AF695427

Higher Structures in Topology, Geometry, and Physics

CONTEMPORARY MATHEMATICS

802

Higher Structures in Topology, Geometry, and Physics

AMS Special Session
Higher Structures in Topology, Geometry, and Physics
March 26–27, 2022
Virtual

Ralph M. Kaufmann
Martin Markl
Alexander A. Voronov
Editors

AMS AMERICAN MATHEMATICAL SOCIETY

EDITORIAL COMMITTEE

Michael Loss, Chair

John Etnyre Angela Gibney Alex Kontorovich

2020 *Mathematics Subject Classification.* Primary 18Mxx, 55Nxx, 81Txx; Secondary 16Txx, 32Gxx, 57Qxx.

Library of Congress Cataloging-in-Publication Data

Names: Kaufmann, Ralph M., editor. | Markl, Martin, 1960- editor. | Voronov, Alexander A., editor. | AMS Special Session on Higher Structures in Topology, Geometry, and Physics (2022 : Online)

Title: Higher structures in topology, geometry, and physics / Ralph M. Kaufmann, Martin Markl, Alexander A. Voronov, editors

Description: Providence, Rhode Island : American Mathematical Society, 2024. | Series: Contemporary mathematics, 0271-4132 ; 802 | "AMS Special Session, Higher Structures in Topology, Geometry, and Physics, Virtual March 26-27, 2022." | Includes bibliographical references.

Identifiers: LCCN 2023048574 | ISBN 9781470471422 (paperback) | ISBN 9781470476427 (ebook)

Subjects: LCSH: Algebraic topology–Congresses. | Quantum field theory–Congresses. | Cohomology operations–Congresses. | Functional analysis–Congresses. | AMS: Algebraic topology – Homology and cohomology theories. | Quantum theory – Quantum field theory; related classical field theories. | Associative rings and algebras – Hopf algebras, quantum groups and related topics. | Several complex variables and analytic spaces – Deformations of analytic structures. | Manifolds and cell complexes – PL-topology.

Classification: LCC QA612 .H54 2024 | DDC 514/.2–dc23/eng/20240402

LC record available at https://lccn.loc.gov/2023048574

Contemporary Mathematics ISSN: 0271-4132 (print); ISSN: 1098-3627 (online)

DOI: https://doi.org/10.1090/conm/802

Color graphic policy. Any graphics created in color will be rendered in grayscale for the printed version unless color printing is authorized by the Publisher. In general, color graphics will appear in color in the online version.

Copying and reprinting. Individual readers of this publication, and nonprofit libraries acting for them, are permitted to make fair use of the material, such as to copy select pages for use in teaching or research. Permission is granted to quote brief passages from this publication in reviews, provided the customary acknowledgment of the source is given.

Republication, systematic copying, or multiple reproduction of any material in this publication is permitted only under license from the American Mathematical Society. Requests for permission to reuse portions of AMS publication content are handled by the Copyright Clearance Center. For more information, please visit `www.ams.org/publications/pubpermissions`.

Send requests for translation rights and licensed reprints to `reprint-permission@ams.org`.

© 2024 by the American Mathematical Society. All rights reserved.
The American Mathematical Society retains all rights
except those granted to the United States Government.
Printed in the United States of America.

∞ The paper used in this book is acid-free and falls within the guidelines
established to ensure permanence and durability.
Visit the AMS home page at `https://www.ams.org/`

10 9 8 7 6 5 4 3 2 1 29 28 27 26 25 24

Contents

Preface

There have been many advances and interactions at the interface of physics and mathematics especially concerning quantum field theory. One common mathematical goal is to give precise meaning to the computation of correlation functions through path integrals in different contexts. The basic assumptions are that fields form a vector space or a linear structure such as a dg–vector space, and that correlation functions are multi-linear functions in these fields. These endow the linear structure with extra operations the most common of which are algebra structures. There is an intricate interplay between the algebra and geometry in the physical computations whose traces one can see in terms of categorical structures and functors between them on the mathematical side. The most well–known may be the formulation of 1+1 dimensional topological field theory through the use of Frobenius algebras following Dijkgraaf. In this case, all the higher correlation functions factor into three–point functions. The geometry is that of surfaces, where this feature translates to pair of pants decomposition. The mathematical formulation following Atiyah then uses functors from a geometric cobordism category to a linear category. More sophisticated versions using gauge theory and dg–structures are of current interest as is the modeling of renormalization and string field theory.

Higher correlators do enter in the case of deformations and non–factorizable structures more generally. In particular, such structures naturally arise in the context of homotopy theory where the homotopy can come from abstract considerations or homotopy transfer. This is also the realm for the appearance of natural algebraic structures such as Hochschild complexes, which in turn describe the geometry of loop spaces. The natural formulation for the correlators and their interrelations is given by operads and their generalizations.

The current volume covers a wide range aspects of such higher structures, such as generalization of operads, models for conformal field theories, string topology, open/closed field theories, actions on Hochschild or related complexes and their geometric and topological aspects.

The aim of the volume is to bring together experts and emerging researchers by sharing new results and research directions.

Ralph M. Kaufmann
Martin Markl
Alexander A. Voronov

Contemporary Mathematics
Volume **802**, 2024
https://doi.org/10.1090/conm/802/16070

On noncommutative differential forms

Boris Tsygan

ABSTRACT. We review the topic of noncommutative differential forms, following the works of Karoubi, Cuntz–Quillen, Cortiñas, Ginzburg–Schedler, and Waikit Yeung. In particular we give a new proof of the theorem of Ginzburg and Schedler that compares extended noncommutative De Rham cohomology to cyclic homology. This theorem is a stronger version of a theorem of Karoubi. We also describe an algebraic structure, namely a category in DG cocategories, that noncommutative forms and other versions of noncommutative calculus are particular cases of.

1. Introduction

The idea of noncommutative differential geometry is to start with an assocaitive algebra A and produce invariants that, in the case when A is commutative, give classical invariants such as differential forms and the De Rham complex. One way of doing this is to use constructions from classical homological algebra, such as Hochschild homology. The relation between these constructions and De Rham complexes had been known since the early 60s [**HKR**], [**R**]. This idea is the basis of cyclic homology theory [**C1**], [**C2**], [**C3**], [**T1**].

On the other hand, one can try to define noncommutative differential forms directly from the algebra A. This idea was present in Connes's approach from the very beginning. It was then advanced by Karoubi and by Cuntz–Quilen in [**K1**], [**CQ**]. The relation between the two approaches, one based on Hochschild homology and the other on noncommutative forms, is strong, deep, and subtle. It had been clarified gradually, starting from the works mentioned above and then in the articles by Ginzburg and Schedler [**GS1**], [**GS2**]. Recent applications include [**WY**] and [**GK**].

Let us start with noncommutative forms. They are defined for any unital algebra A over a commutative unital ring k as a differential graded algebra $(\Omega^\bullet(A), d)$ together with a morphism of graded algebras $A \to \Omega^\bullet(A)$ which is universal with such property. The De Rham cohomology is not interesting, not even in characteristic greater than zero. In fact (noncommutative Poincaré lemma), the morphism $k = \Omega^\bullet(k) \to \Omega^\bullet(A)$ is a quasi-isomorphism. However trivial this fact is, its mere generality has some interesting consequences that we will discuss later. For now,

---

2020 *Mathematics Subject Classification.* Primary 16E40, 16S38; Secondary 16T15.
Key words and phrases. Hochschild and cyclic homology, noncommutative differential forms.

©2024 American Mathematical Society

note one nontrivial way of defining De Rham cohomology, namely the noncommutative De Rham complex

$$\mathrm{DR}^\bullet(A) = \Omega^\bullet(A)/[\Omega^\bullet(A), \Omega^\bullet(A)]. \tag{1.1}$$

Let us next recall the original definition of cyclic homology (for $\mathbb{Q} \subset k$) [**C1**], [**T1**] as the homology of the complex

$$A/\operatorname{im}(1-\tau) \xleftarrow{b} A^{\otimes 2}/\operatorname{im}(1-\tau) \xleftarrow{b} \ldots \xleftarrow{b} A^{\otimes(n+1)}/\operatorname{im}(1-\tau) \xleftarrow{b} \tag{1.2}$$

where

$$\tau(a_0 \otimes \ldots \otimes a_n) = (-1)^n a_n \otimes a_0 \otimes \ldots \otimes a_{n-1}$$

and b is the Hochschild differential. This is a chain complex, as opposed to the noncommutative De Rham complex

$$\mathrm{DR}^0(A) \xrightarrow{d} \mathrm{DR}^1(A) \xrightarrow{d} \ldots \xrightarrow{d} \mathrm{DR}^n(A) \xrightarrow{d}$$

which is a cochain complex. The original observation of Connes was that noncommutative higher traces, i.e., linear functionals

$$\mathrm{DR}^n(A)/d\,\mathrm{DR}^{n-1}(A) \to k,$$

are cyclic cocycles, i.e., linear functionals

$$A^{\otimes(n+1)}/(\operatorname{im}(1-\tau) + \operatorname{im}(b)) \to k.$$

Let us modify the cyclic complex a little. Put

$$C_n(A) = A \otimes \overline{A}^{\otimes n}$$

where $\overline{A} = A/k \cdot 1$; there are two differentials

$$b : C_n(A) \to C_{n-1}(A);\ B : C_n(A) \to C_{n+1}(A)$$

such that $bB + Bb = 0$. The *reduced* cyclic complex of A (i.e., the quotient of (1.2) by the cyclic complex of k) is quasi-isomorphic to

$$A/([A,A] + k \cdot 1) \xleftarrow{b} C_1(A)/BC_0(A) \xleftarrow{b} \ldots \xleftarrow{b} C_n(A)/BC_{n-1}(A) \xleftarrow{b} . \tag{1.3}$$

It turns out that a complex better compatible to the noncommutative De Rham complex is the "dual" cochain complex

$$A/([A,A] + k \cdot 1) \xrightarrow{B} C_1(A)/bC_2(A) \xrightarrow{B} \ldots \xrightarrow{B} C_n(A)/bC_{n+1}(A) \xrightarrow{B} . \tag{1.4}$$

Namely, over the ground ring k of characteristic zero "the original HKR map"

$$\mathrm{HKR}_0(a_0 \otimes a_1 \otimes \ldots \otimes a_n) = \frac{1}{n!} a_0 da_1 \ldots da_n \tag{1.5}$$

is a quasi-isomorphism from (1.4) to the reduced De Rham complex

$$\mathrm{DR}^\bullet(A)/\,\mathrm{DR}^\bullet(k)$$

(cf. [**K1**]).

Now, both complexes (1.3) and (1.4) are quasi-isomorphic to quotients of the periodic cyclic complex

$$\mathrm{CC}^{\mathrm{per}}_\bullet(A) = (C_\bullet(A)((u)), b + uB). \tag{1.6}$$

Those two quotients are described in two diagrams before the beginning of Section 4. (We also take a quotient by the complex of k). This suggests that perhaps the De Rham complex is also a quotient of something larger, and that the De

Rham differential d is also part of a pair of differentials, as is the Hochschild differential b.

This is indeed the case, as shown by Ginzburg and Schedler in [**GS1**], [**GS2**]. Things being subtle, there are two versions of that. Let us start with the extended De Rham complex (Section 4). It is proven in [**GS2**] that it is quasi-isomorphic to its cyclic counterpart. Moreover, the two are quasi-isomorphic as filtered complexes (the filtration given by the bicomplex structure). We will present a new proof of this theorem that, we hope, clarifies the situation and puts it in a more general homological algebraic context.

The reduced version of the extended De Rham complex is quasi-isomorphic to the reduced noncommutative De Rham complex. Therefore we recover Karoubi's theorem. Noncommutative reduced De Rham cohomology gets identified with the cokernel of the periodicity operator S on reduced cyclic homology as in [**K1**].

Another version of a second differential on noncommutative forms is Ginzburg and Schedler's operator ι_Δ. It is actually the first term of the second differential in the extended De Rham complex; it acts on usual, not extended, forms. The pairs (b, B) and (ι_Δ, d) are indeed intertwined by the HKR map (Section 4), but only on the image of the Cuntz–Quillen projection. As for the complement of this image: on it, b and d are acyclic whereas $B = \iota_\Delta = 0$. This allows one to conclude that in the periodic cyclic complex (1.6), one can replace $b + uB$ by $\iota_\Delta + ud$. Other versions of cyclic homology may also be computed in terms of $\iota_\Delta + ud$, in a way that is somewhat more involved.

The last part of the paper (Section 10) is devoted to a general algebraic structure of which noncommutative forms are an example. There are other examples, and it might be that they are all part of a general unified structure. Namely, the structure is of a category in coalgebras (or, more generally, in cocategories), and a module in cocategories over it. This structure provides a package for various algebraic properties of noncommutative versions of forms and multivectors. One example is the Čech–Alexander complex of an associative algebra, both in the classical commutative version from crystalline cohomology theory [**Be**], [**Gr**] and in the noncommutative version of Cortiñas [**Cor1**]. Note that in the noncommutative case, the Čech–Alexander complex projects directly to the complex of noncommutative forms with the De Rham differential. Another example (Section 10.3) is a category in cocategories whose objects are algebras and the cocategory of morphisms between two algebras A and B is constructed from Hochschild cochains of A with values in B. The module in coalgebras puts in correspondence to an algebra A its bar construction.

These examples illuminate how nontrivial algebraic structures appear on the complexes of Hochschild cochains, noncommutative forms, etc. One starts with a rigid structure with many objects and morphisms; everything is rigid up to homotopy. Then one declares all objects to be the same, and all morphisms to be the identity. Now our homotopies—in general connecting two different things—start to connect the same thing to itself; in other words, they become morphisms or/and cocycles. The first example (coming from the classical Čech–Alexander complex) is the bar differential. Other differentials, e.g., B and ι_Δ above, also can be obtained this way. In the example of Section 10.3, this is how one produces the Gerstanhaber bracket on Hochschild cochain complexes.

2. Noncommutative forms

Let A be an associative unital algebra over a commutative unital ring k.

Let $\Omega^\bullet(A)$ be the graded algebra generated by A and by symbols da, $a \in A$, linear in a and subject to relations

(a) $d(ab) = da\, b + adb$;
(b) the unit of A is the unit of $\Omega^\bullet(A)$.

The grading $|a| = 0$, $|da| = 1$ makes $\Omega^\bullet(A)$ a graded algebra. We define the differential $d : \Omega^\bullet(A) \to \Omega^{\bullet+1}(A)$ as the unique graded derivation sending a to da and da to zero for all a in A.

We get a DG algebra $(\Omega^\bullet(A), d)$ together with a morphism of graded algebras $i : A \to \Omega^\bullet(A)$. It is universal, in the sense that for any DG algebra $\Omega^\bullet$ together with a morphism of graded algebras $j : A \to \Omega^\bullet$ there exists unique morphism of DG algebras $f : \Omega^\bullet(A) \to \Omega^\bullet$ such that $j = fi$.

2.1. Noncommutative Poincaré lemma.

LEMMA 2.1. *The embedding*

$$\Omega^\bullet(k) \to \Omega^\bullet(A)$$

is an isomorphism on the cohomology of the differential d.

PROOF. First, note that the naive HKR map

$$A \otimes \bar{A}^{\otimes n} \to \Omega^n(A);\ a_0 \otimes \ldots \otimes a_n \mapsto a_0 da_1 \ldots da_n \tag{2.1}$$

is an isomorphism. Indeed, it is clearly surjective, since every product of factors of the form a_i and db_j may be always transformed into an element of the image of the map above. Then, observe that there is an associative product on $A \otimes \bar{A}^\bullet$; the DGA $\Omega^\bullet(A)$ maps to it by universality, and this map inverts (2.1). We get an isomorphism of algebras which intertwines d with

$$a_0 \otimes \ldots \otimes a_n \mapsto 1 \otimes a_0 \otimes \ldots \otimes a_n.$$

The cohomology of this differential is clearly k. □

REMARK 2.2. One application of the noncommutative Poincaré lemma over $\mathbb{Z}$ is Karoubi's approach to extending De Rham–Sullivan complexes from rational homotopy theory to more general situations [**K2**]–[**K8**].

3. Noncommutative De Rham complex

Define

$$\mathrm{DR}^\bullet(A) = \Omega^\bullet(A)/[\Omega^\bullet(A), \Omega^\bullet(A)] \tag{3.1}$$

and

$$\overline{\mathrm{DR}}^\bullet(A) = \mathrm{DR}^\bullet(A)/\mathrm{DR}^\bullet(k). \tag{3.2}$$

4. The extended noncommutative De Rham complex

Let t be a formal variable of degree one. Define

$$\Omega_t^\bullet(A) = \Omega^\bullet(A) * k[t] \tag{4.1}$$

where $*$ denotes the free product of algebras over k. Put

$$\mathrm{DR}_t^\bullet(A) = \Omega_t^\bullet(A)/[\Omega_t^\bullet(A), \Omega_t^\bullet(A)] \tag{4.2}$$

and

$$\overline{\Omega}_t^\bullet(A) = \Omega^\bullet(A) * k[t]/k[t] \tag{4.3}$$

$$\overline{\mathrm{DR}}_t^\bullet(A) = \overline{\Omega}_t^\bullet(A)/[\overline{\Omega}_t^\bullet(A), \overline{\Omega}_t^\bullet(A)]. \tag{4.4}$$

There is the second grading (by powers of t) on $\Omega_t^\bullet$ as well as on all the spaces above. Therefore, $\Omega_t^\bullet$ is a bi-graded algebra, and all the above spaces are bi-graded. For the first grading, $|d| = 1$ and $|t| = 0$. For the second, $|d| = 0$ and $|t| = 1$. We denote by $\Omega_t^{p,q}$ the component whose degree is p with respect to the first grading and q with respect to the second grading. We get similar decompositions for all the spaces above.

LEMMA 4.1.

$$\mathrm{DR}_t^{n,0} = (\Omega/[\Omega,\Omega])^n; \; \mathrm{DR}_t^{n-1,1} \xrightarrow{\sim} \Omega^{n-1}$$

PROOF. For the component of degree zero in t, this follows directly from the definition. For the component of degree one in t: modulo commutators, every element is uniquely represented by an element ωt where ω is a noncommutative form. □

4.1. The derivation ι_t.

Let $|\omega|$ be the first grading of ω, i.e., $|a|=|t|=0$ and $|da|=1$. Define the graded derivation of degree -1 with respect to this grading by

$$\iota_t(a) = \iota_t(t) = 0; \; \iota_t(da) = [t, a]. \tag{4.5}$$

This is a bi-homogeneous map of degree $(-1, 1)$ satisfying

$$\iota_t^2 = 0.$$

We get complexes

$$\mathrm{DR}_t^{n,0} \xrightarrow{\iota_t} \mathrm{DR}_t^{n-1,1} \xrightarrow{\iota_t} \mathrm{DR}_t^{n-2,2} \xrightarrow{\iota_t} \ldots \xrightarrow{\iota_t} \mathrm{DR}_t^{0,n}. \tag{4.6}$$

They are the columns of the double complex.

$$
\begin{array}{ccccccccc}
A/[A,A] & \xrightarrow{d} & \mathrm{DR}^1(A) & \xrightarrow{d} & \mathrm{DR}^2(A) & \xrightarrow{d} \dots \xrightarrow{d} & \mathrm{DR}^n(A) & \xrightarrow{d} \\
 & & \downarrow{\scriptstyle \iota_t} & & \downarrow{\scriptstyle \iota_t} & & \downarrow{\scriptstyle \iota_t} & \\
 & & At & \longrightarrow & \Omega^1(A)t & \xrightarrow{d} \dots \xrightarrow{d} & \Omega^{n-1}(A)t & \xrightarrow{d} \\
 & & & & \downarrow{\scriptstyle \iota_t} & & \downarrow{\scriptstyle \iota_t} & \\
 & & & & AtAt & \xrightarrow{d} \dots \xrightarrow{d} & \mathrm{DR}_t^{n-2,2}(A) & \longrightarrow \\
 & & & & & \dots & \dots & \longrightarrow \\
 & & & & & & \downarrow & \\
 & & & & & & At\dots At & \xrightarrow{d}
\end{array}
$$

LEMMA 4.2. *The projection*

$$\overline{\mathrm{DR}}_t^\bullet(A) \to \overline{\mathrm{DR}}^\bullet(A)$$

is a quasi-isomorphism.

PROOF. By Lemma 2.1 and the Künneth formula, the rows of the reduced version of the double complex are acyclic, and the spectral sequence converges. □

5. Recollection on cyclic homology

Let

$$C_n(A) = A \otimes \bar{A}^{\otimes n} \tag{5.1}$$

where $\bar{A} = A/k \cdot 1$. Define

$$b : C_n(A) \to C_{n-1};\ B : C_n(A) \to C_{n+1}(A)$$

(5.2)

$$b(a_0 \otimes \dots \otimes a_n) = \sum_{j=0}^{n-1} (-1)^j a_0 \otimes \dots \otimes a_j a_{j+1} \otimes \dots \otimes a_n + (-1)^n a_n a_0 \otimes \dots \otimes a_{n-1}$$

and

$$B(a_0 \otimes \dots \otimes a_n) = \sum_{j=0}^{n} (-1)^{nj} 1 \otimes a_j \otimes \dots \otimes a_n \otimes a_0 \otimes \dots \otimes a_{j-1}. \tag{5.3}$$

One has

$$b^2 = Bb + bB = B^2 = 0. \tag{5.4}$$

This allows us to arrange the terms $C_n(A)$ into a double complex in several different ways. Start with the periodic cyclic complex

$$\mathrm{CC}_\bullet^{\mathrm{per}}(A) = (C_\bullet(A)((u)), b + uB) \tag{5.5}$$

where u is a formal variable of homological degree -2. Define its subcomplex

$$\mathrm{CC}_\bullet^-(A) = (C_\bullet(A)[[u]], b + uB) \tag{5.6}$$

(the negative cyclic complex) and its quotient complex

$$\mathrm{CC}_\bullet(A) = (C_\bullet(A)((u))/uC_\bullet(A)[[u]], b + uB) \tag{5.7}$$

(the cyclic complex). The latter looks like the following.

$$\begin{array}{ccccccccc}
\downarrow{\scriptstyle b} & & \downarrow{\scriptstyle b} & & & & & & \downarrow{\scriptstyle b} \\
A & \xrightarrow{B} & C_1 & \xrightarrow{B} & \dots & & \dots & \xrightarrow{B} & C_n \\
 & & \downarrow{\scriptstyle b} & & & & & & \downarrow{\scriptstyle b} \\
 & & A & \xrightarrow{B} & \dots & & \dots & \xrightarrow{B} & C_{n-1} \\
 & & & & & & & & \dots \\
 & & & & & & \downarrow{\scriptstyle b} & & \downarrow{\scriptstyle b} \\
 & & & & & & A & \xrightarrow{B} & C_1 \\
 & & & & & & & & \downarrow{\scriptstyle b} \\
 & & & & & & & & A
\end{array}$$

The closest to the noncommutative De Rham complex is yet another version that we denote the following by $\mathrm{CD}_\bullet(A)$.

$$\begin{array}{ccccccccccc}
A/[A,A] & \xrightarrow{B} & C_1/bC_2 & \xrightarrow{B} & C_2/bC_3 & \xrightarrow{B} & \dots & \longrightarrow & C_n/bC_{n+1} & \xrightarrow{B} & \\
 & & \downarrow{\scriptstyle b} & & \downarrow{\scriptstyle b} & & & & \downarrow{\scriptstyle b} & & \\
 & & A & \longrightarrow & C_1 & \xrightarrow{B} & \dots & \xrightarrow{B} & C_{n-1} & \xrightarrow{B} & \\
 & & & & \downarrow & & & & \downarrow{\scriptstyle b} & & \\
 & & & & A & \xrightarrow{B} & \dots & \xrightarrow{B} & C_{n-2} & \longrightarrow & \\
 & & & & & & \dots & & \dots & \longrightarrow & \\
 & & & & & & & & \downarrow & & \\
 & & & & & & & & A & \longrightarrow &
\end{array}$$

6. The extended HKR morphism

Define the map

$$\mathrm{HKR}_t : \mathrm{CD}_\bullet(A) \to \mathrm{DR}^\bullet(A) \tag{6.1}$$

as follows. For $k \geq 0$, put

(6.2)

$$a_0 \otimes \dots \otimes a_m \mapsto \frac{1}{(m+k)!} \sum a_0 da_1 \dots da_{j_1} t da_{j_1+1} \dots da_{j_2} t \dots t da_{j_k+1} \dots da_m$$

where the sum is over all $0 \leq j_1 \leq \dots \leq j_k \leq m$. The kth row from above in the (b, B) double complex $\mathrm{CD}_{\bullet,\bullet}(A)$ is mapped to the kth row from above in the (ι_t, d) double complex $\mathrm{DR}_t^{\bullet,\bullet}(A)$ by means of $\mathrm{HKR}^{(k)}$.

THEOREM 6.1. *The extended HKR map* (6.2) *is a quasi-isomorphism for every column and for the total complex. In particular, the homology of the complex* (4.6) *at* $\mathrm{DR}_t^{n-j,j}$ *is isomorphic to* $H_j(A,A)$.

As a consequence we recover Karoubi's theorem.

THEOREM 6.2. *The nth cohomology of the relative noncommutative De Rham complex* $\overline{\mathrm{DR}}^{\bullet}(A)$ *is isomorphic to*

$$\mathrm{Ker}(B : \overline{\mathrm{HC}}_n(A) \to \overline{\mathrm{HH}}_{n+1}(A)).$$

PROOF. Theorem 6.1 establishes quasi-isomorphisms

$$\mathrm{CD}_{\bullet}(A) \xrightarrow{\sim} \mathrm{DR}_t^{\bullet}(A); \quad \overline{\mathrm{CD}}_{\bullet}(A) \xrightarrow{\sim} \overline{\mathrm{DR}}_t^{\bullet}(A).$$

The projection

$$\overline{\mathrm{DR}}_t^{\bullet}(A) \to \overline{\mathrm{DR}}^{\bullet}(A)$$

is a quasi-isomorphism by Lemma 4.2. An n-cycle of $\overline{\mathrm{CD}}_{\bullet}(A)$ is a sum

$$\sum_{j\geq 0} c_{n-2j}; \quad c_{n-2j} \in \overline{C}_{n-2j}(A),\ j>0; \ c_n \in \overline{C}_n(A)/b\overline{C}_{n+1}(A) \tag{6.3}$$

such that

$$(b+B)(\sum c_{n-2j}) = 0$$

in $\oplus_{j\geq 0}\overline{C}_{n-1-2j}(A)$ and

$$Bc_n = 0$$

in $\overline{C}_{n+1}(A)/\overline{C}_{n+2}(A)$. A cycle is homologous to zero if any its representative in $\oplus_{j\geq 0}\overline{C}_{n-2j}(A)$ is in the image of

$$b+B : \oplus_{j\geq 0}\overline{C}_{n+1-2j}(A) \to \oplus_{j\geq 0}\overline{C}_{n-2j}(A).$$

Therefore the nth homology group of $\mathrm{CD}_{\bullet}(A)$ is given by the right-hand side of the formula in Theorem 6.2. □

The proof of Theorem 6.1 will be given in Section 7, and part of it will be based on Theorem 8.4. It proceeds as follows. Consider the extended De Rham complex with the differential ι_t. It is related to Hochschild homology in two ways. First, it is obtained from the tensor algebra of the short bar resolution of the A-bimodule A (cf. (7.2)). We can perform an analogous construction for the *full* bar resolution; we are not sure, however, how to show that the projection to the short version is a quasi-isomorphism. However, if we perform this analogous construction in the dual situation to the DG coalgebra $\mathrm{Bar}(A)$ instead of the algebra A, what we will get is essentially the extended De Rham complex (cf. Lemma 7.2). More precisely: we define a filtration on the extended De Rham complex, and the associated graded module of this filtration is the dual construction discussed above, applied to the *graded, not differential graded* coalgebra $\mathrm{Bar}(A)$ (cf. Remark 7.1). It is known that Hochschild and cyclic homology of any algebra A can be computed in terms of the dual construction, namely Hochschild and cyclic homology of its bar construction ([**Q**], [**FT**]). The Hochschild homology that arises here is not only the usual $H_{\bullet}(A,A)$ but also the higher $H_{\bullet}(A^{\otimes n}, {}_{\alpha}A^{\otimes n})$ for all n, where α is the cyclic permutation viewed as an algebra automorphism of $A^{\otimes n}$. But this is isomorphic to $H_{\bullet}(A,A)$. (Similarly, its cyclic version is isomorphic to the cyclic homology [**Ka**]). We use the dual (coalgebra) version of this to prove Theorem 6.1. The other ingredient in the proof is Theorem 8.4. We outline the proof in Section 7.2; full details are in [**NT**].

6.1. The extended De Rham complex and the Beilinson t-structure. In [**B**], Beilinson defined the t-structure on the filtered derived category. It follows from the definitions that the complex $\mathrm{CD}_\bullet(A)$ coincides with the truncation $\tau^B_{\leq 0}$ with respect to this t-structure, applied to the negative cyclic complex $\mathrm{CC}^-_\bullet(A)$ with the Hodge filtration

$$F^m\,\mathrm{CC}^-_\bullet(A) = u^m C_\bullet(A)[[u]] \tag{6.4}$$

Therefore Theorem 6.1 can be resatted in Corollary 6.3.

COROLLARY 6.3. *There is a natural filtered quasi-isomorphism*

$$\tau^B_{\leq 0}\,\mathrm{CC}^-_\bullet(A) \xrightarrow{\sim} \mathrm{DR}^\bullet_t(A).$$

7. The extended De Rham complex and Hochschild homology

7.1. The extended De Rham complex in terms of the short bar resolution. Recall the standard bar complex:

$$\mathcal{B}_n(A) = A\otimes \overline{A}^{\otimes n}\otimes A;\, b' : \mathcal{B}_n(A)\to \mathcal{B}_{n-1}(A),$$

$$b'(a_0\otimes\ldots\otimes a_{n+1}) = \sum_{j=0}^{n}(-1)^j a_0\otimes\ldots\otimes a_j a_{j+1}\otimes\ldots\otimes a_{n+1}.$$

If we extend this to $\mathcal{B}_{-1}(A) = A$, we get the bimodule resolution of A.

Define

$$\mathcal{B}^{\mathrm{sh}}_1(A) = \Omega^1_A \xrightarrow{\sim} \mathcal{B}_1(A)/\partial\mathcal{B}_2(A), \tag{7.1}$$

$$\mathcal{B}^{\mathrm{sh},(0)}_\bullet(A) = A;$$

$$\mathcal{B}^{\mathrm{sh},(n)}_\bullet(A) = \mathcal{B}^{\mathrm{sh}}_\bullet(A)\otimes_A\ldots\otimes_A \mathcal{B}^{\mathrm{sh}}_\bullet(A),$$

(n-fold tensor product for $n\geq 1$);

$$\mathcal{B}^{\mathrm{sh},(*)}_\bullet(A) = \bigoplus_{n\geq 0}\mathcal{B}^{\mathrm{sh},(n)}_\bullet(A).$$

Then we have

$$(\mathrm{DR}^\bullet_t(A), \iota_t) \xrightarrow{\sim} \mathcal{B}^{\mathrm{sh},(*)}_\bullet(A)/[\mathcal{B}^{\mathrm{sh},(*)}_\bullet(A), \mathcal{B}^{\mathrm{sh},(*)}_\bullet(A)]. \tag{7.2}$$

This construction is used in [**WY**] where it is denoted by $\Upsilon^{(*)}(A)$.

7.2. The extended De Rham complex and the bar construction. Let $\mathrm{DR}^\bullet_{t,+}(A)$ be the subcomplex of $\mathrm{DR}^\bullet_t(A)$ spanned by elements whose degree with respect to t is positive. This subcomplex can be presented in the form that we are going to discuss next.

Let us start with any associative unital differential algebra $(\mathcal{A},\partial)$. View $\mathcal{A}$ as a graded algebra. Introduce a new generator ϵ of degree one and square zero. Consider the cross product algebra

$$\widetilde{\mathcal{A}} = k[\epsilon]\ltimes\mathcal{A} \tag{7.3}$$

generated by ϵ and $\mathcal{A}$ subject to a relation $[\epsilon, a] = \partial a$ for all a in $\mathcal{A}$.

In other words, $\widetilde{\mathcal{A}}$ is generated by the algebra $\mathcal{A}$ and by elements $\underline{a} = \epsilon a$, $a \in \mathcal{A}$, of degree $|a|+1$, linear in a and subject to relations

$$\underline{a} \cdot b = \underline{ab};\ a \cdot \underline{b} = (-1)^{|a|}(\underline{ab} - \partial a \cdot b);\ \underline{a} \cdot \underline{b} = (-1)^{|a|-1}\underline{\partial a \cdot b}. \tag{7.4}$$

Now one can consider the reduced cyclic homology $\overline{\mathrm{HC}}_\bullet(\widetilde{\mathcal{A}})$ of the graded algebra $\widetilde{\mathcal{A}}$. More precisely, we will compute it using the following specific complex defined for any A:

$$\mathrm{CC}'(A) = (\mathrm{Ker}(1-t), b'); \overline{\mathrm{CC}}'(A) = \mathrm{CC}'(A)/\,\mathrm{CC}'(k) \tag{7.5}$$

where $1-t$ and N are as in the standard $(b, b', 1-t, N)$ double complex. (Recall that

$$(\mathrm{Ker}(1-t), b') = (\mathrm{Im}(N), b') \xrightarrow{\sim} (C(A)/\,\mathrm{Ker}(N), b) = (C(A)/\,\mathrm{Im}(1-t), b)$$

and therefore $\mathrm{CC}'(A)$ does compute the cyclic homology).

REMARK 7.1. Let the differential ∂ be equal to zero. Then $\widetilde{\mathcal{A}} \xrightarrow{\sim} A[\epsilon]$; we have

$$C_\bullet^\lambda(A[\epsilon]) \xrightarrow{\sim} C_\bullet^\lambda(A) \oplus \mathcal{B}_\bullet^{(*)}(A)/[\mathcal{B}_\bullet^{(*)}(A), \mathcal{B}_\bullet^{(*)}(A)]_{>0} \tag{7.6}$$

The second summand in the right-hand side is the positively graded part of the same expression as in (7.2) if one replaces $\mathcal{B}_\bullet^{\mathrm{sh}}$ with the full bar resolution $\mathcal{B}_\bullet$. The isomorphism identifies

$$\epsilon a_0^{(1)} \otimes \ldots \otimes a_{m_1}^{(1)} \otimes \epsilon a_0^{(2)} \otimes \ldots \otimes a_{m_2}^{(2)} \otimes \ldots \otimes \epsilon a_0^{(n)} \otimes \ldots \otimes a_{m_n}^{(n)}$$

with

$$(a_0^{(1)} \otimes \ldots \otimes a_{m_1}^{(1)} \otimes 1) \otimes (a_0^{(2)} \otimes \ldots \otimes a_{m_2}^{(2)} \otimes 1) \otimes \ldots \otimes (a_0^{(n)} \otimes \ldots \otimes a_{m_n}^{(n)} \otimes 1).$$

Note that when $\partial \neq 0$ then the terms of the differential that depend on ∂ are not the differential naturally induced by ∂ on the right-hand side of (7.6). The latter preserves the degree indicated by the superscript $(*)$ while the former decreases it by one.

Consider now a dual picture. Let $\mathcal{C}$ be a differential graded counital coalgebra $(\mathcal{C}, \partial)$. For $c \in \mathcal{C}$, let $\underline{c}$ be a formal element of degree $|c|+1$, linear in c. These elements generate the space $\underline{\mathcal{C}}$ which is same as $\mathcal{C}$ but with the grading shifted by one. Let $\widetilde{\mathcal{C}}$ be the graded coalgebra which is a linear direct sum of $\mathcal{C}$ and $\underline{\mathcal{C}}$. The comultiplication is as follows.

$$\Delta c = \sum c^{(1)} \otimes c^{(2)} + \sum (-1)^{|c^{(1)}|} \partial c^{(1)} \otimes \underline{c}^{(2)}, \tag{7.7}$$

$$\Delta \underline{c} = \sum \underline{c}^{(1)} \otimes c^{(2)} + (-1)^{|c^{(1)}|} c^{(1)} \otimes \underline{c}^{(2)} + \sum (-1)^{|c^{(1)}|} \underline{\partial c^{(1)}} \otimes \underline{c}^{(2)}. \tag{7.8}$$

For any counital DG coalgebra C put

$$\mathrm{CC}'(C) = (\mathrm{Coker}(1-t), b'); \overline{\mathrm{CC}}'(C) = \mathrm{Ker}(\mathrm{CC}'(C) \to \mathrm{CC}'(k)). \tag{7.9}$$

LEMMA 7.2. *Let $A = k + \overline{A}$ be the algebra obtained from an algebra $\overline{A}$ by attaching a unit. Then the complex $(\mathrm{DR}_{t,+}^\bullet(A), \iota_t)$ is isomorphic to*

$$\overline{\mathrm{CC}}'(\widetilde{\mathrm{Bar}(\overline{A})})$$

where Bar *stands for the usual bar construction (which is a DG coalgebra).*

PROOF. Take a monomial $\omega_1 t\omega_2 t\ldots\omega_n t$ in $\mathrm{DR}_{t,+}$. Identify it with

$$\alpha_1\otimes\alpha_2\otimes\ldots\otimes\alpha_n \text{ in } \widetilde{\mathrm{CC}'(\mathrm{Bar}(\overline{A}))}$$

where $\alpha_k = (a_0|a_1|\ldots|a_m)$ if $\omega_k = a_0 da_1\ldots da_m$ and $\alpha_k = \underline{(|a_1|\ldots|a_m)}$ if $\omega_k = da_1\ldots da_m$. One checks that this gives an isomorphism of complexes. We will give some more details in Section 7.3. □

7.3. The filtration on the extended De Rham complex. Consider the following filtration on noncommutative forms. Put

$$(d_+\Omega)^n = d\Omega^{n-1},\ n>0;\ (d_+\Omega)^0 = k. \tag{7.10}$$

We say that a monomial

$$\alpha_1 t\ldots t\alpha_N t,\ \alpha_j\in\Omega^\bullet(A), \tag{7.11}$$

lies in $\mathcal{F}^p$ if at least p forms α_j are in $d_+\Omega^\bullet$.

We claim that $\mathrm{gr}^*_{\mathcal{F}}(\mathrm{DR}_t{}^\bullet(A))$ is dual to (7.2) if one replaces the short bar resolution $\mathcal{B}^{\mathrm{sh}}$ by the full bar resolution $\mathcal{B}$ and the algebra A by the graded coalgebra $T(\overline{A}[1])$, the free coalgebra of $\overline{A}[1]$ where $\overline{A} = A/k\cdot 1$. (This may sound a bit strange since full $\mathcal{B}_\bullet$ is larger than $\mathcal{B}^{\mathrm{sh}}_\bullet$ and $T(\overline{A}[1])$ is larger than A. Such is life in Hilbert's hotel).

More precisely, for any graded counital coalgebra C, denote

$$\bar{C}^\bullet_{\mathrm{II}}(C)^{(0)} = \overline{\mathrm{CC}}^\bullet_{\mathrm{II}}(C) = (\bar{C}^{\otimes(\bullet+1)})_{C_{\bullet+1}} \tag{7.12}$$

with the differential dual to b (or take invariants with the differential dual to b'). For $n>0$

$$C^\bullet_{\mathrm{II}}(C)^{(n)} = (\bar{C}[-1]^{\otimes\bullet}\otimes C\otimes\ldots\otimes\bar{C}[-1]^{\otimes\bullet}\otimes C)^{C_n} \tag{7.13}$$

(the tensor product on the right is n-fold); the differential sends a monomial

$$c_1^{(1)}\otimes\ldots c_{m_1}^{(1)}\otimes x_1\otimes\ldots\otimes c_1^{(n)}\otimes\ldots c_{m_n}^{(n)}\otimes x_n \tag{7.14}$$

to

$$\sum_{i=1}^{n}\sum_{j-1}^{m_i}\pm\ldots\otimes\Delta(c_j^{(i)})\otimes\ldots+\sum_{j=1}^{n}(\ldots\otimes\Delta'(x_j)\ldots+\ldots\otimes\Delta''(x_j)\ldots)$$

where we use the following notation. First,

$$\Delta' : C\to\bar{C}[-1]\otimes C;\ \Delta'' : C\to C\otimes\bar{C}[-1] \tag{7.15}$$

is the comultiplication followed by the projection. Furthermore, the second half of the $j=n$ term is by definition

$$\pm x_n^{(2)}\otimes c_1^{(1)}\otimes\ldots c_{m_1}^{(1)}\otimes x_1\otimes\ldots\otimes c_1^{(n)}\otimes\ldots c_{m_n}^{(n)}\otimes x_n^{(1)}$$

where

$$\Delta(x) = \sum x^{(1)}\otimes x^{(2)}.$$

LEMMA 7.3.

$$\mathrm{gr}^*_{\mathcal{F}}(\mathrm{DR}_t{}^\bullet(A)) \xrightarrow{\sim} \bar{C}^\bullet_{\mathrm{II}}(C)^{(0)}\oplus\oplus_{n>0}C^\bullet_{\mathrm{II}}(C)^{(n)}$$

where $C = T(\bar{A}[1])$.

PROOF. Take a monomial

$$\alpha_1^{(1)}t\ldots\alpha_{m_1}^{(1)}t\beta_1 t\ldots t\alpha_1^{(n)}t\ldots\alpha_{m_n}^{(n)}t\beta_n \tag{7.16}$$

in $\mathrm{gr}^n(\mathrm{DR}_t^+)$, with β_j in $d_+\Omega$. Associate to it a monomial (7.14) by the following rule: to a form $a_0da_1\ldots da_k$, associate an element $(a_0|\ldots|a_k)$ of $T(\bar{A}[1])$; to a form $db_1\ldots db_l$, associate an element $\underline{(b_1|\ldots|b_l)}$ for $l>0$ or 1 for $l=0$. Observe that:

(1)

$$\iota_t(a_0da_1\ldots da_n)=\sum_{k=1}^{n}(-1)^{k-1}(a_0da_1\ldots da_{k-1})t(a_kda_{k+1}\ldots da_n)$$
$$\mod(d_+\Omega t\Omega+\Omega td_+\Omega).$$

(2)

$$\iota_t(da_1\ldots da_n)=\sum_{k=1}^{n}(-1)^k(a_1da_2\ldots da_k)t(da_{k+1}da_{k+2}\ldots da_n)+$$
$$\sum_{k=1}^{n}(-1)^{k-1}(da_1\ldots da_{k-1})t(a_kda_{k+1}\ldots da_n)\mod(d_+\Omega td_+\Omega).$$

The statement follows. ∎

REMARK 7.4. In other words, the accociated graded module of the filtration $\mathcal{F}$ is the dual construction of the one described in Remark 7.1, with an algebra A replaced by the graded coalgebra $T(\bar{A}[1])$. The differential in the spectral sequence is related to but not induced by the bar differential.

Now compute the spectral sequence of the filtration $\mathcal{F}$. The first term is the reduced cyclic cohomology of $T(\bar{A}[1])$. Since the coalgebra is cofree,

$$\overline{\mathrm{HC}}^\bullet_{\mathrm{II}}(T(\bar{A}[1]))\xrightarrow{\sim}\bigoplus_{n=1}^{\infty}(T^n(\bar{A}[1]))^{C_n}. \tag{7.17}$$

By the dual version of [**Ka**], all homologies of $\mathrm{gr}^n_{\mathcal{F}}$ for $n>0$ are isomophic to the Hochschild cohomology $\mathrm{HH}^\bullet_{\mathrm{II}}(T(\bar{A}[1]))$, which is computed by the short Hochschild complex

$$C^\bullet_{\mathrm{sh}}(T(\bar{A}[1]))=(T(\bar{A}[1])\xrightarrow{b}T(\bar{A}[1])\otimes\bar{A}[1]). \tag{7.18}$$

7.4. The spectral sequence of the filtration $\mathcal{F}$. We claim that the differential $\mathrm{gr}^k_{\mathcal{F}}\Omega_t^m\to\mathrm{gr}^{k+1}_{\mathcal{F}}\Omega_t^{m-1}$ acts on (7.18) as follows: on $T(\bar{A}[1])$, by b'; on $\bar{A}[1]\otimes T(\bar{A}[1])$, by b. To see that, we have to compare the short complex to the higher full complexes. Namely: we have to compute the compositions

$$C^\bullet_{\mathrm{sh}}(T(\bar{A}[1]))\to C^\bullet_{\mathrm{II}}(T(\bar{A}[1]))\to C^\bullet_{\mathrm{II}}(T(\bar{A}[1]))^{(n)} \tag{7.19}$$

and

$$C_{\mathrm{II}}(T(\bar{A}[1]))^{(n)}\to C^\bullet_{\mathrm{II}}(T(\bar{A}[1]))\to C^\bullet_{\mathrm{sh}}(T(\bar{A}[1])). \tag{7.20}$$

More precisely, we know that, for a free algebra $T(V)$, each $\mathcal{B}^{(n)}_\bullet(T(V))_{C_n}$ computes $\mathrm{HH}_\bullet(T(V))$. In fact there is an explicit homotopy equivalence between $\mathcal{B}^{(n)}(T(V))_{C_n}$ and $\mathcal{B}^{(1)}_\bullet(T(V))$, the latter being the standard Hochschild complex $C_\bullet(T(V))$. Actually, $\mathcal{B}^{(n)}_\bullet(T(V))$ is identical to

$$(\mathcal{B}_\bullet(A)\otimes_A\ldots\otimes_A\mathcal{B}_\bullet(A))\otimes_{A\otimes A^{\mathrm{op}}}A;$$

therefore a homotopy equivalence between $\mathcal{B}_\bullet^{(n)}(T(V))$ and $\mathcal{B}_\bullet^{(1)}(T(V))$ is straightforward (a standard comparison between the complexes computing the Tor functors using two different resolutions). This homotopy equivalence is easily modified to descend to coinvariants of C_n.

Furthermore, there is an explicit homotopy equivalence between $C_\bullet(T(V))$ and $C_\bullet^{\mathrm{sh}}(T(V))$.

A dual version of the construction above, with V replaced by $A[1]$ and $T(V)$ by the coalgebra $T(\bar{A}[1])$, is what (7.19) and (7.20) refer to. Now the differentials can be computed explicitly. This computation is performed in detail in [**NT**].

This argument computes the E^2 term of the spectral sequence for all but two leftmost columns. As soon as we compute it for those two, we will have the theorem proven. Indeed, we will know that the extended HKR map (6.1) is a quasi-isomorphism. Indeed, the truncated Hochschid complex

$$C_n(A)/bC_{n+1}(A) \to C_{n-1}(A) \to \ldots \to C_0(A) \tag{7.21}$$

has its own filtration

$$\mathcal{F}^{n-j}C_j(A) = C_j(A);\ \mathcal{F}^{n-j+1}C_j(A) = 1 \otimes \bar{A}^{\otimes j}; \mathcal{F}^{n-j+2}C_j(A) = 0 \tag{7.22}$$

It is straightforward that the extended HKR map preserves the filtration. Using Theorem 8.4, this finishes the proof of Theorem 6.1.

8. The (ι_Δ, d) double complex

8.1. The differential ι_Δ.

DEFINITION 8.1. Put

$$\iota_\Delta(a_0da_1\ldots da_n) = \sum_{j=1}^{n}(-1)^{n(j-1)}[a_j, da_{j+1}\ldots da_n a_0 da_1\ldots da_{j-1}].$$

This is just the composition

$$\Omega^n(A) \to \mathrm{DR}^n(A) \xrightarrow{\iota_t} \mathrm{DR}^{n-1,1}(A) \xrightarrow{\sim} \Omega^{n-1}(A).$$

We have

$$\iota_\Delta^2 = [d, \iota_\Delta] = d^2 = 0. \tag{8.1}$$

8.2. The Karoubi operator κ. First look at the component of HKR_t at level one. We get from (6.2)

$$\mathrm{HKR}^{(1)} = \frac{1}{(n+1)!}(1 + \kappa + \ldots \kappa^n) \tag{8.2}$$

where

$$\kappa(a_0da_1\ldots da_n) = (-1)^{n-1}da_n \cdot a_0da_1\ldots da_{n-1}. \tag{8.3}$$

LEMMA 8.2. *On Ω^n one has*

$$\iota_\Delta = (1 + \kappa + \ldots + \kappa^{n-1})b.$$

This follows from HKR_t being a morphism of double complexes.

8.3. The Cuntz–Quillen projection P.

LEMMA 8.3.

$$(\kappa^n - 1)(\kappa^{n+1} - 1) = 0$$

on Ω^n.

This is proven in [**CQ**], [**GS1**]. Let

$$F(x) = \frac{(x^n - 1)(x^{n+1} - 1)}{(x-1)^2}. \tag{8.4}$$

We see that $F(x)$ is a polynomial coprime to $x - 1$, and we have

$$(1 - \kappa)^2 F(\kappa) = 0.$$

Therefore we can define P to be the projection upon $\mathrm{Ker}(1-\kappa)^2$ along the sum of generalized eigenspaces of $F(\kappa)$ corresponding to generalized eigenvalues other than one. This projection is a polynomial in κ. It is the Cuntz–Quillen projection. Put also $P^\perp = 1 - P$.

8.4. The main result. We will use the identification

$$C_\bullet(A) \xrightarrow{\sim} \Omega^\bullet(A); a_0 \otimes \ldots \otimes a_n \mapsto a_0 da_1 \ldots da_n \tag{8.5}$$

(the crude HKR). We get two pairs of commuting differentials on $\Omega^\bullet(A)$: one is (b, B) and the other is (ι_Δ, d).

Here we summarize the main result from [**GS2**]. The comparison between the (b, B) and (ι_Δ, d) double complexes is given by Theorem 8.4.

THEOREM 8.4.

(1) *The projections P and $P^\perp$ commute with b, B, ι_Δ, and d.*
(2)
$$\Omega^\bullet(A) = P\Omega^\bullet(A) \oplus P^\perp\Omega^\bullet(A).$$
(3) *On $P\Omega^\bullet(A)$: Let $(\mathcal{N}!)^{-1}$ be the operator whose restriction to $\Omega^n(A)$ is $\frac{1}{n!}\mathrm{Id}$. Then $(\mathcal{N}!)^{-1}$ intertwines b with ι_Δ and B with d.*
(4) *On $P^\perp\Omega^\bullet(A)$: $B = 0$; $\iota_\Delta = 0$; both b and d are contractible.*

Another way to express (3): $P \circ \mathrm{HKR}^{(0)}$ intertwines b with ι_Δ and B with d where

$$\mathrm{HKR}^{(0)}(a_0 \otimes \ldots \otimes a_n) = \frac{1}{n!} a_0 da_1 \ldots da_n. \tag{8.6}$$

In particular we get a theorem from [**GS1**].

THEOREM 8.5. *The periodic cyclic complex $(C_\bullet(A)((u)), b + uB)$ is quasi-isomorphic to $(\Omega^\bullet(A)((u)), \iota_\Delta + ud)$.*

9. The extended noncommutative De Rham complex and the representation scheme

Here we follow [**GS1**], [**GS2**], and [**WY**]. The following was the main motivation for Ginzburg and Schedler to introduce the extended De Rham complex.

For an associative algebra A and a natural number n, let $\mathcal{O}_{\mathrm{Rep}_n(A)}$ be the algebra generated by elements $\rho_{ij}(a)$, k-linear in $a \in A$, $1 \leq i, j \leq n$, subject to relations

$$\rho_{ij}(ab) = \sum_{l=1}^{n} \rho_{il}(a)\rho_{lj}(b); \quad \rho_{ij}(1) = \delta_{ij}.$$

In other words, points of the scheme $\mathrm{Rep}_n(A) = \mathrm{Spec}(\mathcal{O}_{\mathrm{Rep}_n(A)})$ are n-dimensional representations of A.

Noncommutative forms are closely related to geometry of the representation scheme. To start with, an element of $A/[A,A]$ defines an algebraic function on $\mathrm{Rep}_n(A)$ for any n, namely

$$\mathrm{tr}\,\rho(a) = \sum_{j=1}^{n} \rho_{jj}(a). \tag{9.1}$$

More generally, let $(\Omega^\bullet_{\mathrm{Rep}_n(A)}, d)$ be the algebraic De Rham complex of $\mathcal{O}_{\mathrm{Rep}_n(A)}$. Then there is a morphism of complexes

$$\mathrm{DR}^\bullet(A) \to (\Omega^\bullet_{\mathrm{Rep}(A)}, d) \tag{9.2}$$

defined as follows. Define a morphism of DG algebras

$$\Omega^\bullet(A) \to M_n(\Omega^\bullet_{\mathrm{Rep}(A)}) \tag{9.3}$$

by

$$a \mapsto \rho(a);\ da \mapsto d\rho(a), \tag{9.4}$$

and compose it with the trace map

$$\mathrm{tr} : M_n(\Omega^\bullet_{\mathrm{Rep}(A)}) \to \Omega^\bullet_{\mathrm{Rep}(A)}. \tag{9.5}$$

This composition is zero on commutators and therefore produces a morphism (9.2).

The group GL_n acts on the representation scheme. Define the Cartan model of the equivariant De Rham complex

$$\Omega^{\bullet,\mathrm{GL}_n}_{\mathrm{Rep}_n(A)} = ((\Omega^\bullet_{\mathrm{Rep}_n(A)}[\mathfrak{gl}^*_n])^{\mathrm{GL}_n}, d + \iota_{\mathfrak{gl}_n}). \tag{9.6}$$

Then (9.2) can be extended to a morphism of complexes

$$\mathrm{DR}^\bullet_t(A) \to \Omega^{\bullet,\mathrm{GL}_n}_{\mathrm{Rep}_n(A)} \tag{9.7}$$

defined as follows. Extend (9.3) to a morphism

$$\Omega^\bullet_t(A) \to k[\mathfrak{gl}^*_n] \otimes M_n(\Omega^\bullet_{\mathrm{Rep}(A)}) \tag{9.8}$$

by

$$t \mapsto \sum_{j,k} E^*_{jk} \otimes \rho_{jk}(1); \tag{9.9}$$

this intertwines d with d and ι_t with $\iota_{\mathfrak{gl}_n}$. Indeed,

$$\iota_{\mathfrak{gl}_n}(\rho(a)) = 0;$$

$$\iota_{\mathfrak{gl}_n}(d\rho(a)) = \sum_{i,j} E^*_{ij} \otimes [E_{ij}, \rho(a)] = [\sum_{i,j} E^*_{ij} \otimes \rho_{ij}(1), \rho(a)].$$

Follow this with the trace map (9.5) from matrix-valued forms to forms. We get a morphism

$$\omega_0 t\omega_1 \dots t\omega_n \mapsto (X \mapsto \mathrm{tr}(\omega_0 X\omega_1 \dots X\omega_n)) \tag{9.10}$$

($X \in \mathfrak{gl}_n$). The image of this map is obviously inside the submodule of GL_n-invariants. Thus we get the morphism (9.6).

This morphism, and Waikit Yeung's version for multivector fields, provides a bridge between noncommutative geometry of an algebra and classical geometry of its representation scheme.

10. Noncommutative Poincaré lemma and the categorical nature of the differentials

LEMMA 10.1. *Let $f, g : A \to B$ be two morphism of algebras. Then f and g induce homotopic morphisms $(\Omega^\bullet(A), d) \to (\Omega^\bullet(B), d)$.*

PROOF. This follows immediately from the noncommutative Poincaré lemma (Lemma 2.1) (when A is flat over k. In fact a homotopiy can be easily constructed explicitly, and in more than one way. Let

$$\Omega^n(A) \to \oplus_{j=0}^{n-1} \Omega^j(A) t \Omega^{n-1-j}(A) \tag{10.1}$$

be the differential ι_t (4.5) restricted to $\Omega^n(A)$ (i.e., to the homogenous part of degree zero in t of $\Omega_t^\bullet(A)$). We have two morphisms

$$\mu_1, \mu_2 : \Omega^k(A) t \Omega^l(A) \to \Omega^{k+l}(B), \tag{10.2}$$

$$\mu_1(\omega_1 t \omega_2) = f(\omega_1) g(\omega_2);\ \mu_2(\omega_1 t \omega_2) = (-1)^{kl} g(\omega_2) f(\omega_1). \tag{10.3}$$

Both μ_1 and μ_2 are homotopies between $f, g : \Omega^\bullet(A) \to \Omega^\bullet(B)$. This is shown by a direct computation. □

Now let $A = B$ and $f = g = \mathrm{id}_A$. It follows from

$$[d, \mu_j] = f - g$$

that in this case

$$[d, \mu_j] = 0.$$

We have

$$\mu_1 = 0;\ \mu_2 = \iota_\Delta.$$

Is there a categorical reason for $\iota_\Delta^2 = 0$? Let us first tighten the structure. Namely, for any $f_1, \ldots, f_m : A \to B$ define

$$\mu_1(f_1, \ldots, f_m) : \Omega^\bullet(A) \to \Omega^{\bullet - m + 1}(B) \tag{10.4}$$

as follows. Define

$$\iota_t^{(m)} : \Omega^\bullet(A) \to \Omega^\bullet(A) t \ldots t \Omega^\bullet(A)[1 - m], \tag{10.5}$$

$$a_0 da_1 \ldots da_n \mapsto \sum_{1 \le j_1 < \ldots < j_m \le m} (-1)^{j_1 + \ldots + j_m - m} a_0 da_1 \ldots [t, a_{j_1}] \ldots [t, a_{j_m}] \ldots da_n.$$

Also put $\mu_1(f) = f$.

LEMMA 10.2.

$$\sum_{j=2}^{m} (-1)^{j-1} \mu_1(f_1, \ldots, \widehat{f_j}, \ldots, f_{m+1}) + [d, \mu_1(f_1, \ldots, f_{m+1})] = 0.$$

The proof is by direct verification.

Let $\mathbf{c}(A, B)$ be the category whose objects are morphisms $f : A \to B$ and there is one morphism between any f and g. Lemma 10.2 states that morphisms μ_1 define a pairing

$$\Omega^\bullet(A) \otimes \mathbb{Z}\mathrm{N}(\mathbf{c}(A, B)) \to \Omega^\bullet(B). \tag{10.6}$$

Here N stands for the nerve, and $\mathbb{Z}\mathrm{N}$ for the simplicial free Abelian group generated by the nerve and viewed as a chain complex.

There is also the product

$$\mathbb{Z}\,\mathrm{N}(\mathbf{c}(A,B))\otimes\mathbb{Z}\,\mathrm{N}(\mathbf{c}(B,C))\to\mathbb{Z}\,\mathrm{N}(\mathbf{c}(A,C)). \tag{10.7}$$

It is defined as follows. For $f_j : A\to B$ and $g_k : B\to C$,

$$(f_1|\ldots|f_m)(g_1|\ldots|g_n)=\sum\pm(\ldots|g_kf_j|\ldots) \tag{10.8}$$

where the sum is taken over all paths from g_1f_1 to g_nf_m using the following two moves:

$$\text{from } g_kf_j \text{ to } g_{k+1}f_j \text{ or to } g_kf_{j+1}$$

Every permutation of f_j and g_k introduces a factor -1. For example,

$$(f_1|f_2)(g_1|g_2)=(g_1f_1|g_1f_2|g_2f_2)-(g_1f_1|g_2f_1|g_2f_2).$$

Lemma 10.3. *The product* (10.7) *is associative and compatible with the action* (10.6).

We do not know at the moment if μ_2 (and its partial case ι_Δ) is part of the structure described above. We know this for another related structure that we are going to describe next.

10.1. The Čech–Alexander complex. The noncommutative complex $\Omega^\bullet(A)$ is a quotient of another important complex:

$$A^{(n)}=A^{\otimes(n+1)};\ \check{\partial}:A^{(n)}\to A^{(n+1)};\ \check{\partial}=\sum_{j=0}^{n+1}(-1)^j\partial_j; \tag{10.9}$$

$$\partial_j(a_0\otimes\ldots\otimes a_n)=a_0\otimes\ldots\otimes 1\otimes a_j\otimes\ldots\otimes a_n \tag{10.10}$$

Define

$$\mu(f_1,\ldots,f_m)(a_0\otimes\ldots\otimes a_n)=\sum\pm f_1(a_0)\otimes\ldots\otimes f_1(a_{j_1})f_2(a_{j_1+1})$$
$$\otimes\ldots\otimes f_{m-1}(a_{j_{m-1}})f_m(a_{j_{m-1}+1})\otimes\ldots\otimes f_m(a_n).$$

The sum is taken over all $0\leq j_1<\ldots<j_m\leq n-1$. We get the pairing

$$A^{(\bullet)}\otimes\mathbb{Z}\,\mathrm{N}(\mathbf{c}(A,B))\to B^{(\bullet)}. \tag{10.11}$$

Lemma 10.4. *The product* (10.7) *is compatible with the action* (10.11).

There is an extra piece of structure here. Namely, the deconcatenation operations

$$(f_1,\ldots,f_m)\mapsto\sum_{j=1}^{m-1}(f_1,\ldots,f_j)\otimes(f_{j+1},\ldots,f_m) \tag{10.12}$$

and

$$a_1\otimes\ldots\otimes a_n\mapsto\sum_{j=1}^{n-1}(a_1\otimes\ldots\otimes a_j)\otimes(a_{j+1}\otimes\ldots\otimes a_n) \tag{10.13}$$

turn both $\mathbb{Z}\,\mathrm{N}(\mathbf{c}(A,B))$ and

$$\bar{A}^{(\bullet)}[1]=(A^{(\bullet)}/k^{(\bullet)})[1] \tag{10.14}$$

into differential graded coalgebras.

PROPOSITION 10.5. *The product* (10.7) *and the pairing*

$$\bar{A}^{(\bullet)}[1] \otimes \mathbb{Z}\,\mathrm{N}(\mathbf{c}(A,B)) \to \bar{B}^{(\bullet)}[1]$$

(induced by (10.11)*) are morphisms of differential graded coalgebras.*

When $A = B$ and $f = g = \mathrm{id}_A$ then we recover the bar differential

$$\mu(\mathrm{id},\mathrm{id})(a_0 \otimes \ldots \otimes a_n) = \sum_{j=0}^{n-1} (-1)^j a_0 \otimes \ldots \otimes a_j a_{j+1} \otimes \ldots \otimes a_n. \tag{10.15}$$

The identity $\mu(\mathrm{id},\mathrm{id})^2 = 0$ follows from (10.4) though of course it can be checked directly.

We have constructed a category in differential graded coalgebras; Čech–Alexander complexes form "a module in coalgebras" over it. This provides a strong form of the statement that all morphisms define the same morphism on Čech–Alexander cohomology.

REMARK 10.6. Of course the last sentence above is obviously true because the embedding $k^{(\bullet)} \to A^{(\bullet)}$ is a quasi-isomorphism. (In fact, for any section $s : A \to k$ of the embedding $k \to A$, the map

$$a_0 \otimes \ldots \otimes a_n \mapsto s(a_0) a_1 \otimes \ldots \otimes a_n \tag{10.16}$$

is a homotopy between that embedding and the identity). There are cases, however, when we can modify the complex and then its cohomology becomes interesting. For example, let A be a finitely generated commutative algebra. Let $\pi : P \to A$ be an epimorphism where P is a polynomial algebra in finitely many variables. Let I be the kernel of π. Denote by $\widehat{P}^{(n)}$ the $I^{(n)}$-adic completion of $P^{\otimes(n+1)}$ where

$$I^{(n)} = \mathrm{Ker}(P^{\otimes(n+1)} \to A).$$

In characteristic p one can also take the completed divided powers envelope. Then the differential $\check{\partial}$ and the maps $\mu(f_1, \ldots, f_m)$ extend to $P^{(\bullet)}$ while the homotopy (10.16) does not. For every $\pi : P \to A$ and $\rho : Q \to B$ and for every $f : A \to B$ there is a morphism $\widetilde{f} : P \to Q$ such that $\rho\widetilde{f} = f\pi$. Therefore the complex $P^{(\bullet)}$ depends, essentially, only on A. Lemma 10.4 and Proposition 10.5 also provide a recipe for gluing the complexes $P^{(\bullet)}$ for a sheaf of commutative algebras. This gives a definition of crystalline cohomology [**Be**], [**Gr**].

10.2. Noncommutative Čech–Alexander complex. Now let R be an associative algebra. Let $R^{(n)}$ be the $(n+1)$-fold *free* product of R with itself. (This is the $(n+1)$-fold coproduct in the category of associative rather than commutative algebras). One defines the coface maps

$$\partial_j : R^{(n-1)} \to R^{(n)},\ 0 \le j \le n \tag{10.17}$$

as follows. We denote by R_k the kth copy of R in the free product. Then ∂_j is the morphism of algebras that sends R_k identically to R_k for $k < j$ and to R_{k+1} for $k \le j$.

Now let R and S be two associative algebras. Let $f_1, \ldots, f_m : R \to S$ be morphisms of algebras. The maps

$$\mu(f_1, \ldots, f_m) : R^{(n)} \to S^{(n-m+1)}$$

are defined analogously to the above. Namely, for any nondecreasing surjection $\sigma : \{0, \dots, n\} \to \{0, \dots, n-m+1\}$ let

$$h(\sigma) : R^{(n)} \to S^{(n-m+1)}$$

be the algebra morphism that sends R_k to $S_{\sigma(k)}$ by the morphism $f_{k-\sigma(k)+1}$. Then

$$\mu(f_1, \dots, f_m) = \sum (-1)^{p(\sigma)} h(\sigma)$$

where

$$p(\sigma) = \sum_{l=0}^{n-m+1} l \operatorname{Card}(\sigma^{-1}(\{l\})).$$

The coalgebra structure on the Čech–Alexander complexes also lifts from the commutative case. Define for $0 \leq j < n$

$$\Delta_{j,n} : R^{(n)} \to R^{(j)} \otimes R^{(n-1-j)}$$

to be the algebra morphism that sends R_k identically to $R_k \otimes 1$ for $k \leq j$ and to $1 \otimes R_{k-j-1}$ for $k > j$. Then

$$\Delta | R^{(n)} = \sum_{j=0}^{n-1} \Delta_{j,n} \tag{10.18}$$

defines a DG coalgebra structure on $R^{(\bullet)}[1]$.

PROPOSITION 10.7. *The product* (10.7) *and the pairing*

$$\bar{R}^{(\bullet)}[1] \otimes \mathbb{Z}\,\mathrm{N}(\mathbf{c}(R, S)) \to \bar{S}^{(\bullet)}[1]$$

(induced by (10.11)*) are morphisms of differential graded coalgebras. They satisfy the associativity condition when three algebras A, B, C are given.*

10.2.1. *Noncommutative crystalline cohomology in characteristic zero.* Here we follow Cortiñas [**Cor1**]. Let A be an associative algebra. Let $R \to A$ be an epimorphism where R is a free algebra. Let $J^{(n)} = \operatorname{Ker}(R^{(n)} \to A)$. Let $R^{(\bullet)}/[R^{(\bullet)}, R^{(\bullet)}]^{\wedge}$ be the completion of $R^{(\bullet)}/[R^{(\bullet)}, R^{(\bullet)}]$ with respect to the filtration induced by powers of $J^{(\bullet)}$.

THEOREM 10.8 ([**Cor1**]). *The complex*

$$R_\natural{}^{(\bullet)\wedge} = R^{(\bullet)}/[R^{(\bullet)}, R^{(\bullet)}]^{\wedge}, \check{\partial}) \tag{10.19}$$

computes the periodic cyclic homology of A.

All the morphisms $\mu(f_1, \dots, f_m)$ extend to the completion. The coalgebra structure descends to the quotient by commutators. Also, denote

$$\bar{R}_\natural^{(\bullet)\wedge} = R_\natural{}^{(\bullet)\wedge}/k_\natural{}^{(\bullet)\wedge}$$

We get a stronger form of the statement [**Cor1**] that the cohomology is independent of the choice of R.

PROPOSITION 10.9. *The product* (10.7) *and the pairing*

$$\bar{R}_\natural^{(\bullet)\wedge} \otimes \mathbb{Z}\,\mathrm{N}(\mathbf{c}(R, S)) \to \bar{S}_\natural^{(\bullet)\wedge}$$

(induced by (10.11)*) are morphisms of differential graded coalgebras that satisfy the associativity condition.*

Let us say a few words about the proof of Theorem 10.19. In the commutative case, the classical proof starts with the Čech–Alexander–De Rham double complex and shows that its total complex is quasi-isomorphic to both Čech–Alexander and De Rham complexes. In the noncommutative case, the Čech–Alexander complex already *is* a version of the De Rham complex. This is shown by expressing the algebras $R^{(\bullet)}$ in terms of noncommutative forms (generalizing the case $\bullet = 1$ due to Cuntz and Quillen). After that, the theorem follows from Karoubi's theorem (Theorem 6.2) and Goodwillie's theorem [**Go**].

10.3. Bar construction and Hochschild cochains. There is another, possibly related, example of the structure discussed above. Namely, replace the additive category $\mathbb{Z}\mathbf{c}(A,B)$ by the differential graded category $\mathbf{C}^\bullet(A,B)$ as follows. Its objects are morphisms $f : A \to B$. The complex of morphisms is

$$\mathbf{C}^\bullet(A,B)(f,g) = C^\bullet(A, {}_fB_g) = \mathrm{Hom}_k(A^{\otimes\bullet}, B), \tag{10.20}$$

the Hochschild cochain complex of A with coefficients in B viewed as an A-bimodule with the action $a_1\cdot b\cdot a_2 = f(a_1)bg(a_2)$. The composition is given by the cup product.

There are morphisms of DG coalgebras

$$\mathrm{Bar}(\mathbf{C}^\bullet(A,B)) \otimes \mathrm{Bar}(\mathbf{C}^\bullet(B,C)) \to \mathrm{Bar}(\mathbf{C}^\bullet(A,B)), \tag{10.21}$$

$$\mathrm{Bar}(A) \otimes \mathrm{Bar}(\mathbf{C}^\bullet(B,C)) \to \mathrm{Bar}(B). \tag{10.22}$$

The first one is associative and the two agree with each other. The morphisms are a categorized version of the morphisms based on brace operations [**GV**], [**GJ**]. The details can be found in [**T2**] and [**NT**].

This is one way of saying that algebras form a two-category, i.e., a category in DG categories. Namely, the above says that algebras form a (strict) category in DG cocategories; one can apply the bar construction and define a version of a 2-category up to homotopy.

REMARK 10.10. One gets a feeling that the algebraic structures described in (10.6), Propositions 10.5 and 10.9, (10.21), and (10.22) are parts of the same structure. More precisely: the bar differential appears as a differential in Proposition 10.9 and as the action of the morphism (id, id) in Proposition 10.5; Čech–Alexander cochains project to forms (via the naive HKR map); and the cocategories (coalgebras) $\mathbb{Z}[N\mathbf{c}(A,B), \mathbf{C}(A,B)]$ are clearly related.

Acknowledgments

The author is grateful to P. Bressler, A. Gorokhovsky, D. Kaledin, R. Kaufmann, M. Rivera, and R. Nest for fruitful discussions, and to E. Getzler, V. Ginzburg, A. Goncharov, M. Kontsevich, T. Schedler, and Waikit Yeung for patiently explaining to him various aspects of their work on noncommutative forms and related topics. The author is also grateful to the referee whose careful reading and many insights and suggestions really improved the paper.

References

[B] A. A. Beĭlinson, *On the derived category of perverse sheaves*, K-theory, arithmetic and geometry (Moscow, 1984), Lecture Notes in Math., vol. 1289, Springer, Berlin, 1987, pp. 27–41, DOI 10.1007/BFb0078365. MR923133

[Be] P. Berthelot, *Cohomologie cristalline des schémas de caractéristique* $p > 0$ (French), Lecture Notes in Mathematics, Vol. 407, Springer-Verlag, Berlin-New York, 1974. MR384804

[C1] H. Takai and T. Natsume, *A. Connes' noncommutative differential geometry* (Japanese), Sūgaku **35** (1983), no. 2, 97–112. MR732432

[C2] A. Connes, *Introduction to noncommutative differential geometry*, Workshop Bonn 1984 (Bonn, 1984), Lecture Notes in Math., vol. 1111, Springer, Berlin, 1985, pp. 3–16, DOI 10.1007/BFb0084582. MR797413

[C3] A. Connes, *Noncommutative differential geometry*, Inst. Hautes Études Sci. Publ. Math. **62** (1985), 257–360. MR823176

[Cor1] G. Cortiñas, *Periodic cyclic homology as sheaf cohomology*, K-Theory **20** (2000), no. 2, 175–200, DOI 10.1023/A:1007867728238. Special issues dedicated to Daniel Quillen on the occasion of his sixtieth birthday, Part II. MR1798825

[Cor2] G. Cortiñas, *Infinitesimal K-theory*, J. Reine Angew. Math. **503** (1998), 129–160, DOI 10.1515/crll.1998.094. MR1650347

[CQ] J. Cuntz and D. Quillen, *Operators on noncommutative differential forms and cyclic homology*, Geometry, topology, & physics, Conf. Proc. Lecture Notes Geom. Topology, IV, Int. Press, Cambridge, MA, 1995, pp. 77–111. MR1358613

[DK1] W. G. Dwyer and D. M. Kan, *Three homotopy theories for cyclic modules*, Proceedings of the Northwestern conference on cohomology of groups (Evanston, Ill., 1985), J. Pure Appl. Algebra **44** (1987), no. 1-3, 165–175, DOI 10.1016/0022-4049(87)90022-3. MR885102

[DK2] W. G. Dwyer, M. J. Hopkins, and D. M. Kan, *The homotopy theory of cyclic sets*, Trans. Amer. Math. Soc. **291** (1985), no. 1, 281–289, DOI 10.2307/1999909. MR797060

[DK3] W. G. Dwyer and D. M. Kan, *Normalizing the cyclic modules of Connes*, Comment. Math. Helv. **60** (1985), no. 4, 582–600, DOI 10.1007/BF02567433. MR826872

[FT] B. L. Feĭgin and B. L. Tsygan, *Cyclic homology of algebras with quadratic relations, universal enveloping algebras and group algebras*, K-theory, arithmetic and geometry (Moscow, 1984), Lecture Notes in Math., vol. 1289, Springer, Berlin, 1987, pp. 210–239, DOI 10.1007/BFb0078369. MR923137

[GV] M. Gerstenhaber and A. A. Voronov, *Homotopy G-algebras and moduli space operad*, Internat. Math. Res. Notices **3** (1995), 141–153, DOI 10.1155/S1073792895000110. MR1321701

[GJ] E. Getzler and J. D. S. Jones, *Operads, homotopy algebra, and iterated integrals for double loop spaces*, hep-th/9305013, 1–70 (1993).

[GS1] V. Ginzburg and T. Schedler, *Free products, cyclic homology, and the Gauss-Manin connection*, Adv. Math. **231** (2012), no. 3-4, 2352–2389, DOI 10.1016/j.aim.2012.06.016. MR2964640

[GS2] V. Ginzburg and T. Schedler, *A new construction of cyclic homology*, Proc. Lond. Math. Soc. (3) **112** (2016), no. 3, 549–587, DOI 10.1112/plms/pdw001. MR3474483

[GK] A. Goncharov and M. Kontsevich, *Spectral description of noncommutative local systems on surfaces and noncommutative cluster varieties*, arXiv:2108.04168, (2004).

[Go] T. G. Goodwillie, *Cyclic homology, derivations, and the free loopspace*, Topology **24** (1985), no. 2, 187–215, DOI 10.1016/0040-9383(85)90055-2. MR793184

[Gr] A. Grothendieck, *Crystals and the de Rham cohomology of schemes*, Dix exposés sur la cohomologie des schémas, Adv. Stud. Pure Math., vol. 3, North-Holland, Amsterdam, 1968, pp. 306–358. Notes by I. Coates and O. Jussila. MR269663

[HKR] G. Hochschild, B. Kostant, and A. Rosenberg, *Differential forms on regular affine algebras*, Trans. Amer. Math. Soc. **102** (1962), 383–408, DOI 10.2307/1993614. MR142598

[Ka] D. B. Kaledin, *Cartier isomorphism for unital associative algebras*, Proc. Steklov Inst. Math. **290** (2015), no. 1, 35–51, DOI 10.1134/S0081543815060048. Published in Russian in Tr. Mat. Inst. Steklova **290** (2015), 43–60. MR3488779

[K1] M. Karoubi, *Homologie cyclique et K-théorie* (French, with English summary), Astérisque **149** (1987), 147. MR913964

[K2] M. Karoubi, *Correspondance de Dold-Kan et formes differentielles* (French), J. Algebra **198** (1997), no. 2, 618–626, DOI 10.1006/jabr.1997.7168. MR1489915

[K3] M. Karoubi, *Formes différentielles non commutatives et opérations de Steenrod* (French), Topology **34** (1995), no. 3, 699–715, DOI 10.1016/0040-9383(94)00043-K. MR1341815

[K4] M. Karoubi, *Formes différentielles non commutatives et cohomologie à coefficients arbitraires* (French), Trans. Amer. Math. Soc. **347** (1995), no. 11, 4277–4299, DOI 10.2307/2155038. MR1316853

[K5] M. Karoubi, *Formes topologiques non commutatives* (French, with English summary), Ann. Sci. École Norm. Sup. (4) **28** (1995), no. 4, 477–492. MR1334610

[K6] M. Karoubi, *Méthodes quantiques en topologie algébrique* (French, with English and French summaries), C. R. Acad. Sci. Paris Sér. I Math. **328** (1999), no. 9, 755–758, DOI 10.1016/S0764-4442(99)80266-3. MR1694110

[K7] M. Karoubi, *Quantum methods in algebraic topology*, Topology, geometry, and algebra: interactions and new directions (Stanford, CA, 1999), Contemp. Math., vol. 279, Amer. Math. Soc., Providence, RI, 2001, pp. 177–193, DOI 10.1090/conm/279/04560. MR1850747

[K8] M. Karoubi, *Braiding of differential forms and homotopy types* (English, with English and French summaries), C. R. Acad. Sci. Paris Sér. I Math. **331** (2000), no. 10, 757–762, DOI 10.1016/S0764-4442(00)01693-1. MR1807185

[L] J.-L. Loday, *Cyclic homology*, Grundlehren der mathematischen Wissenschaften [Fundamental Principles of Mathematical Sciences], vol. 301, Springer-Verlag, Berlin, 1992. Appendix E by María O. Ronco, DOI 10.1007/978-3-662-21739-9. MR1217970

[NT] R. Nest and B. Tsygan, *Cyclic Homology*, book in progress, available at: `https://sites.math.northwestern.edu/~tsygan/Part1.pdf`

[Q] D. Quillen, *Algebra cochains and cyclic cohomology*, Inst. Hautes Études Sci. Publ. Math. **68** (1988), 139–174 (1989). MR1001452

[R] G. S. Rinehart, *Differential forms on general commutative algebras*, Trans. Amer. Math. Soc. **108** (1963), 195–222, DOI 10.2307/1993603. MR154906

[T1] B. L. Tsygan, *Homology of matrix Lie algebras over rings and the Hochschild homology* (Russian), Uspekhi Mat. Nauk **38** (1983), no. 2(230), 217–218. MR695483

[T2] B. Tsygan, *Noncommutative calculus and operads*, Topics in noncommutative geometry, Clay Math. Proc., vol. 16, Amer. Math. Soc., Providence, RI, 2012, pp. 19–66. MR2986860

[WY] Wai-Kit Yeung, *Pre-Calabi-Yau structures and moduli of representations*, `arXiv:1802.05398`, (2022).

Department of Mathematics, Northwestern University, Evanston, Illinois 60208
Email address: `b-tsygan@northwestern.edu`

Contemporary Mathematics
Volume **802**, 2024
https://doi.org/10.1090/conm/802/16071

An algebraic model for the free loop space

Manuel Rivera

ABSTRACT. We describe an algebraic chain level construction that models the passage from an arbitrary topological space to its free loop space. The input of the construction is a categorical coalgebra, i.e., a curved coalgebra satisfying certain properties, and the output is a chain complex. The construction is a modified version of the co-Hochschild complex of a differential graded (dg) coalgebra. When applied to the chains on an arbitrary simplicial set X, appropriately interpreted, it yields a chain complex that is naturally quasi-isomorphic to the singular chains on the free loop space of the geometric realization of X. We relate this construction to a twisted tensor product model for the free loop space constructed using the adjoint action of a dg Hopf algebra model for the based loop space.

1. Introduction

To any space Y one can naturally associate a new space LY defined as the set of continuous maps from the circle S^1 to Y equipped with the compact-open topology. The space LY is called the *free loop space* of Y and may be equipped with a natural S^1-action given by rotation of loops. In this article, we present an algebraic chain level model for the passage $Y \mapsto LY$ that does not assume any hypothesis on the underlying space Y. The construction requires a small amount of data and, consequently, is potentially useful for calculations and for studying how the algebraic topology of a geometric space manifests at the level of the free loop space.

The main idea is inspired by the following picture. Suppose X is a simplicial set. For the purposes of capturing the intuitive idea, the reader may assume X arises from a simplicial complex equipped with a total ordering of its vertices. We consider ordered sequences $(\sigma_0, \cdots, \sigma_p)$ of simplicies in X such that the last vertex of σ_i is the first vertex of σ_{i+1} for $i = 0, \ldots, p-1$ and the last vertex of σ_p is the first vertex of σ_0. These ordered sequences of simplices were called *closed necklaces* in [**RS18**].

We want to think of a closed necklace in X as a family of free loops in $|X|$, the geometric realization of X, parameterized by a cube of an appropriate dimension. More precisely, adapting a classical construction of Adams described in [**Ada56**], it is possible to decompose a closed necklace $(\sigma_0, \cdots, \sigma_p)$ into a family of free

2020 *Mathematics Subject Classification.* Primary 18Mxx.

This research was supported by NSF Grant 210554 and the Karen EDGE Fellowship.

©2024 American Mathematical Society

loops in $|X|$ parameterized by a particular subdivision of a cube of dimension $|\sigma_0| + \cdots + |\sigma_p| - p$ having the following properties:

(i) the base points of every loop in this family always lie inside the special simplex σ_0, and
(ii) the boundary of such family may be described in terms of all "sub-closed necklaces" of codimension 1.

In [**RS18**], this idea was used to construct a *combinatorial* model for $L|X|$ given by gluing a set of polyhedra indexed by closed necklaces in X. In the present paper, we are concerned with an *algebraic* version of this construction. Namely, for an arbitrary commutative ring $\mathbf{k}$, we describe a functorial construction that produces a $\mathbf{k}$-chain complex, directly from the natural algebraic structure of the normalized $\mathbf{k}$-chains $C_*(X)$ suitable interpreted, that computes the $\mathbf{k}$-homology of $L|X|$. We highlight four essential observations that are used in this algebraic construction.

(1) The first observation is that the graded $\mathbf{k}$-module freely generated by closed necklaces in X may be described algebraically in terms of the *cotensor product* for a bicomodule structure of the normalized chains $C_*(X)$ over the coalgebra $C_0(X) = \mathbf{k}[X_0]$ generated by the set of vertices of X. This bicomodule structure on $C_*(X)$ is determined by projecting a simplex to its first or last vertex. Then for any pair of simplices σ_0 and σ_1 in X, requiring the last vertex of σ_0 to be the first vertex of σ_1 is equivalent to requiring the tensor $\sigma_0 \otimes \sigma_1 \in C_*(X) \otimes_{\mathbf{k}} C_*(X)$ to lie inside the sub-$\mathbf{k}$-module

$$C_*(X) \square_{C_0(X)} C_*(X) \subseteq C_*(X) \otimes_{\mathbf{k}} C_*(X),$$

where $\square_{C_0(X)}$ denotes the cotensor product of $C_0(X)$-bicomodules.

(2) The second observation is that the boundary of a closed necklace representing a family (or "chain") of free loops may be described algebraically in terms of certain simplicial face maps and the Alexander-Whitney coproduct

$$\Delta \colon C_*(X) \to C_*(X) \square_{C_0(X)} C_*(X).$$

The resulting description of the boundary is reminiscent of the differential of the *co-Hochschild complex* of a differential graded (dg) coalgebra as studied in [**Doi81**], [**HPS09**], and other articles. However, now we are in the different context of comonoids in the category of bicomodules over a coalgebra with cotensor product as monoidal structure.

(3) The third observation is that the relevant structure of $C_*(X)$ for our purposes may be packaged as a curved coalgebra satisfying certain properties, which we call a *categorical coalgebra*. This notion is inspired by Holstein and Lazarev's categorical Koszul duality theory [**HL22**]. Any categorical coalgebra C gives rise to a comonoid in the monoidal category of C_0-bicomodules with cotensor product. Furthermore, any categorical coalgebra gives rise to a dg category through a many object version of the cobar construction. The main construction of the article is then a version of the co-Hochschild chain complex for categorical coalgebras that coincides with the classical co-Hochschild complex when restricted to connected dg coalgebras. It has a categorical coalgebra as input and a mixed complex, i.e. a chain complex equipped with an additional degree $+1$ operator squaring to zero, as output. The co-Hochschild complex is invariant with respect to a suitable notion of weak equivalence drawn from Koszul duality theory.

(4) The fourth observation is that in order to recover a homological model for the free loop space $L|X|$ for an arbitrary simplicial set X, certain formal localization must be performed at the 1-simplices of X. This step is not necessary if all the 1-simplices in X already have inverses up to homotopy (e.g., X is a Kan complex). This localization may be described in purely algebraic terms. In order to extract the set of elements to be localized, we consider categorical coalgebras equipped with an additional dg coalgebra enrichment on their associated dg category. We call these *B_∞-categorical coalgebras*. We define an extended version of the co-Hochschild complex that takes a B_∞-categorical coalgebra and produces a chain complex by formally inverting a particular set (extracted by applying the *set-like elements functor* to the dg coalgebra enrichment of the associated cobar dg category) in the co-Hochschild complex of the underlying categorical coalgebra.

Our main result, informally stated, is the following.

THEOREM 1.1. *For an arbitrary simplicial set X, the extended co-Hochschild complex of $C_*(X)$, a B_∞-categorical coalgebra model for the normalized chains on X, is naturally quasi-isomorphic to $C_*^{\text{sing}}(L|X|)$, the singular chains on the free loop space.*

As our proof will reveal, this result may be understood as a simplification of a theorem proved by Goodwillie in [**Goo85**] and independently by Burghelea and Fedorowicz in [**BF86**] saying the following. For any path-connected pointed topological space (Y, b), the Hochschild chain complex of the Pontryagin dg algebra $C_*^{\text{sing}}(\Omega_b Y)$ of singular chains on the based (Moore) loop space of Y at b is naturally quasi-isomorphic to $C_*^{\text{sing}}(LY)$. Our streamlined model is essentially deduced from this result using (a generalization of) the fact that for any conilpotent dg coalgebra C we have two resolutions for the dg algebra $A = \text{Cobar}(C)$ as an A-bimodule: 1) the classical two-sided bar resolution $\text{Bar}(A, A, A)$, and 2) a smaller resolution $\mathcal{Q}(A, C, A)$ with underlying module $A \otimes C \otimes A$. The first one is used when defining the Hochschild complex of A, while the second one is used when defining the co-Hochschild complex of C. We describe an explicit natural quasi-isomorphism of A-bimodules

$$\mathcal{Q}(A, C, A) \xrightarrow{\simeq} \text{Bar}(A, A, A),$$

see Proposition 5.2.

Finally, we establish a relationship between the extended co-Hochschild complex model for the free loop space and Brown's twisted tensor product model for a fibration. This involves proving that a natural dg bialgebra structure constructed on the extended cobar construction of a reduced simplicial set is in fact a dg Hopf algebra. We then model the holonomy of the free loop space fibration in terms of the adjoint action of such dg Hopf algebra. This relationship with Brown's twisted tensor product may be used to give an algebraic model of the inclusion $Y \hookrightarrow LY$ of points as constant loops in terms of the co-Hochschild complex. We expect this to be useful in studying and computing the string topology of non-simply connected manifolds.

2. Preliminaries

Fix a commutative ring with unit $\mathbf{k}$. We assume familiarity with the notions of differential graded (dg) $\mathbf{k}$-modules, $\mathbf{k}$-algebras, $\mathbf{k}$-coalgebras and $\mathbf{k}$-categories. For generalities about dg categories and their homotopy theory we refer to [**Tab05**],

[**Tab10**], and [**Toë11**]. All (co)algebras in this article will be (co)associative and (co)unital. All differentials will have degree -1. We denote by $\mathsf{Ch}_{\mathbf{k}}$ the category of dg $\mathbf{k}$-modules (i.e., $\mathbf{k}$-chain complexes) and $\mathsf{Ch}_{\mathbf{k}}^{\geq 0}$ its full sub-category of non-negatively graded objects. Denote by $\mathsf{dgAlg}_{\mathbf{k}}$, $\mathsf{dgCoalg}_{\mathbf{k}}$, and $\mathsf{dgCat}_{\mathbf{k}}$ the categories of dg algebras, dg coalgebras, and small dg categories, respectively. A additional subscript of "≥ 0" in the notation for these categories will also mean the full sub-category of non-negatively graded objects. Whenever we write $\otimes$ we mean $\otimes_{\mathbf{k}}$, unless noted otherwise. All signs in this article are determined by the Koszul sign convention.

For any graded algebra A, we denote by A^{op} the graded algebra with A as underlying $\mathbf{k}$-module and multiplication defined by $\mu_A \circ t$, where $t : A \otimes A \to A \otimes A$ is given by $t(a \otimes b) = (-1)^{|a||b|} b \otimes a$ and $\mu\colon A \otimes A \to A$ is the multiplication of A. Similarly, for any graded coalgebra C, we denote by C^{op} the graded coalgebra with X as underlying $\mathbf{k}$-module and coproduct defined by $t \circ \Delta$ where $\Delta\colon C \to C \otimes C$ is the coproduct of C.

For any set S we denote by $\mathbf{k}[S]$ the $\mathbf{k}$-coalgebra whose underlying $\mathbf{k}$-module is freely generated by S and whose coproduct $\Delta\colon \mathbf{k}[S] \to \mathbf{k}[S] \otimes \mathbf{k}[S]$ is determined by $\Delta(s) = s \otimes s$ for any $s \in S$. The counit $\varepsilon\colon \mathbf{k}[S] \to \mathbf{k}$ is detrmined by $\varepsilon(s) = 1_{\mathbf{k}}$ for any $s \in S$.

2.1. Cotensor product. Let C be a dg $\mathbf{k}$-coalgebra, which is flat as a $\mathbf{k}$-module. Let M and N be dg right and left C-comodules, respectively, with coaction maps $\rho_M\colon M \to M \otimes C$ and $\rho_N\colon N \to C \otimes N$. The *cotensor product of M and N over C* is defined as

$$M \underset{C}{\square} N = \ker(\rho_M \otimes \mathrm{id}_N - \mathrm{id}_M \otimes \rho_N\colon M \otimes N \to M \otimes C \otimes N). \tag{2.1}$$

The category of dg C-bicomodules, denoted by C-$\mathsf{biComod}$, becomes a monoidal category when equipped with the cotensor product $\square_C$ and unit object C. Suppose A is a monoid in this category, namely, a dg C-bicomodule equipped with an associative product $A \square_C A \to A$ and unit $u : C \to A$. Let E and F be dg right and left C-comodules, respectively. Suppose that E and F are further equipped with right and left dg A-module structures respectively, namely, we have action maps $\rho_E : E \square_C A \to E$ and $\rho_F : A \square_C F \to F$ satisfying the usual compatibilities. Define the *tensor product of E and F over A* to be the dg $\mathbf{k}$-module

$$E \underset{A}{\bigotimes} F = \operatorname{coker}(\rho_E \underset{C}{\square} \mathrm{id}_F - \mathrm{id}_E \underset{C}{\square} \rho_F\colon E \underset{C}{\square} A \underset{C}{\square} F \to E \underset{C}{\square} F). \tag{2.2}$$

2.2. DG categories as monoids. Given any dg category $\mathsf{A} \in \mathsf{dgCat}_{\mathbf{k}}$ with object set $\mathcal{O}_{\mathsf{A}}$ define a monoid $\mathcal{M}(\mathsf{A})$ in the monoidal category of dg $\mathbf{k}[\mathcal{O}_{\mathsf{A}}]$-bicomodules equipped with the cotensor product $\square_{\mathbf{k}[\mathcal{O}_{\mathsf{A}}]}$ as follows. The underlying dg $\mathbf{k}$-module of $\mathcal{M}(\mathsf{A})$ is given by the direct sum

$$\bigoplus_{x,y \in \mathcal{O}_{\mathsf{A}}} \mathsf{A}(x,y).$$

The $\mathbf{k}[\mathcal{O}_{\mathsf{A}}]$-bicomodule structure maps

$$\mathcal{M}(\mathsf{A}) \to \mathbf{k}[\mathcal{O}_{\mathsf{A}}] \otimes \mathcal{M}(\mathsf{A})$$

and

$$\mathcal{M}(\mathsf{A}) \to \mathcal{M}(\mathsf{A}) \otimes \mathbf{k}[\mathcal{O}_{\mathsf{A}}]$$

are induced by the source and target maps in A, respectively. The monoid structure

$$\mathcal{M}(\mathsf{A}) \underset{\mathbf{k}[\mathcal{O}_\mathsf{A}]}{\square} \mathcal{M}(\mathsf{A}) \to \mathcal{M}(\mathsf{A})$$

is induced by the composition of morphisms in A and the unit map $\mathbf{k}[\mathcal{O}_\mathsf{A}] \to \mathcal{M}(\mathsf{A})$ is determined by $x \mapsto \mathrm{id}_x \in \mathsf{A}(x,x)_0$ for all $x \in \mathcal{O}_\mathsf{A}$.

2.3. Two sided bar construction. Let A be a dg category and, for simplicity, denote by A_0 the coalgebra $\mathbf{k}[\mathcal{O}_\mathsf{A}]$. Let M and N be right and left dg modules over $\mathcal{M}(\mathsf{A})$, respectively, in the monoidal category $(\mathsf{A}_0\text{-biComod}, \square_{\mathsf{A}_0})$. This means that M and N are dg A_0-bicomodules equipped dg maps

$$\mathsf{M}\square_{\mathsf{A}_0}\mathcal{M}(\mathsf{A}) \to \mathsf{M} \text{ and } \mathcal{M}(\mathsf{A})\square_{\mathsf{A}_0}\mathsf{N} \to \mathsf{N}$$

defining right and left dg $\mathcal{M}(\mathsf{A})$-actions, respectively. The *two-sided bar construction of* M *and* N *over* A is the dg A_0-bicomodule

$$\mathrm{Bar}_{\mathsf{A}_0}(\mathsf{M}, \mathsf{A}, \mathsf{N})$$

defined as follows. The underlying graded $\mathbf{k}$-module is defined to be

$$\bigoplus_{i=0}^{\infty} \Big(\mathsf{M}\underset{\mathsf{A}_0}{\square}(s^{+1}\overline{\mathcal{M}}(\mathsf{A}))^{\square i}\underset{\mathsf{A}_0}{\square}\mathsf{N}\Big),$$

where $\overline{\mathcal{M}}(\mathsf{A}) = \mathcal{M}(\mathsf{A})/u(\mathsf{A}_0)$, where $u : \mathsf{A}_0 \to \overline{\mathcal{M}}(\mathsf{A})$ is the unit map.

We will use the classical "bar" notation

$$m[a_1|\cdots|a_p]n$$

to denote a generator

$$m\square s^{+1}a_1\square\cdots\square s^{+1}a_p\square n,$$

where $m \in \mathsf{M}$, $n \in \mathsf{N}$, and $a_i \in \mathcal{M}(\mathsf{A})$ for $i = 1, \ldots, p$. The differential

$$\partial_{\mathsf{M},\mathsf{A},\mathsf{N}} \colon \bigoplus_{i=0}^{\infty} \Big(M\underset{\mathsf{A}_0}{\square}(s^{+1}\overline{\mathcal{M}}(\mathsf{A}))^{\square i}\underset{\mathsf{A}_0}{\square}N\Big) \to \bigoplus_{i=0}^{\infty} \Big(M\underset{\mathsf{A}_0}{\square}(s^{+1}\overline{\mathcal{M}}(\mathsf{A}))^{\square i}\underset{\mathsf{A}_0}{\square}N\Big)$$

is defined as the sum of linear maps

$$\partial_{\mathsf{M},\mathsf{A},\mathsf{N}} = d_\mathsf{M}\square\,\mathrm{id}_{\mathcal{M}(\mathsf{A})}\square\,\mathrm{id}_\mathsf{N} + \mathrm{id}_\mathsf{M}\square D_{\mathcal{M}(\mathsf{A})}\square\,\mathrm{id}_\mathsf{N} + \mathrm{id}_\mathsf{M}\square\,\mathrm{id}_{\mathcal{M}(\mathsf{A})}\square d_\mathsf{N} + \theta,$$

where $d_\mathsf{M}, d_\mathsf{N}$, and $D_{\mathcal{M}(\mathsf{A})}$ are the differentials of M, N, and $\mathcal{M}(\mathsf{A})$, respectively, and θ is given by the following formula

$$\theta(m[a_1|\cdots|a_p]n)$$

$$= m\cdot a_1[a_2|\cdots|a_p]n + \sum_{i=1}^{p-1} \pm m[a_1|\cdots|a_i\cdot a_{i+1}|\cdots|a_p]n \pm m[a_1|\cdots|a_{p-1}]a_p\cdot m.$$

The associativity of the monoid structure of $\mathcal{M}(\mathsf{A})$, the compatibilities of the differentials with the products and actions, and $d_M^2 = d_N^2 = D^2_{\mathcal{M}(\mathsf{A})} = 0$ all together imply that $\partial^2_{\mathsf{M},\mathsf{A},\mathsf{N}} = 0$. The A_0-bicomodule structure on

$$\bigoplus_{i=0}^{\infty} \Big(M\underset{\mathsf{A}_0}{\square}(s^{+1}\overline{\mathcal{M}}(\mathsf{A}))^{\square i}\underset{\mathsf{A}_0}{\square}N\Big)$$

is given by the left and right A_0-comodule structures of M and N, respectively.

2.4. The Hochschild complex. For any dg category A, the chain complex

$$\mathrm{Bar}_{\mathsf{A}_0}(\mathcal{M}(\mathsf{A}), \mathsf{A}, \mathcal{M}(\mathsf{A}))$$

has a natural dg $\mathcal{M}(\mathsf{A})$-bimodule structure in the category of A_0-bicomodules with cotensor product $\square_{\mathsf{A}_0}$. This construction is clearly functorial with respect to morphisms of dg categories. We recall the definition of the Hochschild chain complex.

DEFINITION 2.1. Define a functor

$$\mathcal{CH}_*\colon \mathsf{dgCat}_{\mathbf{k}}^{\geq 0} \to \mathsf{Ch}_{\mathbf{k}}^{\geq 0},$$

called the *Hochschild complex*, as follows. For any $\mathsf{A} \in \mathsf{dgCat}_{\mathbf{k}}^{\geq 0}$, the underlying dg $\mathbf{k}$-module of $\mathcal{CH}_*(\mathsf{A})$ is defined by

$$\mathrm{Bar}_{\mathsf{A}_0}(\mathcal{M}(\mathsf{A}), \mathsf{A}, \mathcal{M}(\mathsf{A})) \bigotimes_{\mathcal{M}(\mathsf{A})\otimes\mathcal{M}(\mathsf{A})^{op}} \mathcal{M}(\mathsf{A}),$$

see Section 2.2 for notation.

The generators of $\mathcal{CH}_*(\mathsf{A})$ may be written as $[a_1|\cdots|a_p]a_{p+1}$, where $a_{p+1} \in \mathcal{M}(\mathsf{A})$, $a_i \in \overline{\mathcal{M}}(\mathsf{A})$ with $\mathsf{s}(a_i) = \mathsf{t}(a_{i+1})$ for $i = 1\cdots p$, and $\mathsf{t}(a_{p+1}) = \mathsf{s}(a_p)$. Using this notation, the differential

$$\partial_{\mathsf{A}}\colon \mathcal{CH}_*(\mathsf{A}) \to \mathcal{CH}_{*-1}(\mathsf{A})$$

is given by the same formula as the differential for the Hochschild complex of a dg algebra. One may equip this construction with a mixed complex structure via Connes' operator

$$B : \mathcal{CH}_*(\mathsf{A}) \to \mathcal{CH}_{*+1}(\mathsf{A}).$$

In this setting, B is given by

$$B([a_1|\cdots|a_p]a_{p+1}) = \sum_{i=1}^{p+1} \pm[a_i|\cdots|a_{p+1}|a_1|\cdots|a_{i-1}]\,\mathrm{id}_{\mathsf{s}(a_i)}\, s.$$

Just as in the classical case of the Hochschild complex of a dg algebra, one may check that $(\bigoplus_{n=0}^{\infty} \mathcal{CH}_n(\mathsf{A}), \partial_{\mathsf{A}}, B)$ is a non-negatively graded mixed complex functorially associated to any $\mathsf{A} \in \mathsf{dgCat}_{\mathbf{k}}^{\geq 0}$.

3. Categorical coalgebras and the cobar construction

In this section we define the notion of categorical coalgebras. This is a version of a curved coalgebra over a set of "objects" or "points" satisfying certain properties. Any categorical coalgebra gives rise to a dg category through a many object version of the cobar construction. These notions have been adapted from [**HL22**] in order to be applied to the algebraic topology setting and to work over an arbitrary commutative ring $\mathbf{k}$. The corresponding notion in [**HL22**] is that of a "pointed curved coalgebra".

3.1. Categorical coalgebras.

DEFINITION 3.1. A *categorical* $\mathbf{k}$*-coalgebra* consists of the data $C = (C, \Delta, d, h)$ such that

(1) $C = \bigoplus_{i=0}^{\infty} C_i$ is a nonnegatively graded $\mathbf{k}$-module.
(2) $\Delta : C \to C \otimes C$ is a degree 0 coassociative counital coproduct with counit $\varepsilon : C \to \mathbf{k}$

(3) The set

$$\mathcal{S}(C) := \{x \in C : \Delta(x) = x \otimes x, \varepsilon(x) = 1_{\mathbf{k}}\}$$

of "set-like" elements in C is nonempty and

$$C_0 \cong \mathbf{k}[\mathcal{S}(C)].$$

(4) $d : C \to C$ is a linear map of degree -1 which is a graded coderivation of Δ.

(5) The projection map $\epsilon : C \to C_0$ satisfies $\epsilon \circ d = 0$. In other words, $d : C_1 \to C_0$ is the zero map.

(6) $h : C \to \mathbf{k}$ is a linear map of degree -2 satisfying $h \circ d = 0$ and

$$d \circ d = (h \otimes \mathrm{id}) \circ (\Delta - \Delta^{op}) \tag{3.1}$$

where $\Delta^{op} = t \circ \Delta$ for $t(x \otimes y) = (-1)^{|x||y|} y \otimes x$. The right-hand side of the above equation is being considered as a map $C \to \mathbf{k} \otimes C \cong C$. The map h is called the *curvature* of C. Equation 3.1 may be rewritten as

$$d^2(x) = \sum_{(x)} h(x')x'' + x'h(x'').$$

Any categorical coalgebra C has a natural C_0-bicomodule structure with coaction maps

$$\rho_l : C \xrightarrow{\Delta} C \otimes C \xrightarrow{\epsilon \otimes \mathrm{id}_C} C_0 \otimes C$$

and

$$\rho_r : C \xrightarrow{\Delta} C \otimes C \xrightarrow{\mathrm{id}_C \otimes \epsilon} C \otimes C_0.$$

Furthermore, the coassociativity of $\Delta : C \to C \otimes C$ implies that $\Delta : C \to C \otimes C$ factors as $C \to C \square_{C_0} C \to C \otimes C$; so C may be regarded as a comonoid in the category of graded C_0-bicomodules with monoidal structure given by the cotensor product $\square_{C_0}$.

REMARK 3.2. Note that it is possible for a categorical coalgebra to have nonzero cuvature and for $d : C \to C$ to square zero.

DEFINITION 3.3. A *morphism of categorical coalgebras* $C = (C, \Delta, d, h)$ and $C' = (C', \Delta', d', h')$ consists of a pair (f_0, f_1) where

(1) $f_0 : (C, \Delta) \to (C', \Delta')$ is a morphism of graded $\mathbf{k}$-coalgebras,

(2) $f_1 : C \to C'_0$ is a C'_0-bicomodule map of degree -1 such that the composition $\bar{f_1} = \varepsilon \circ f_1$, where ε' is the counit of C', satisfies

$$f_0 \circ d = d' \circ f_0 + (\bar{f_1} \otimes f_0) \circ (\Delta - \Delta^{op}) \tag{3.2}$$

and

$$h' \circ f_0 = h + \bar{f_1} \circ d + (\bar{f_1} \otimes \bar{f_1}) \circ \Delta. \tag{3.3}$$

The composition of two morphisms of categorical coalgebras is defined by

$$(g_0, g_1) \circ (f_0, f_1) = (g_0 \circ f_0, g_1 \circ f_0 + g_0 \circ f_1). \tag{3.4}$$

Denote by $\mathsf{cCoalg}_{\mathbf{k}}$ the category of categorical coalgebras.

3.2. The cobar functor. Working over a field and with unbounded complexes, Holstein, and Lazarev define in [**HL22**] a functor from pointed curved coalgebras to dg categories extending the classical cobar functor from conilpotent dg coalgebras to augmented dg algebras. The same construction can be defined for categorical coalgebras over an arbitrary ring. We now describe this construction in our setting.

DEFINITION 3.4. Define a functor

$$\mathbf{\Omega}\colon \mathsf{cCoalg}_{\mathbf{k}} \to \mathsf{dgCat}_{\mathbf{k}}^{\geq 0},$$

called the *cobar functor*, as follows. Given any $C = (C, d, \Delta, h) \in \mathsf{cCoalg}_{\mathbf{k}}$, the objects of $\mathbf{\Omega}(C)$ are the elements of the set $\mathcal{S}(C)$ of set-like elements in C.

For any $x \in \mathcal{S}(C)$ denote by $i_x : \mathbf{k} \to C_0 = \mathbf{k}[\mathcal{S}(C)]$ the map determined by $i_x(1_{\mathbf{k}}) = x$. The map i_x gives rise to a C_0-bicomodule structure on $\mathbf{k}$ through the maps

$$\mathbf{k} \cong \mathbf{k} \otimes \mathbf{k} \xrightarrow{i_x \otimes \mathrm{id}_{\mathbf{k}}} C_0 \otimes \mathbf{k}$$

and

$$\mathbf{k} \cong \mathbf{k} \otimes \mathbf{k} \xrightarrow{\mathrm{id}_{\mathbf{k}} \otimes i_x} \mathbf{k} \otimes C_0.$$

We denote this C_0-bicomodule by $\mathbf{k}_x$ and its generator by id_x.

Write $C = \overline{C} \oplus C_0$. Denote by $s^{-1}\overline{C}$ the graded $\mathbf{k}$-module obtained by applying the shifting $\overline{C}$ by -1. We have the following three degree -1 maps:

(1) $\overline{d}\colon s^{-1}\overline{C} \to s^{-1}\overline{C}$
(2) $\overline{\Delta}\colon s^{-1}\overline{C} \to s^{-1}\overline{C} \otimes s^{-1}\overline{C}$, and
(3) $\overline{h}\colon s^{-1}\overline{C} \xrightarrow{s^{+1}} \overline{C} \xrightarrow{\rho_r} C \otimes C_0 \xrightarrow{h \otimes \mathrm{id}} \mathbf{k} \otimes C_0 \cong C_0$.

For any two $x, y \in \mathcal{S}(C)$ define a nonnegatively graded $\mathbf{k}$-module by

$$\mathbf{\Omega}(C)(x,y) = \bigoplus_{i=0}^{\infty} \mathbf{k}_x \underset{C_0}{\square} (s^{-1}\overline{C})^{\square i} \underset{C_0}{\square} \mathbf{k}_y,$$

where $(s^{-1}\overline{C})^{\square i}$ denotes the i-fold cotensor product of C_0-bicomodules and

$$(s^{-1}\overline{C})^{\square 0} = C_0.$$

We will use the notation

$$\{c_1|\cdots|c_p\}$$

to denote a generator

$$\mathrm{id}_x \square s^{-1}c_1 \square \cdots \square s^{-1}c_p \square \mathrm{id}_y \in \mathbf{\Omega}(C)(x,y).$$

We say the monomial $\{c_1|\cdots|c_p\}$ has *length* p. In particular, note

$$\mathbf{\Omega}(C)(x,x)_0 = \mathbf{k}_x \oplus \mathbf{k}_x \square s^{-1}C_1 \square \mathbf{k}_x.$$

The differential

$$D_{x,y} : \mathbf{\Omega}(C)(x,y)_k \to \mathbf{\Omega}(C)(x,y)_{k-1}$$

is defined by extending

$$\overline{h} + \overline{d} + \overline{\Delta}\colon \mathbf{k}_x \square s^{-1}\overline{C} \square \mathbf{k}_y \to \mathbf{k}_x \square C_0 \square \mathbf{k}_y \oplus \mathbf{k}_x \square s^{-1}\overline{C} \square \mathbf{k}_y \oplus \mathbf{k}_x \square (s^{-1}\overline{C})^{\square 2} \square \mathbf{k}_y$$

as a "derivation" to monomials of arbitrary length. It follows directly from (2), (4), and (6) in Definition 3.1 that $D_{x,y} \circ D_{x,y} = 0$. The composition in $\mathbf{\Omega}(C)$ is given by concatenation of monomials. For every $x \in \mathcal{S}(C)$, $1_{\mathbf{k}} \in \mathbf{k}_x \cong \mathbf{k}_x \square C_0 \square \mathbf{k}_x \subset \mathbf{\Omega}(C)(x,x)_0$ is the identity morphism.

Given a morphism $(f_0, f_1) : C \to C'$ between categorical coalgebras, define a morphism

$$\mathbf{\Omega}(f_0, f_1)\colon \mathbf{\Omega}(C) \to \mathbf{\Omega}(C')$$

of dg categories as follows. Since $f_0 : C \to C'$ is a map of coalgebras, f_0 restricts to a map of sets $\mathcal{S}(C) \to \mathcal{S}(C')$, which defines the functor $\mathbf{\Omega}(f_0, f_1)$ on objects. For any two $x, y \in \mathcal{S}(C)$ define

$$\mathbf{\Omega}(f_0, f_1)_{x,y}\colon \mathbf{\Omega}(C)(x, y) \to \mathbf{\Omega}(C')(f_0(x), f_0(y))$$

by extending the map

$$\mathbf{k}_x \square s^{-1}\overline{C}\square \mathbf{k}_y \longrightarrow \mathbf{k}_{f_0(x)}\square s^{-1}\overline{C}\square \mathbf{k}_{f_0(y)} \oplus \mathbf{k}_{f_0(x)}\square C_0'\square \mathbf{k}_{f_0(y)} \tag{3.5}$$

$$\{c\} \longmapsto \{f_0(c)\} + \mathrm{id}_{f_0(x)}\square f_1(c)\square \mathrm{id}_{f_0(y)} \tag{3.6}$$

"multiplicatively" to monomials $\{c_1|\cdots|c_p\}$ of arbitrary length. Note that

$$\mathbf{k}_{f_0(x)}\square C_0'\square \mathbf{k}_{f_0(y)}$$

is a nontrivial $\mathbf{k}$-module if and only if $f_0(x) = f_0(y)$, in which case it is isomorphic to $\mathbf{k}$. Hence, $\mathrm{id}_{f_0(x)}\square f_1(c)\square \mathrm{id}_{f_0(y)}$ may be identified with a scalar. It follows directly from (3.2) and (3.3) that $\mathbf{\Omega}(f_0, f_1)_{x,y}$ is a chain map for each $x, y \in \mathcal{O}_C$ and from (3.4) that compositions are compatible.

REMARK 3.5. When $\mathbf{k}$ is a field, a categorical $\mathbf{k}$-coalgebra is a pointed curved $\mathbf{k}$-coalgebra C (as defined in [**HL22**]) that is non-negatively graded and whose coradical is exactly the degree zero summand $C_0 \subseteq C$ (which is assumed to be non-trivial). In this case, the splitting map (which is part of the structure in the definition of a pointed curved coalgebra) is precisely the projection map $C \to C_0$. In particular, by the definition of a pointed curved coalgebra, C_0 is generated by the set-like (sometimes called "group-like") elements of C. Our notion of categorical coalgebra allows for the definition of a cobar functor landing in differential non-negatively graded categories even when working over an arbitrary ring.

DEFINITION 3.6. A *B_∞-categorical coalgebra* is a categorical coalgebra C equipped with degree 0 coassociative coproducts

$$\nabla_{x,y} : \mathbf{\Omega}(C)(x, y) \to \mathbf{\Omega}(C)(x, y) \otimes \mathbf{\Omega}(C)(x, y)$$

for all $x, y \in \mathcal{S}(C)$ making $\mathbf{\Omega}(C)$ into a category enriched over $(\mathsf{dgCoalg}_{\mathbf{k}}^{\geq 0}, \otimes)$, the monoidal category of differential nonnegatively graded coassociative counital $\mathbf{k}$-coalgebras. B_∞-categorical coalgebras form a category when equipped with maps of categorical coalgebras that preserve the additional structure. We denote this category by $\mathsf{B}_\infty\text{-}\mathsf{cCoalg}_{\mathbf{k}}$. This notion has also been considered in [**MRZ23**].

3.3. The extended cobar functor. We define a new version of the cobar construction by formally inverting set-like elements in the dg coalgebra of morphisms of the cobar construction of a B_∞-categorical coalgebra, generalizing a construction of [**HT10**]. This will give rise to a functorial construction that recovers the dg category of paths of the geometric realization of a simplicial set X when applied to a B_∞-categorical coalgebra of chains on X, as it will be discussed in Section 4.

Let

$$\mathcal{Z}\colon \mathsf{B}_\infty\text{-}\mathsf{cCoalg}_{\mathbf{k}} \to \mathsf{Cat}$$

be the functor defined as follows. For any $C \in \mathsf{B}_\infty\text{-cCoalg}_{\mathbf{k}}$, the set of objects of $\mathcal{Z}(C)$ is $\mathcal{S}(C)$. For any two objects x and y

$$\mathcal{Z}(C)(x,y) = \{f \in \mathbf{\Omega}(C)(x,y) | \nabla_{x,y}(f) = f \otimes f \text{ and } \varepsilon_{x,y}(f) = 1\}.$$

In other words, $\mathcal{Z}(C)(x,y) = \mathcal{S}(\mathbf{\Omega}(C)(x,y), \nabla_{x,y})$. Since the composition in $\mathbf{\Omega}(C)$ is compatible with the dg coalgebra structures on the morphisms and identity morphisms are set-like, it follows that $\mathcal{Z}(C)$ becomes a category with composition induced by that of $\mathbf{\Omega}(C)$. The functoriality of the construction follows since $\mathbf{\Omega}$ is a functor and taking set-like elements in a coalgebra is functorial.

DEFINITION 3.7. Define a functor

$$\widehat{\mathbf{\Omega}} \colon \mathsf{B}_\infty\text{-cCoalg}_{\mathbf{k}} \to \mathsf{dgCat}_{\mathbf{k}}^{\geq 0},$$

called the *extended cobar functor*, by letting

$$\widehat{\mathbf{\Omega}}(C) = \mathbf{\Omega}(C)[\mathcal{Z}(C)^{-1}],$$

namely, by formally (strictly) inverting the set of 0-cycles determined by the morphisms of $\mathcal{Z}(C)$ inside $\mathbf{\Omega}(C)$.

REMARK 3.8. In practice, we will consider the above construction when the natural map

$$\mathbf{k}[\mathcal{Z}(C)] \to \mathbf{\Omega}(C)$$

is a cofibration of cofibrant and locally $\mathbf{k}$-flat dg categories, where $\mathbf{k}[\mathcal{Z}(C)]$ denotes the dg category obtained by linearizing the morphisms of $\mathcal{Z}(C)$ and defining each differential to be trivial.

In this case, the strict localization

$$\widehat{\mathbf{\Omega}}(C) = \mathbf{\Omega}(C)[\mathcal{Z}(C)^{-1}]$$

is a homotopical localization. In other words, under these hypotheses, an $\mathbf{\Omega}$-quasi-equivalence $f : C \to C'$ between categorical coalgebras induces a quasi-equivalence $\widehat{\mathbf{\Omega}}(f) \colon \widehat{\mathbf{\Omega}}(C) \to \widehat{\mathbf{\Omega}}(C')$ of dg categories. This follows since we may interpret the extended cobar functor as a pushout of dg categories

$$\widehat{\mathbf{\Omega}}(C) = \mathbf{\Omega}(C) \bigsqcup_{\mathbf{k}[\mathcal{S}(C)]} \mathbf{k}[\mathcal{Z}(C)][\mathcal{Z}(C)^{-1}].$$

This pushout is a homotopy pushout in Tabuada's model structure on $\mathsf{dgCat}_{\mathbf{k}}$ when $\mathbf{k}[\mathcal{Z}(C)] \to \mathbf{\Omega}(C)$ is a cofibration of dg categories, since both $\mathbf{k}[\mathcal{Z}(C)]$ and $\mathbf{k}[\mathcal{Z}(C)][\mathcal{Z}(C)^{-1}]$ are locally $\mathbf{k}$-flat dg categories and consequently left proper objects, see [**Hol14**].

4. Chains on a simplicial set and the dg category of paths

The first goal of this section is to describe a version of the normalized simplicial chains as a functor

$$C_* \colon \mathsf{sSet} \to \mathsf{B}_\infty\text{-cCoalg}_{\mathbf{k}}.$$

Then we show that, for any simplicial set X, the extended cobar construction applied to the B_∞-categorical coalgebra of chains $C_*(X)$ yields a model for the dg category of paths on the topological space $|X|$.

4.1. Chains as a categorical coalgebra. For any simplicial set X, denote by $(\overline{C}_*(X), \partial)$ the dg $\mathbf{k}$-module of normalized simplicial chains. Recall the Alexander–Whitney coproduct, given on any simplex $\sigma \in X_n$ by

$$\Delta(\sigma) = \sum_{i=0}^{n} \sigma(0, \dots, i) \otimes \sigma(i, \dots, n),$$

induces a coassociative coproduct

$$\Delta : \overline{C}_*(X) \to \overline{C}_*(X) \otimes \overline{C}_*(X)$$

of degree 0. In the above formula, $\sigma(0, \dots, i)$ and $\sigma(i, \dots, n)$ denote the first i-th and last $(n-i)$-th faces of σ, respectively. This construction gives rise to a functor

$$C_*^{\Delta} : \mathsf{sSet} \to \mathsf{dgCoalg}_{\mathbf{k}}^{\geq 0}$$

given by

$$C_*^{\Delta}(X) = (\overline{C}_*(X), \partial, \Delta)$$

For any two simplicial sets X and Y, the natural Eilenberg-Zilber shuffle map

$$\mathrm{EZ}_{X,Y} : \overline{C}_*(X) \otimes \overline{C}_*(Y) \to \overline{C}_*(X \times Y)$$

is a map of dg coalgebras and consequently makes C_*^{Δ} into a lax monoidal functor, as explained in [**EM66**, 17.6].

The projection map $\epsilon : \overline{C}_*(X) \to \overline{C}_0(X)$ does not satisfy $\epsilon \circ \partial = 0$. However, as suggested in [**HL22**], the differential ∂ may be modified to obtain a categorical coalgebra as follows.

DEFINITION 4.1. For any $X \in \mathsf{sSet}$ define a categorical coalgebra $\widetilde{C}_*^{\Delta}(X) \in \mathsf{cCoalg}_{\mathbf{k}}$ as follows. The underlying graded $\mathbf{k}$-module of $\widetilde{C}_*^{\Delta}(X)$ is exactly $\overline{C}_*(X)$, which is given by $\overline{C}_n(X) = \mathbf{k}[X_n]/D(X_n)$, where $D(X_n) \subseteq \mathbf{k}[X_n]$ is the sub $\mathbf{k}$-module generated by degenerate n-simplices.

Let $e : \mathbf{k}[X_1] \to \mathbf{k}$ be the linear map sending degenerate 1-simplices to $0 \in \mathbf{k}$ and non-degenerate 1-simplices to $1 \in \mathbf{k}$. The map e induces a linear map $\widetilde{e} : \overline{C}_1(X) \to \mathbf{k}$. Define a new differential

$$\widetilde{\partial} : \overline{C}_*(X) \to \overline{C}_{*-1}(X)$$

by

$$\widetilde{\partial} = \partial - (\mathrm{id} \otimes \widetilde{e} - \widetilde{e} \otimes \mathrm{id}) \circ \Delta.$$

The map $\widetilde{\partial}$ is a coderivation of Δ and the projection map $\epsilon : \overline{C}_*(X) \to \overline{C}_0(X)$ now satisfies $\epsilon \circ \widetilde{\partial} = 0$. Finally, define $h : \overline{C}_2(X) \to \mathbf{k}$ by

$$h = (\widetilde{e} \otimes \widetilde{e}) \circ \Delta + \widetilde{e} \circ \partial.$$

A routine check yields that

$$\widetilde{C}_*^{\Delta}(X) = (\overline{C}_*(X), \widetilde{\partial}, \Delta, h)$$

defines an object in $\mathsf{cCoalg}_{\mathbf{k}}$. Furthermore, this construction gives rise to a functor

$$\widetilde{C}_*^{\Delta} : \mathsf{sSet} \to \mathsf{cCoalg}_{\mathbf{k}}.$$

The following result establishes a connection between the cobar functor from categorical coalgebras to dg categories and the dg nerve functor originally defined in [**Lur17**].

THEOREM 4.2. *The composition of functors*

$$\mathbf{\Omega} \circ \widetilde{C}_*^{\Delta} \colon \mathsf{sSet} \to \mathsf{dgCat}_{\mathbf{k}}$$

is naturally isomorphic to the (1-categorical) left adjoint of the dg nerve functor

$$\mathrm{N}_{dg} \colon \mathsf{dgCat}_{\mathbf{k}} \to \mathsf{sSet}.$$

PROOF. This result was proved in [**HL22**, Section 4]. The one object case was proved in [**RZ18**]. □

The dg nerve and its left adjoint also fit into a Quillen adjunction as we now record. Denote the left adjoint of N_{dg} by

$$\mathbf{\Lambda} \colon \mathsf{sSet} \to \mathsf{dgCat}_{\mathbf{k}}.$$

THEOREM 4.3. *The adjunction*

$$\mathbf{\Lambda} \colon \mathsf{sSet} \rightleftarrows \mathsf{dgCat}_{\mathbf{k}} \colon \mathrm{N}_{dg}$$

is a Quillen adjunction of model categories when sSet *is equipped with Joyal's model structure and* $\mathsf{dgCat}_{\mathbf{k}}$ *with Tabuada's model structure. In particular,* $\mathbf{\Lambda} \cong \mathbf{\Omega} \circ \widetilde{C}_*^{\Delta}$ *sends categorical equivalences of simplicial sets to quasi-equivalences of dg categories.*

PROOF. This is [**Lur17**, Proposition 1.3.1.20]. The second statement follows since all simiplicial sets are cofibrant in Joyal's model structure. □

REMARK 4.4. In general, the functor $\mathbf{\Lambda}$ does not send weak homotopy equivalences of simplicial sets to quasi-equivalences of dg categories. However, if $f \colon X \to X'$ is a weak homotopy equivalence and X and X' are "group-like", namely, their homotopy categories are groupoids, then $\mathbf{\Lambda}(f)$ is a quasi-equivalence of dg categories.

4.2. Chains as a B_∞-categorical coalgebra. We now describe a natural lift of $\widetilde{C}_*^{\Delta}$ to the category $\mathsf{B}_\infty\text{-}\mathsf{cCoalg}_{\mathbf{k}}$. For simplicity we will denote this lift by

$$C_* \colon \mathsf{sSet} \to \mathsf{B}_\infty\text{-}\mathsf{cCoalg}_{\mathbf{k}}.$$

To define this lift we use the factorization of $\mathbf{\Lambda}$, constructed in [**RZ18**, Section 6], through the category of small categories enriched over cubical sets (with connections). More precisely, in [**RZ18**] we constructed a functor

$$\mathfrak{C}_{\square_c} \colon \mathsf{sSet} \to \mathsf{Cat}_{\square_c}$$

from simplicial sets to the category of small categories enriched over the monoidal category of cubical sets (with connections) with Day convolution product. Then we showed that $\mathbf{\Lambda}$ is naturally isomorphic to the composition

$$\mathsf{sSet} \xrightarrow{\mathfrak{C}_{\square_c}} \mathsf{Cat}_{\square_c} \xrightarrow{\mathfrak{Q}_{\square_c}} \mathsf{dgCat}_{\mathbf{k}},$$

where $\mathfrak{Q}_{\square_c}$ is the functor obtained by applying the monoidal functor Q_* of normalized cubical chains at the level of morphisms.

The chain complex of normalized cubical chains $Q_*(K)$ on a cubical set K, with or without the extra data of connections, has a natural coproduct structure

$$\nabla \colon Q_*(K) \to Q_*(K) \otimes Q_*(K)$$

making $Q_*(K)$ into a dg coalgebra. It is completely determined by its action on the standard 0-cube and 1-cube given by the formulas

$$\nabla[0] = [0] \otimes [0]$$

and

$$\nabla\colon [0,1] \mapsto [0] \otimes [0,1] + [0,1] \otimes [1],$$

respectively.

Hence, the natural isomorphism $\Lambda \cong \mathfrak{Q}_{\square_c} \circ \mathfrak{C}_{\square_c}$ together with this cubical coproduct provides a natural lift of $\mathbf{\Lambda}\colon \mathsf{sSet} \to \mathsf{dgCat}_{\mathbf{k}}$ to the category of small categories enriched over the monoidal category of dg $\mathbf{k}$-coalgebras. Using the identification $\mathbf{\Lambda} \cong \mathbf{\Omega} \circ \widetilde{C}_*^{\Delta}$, we may interpret this additional structure as a functor

$$C_*\colon \mathsf{sSet} \to \mathsf{B}_\infty\text{-}\mathsf{cCoalg}_{\mathbf{k}}$$

lifting

$$\widetilde{C}_*^{\Delta}\colon \mathsf{sSet} \to \mathsf{cCoalg}_{\mathbf{k}}.$$

In particular, if X is a simplicial set with one vertex (i.e., a reduced simplicial set) then this construction provides the dg algebra $\mathbf{\Omega}(C_*^{\Delta}(X))$ with a natural dg bialgebra structure.

In the one vertex case, a version of this construction has been studied in detail in [**Bau80**], [**Bau81**] and, more recently, in [**MMR21**]. In the many vertex case, this version of the chains functor has been also discussed in [**MRZ23**]. We refer the reader to these references for more details.

4.3. The extended cobar construction as a model for the path category. For any topological space Y denote by $\mathcal{P}Y$ the topologically enriched category whose objects are the points of Y and morphisms $\mathcal{P}Y(x,y)$ are given by the space (with compact-open topology) of pairs (r,γ) where r is a non-negative real number (which we call the "parameter") and $\gamma\colon [0,r] \to Y$ a continuous path with $\gamma(0) = x$ and $\gamma(r) = y$. Composition is given by concatenation of paths and adding the corresponding parameters. Identities are constant paths with parameter $r = 0$. We call $\mathcal{P}Y$ the *path category* of Y.

Denote by $C_*^{\mathrm{sing}}(\mathcal{P}Y)$ the dg $\mathbf{k}$-category obtained by applying the normalized singular chains functor (equipped with the Elienberg–Zilber lax structure) on the morphisms of the topologically enriched category $\mathcal{P}Y$. This gives rise to a functorial construction

$$\mathsf{Top} \to \mathsf{dgCat}_{\mathbf{k}}$$
$$Y \mapsto C_*^{\mathrm{sing}}(\mathcal{P}Y)$$

that sends weak homotopy equivalences of spaces to quasi-equivalence of dg categories.

THEOREM 4.5. *For any simplicial set X, the dg categories $\widehat{\mathbf{\Omega}}(C_*(X))$ and $C_*^{\mathrm{sing}}(\mathcal{P}|X|)$ are naturally quasi-equivalent.*

PROOF. The natural map

$$\mathbf{k}[\mathcal{Z}(C_*(X))] \to \mathbf{\Omega}(C_*(X))$$

is a cofibration between cofibrant dg categories, since, by Theorem 4.2, it may be identified with

$$\mathbf{\Lambda}(i)\colon \mathbf{\Lambda}(\mathrm{sk}_1 X) \to \mathbf{\Lambda}(X),$$

where $\mathbf{\Lambda}\colon \mathsf{sSet} \to \mathsf{dgCat}_{\mathbf{k}}$ is the left adjoint of the dg nerve functor and $i\colon \mathrm{sk}_1 X \hookrightarrow X$ the inclusion of the 1-skeleton of X into X. By Remark 3.8, the (ordinary) pushout $\widehat{\mathbf{\Omega}}(C_*(X)) \cong \mathbf{\Lambda}(X)[X_1^{-1}]$ is actually a homotopy pushout of the maps $\mathbf{\Lambda}(i)\colon \mathbf{\Lambda}(\mathrm{sk}_1 X) \to \mathbf{\Lambda}(X)$ and $\mathbf{\Lambda}(\mathrm{sk}_1 X) \to \mathbf{\Lambda}(\mathrm{sk}_1 X)[X_1^{-1}]$, where $\mathbf{\Lambda}(\mathrm{sk}_1 X)[X_1^{-1}]$ is the dg category obtained by linearizing the free groupoid generated by the quiver $X_1 \rightrightarrows X_0$ determined by the first two simplicial face maps. Let

$$\mathcal{K}\colon \mathsf{sSet} \to \mathsf{sSet}$$

be a Kan replacement functor so that there is a natural quasi-equivalence of dg categories

$$\mathbf{\Lambda}(\mathrm{sk}_1 X)[X_1^{-1}] \simeq \mathbf{\Lambda}(\mathcal{K}(\mathrm{sk}_1 X)).$$

By Theorem 4.3, $\mathbf{\Lambda}\colon \mathsf{sSet} \to \mathsf{dgCat}_{\mathbf{k}}$ is a left Quillen functor between Joyal's model structure on simplicial sets and Tabuada's model structure on dg categories, thus $\mathbf{\Lambda}$ preserves homotopy pushouts. Hence, we have natural quasi-equivalence of dg categories

$$\widehat{\mathbf{\Omega}}(C_*(X)) \cong \mathbf{\Lambda}(X)[X_1^{-1}] \simeq \mathbf{\Lambda}(X \coprod_{\mathrm{sk}_1 X} \mathcal{K}(\mathrm{sk}_1 X)).$$

Note the map

$$X \to X \coprod_{\mathrm{sk}_1 X} \mathcal{K}(\mathrm{sk}_1 X)$$

is a weak homotopy equivalence of simplicial sets and the homotopy category of $X \coprod_{\mathrm{sk}_1 X} \mathcal{K}(\mathrm{sk}_1 X)$ is a groupoid. It follows that $\mathbf{\Lambda}(X \coprod_{\mathrm{sk}_1 X} \mathcal{K}(\mathrm{sk}_1 X))$ is naturally quasi-equivalent to the dg category $C_*^{\mathrm{sing}}(\mathcal{P}|X \coprod_{\mathrm{sk}_1 X} \mathcal{K}(\mathrm{sk}_1 X)|)$ and consequently to $C_*^{\mathrm{sing}}(\mathcal{P}|X|)$, as desired. □

5. The co-Hochschild complex of a categorical coalgebra

We define a version of the co-Hochschild complex for categorical coalgebras. Then we establish a relationship with the Hochschild complex of a dg category. The co-Hochschild complex in the case of connected dg coalgebras has been sutided in [**HPS09**] and [**HS21**].

5.1. The co-Hochschild complex.

We construct a functor

$$co\mathcal{CH}_*\colon \mathsf{cCoalg}_{\mathbf{k}} \to \mathsf{Ch}_{\mathbf{k}}^{\geq 0},$$

called the *co-Hochschild complex*, as follows. For any categorical coalgebra C, the underlying graded $\mathbf{k}$-module of $co\mathcal{CH}_*(C)$ is defined by

$$C \underset{C_0 \otimes C_0^{op}}{\square} \mathcal{M}(\mathbf{\Omega}(C)) := (C \underset{C_0}{\square} \mathcal{M}(\mathbf{\Omega}(C))) \bigcap (C \underset{C_0^{op}}{\square} \mathcal{M}(\mathbf{\Omega}(C)))$$

Explicitly, this notation is saying that $co\mathcal{CH}_k(C)$ is generated as a graded $\mathbf{k}$-module by monomials

$$x = x_0 \square s^{-1} x_1 \square \cdots \square s^{-1} x_p = x_0\{x_1|\cdots|x_p\},$$

where $x_0 \in C$, $x_i \in \overline{C}$ for $i = 1, \ldots, p$, $x_p \otimes x_0 \in C \underset{C_0}{\square} C$, and $|x_0| + |x_1| + \cdots + |x_p| - p = k$. The differential

$$\partial\colon co\mathcal{CH}_k(C) \to co\mathcal{CH}_{k-1}(C)$$

is defined by

$$\begin{aligned}\partial(x) &= d_C x_0\{x_1|\cdots|x_p\} + \sum_{i=1}^{p} \pm x_0\{x_1|\cdots|d_C x_i|\cdots|x_p\}\\ &+ \sum_{i=1}^{p} \pm x_0\{x_1|\cdots|\overline{h}(x_i)|\cdots|x_p\} + \sum_{(x_0)} \pm x_0'\{x_0''|x_1|\cdots|x_p\}\\ &+ \sum_{i=1}^{p}\sum_{(x_i)} \pm x_0\{x_1|\cdots|x_i'|x_i''|\cdots|x_p\} + \sum_{(x_0)} \pm x_0''\{x_1|\cdots|x_p|x_0'\}.\end{aligned}$$

The signs are determined, as usual, from the Koszul sign convention. One may also equip the chain complex $co\mathcal{CH}_*(C)$ with the further structure of a mixed complex by defining a degree $+1$ operator

$$P\colon co\mathcal{CH}_*(C) \to co\mathcal{CH}_{*+1}(C)$$

through the formula

$$P(x_0\{x_1|\cdots|x_p\}) = \sum_{i=1}^{p} \pm\varepsilon(x_0)x_i\{x_{i+1}|\cdots|x_p|x_1|\cdots|x_{i-1}\},$$

where $\varepsilon : C \to \mathbf{k}$ denotes the counit of C. A straightforward computation yields that

$$(co\mathcal{CH}_*(C), \partial, P)$$

is a non-negatively graded mixed $\mathbf{k}$-complex. This construction is clearly functorial with respect to maps of categorical coalgebras.

5.2. The extended co-Hochschild complex. Define the *extended Hochschild complex* as the functor

$$\widehat{co\mathcal{CH}_*}\colon \mathsf{B}_\infty\text{-}\mathsf{cCoalg}_{\mathbf{k}} \to \mathsf{Ch}_{\mathbf{k}}^{\geq 0}$$

given by

$$C \underset{C_0\otimes C_0^{op}}{\square} \mathcal{M}(\widehat{\mathbf{\Omega}}(C)).$$

5.3. Relationship with the Hochschild chain complex. Let C be a categorical coalgebra and M and N right and left dg modules over $\mathcal{M}(\mathbf{\Omega}(C))$, respectively, as in Section 2.3. Define a graded $\mathbf{k}$-module

$$\mathcal{Q}(\mathsf{M}, C, \mathsf{N}) := \mathsf{M}\underset{C_0}{\square}C\underset{C_0}{\square}\mathsf{N}$$

and consider the linear map

$$\partial_{\mathcal{Q}}\colon \mathcal{Q}(\mathsf{M}, C, \mathsf{N}) \to \mathcal{Q}(\mathsf{M}, C, \mathsf{N})$$

of degree -1 defined by

$$\partial_{\mathcal{Q}} = d_{\mathsf{M}}\square\,\mathrm{id}_C\,\square\,\mathrm{id}_{\mathsf{N}} + \mathrm{id}_{\mathsf{M}}\,\square d_C\square\,\mathrm{id}_{\mathsf{N}} + \mathrm{id}_{\mathsf{M}}\,\square\,\mathrm{id}_C\,\square d_{\mathsf{N}} + \theta',$$

where

$$\theta'(m\square c\square n) = \pm(m\cdot\{c'\})\square c''\square n \pm m\square c'\square(\{c''\}\cdot n).$$

REMARK 5.1. Note that $\partial_{\mathcal{Q}}$ may not square to zero, since d_C may not square to zero in a categorical coalgebra. When C is a connected dg coalgebra then $\mathcal{Q}(\mathsf{M}, C, \mathsf{N})$ is indeed a chain complex.

We will now define a map H of degree $+1$ and two maps π and α of degree 0 fitting in the diagram

$$H \circlearrowright \mathrm{Bar}_{C_0}(\mathsf{M}, \mathbf{\Omega}(C), \mathsf{N}) \underset{\alpha}{\overset{\pi}{\rightleftarrows}} \mathcal{Q}(\mathsf{M}, C, \mathsf{N}).$$

These maps will satisfy three equations given in Proposition 5.2.

(1) Define

$$\pi\colon \mathrm{Bar}_{C_0}(\mathsf{M}, \mathbf{\Omega}(C), \mathsf{N}) \to \mathcal{Q}(\mathsf{M}, C, \mathsf{N})$$

on any generator $m[a_1|\cdots|a_p]n$ by letting

$$\pi(m[a_1|\cdots|a_p]n) = 0 \text{ if } p > 1 \ ,$$

and when $p = 1$, writing $a_1 = \{c_1|\cdots|c_q\}$, define

$$\pi(m[\{c_1|\cdots|c_q\}])n = \sum_{i=1}^{q} m \cdot \{c_1|\cdots|c_{i-1}\}\square[\{c_i\}]\square\{c_{i+1}|\cdots|c_q\} \cdot n$$

if $q > 0$, and

$$\pi(m[x]n) = m\square x \square n$$

if $q = 0$ and $x \in \mathcal{S}(C) \subset C_0$.

(2) Define

$$\alpha\colon \mathcal{Q}(\mathsf{M}, C, \mathsf{N}) \to \mathrm{Bar}_{C_0}(\mathsf{M}, \mathbf{\Omega}(C), \mathsf{N})$$

by

$$\alpha(m\square c\square n) = m[\{c\}]n + \sum m[\{c'\}|\{c''\}]n + \sum m[\{c'\}|\{c''\}|\{c'''\}]n + \cdots$$

Note α is well defined since the induced coproduct $\Delta\colon \overline{C} \to \overline{C} \otimes \overline{C}$ is of degree 0 and $\overline{C}$ is concentrated on positive degrees.

(3) Define

$$H : \mathrm{Bar}_{C_0}(\mathsf{M}, \mathbf{\Omega}(C), \mathsf{N}) \to \mathrm{Bar}_{C_0}(\mathsf{M}, \mathbf{\Omega}(C), \mathsf{N})$$

to be a degree $+1$ linear map given on a generator $m[a_1|\cdots|a_p]n$ as follows. Write $a_1 = \{c_1|\cdots|c_m\}$ and let

$$H(m[\{c_1|\cdots|c_m\}|a_2|\cdots|a_p]n) = 0 \text{ if } m < 2 \ ,$$

$$\begin{aligned} H(m[\{c_1|c_2\}|a_2|\cdots|a_p]n) &= m[\{c_1\}|\{c_2\}|a_2|\cdots|a_p]n \\ &+ \sum [\{c_1'\}|\{c_1''\}|\{c_2\}|a_2|\cdots|a_p]n \\ &+ \sum [\{c_1'\}|\{c_1''\}|\{c_1'''\}|\{c_2\}|a_2|\cdots|a_p]n + \cdots \text{ if } m = 2, \end{aligned}$$

and, if $m > 2$, let

$$\begin{aligned} &H(m[\{c_1|\cdots|c_m\}|a_2|\cdots|a_p]n) \\ &= \sum_{i=1}^{m} m \cdot \{c_1|\cdots|c_{i-1}\}[\{c_i\}|\{c_{i+1}|\cdots|c_m\}|a_2|\cdots|a_p]n \\ &+ \sum_{i=1}^{m} m \cdot \{c_1|\cdots|c_{i-1}\}[\{c_i'\}|\{c_i''\}|\{c_{i+1}|\cdots|c_m\}|a_2|\cdots|a_p]n \\ &+ \sum_{i=1}^{m} m \cdot \{c_1|\cdots|c_{i-1}\}[\{c_i'\}|\{c_i''\}|\{c_i'''\}|\{c_{i+1}|\cdots|c_m\}|a_2|\cdots|a_p]n + \cdots \end{aligned}$$

A tedious but straightforward computation verifies the following equations hold.

PROPOSITION 5.2. *The maps π, α, and H defined above satisfy the equations*

$$\pi \circ \partial_{\mathsf{M},C,\mathsf{N}} = \partial_{\mathcal{Q}} \circ \pi, \tag{5.1}$$

$$\alpha \circ \partial_{\mathcal{Q}} = \partial_{\mathsf{M},C,\mathsf{N}} \circ \alpha, \text{ and} \tag{5.2}$$

$$H \circ \partial_{\mathsf{M},C,\mathsf{N}} + \partial_{\mathsf{M},C,\mathsf{N}} \circ H = \alpha \circ \pi - \mathrm{id}_{\mathrm{Bar}_{C_0}(\mathsf{M},\mathbf{\Omega}(C),\mathsf{N})}. \tag{5.3}$$

We avoid using the terminology "chain contraction" in the above proposition precisely because $\partial_{\mathcal{Q}}$ might not square to zero. In any case, as our main application we consider the case $\mathsf{M} = \mathsf{N} = \mathcal{M}(\mathbf{\Omega}(C))$. In this particular case,

$$\partial_{\mathcal{Q}} : \mathcal{Q}(\mathcal{M}(\mathbf{\Omega}(C)), C, \mathcal{M}(\mathbf{\Omega}(C))) \to \mathcal{Q}(\mathcal{M}(\mathbf{\Omega}(C)), C, \mathcal{M}(\mathbf{\Omega}(C)))$$

does square to zero and so $(\mathcal{Q}(\mathcal{M}(\mathbf{\Omega}(C)), C, \mathcal{M}(\mathbf{\Omega}(C))), \partial_{\mathcal{Q}})$ defines a dg $\mathbf{k}$-module. In fact, it follows from the definition of a categorical coalgebra that the two terms

$$\pm m \square d_C^2(c) \square n$$

and

$$\pm (m \cdot \overline{h}\{c'\}) \square c'' \square n \pm m \square c' \square (\overline{h}\{c''\} \cdot n)$$

in $\partial_{\mathcal{Q}}^2(m \square c \square n)$ cancel each other.

THEOREM 5.3. *For any categorical coalgebra C, there is a natural chain contraction of dg $\mathbf{k}$-modules*

$$\overline{H} \circlearrowright \mathcal{CH}_*(\mathbf{\Omega}(C)) \underset{\overline{\alpha}}{\overset{\overline{\pi}}{\rightleftarrows}} co\mathcal{CH}_*(C).$$

If C is a B_∞-categorical coalgebra, then there is a natural chain contraction of dg $\mathbf{k}$-modules

$$\widehat{H} \circlearrowright \mathrm{Bar}_{C_0}(\mathcal{M}(\widehat{\mathbf{\Omega}}(C)), \mathbf{\Omega}(C), \mathcal{M}(\widehat{\mathbf{\Omega}}(C))) \bigotimes_{\mathcal{M}(\widehat{\mathbf{\Omega}}(C)) \otimes \mathcal{M}(\widehat{\mathbf{\Omega}}(C))^{op}} \widehat{\mathbf{\Omega}}(C) \underset{\widehat{\alpha}}{\overset{\widehat{\pi}}{\rightleftarrows}} \widehat{co\mathcal{CH}_*}(C).$$

PROOF. Recall

$$\mathcal{CH}_*(\mathbf{\Omega}(C)) = \mathrm{Bar}_{C_0}(\mathcal{M}(\mathbf{\Omega}(C)), \mathbf{\Omega}(C), \mathcal{M}(\mathbf{\Omega}(C))) \bigotimes_{\mathcal{M}(\mathbf{\Omega}(C)) \otimes \mathcal{M}(\mathbf{\Omega}(C))^{op}} \mathcal{M}(\mathbf{\Omega}(C)).$$

Now we observe there is a natural isomorphism of dg $\mathbf{k}$-modules

$$co\mathcal{CH}_*(C) \cong \mathcal{Q}(\mathcal{M}(\mathbf{\Omega}(C)), C, \mathcal{M}(\mathbf{\Omega}(C))) \bigotimes_{\mathcal{M}(\mathbf{\Omega}(C)) \otimes \mathcal{M}(\mathbf{\Omega}(C))^{op}} \mathcal{M}(\mathbf{\Omega}(C)).$$

Using the notation of Proposition 5.2, define

$$\overline{\pi} = \pi \bigotimes_{\mathcal{M}(\mathbf{\Omega}(C)) \otimes \mathcal{M}(\mathbf{\Omega}(C))^{op}} \mathrm{id}_{\mathcal{M}(\mathbf{\Omega}(C))},$$

$$\overline{\alpha} = \alpha \bigotimes_{\mathcal{M}(\mathbf{\Omega}(C)) \otimes \mathcal{M}(\mathbf{\Omega}(C))^{op}} \mathrm{id}_{\mathcal{M}(\mathbf{\Omega}(C))}, \text{ and}$$

$$\overline{H} = H \bigotimes_{\mathcal{M}(\mathbf{\Omega}(C)) \otimes \mathcal{M}(\mathbf{\Omega}(C))^{op}} \mathrm{id}_{\mathcal{M}(\mathbf{\Omega}(C))}.$$

It follows from Proposition 5.2 that $\overline{\pi}$, $\overline{\alpha}$ and $\overline{H}$ define the data of a natural chain contraction. The proof of the second statement is similar. □

5.4. Invariance of the co-Hochschild complex. Recall a morphism of dg categories $f\colon \mathsf{A} \to \mathsf{A}'$ is a *quasi-equivalence* if

(1) for any two objects x and y in A, the induced map $f_{x,y}\colon \mathsf{A}(x,y) \to \mathsf{A}'(f(x), f(y))$ is a quasi-isomorphism of dg $\mathbf{k}$-modules, and
(2) if the induced map on homotopy categories $H_0(f)\colon H_0(\mathsf{A}) \to H_0(\mathsf{A}')$ is essentially surjective.

DEFINITION 5.4. A map $f\colon C \to C'$ between categorical coalgebras is called an *$\boldsymbol{\Omega}$-quasi-equivalence* if $\boldsymbol{\Omega}(f)\colon \boldsymbol{\Omega}(C) \to \boldsymbol{\Omega}(C')$ is a quasi-equivalence of dg categories.

The co-Hochschild complex is invariant under $\boldsymbol{\Omega}$-quasi-equivalences assuming suitable hypotheses, as shown next.

PROPOSITION 5.5. *If $f\colon C \to C'$ is an $\boldsymbol{\Omega}$-quasi-equivalence between categorical coalgebras C and C' that are flat as $\mathbf{k}$-modules, then*

$$co\mathcal{CH}_*(f)\colon co\mathcal{CH}_*(C) \to co\mathcal{CH}_*(C')$$

is a quasi-isomorphism of dg $\mathbf{k}$-modules.

PROOF. If C and C' are flat as $\mathbf{k}$-modules then the dg categories $\boldsymbol{\Omega}(C)$ and $\boldsymbol{\Omega}(C')$ are locally $\mathbf{k}$-flat. Hence, the quasi-equivalence

$$\boldsymbol{\Omega}(f)\colon \boldsymbol{\Omega}(C) \to \boldsymbol{\Omega}(C')$$

induces a quasi-isomorphism between Hochschild complexes

$$\mathcal{CH}_*(\boldsymbol{\Omega}(f))\colon \mathcal{CH}_*(\boldsymbol{\Omega}(C)) \to \mathcal{CH}_*(\boldsymbol{\Omega}(C')).$$

By Corollary 5.3, we have natural quasi-isomorphisms

$$\mathcal{CH}_*(\boldsymbol{\Omega}(C)) \simeq co\mathcal{CH}_*(C)$$

and

$$\mathcal{CH}_*(\boldsymbol{\Omega}(C')) \simeq co\mathcal{CH}_*(C').$$

It follows that the induced map

$$co\mathcal{CH}_*(f)\colon co\mathcal{CH}_*(C) \to co\mathcal{CH}_*(C')$$

is a quasi-isomorphism. □

6. The co-Hochschild complex as a model for the free loop space

We establish a relationship between the extended co-Hochschild complex and the free loop space. For any topological space Y, we denote by LY the *free loop space* of Y modeled as the space (with compact-open topology) of pairs (r, γ) where r is a non-negative real number and $\gamma\colon [0, r] \to Y$ a continous map with $\gamma(0) = \gamma(r)$. Denote by $C_*^{\text{sing}}(LY)$ the dg $\mathbf{k}$-module of normalized singular chains on LY. Let $\mathbf{k}$ be a commutative ring and for any simplicial set X denote by $C_*(X)$ its B_∞-categorical $\mathbf{k}$-coalgebra of chains. Note that the S^1-action on $L|X|$ given by rotating loops gives rise to an operator $R : C_*^{\text{sing}}(L|X|) \to C_{*+1}^{\text{sing}}(L|X|)$ which gives the chain complex $C_*^{\text{sing}}(L|X|)$ the extra structure of a mixed complex.

THEOREM 6.1. *For any simplicial set X, the dg $\mathbf{k}$-modules $\widehat{co\mathcal{CH}}_*(C_*(X))$ and $C_*^{\text{sing}}(L|X|)$ are naturally quasi-isomorphic.*

PROOF. For simplicity write $C = C_*(X)$. By Theorem 5.3, there is a natural chain homotopy equivalence

$$\widehat{co\mathcal{CH}_*}(C) \simeq \mathrm{Bar}_{C_0}(\mathcal{M}(\widehat{\mathbf{\Omega}}(C)), \mathbf{\Omega}(C), \mathcal{M}(\widehat{\mathbf{\Omega}}(C))) \bigotimes_{\mathcal{M}(\widehat{\mathbf{\Omega}}(C))\otimes\mathcal{M}(\widehat{\mathbf{\Omega}}(C))^{op}} \widehat{\mathbf{\Omega}}(C).$$

Note that $\mathbf{\Omega}(C)$ is a cofibrant dg category being naturally isomorphic to $\mathbf{\Lambda}(X)$ where $\mathbf{\Lambda}\colon \mathsf{sSet} \to \mathsf{dgCat}_{\mathbf{k}}$ denotes the left adjoint of the dg nerve functor. Hence, the natural map

$$\mathbf{\Omega}(C) \cong \mathbf{\Lambda}(X) \to \mathbf{\Lambda}(X)[X_1^{-1}] \cong \widehat{\mathbf{\Omega}}(C)$$

induces a quasi-isomorphism after applying bar constructions. This is a classical fact, for instance see [**FM94**, Section Q5]. More precisely, the natural map

$$\mathrm{Bar}_{C_0}(\mathcal{M}(\widehat{\mathbf{\Omega}}(C)), \mathbf{\Omega}(C), \mathcal{M}(\widehat{\mathbf{\Omega}}(C))) \to \mathrm{Bar}_{C_0}(\mathcal{M}(\widehat{\mathbf{\Omega}}(C)), \widehat{\mathbf{\Omega}}(C), \mathcal{M}(\widehat{\mathbf{\Omega}}(C)))$$

is a quasi-isomorphism. Consequently, we have a natural quasi-isomorphism

$$\begin{aligned}\widehat{co\mathcal{CH}_*}(C) &\xrightarrow{\simeq} \mathrm{Bar}_{C_0}(\mathcal{M}(\widehat{\mathbf{\Omega}}(C)), \widehat{\mathbf{\Omega}}(C), \mathcal{M}(\widehat{\mathbf{\Omega}}(C))) \bigotimes_{\mathcal{M}(\widehat{\mathbf{\Omega}}(C))\otimes\mathcal{M}(\widehat{\mathbf{\Omega}}(C))^{op}} \widehat{\mathbf{\Omega}}(C) \\ &= \mathcal{CH}_*(\widehat{\mathbf{\Omega}}(C)).\end{aligned}$$

The invariance of the Hochschild complex with respect to quasi-equivalences between locally $\mathbf{k}$-flat dg categories, together with Theorem 4.5, implies that $\mathcal{CH}_*(\widehat{\mathbf{\Omega}}(C))$ is naturally quasi-isomorphic to $\mathcal{CH}_*(C_*^{\mathrm{sing}}(\mathcal{P}|X|))$. By a result proved by Goodwillie [**Goo85**] and also (independently) by Burghelea and Fiedorowicz [**BF86**], there is a natural quasi-isomorphism

$$\mathcal{CH}_*(C_*^{\mathrm{sing}}(\mathcal{P}|X|)) \xrightarrow{\simeq} C_*^{\mathrm{sing}}(L|X|),$$

as desired. □

REMARK 6.2. If the simplicial set X has the property of being "group-like", namely, if the homotopy category of X is a groupoid, then the (non-extended) co-Hochschild complex $co\mathcal{CH}_*(C_*(X))$ of the underlying categorical coalgebra of $C_*(X)$ is already naturally quasi-isomorphic to $C_*^{\mathrm{sing}}(L|X|)$. In other words, there is no need to localize if every 1-simplex in X is invertible up to homotopy. In this case, $co\mathcal{CH}_*(C_*(X))$ and $C_*^{\mathrm{sing}}(L|X|)$ are quasi-isomorphic as mixed complexes. This follows since the quasi-isomorphism $\overline{\pi}$ in Theorem 5.3 is a morphism of mixed complexes (i.e. it intertwines the operators B and P) together with the fact that the quasi-isomorphism

$$\mathcal{CH}_*(C_*^{\mathrm{sing}}(\mathcal{P}|X|)) \xrightarrow{\simeq} C_*^{\mathrm{sing}}(L|X|)$$

constructed in [**Goo85**] and [**BF86**] is also a morphism of mixed complexes (i.e. intertwines the operators B and R). A theory of cyclic homology for categorical coalgebras will be developed by Daniel Tolosa in his PhD thesis.

7. dg Hopf algebras and the adjoint action

This section has two goals. The first goal is to clarify the relationship between the co-Hochschild complex model for the free loop space and Brown's twisted tensor product model for the total space of a fibration. The latter uses the conjugation action of (a topological group model of) the based loop space on itself when defining the twisted differential, while the first does not use any antipode map (or inverses) when defining the co-Hochschild complex differential.

The second goal is to describe a natural chain map

$$C \to \widehat{co\mathcal{CH}_*}(C)$$

modeling the continuous map $Y \to LY$ sending each point of a space Y to its corresponding constant loop in the free loop space. We expect this to be useful in string topology of nonsimply connected manifolds, where constant loops play a delicate role.

For simplicity, in this section we work with connected dg **k**-coalgebras instead of categorical **k**-coalgebras, namely, we work in the one object case. For homotopy theoretic applications this is not a strong hypothesis since pointed connected homotopy types may be modeled by reduced simplicial sets (i.e., simplicial sets with a single vertex). For any reduced simplicial set X, we denote by $C_*(X)$ the normalized simplicial chains on X, which is a connected dg coalgebra when equipped with the Alexander-Whitney coproduct. It follows from Section 4.2 that the dg algebra (or dg category with one object) $\widehat{\mathbf{\Omega}}(C_*(X))$ has a natural dg bialgebra structure naturally quasi-isomorphic to the dg bialgebra structure on the singular chain complex $C_*^{\text{sing}}(\Omega_b|X|)$ of based (Moore) loops in $|X|$ at b. One of the main technical steps in this section is showing that the dg bialgebra $\widehat{\mathbf{\Omega}}(C_*(X))$ has the property of being a dg Hopf algebra. This means there is a map of dg **k**-modules

$$s\colon \widehat{\mathbf{\Omega}}(C_*(X)) \to \widehat{\mathbf{\Omega}}(C_*(X))$$

satisfying

$$\mu \circ (s \otimes \mathrm{id}) \circ \nabla = \eta \circ \varepsilon = \mu \circ (\mathrm{id} \otimes s) \circ \nabla, \tag{7.1}$$

where $\nabla\colon \widehat{\mathbf{\Omega}}(C_*(X)) \to \widehat{\mathbf{\Omega}}(C_*(X)) \otimes \widehat{\mathbf{\Omega}}(C_*(X))$ is the coproduct, $\mu\colon \widehat{\mathbf{\Omega}}(C_*(X)) \otimes \widehat{\mathbf{\Omega}}(C_*(X)) \to \widehat{\mathbf{\Omega}}(C_*(X))$ the product, $\varepsilon\colon \widehat{\mathbf{\Omega}}(C_*(X)) \to \mathbf{k}$ the counit, and $\eta\colon \mathbf{k} \to \widehat{\mathbf{\Omega}}(C_*(X))$ the unit.

7.1. Hochschild chain complex and adjoint action. Given a map $f\colon A \to B$ of dg **k**-algebras, denote by f^*B the left dg $A \otimes A^{op}$-module whose underlying dg **k**-module is B and the $A \otimes A^{op}$-right action is induced through f. Denote by

$$\mathcal{CH}_*(A, f^*B) := \mathrm{Bar}(A, A, A) \otimes_{A \otimes A^{op}} f^*B$$

the (normalized) Hochschild chain complex of A with coefficients in $f^*(B)$. This has

$$\bigoplus_{p=0}^{\infty} (s^{+1}\overline{A})^{\otimes p} \otimes f^*B$$

as underlying graded **k**-module and we write generators $[a_1|\cdots|a_p]b$ as usual.

Now suppose $f : A \to B$ is a map of dg **k**-*bialgebras*. Furthermore, suppose the dg bialgebra B has the property of being a dg Hopf algebra. Denote by $f^*_{ad}(B)$ the left A-module whose underlying dg **k**-module is B and left A-action given by

$$a \cdot b = \sum_{(a)} (-1)^{|a'|(|a''|+|b|)} f(a'')bs(f(a')),$$

for any $a \in A$, $b \in B$, where $s : B \to B$ denotes the antipode of B. We call this the *adjoint action of A on B via f*.

The counit $\epsilon_A : A \to \mathbf{k}$ of the dg Hopf algebra A makes $\mathbf{k}$ into a right dg A-bimodule. Define a dg $\mathbf{k}$-module

$$\mathcal{C}_*(A, f^*_{ad}B) := \mathrm{Bar}(\mathbf{k}, A, f^*_{ad}B).$$

PROPOSITION 7.1. *Let $f : A \to B$ be a dg $\mathbf{k}$-bialgebra map, where B has the property of being a dg Hopf algebra. There is a natural isomorphism of dg $\mathbf{k}$-modules*

$$\mathcal{CH}_*(A, f^*B) \cong \mathcal{C}_*(A, f^*_{ad}B).$$

PROOF. Define

$$\varphi\colon \mathcal{CH}_*(A, f^*B) \to \mathcal{C}_*(A, f^*_{ad}B) \tag{7.2}$$

$$\varphi([a_1|\cdots|a_p]b) := \sum \pm[a_1''|\cdots|a_p'']bf(a_1'\cdots a_p'). \tag{7.3}$$

The antipode compatibility (equation 7.1) in the definition of a dg Hopf algebra implies that φ is a chain map with (strict) inverse given by the chain map

$$\varphi^{-1}([a_1|\cdots|a_p]b) = \sum \pm[a_1''|\cdots|a_p'']bs(f(a_1'\cdots a_p')), \tag{7.4}$$

where $s : B \to B$ denotes the antipode of B. □

7.2. The cobar construction as a dg Hopf algebra. We wish to apply the above discussion to a dg Hopf algebra model for the based loop space. Recall that, as discussed in Section 4.2, for any reduced simplicial set X, there is a natural coproduct

$$\nabla\colon \widehat{\boldsymbol{\Omega}}(C_*(X)) \to \widehat{\boldsymbol{\Omega}}(C_*(X)) \otimes \widehat{\boldsymbol{\Omega}}(C_*(X))$$

making $(\widehat{\boldsymbol{\Omega}}(C_*(X)), D, \otimes, \nabla)$ into a dg bialgebra, which turns out to be a dg Hopf algebra, as shown next.

THEOREM 7.2. *For any reduced simplicial set X, the dg bialgebra $(\widehat{\boldsymbol{\Omega}}(C_*(X)), D, \otimes, \nabla)$ has the property of being a dg Hopf algebra.*

PROOF. We must show the existence of an antipode map

$$s\colon \widehat{\boldsymbol{\Omega}}(C_*(X)) \to \widehat{\boldsymbol{\Omega}}(C_*(X))$$

that is also a map of dg $\mathbf{k}$-modules. First note that the $\mathbf{k}$-bialgebra $\widehat{\boldsymbol{\Omega}}(C_*(X))_0$ on degree 0 is a Hopf algebra being isomorphic to the group algebra $\mathbf{k}[G(X_1)]$, where $G(X_1)$ denotes the free group generated by the set of 1-simplices. The antipode

$$s_0\colon \widehat{\boldsymbol{\Omega}}(C_*(X))_0 \to \widehat{\boldsymbol{\Omega}}(C_*(X))_0$$

is determined by sending a group-like element to its inverse. Recall that $\widehat{\boldsymbol{\Omega}}(C_*(X))$ if and only if the identity map $\mathrm{id}\colon \widehat{\boldsymbol{\Omega}}(C_*(X)) \to \widehat{\boldsymbol{\Omega}}(C_*(X))$ is invertible as an element in the convolution algebra $(\mathrm{Hom}(\widehat{\boldsymbol{\Omega}}(C_*(X)), \widehat{\boldsymbol{\Omega}}(C_*(X))), \star)$. So far we know that the restriction

$$\mathrm{id}\,|_{\widehat{\boldsymbol{\Omega}}(C_*(X))_0}\colon \widehat{\boldsymbol{\Omega}}(C_*(X))_0 \to \widehat{\boldsymbol{\Omega}}(C_*(X))$$

is an invertible element in the convolution algebra. We now explain why id is invertible following a classical argument of Takeuchi adapted to the dg setting. First recall that Hess and Tonks constructed in [**HT10**] a chain homotopy equivalence between $\widehat{\boldsymbol{\Omega}}(C_*(X))$ and $G(X)$, the simplicial group functorially associated to X known as Kan's loop group construction. In particular, this involved constructing two maps of dg algebras

$$\phi\colon \widehat{\boldsymbol{\Omega}}(C_*(X)) \to C_*(G(X))$$

and

$$\psi\colon C_*(G(X)) \to \widehat{\mathbf{\Omega}}(C_*(X))$$

that are chain homotopy inverses to each other, see [**HT10**, Theorem 15]. Since $G(X)$ is a simplicial *group*, the chain complex $C_*(G(X))$ may be equipped with a natural dg Hopf algebra structure with product induced by that of $G(X)$ (together with the Eilenberg–Zilber map) and Alexander–Whitney coproduct. The antipode

$$s_{G(X)}\colon C_*(G(X)) \to C_*(G(X))$$

is induced by applying the normalized chains functor to the inverse map

$$(-)^{-1}\colon G(X) \to G(X).$$

Furthermore, the map $g\colon \widehat{\mathbf{\Omega}}(C_*(X)) \to \widehat{\mathbf{\Omega}}(C_*(X))$ defined by $g := \psi \circ s_{G(X)} \circ \phi$ is a map of dg $\mathbf{k}$-modules extending s_0. This map might not be the inverse of id in the convolution algebra but the proof of [**Tak71**, Lemma 14] explains how one may obtain such an inverse from g. □

REMARK 7.3. [**Tak71**, Lemma 14] implies that a graded bialgebra B, over a field, is a Hopf algebra if and only if the degree 0 bialgebra B_0 is a Hopf algebra. However, we cannot apply this result right away to our context because the proof uses the existence of an arbitrary linear map extending the antipode in degree 0, a fact we do not immediately have in the dg setting (even over a field) since the arbitrary extension might not preserve differentials.

7.3. co-Hochschild complex and Brown's twisted tensor product. Recall the following classical construction of Ed Brown.

DEFINITION 7.4. Let (C, d_C, Δ_C) be a dg coalgebra, (A, d_A, μ_A) an dg algebra, and (M, d_M) a left dg A-module with action $\mu_M : A \otimes M \to M$. Suppose $\tau : C \to A$ is a twisting cochain, that is, a map of degree -1 satisfying

$$d_A \circ \tau + \tau \circ d_C + \mu_A \circ (\tau \otimes \tau) \circ \Delta_C = 0.$$

The *twisted tensor product of C and M over τ* is defined to be the chain complex $C \otimes_\tau M := (C \otimes M, d_\tau)$ where

$$d_\tau = d_C \otimes \mathrm{id}_M + \mathrm{id}_C \otimes d_M + (\mathrm{id}_C \otimes \mu_M) \circ (\mathrm{id}_C \otimes \tau \otimes \mathrm{id}_M) \circ (\Delta_C \otimes \mathrm{id}_M).$$

Brown proved that the above construction gives rise to a chain model for the total space E of a fibration $F \to E \to B$ using a canonical twisting cochain $\tau\colon C_*(B) \to C_{*-1}(\Omega B)$ (following a geometric construction of F. Adams) and the holonomy chain map $\mu_F\colon C_*(\Omega B) \otimes C_*(F) \to C_*(F)$.

We now compare the extended co-Hochschild complex of the chains on a reduced simplicial set to a twisted tensor product constructed via the adjoint action on the dg Hopf algebra structure of the cobar construction. The adjoint action is an algebraic model for the conjugation action of based loops and consequently encodes the holonomy of the free loop fibration $\Omega B \to LB \to B$.

For any connected dg coalgebra C denote by

$$\iota\colon C \to \mathbf{\Omega}(C)$$

the twisting cochain given by

$$\iota(x) = \{x\}.$$

The twisting cochain ι is called the *universal twisting cochain of C*.

Given any reduced simplicial set X, let $i\colon \mathbf{\Omega}(C_*(X)) \hookrightarrow \widehat{\mathbf{\Omega}}(C_*(X))$ the canonical inclusion map, and $i^*_{ad}\widehat{\mathbf{\Omega}}(C_*(X))$ the left dg $\mathbf{\Omega}(C_*(X))$-module determined by the action action, as defined in Section 7.1 using the dg Hopf algebra structure of Theorem 7.2.

THEOREM 7.5. *For any reduced simplicial set X there is a natural chain homotopy equivalence of dg $\mathbf{k}$-modules*

$$C_*(X) \otimes_\iota i^*_{ad}\widehat{\mathbf{\Omega}}(C_*(X)) \leftrightarrows \widehat{co\mathcal{CH}_*}(C_*(X)).$$

PROOF. For simplicity, we denote the connected dg coalgebra $C_*(X)$ by C, the augmented dg algebra $\mathbf{\Omega}(C_*(X))$ by A, and the dg Hopf algebra $\widehat{\mathbf{\Omega}}(C_*(X))$ by $\widehat{A}$. Let $i\colon A \to \widehat{A}$ be the natural inclusion map. As in Section 7.1, we denote by $i^*_{ad}\widehat{A}$ to be $\widehat{A}$ equipped with the right dg A-module structure given by the adjoint action and $i^*(A)$ when equipped with the left $(A \otimes A^{op})$-action.

Note there is a natural isomorphism of dg $\mathbf{k}$-modules

$$C \otimes_\iota i^*_{ad}\widehat{A} \cong \mathcal{Q}(\mathbf{k}, C, A) \otimes_A i^*_{ad}\widehat{A}, \tag{7.5}$$

where $\mathbf{k}$ is thought of as a right dg A-module via the augmentation map.

By Proposition 5.2, there is a natural chain homotopy equivalence

$$\mathcal{Q}(\mathbf{k}, C, A) \otimes_A i^*_{ad}\widehat{A} \leftrightarrows \mathrm{Bar}(\mathbf{k}, A, A) \otimes_A i^*_{ad}\widehat{A} \cong \mathcal{C}_*(A, i^*_{ad}\widehat{A}). \tag{7.6}$$

By Proposition 7.1, there is a natural isomorphism of dg $\mathbf{k}$-modules

$$\mathcal{C}_*(A, i^*_{ad}\widehat{A}) \cong \mathcal{CH}_*(A, i^*\widehat{A}) = \mathrm{Bar}(A, A, A) \otimes_{A \otimes A^{op}} i^*\widehat{A} \tag{7.7}$$

Again by Proposition 5.2, there is a natural chain homotopy equivalence

$$\mathrm{Bar}(A, A, A) \otimes_{A \otimes A^{op}} i^*\widehat{A} \leftrightarrows \mathcal{Q}(A, C, A) \otimes_{A \otimes A^{op}} i^*\widehat{A}. \tag{7.8}$$

Now observe there is a natural isomorphism of dg $\mathbf{k}$-modules

$$\mathcal{Q}(A, C, A) \otimes_{A \otimes A^{op}} i^*\widehat{A} \cong \widehat{co\mathcal{CH}_*}(C). \tag{7.9}$$

Putting together equations (7.5), (7.6), (7.7), (7.8), and (7.9) yields the desired result. □

REMARK 7.6. Note that the $\mathbf{k}$-linear map $C_*(X) \to \widehat{co\mathcal{CH}_*}(C_*(X))$ given by sending $\sigma \mapsto \sigma \otimes 1_{\mathbf{k}}$ is *not* in general a chain map (it would be if $C_*(X)$ was strictly cocommutative, which is not). However, we do have a chain map

$$C_*(X) \to C_*(X) \otimes_\iota i^*_{ad}\widehat{\mathbf{\Omega}}(C_*(X))$$

given by the same formula

$$\sigma \mapsto \sigma \otimes 1_{\mathbf{k}}.$$

Composing this map with the chain map

$$C_*(X) \otimes_\iota i^*_{ad}\widehat{\mathbf{\Omega}}(C_*(X)) \to \widehat{co\mathcal{CH}_*}(C_*(X))$$

given in Theorem 7.5, we obtain a chain map

$$C_*(X) \to \widehat{co\mathcal{CH}_*}(C_*(X))$$

modeling the continuous map $X \to LX$ that sends a point b in X to the constant loop at b. We expect this map to be useful in studying the Goresky–Hingston coproduct and the corresponding Lie cobracket in the S^1-equivariant setting in the string topology of nonsimply connected manifolds.

Acknowledgments

The author would like to thank Julian Holstein, Ralph Kaufmann, Andrey Lazarev, Anibal Medina, Yang Mo, Maximilien Peroux, Dennis Sullivan, Daniel Tolosa, and Mahmoud Zeinalian for fruitful exchanges.

References

[Ada56] J. F. Adams, *On the cobar construction*, Proc. Nat. Acad. Sci. U.S.A. **42** (1956), 409–412, DOI 10.1073/pnas.42.7.409. MR79266

[Bau80] H. J. Baues, *Geometry of loop spaces and the cobar construction*, Mem. Amer. Math. Soc. **25** (1980), no. 230, ix+171, DOI 10.1090/memo/0230. MR567799

[Bau81] H. J. Baues, *The double bar and cobar constructions*, Compositio Math. **43** (1981), no. 3, 331–341. MR632433

[BF86] D. Burghelea and Z. Fiedorowicz, *Cyclic homology and algebraic K-theory of spaces. II*, Topology **25** (1986), no. 3, 303–317, DOI 10.1016/0040-9383(86)90046-7. MR842427

[Doi81] Yukio Doi, *Homological coalgebra*, J. Math. Soc. Japan **33** (1981), no. 1, 31–50, DOI 10.2969/jmsj/03310031. MR597479

[EM66] Samuel Eilenberg and John C. Moore, *Homology and fibrations. I. Coalgebras, cotensor product and its derived functors*, Comment. Math. Helv. **40** (1966), 199–236, DOI 10.1007/BF02564371. MR203730

[FM94] Eric M. Friedlander and Barry Mazur, *Filtrations on the homology of algebraic varieties*, Mem. Amer. Math. Soc. **110** (1994), no. 529, x+110, DOI 10.1090/memo/0529. With an appendix by Daniel Quillen. MR1211371

[Goo85] Thomas G. Goodwillie, *Cyclic homology, derivations, and the free loopspace*, Topology **24** (1985), no. 2, 187–215, DOI 10.1016/0040-9383(85)90055-2. MR793184

[HL22] J. Holstein and A. Lazarev, *Categorical Koszul duality*. part B, Adv. Math. **409** (2022), no. part B, Paper No. 108644, 52, DOI 10.1016/j.aim.2022.108644. MR4477015

[Hol14] Julian V. S. Holstein, *Properness and simplicial resolutions in the model category dgCat*, Homology Homotopy Appl. **16** (2014), no. 2, 263–273, DOI 10.4310/HHA.2014.v16.n2.a14. MR3272303

[HPS09] Kathryn Hess, Paul-Eugène Parent, and Jonathan Scott, *CoHochschild homology of chain coalgebras*, J. Pure Appl. Algebra **213** (2009), no. 4, 536–556, DOI 10.1016/j.jpaa.2008.08.001. MR2483836

[HS21] Kathryn Hess and Brooke Shipley, *Invariance properties of co-Hochschild homology*, J. Pure Appl. Algebra **225** (2021), no. 2, Paper No. 106505, 27, DOI 10.1016/j.jpaa.2020.106505. MR4128640

[HT10] Kathryn Hess and Andrew Tonks, *The loop group and the cobar construction*, Proc. Amer. Math. Soc. **138** (2010), no. 5, 1861–1876, DOI 10.1090/S0002-9939-09-10238-1. MR2587471

[Lur17] Jacob Lurie, *Higher algebra*, 2017.

[MMR21] Anibal M. Medina-Mardones and Manuel Rivera, *Adams' cobar construction as a monoidal e_∞-coalgebra model of the based loop space*, 2021.

[MRZ23] Emilio Minichiello, Manuel Rivera, and Mahmoud Zeinalian, *Categorical models for path spaces*, Adv. Math. **415** (2023), Paper No. 108898, 48, DOI 10.1016/j.aim.2023.108898. MR4544567

[RS18] Manuel Rivera and Samson Saneblidze, *A combinatorial model for the free loop fibration*, Bull. Lond. Math. Soc. **50** (2018), no. 6, 1085–1101, DOI 10.1112/blms.12202. MR3891945

[RZ18] Manuel Rivera and Mahmoud Zeinalian, *Cubical rigidification, the cobar construction and the based loop space*, Algebr. Geom. Topol. **18** (2018), no. 7, 3789–3820, DOI 10.2140/agt.2018.18.3789. MR3892231

[Tab05] Goncalo Tabuada, *Une structure de catégorie de modèles de Quillen sur la catégorie des dg-catégories* (French, with English and French summaries), C. R. Math. Acad. Sci. Paris **340** (2005), no. 1, 15–19, DOI 10.1016/j.crma.2004.11.007. MR2112034

[Tab10] Gonçalo Tabuada, *Differential graded versus simplicial categories*, Topology Appl. **157** (2010), no. 3, 563–593, DOI 10.1016/j.topol.2009.10.015. MR2577490

[Tak71] Mitsuhiro Takeuchi, *Free Hopf algebras generated by coalgebras*, J. Math. Soc. Japan **23** (1971), 561–582, DOI 10.2969/jmsj/02340561. MR292876

[Toë11] Bertrand Toën, *Lectures on dg-categories*, Topics in algebraic and topological K-theory, Lecture Notes in Math., vol. 2008, Springer, Berlin, 2011, pp. 243–302, DOI 10.1007/978-3-642-15708-0. MR2762557

Department of Mathematics, Purdue University, 150 N. University Street, West Lafayette, Indiana 47907-2067

Email address: `manuelr@purdue.edu`

Contemporary Mathematics
Volume **802**, 2024
https://doi.org/10.1090/conm/802/16072

The structure of simply colored coalgebras

Yang Mo

Abstract. A simply colored coalgebra is a coassociative counital coalgebra C over an arbitrary ring R, which can be decomposed into a direct sum of two R-modules: one generated by set-like elements and another consisting of conilpotent elements. Our main result is the equivalence between simply colored coalgebras over a field k and pointed coalgebras with a choice of splitting of its coradical. Additionally, we also prove that the category of simply colored coalgebras is both complete and cocomplete.

Introduction

The aim of this paper is to delve into the subtleties in [**KM22a**] and to extend its results. To begin, we briefly discuss the initial motivation for this collaborative work with R. Kaufmann.

In [**MP12**], Peter May constructs an component coalgebra in the following sense: Let $C_g = Rg \oplus \overline{C}_g$ and $\overline{C}_g = \{x|\Delta(x) = x \otimes g + x' \otimes x'' + g \otimes x, \deg(x') > 0 \text{ and } \deg(x'') > 0\}$ and g is a set-like[1] element in C_g. In essence, C_g is a connected graded coalgebra over R that lacks primitive elements and is indexed by the unique set-like element in g. Given such a family $\{C_g\}$ of them, we define the component coalgebra as the direct sum of coalgebras over the family, i.e.,

$$C = \oplus_g C_g$$

This construction is motivated from the study of loop spaces. It should be noted that each g is also a set-like element in the direct sum C. A component bialgebra refers to a bialgebra in which its coalgebra is a component coalgebra. If the bialgebra is group-like, meaning every set-like element is invertible under the multiplication, then the bialgebra becomes a Hopf algebra.

At a certain point in our study of Hopf algebras obtained from Feynman categories, we realized the need for a more general formulation. In a broader sense, the components should be indexed by two potentially distinct set-like elements, denoted as g and h, within C. In other words, we have

$$C = \oplus_{g,h} C_{g,h}.$$

2020 *Mathematics Subject Classification.* Primary 16T15; Secondary 18M70.

[1] In this paper, we adopt the unconventional term **"set-like"** instead of the commonly used term **"group-like"** found in [**MP12**] [**Mon93**], [**Rad12**], [**Swe69**], in order to maintain consistency.

©2024 American Mathematical Society

The good news is that pointed coalgebras (not necessarily graded) fulfill the required property, as shown in [**Rad12**]. Although each $C_{g,h}$ may not be a coaugmented coalgebra, the existence of an antipode for a pointed coalgebra C remains the same: every set-like element is invertible. However, the bad news is that the notion of pointed coalgebras is not well-defined when the ground ring is not a field. This difficulty arises from the fact that the notion of subcoalgebras may not be well-behaved.

To address this challenge, we introduce the concept of a simply colored coalgebra. It is defined as a triple (C, G, δ), where C is a coalgebra, G is a set of set-like elements in C (this will be called colors), and $\delta : C \to C[G]$ is a coalgebra retraction that maps C onto the set-like coalgebra $C[G]$. This decomposition allows us to express C as the direct sum $C = C[G] \oplus \ker \delta$.

The simply colored coalgebra can be defined over any commutative rings with unity, such as $\mathbb{Z}$. Furthermore, we demonstrate that any non-negatively graded coalgebra with its zero degree freely spanned by set-like elements can be regarded as a simply colored coalgebra. In particular, a connected graded coalgebra is simply colored.

This framework enables us to generalize the results presented in [**GCKT20a**] and [**GCKT20b**]. For further details, We refer readers to [**KM22a**], which uses different terminology and introduces additional variants.

In this paper, we present the following results to expand upon the topics discussed in [**KM22a**]. To be more precise,

(1) The reduced comultiplication can be defined more simply in terms of a retraction (Theorem 2.5). This allows us to extend [**KM22a**, Lemma 2.19] to any comonoid in an monoidal abelian category. This demonstrates that the use of set-like coalgebras is not necessary in the argument and that the coassociativity of the reduced comultiplication can be proved without it.

(2) A pointed coalgebra with a splitting is a pointed coalgebra C equipped with an additional coalgebra homomorphism δ onto its coradical, such that C can be decomposed as $C = C_0 \oplus \ker \delta$. We show that a simply colored coalgebra over a field is equivalent to a pointed coalgebra with a splitting. More precisely,

    THEOREM (2.16). *Let k be a field,*

    (a) *Every simply colored coalgebra (C, G, δ) over k can be viewed as a pointed coalgebra with a splitting (C, δ).*

    (b) *Conversely, if (C, δ) is a pointed coalgebra with a splitting, it naturally gives rise to a simply colored coalgebra structure (C, G, δ), where G is the set of all set-like elements in C.*

    In particular, it is worth noting that coaugmented conilpotent coalgebras can be precisely characterized as pointed coalgebras with a splitting where G consists of a single element. While this fact is likely known in the literature (see, for example, [**HL22**]), the detailed argument supporting this characterization is not commonly found.

(3) In the last section, we show the category of pointed coalgebras with a splitting is both complete and cocomplete. It is worth mentioning that this result is independently proven in [**HL22**]. However, we will present a different approach that does not rely on pseudocompact algebras.

We should highlight that the characterization of reduced comultiplication (as discussed in Theorem 2.5 and Section 2.1.2) enables us to define a "pointed-co-algebra-like" cooperad. It provides a generalization of the existing notion of "conil-potent coaugmented-like" conilpotent coaugmented cooperads. As a consequence, it could potentially pave the way for a broader generalization of Koszul duality of operads, similar to the approach taken in [**HL22**], which may serve as a subject for future research.

Conventions and notations. We assume that readers are familiar with the definitions and basic examples of coalgebras, bialgebras, and Hopf algebras. Throughout this paper, we work over a unital commutative associative ring R or a field k. Given the commutativity of R, an R-module naturally becomes a bimodule. The tensor product $\otimes$ is considered to be over the ground ring R, unless otherwise stated. We use ρ_l (and ρ_r) to denote a left coaction (and right coaction). The symbols ω_l and ω_r are reserved for coactions induced via retraction, which we will discuss later in the paper. It $M = N \oplus N'$ can be decomposed, then the notation $\pi_N : M \to N$ is used to represent the canonical projection onto the submodule N. If the submodule is clear from the context, we omit the subscript B from π_N. Finally, in an unconventional manner, we will use the term **set-like** elements to refer to what is commonly known as group-like elements in most literature.

1. Preliminaries

Most of the material in this section can be found in textbooks such as [**Rad12**], [**Mon93**], and [**Swe69**]. First, we will provide a brief overview of the fundamental concepts

DEFINITION 1.1.

(1) A *cosimple* coalgebra is a coalgebra whose subcoalgebras are exactly zero and itself.
(2) A *coradical* of coalgebra is a sum of all its cosimple subcoalgbras.
(3) A *cosemisimple* coalgebra is a coalgebra which equals to its own coradical.

According to the fundamental theorem of coalgebras, any cosimple coalgebra must be finite-dimensional. Its dual algebra corresponds precisely to a simple algebra (see [**Mon93**, Chapter 1]). While it is commonly referred to as a simple coalgebra in most literature, we will retain the "co-" prefix for consistency. Additionally, it can be proven that the sum of cosimple coalgebras must be a direct sum (see for example [**Mon93**]).

EXAMPLE 1.2 ([**Rad12**]). Here is an cosimple coalgebra over a field. We consider the (standard) matrix coalgebra $M_n^c(k)$ over a field k. As a vector space, it is spanned by elements e_{ij} endowed with the following coalgebra structure.

$$\Delta(e_{ij}) = \sum_{1\leq t\leq n} e_{it} \otimes e_{tj}$$
$$\epsilon(e_{ij}) = \delta_{ij}$$

$M_n^c(k)$ as a vector space is finite dimensional, so its linear dual $(M_n^c(k))^*$ is a algebra over k. In fact, one can check it is isomorphic to the standard matrix algebra $M_n(k)$ as

$$e_{ij}^* e_{kl}^* = \delta_{jk} e_{il}^*.$$

Since the matrix algebra $M_n(k)$ is simple (the only ideal is 0 and itself), by duality the only subcoalgebra is 0 and itself. By definition, $M_n^c(k)$ is a simple coalgebra. $M_n^c(k)$ is pointed if and only if $n = 1$, since the dimension of simple pointed coalgebra can only have dimension of 1 and by comparison the dimension of the matrix coalgebra $M_n^c(k)$ is n^2 ($n^2 > 1$ for $n > 1$).

DEFINITION 1.3. Let C be a coalgebra over a field k and X, Y subspaces. Define the *wedge product* $X \wedge Y$ of X and Y to be the kernel of the composition

$$C \xrightarrow{\quad \Delta \quad} C \otimes C \xrightarrow{\quad \pi\otimes\pi \quad} C/X \otimes C/Y.$$

There are two important observations from the definition, which were left as exercises in [**Swe69**].

PROPOSITION 1.1 ([**Swe69**]).

(1) $(X \wedge Y) \wedge Z = X \wedge (Y \wedge Z)$
(2) $X \wedge Y = \Delta^{-1}(C \otimes Y + X \otimes C)$

The first assertion follows from coassociativity $(\Delta \otimes id) \circ \Delta = (id \otimes \Delta) \circ \Delta$ that the kernel of

$$C \xrightarrow{(\Delta\otimes id)\circ\Delta} C \otimes C \otimes C \xrightarrow{\quad \pi\otimes\pi\otimes\pi \quad} C/X \otimes C/Y \otimes C/Z$$

is the same as the kernel of

$$C \xrightarrow{(id\otimes\Delta)\circ\Delta} C \otimes C \otimes C \xrightarrow{\quad \pi\otimes\pi\otimes\pi \quad} C/X \otimes C/Y \otimes C/Z.$$

As a result, it make senses to use the notation $\wedge^n X = (\wedge^{n-1} X) \wedge X$ without attention to the order. The second assertion follows form the following linear algebra exercise.

LEMMA 1.4. *If $f : V_1 \to W_1$ and $g : V_2 \to W_2$ are linear maps between vector spaces. Then*

$$\ker(f \otimes g) = \ker(f) \otimes V_2 + V_1 \otimes \ker(g).$$

PROOF. Every monomorphism of vector spaces splits. Thus the image of $f \otimes g$ is $f(V_1) \otimes f(V_2) \subset W_1 \otimes W_2$. Without loss of generality, assume both f and g surjective. In this case, since the short exact sequence splits, f induces a splitting as $V_1 = \ker f \oplus V_1'$ s.t. $V_1' \cong W_1$ via f. Similarly, g induces a splitting as $V_2 = \ker g \oplus V_2'$ s.t. $V_2' \cong W_2$ via f Then

$$V_1 \otimes V_2 = \ker f \otimes \ker g \oplus \ker f \otimes V_2' \oplus V_1' \otimes \ker g \oplus V_1' \otimes V_2'.$$

Note if $s \circ f = id$ and $t \circ g = id$, then $(s \otimes t) \circ (f \otimes g) = (s \circ f) \otimes (t \circ g) = id \otimes id = id$. This implies given the splitting we chose for f and g, $im(f \otimes g) = V_1' \otimes V_2'$ and

$$\ker(f \otimes g) = \ker f \otimes \ker g \oplus \ker f \otimes V_2' \oplus V_1' \otimes \ker g = \ker(f) \otimes V_2 + V_1 \otimes \ker(g). \quad \square$$

Let C_0 be the coradical for an coalgebra C. It is known ([**Rad12**, p. 125]) the sequence of subspaces $C_n = \wedge^n C_0$ forms a exhaustive coalgebra filtration; it is called *coradical filtration* of C.

DEFINITION 1.5. A *pointed* coalgebra over a field is defined to be a coalgebra in which every cosimple subcoalgebras is one dimensional.

Alternatively, a pointed coalgebra can be defined as a coalgebra whose coradical is generated freely by set-like elements, since every one-dimensional coalgebra is necessarily generated by a set-like element. It is essential to note that the set of set-like elements is linearly independent (refer to [**Rad12**, Lemma 2.1.12] for further details).

EXAMPLE 1.6.

(1) The simplest form of a pointed coalgebra is given by kg, where g is a set-like element. By duality, every one-dimensional cosimple coalgebra is isomorphic to such a coalgebra kg. The element g in this case is referred to as a *set-like* element. Furthermore, the sum of one-dimensional cosimple coalgebras is also pointed.
(2) According to [**Swe69**], a cocommutative coalgebra over an algebraically closed field is pointed.
(3) As stated in [**Rad12**], a graded coalgebra with its degree zero component spanned by group-like elements is pointed. This follows from the observation that the coradical must be contained within the degree one component ([**Rad12**, Proposition 4.1.2]).
(4) Every path coalgebra associated with a quiver is a pointed coalgebra. This can be verified either directly or by utilizing property (2) and noting that the path coalgebra is graded, with its zero component being spanned by set-like elements (vertices). The path coalgebra is coradically graded. Further information can be found in [**Chi03**].

We know that the path coalgebra associated with a quiver is pointed. However, the converse remains an open problem. Nonetheless, we can establish the following weaker converse.

PROPOSITION 1.2 ([**D97**]). *Let C be a pointed coalgebra. Then there exists a quiver such that C is isomorphic to an admissible subcoalgebra of the path coalgebra of the quiver.*

It has been shown in [**Mon93**] that every pointed coalgebra admits a splitting of its coradical. However, the choice of splitting is not unique. To record the information of the splitting δ, we come up with the following notion.

DEFINITION 1.7. A *pointed coalgebra with a splitting* is a pair (C, δ) consisting of a pointed coalgebra C and a surjective coalgebra homomorphism δ from C onto its coradical C_0.

REMARK 1.8. The splitting of coradical is a part of structure, and it is not property of pointed coalgebras.

Note such a δ induces a splitting $C = C_0 \oplus I$ where $I = \ker \delta$. Conversely, if we are given a pointed coalgebra with $C = C_0 \oplus I$ where I is a coideal, then the canonical projection $\pi_{C_0} : C \to C_0$ naturally provides a splitting. Hence, we occasionally use the notation (C, I) to denote a pointed coalgebra with a splitting.

Lastly, we demonstrate that the category of pointed coalgebras is complete and cocomplete. We begin by presenting the following observations:

LEMMA 1.9.

(1) *Let $f : C \to D$ be a surjective coalgebra homomorphism. If I is a coideal in C, then $f(I)$ is also a coideal in D.*

(2) *If $f : C \to D$ is a surjective coalgebra homomorphism, where C is pointed, then D is also pointed, and $f(C_0) = D_0$.*
(3) *The category of coalgebras over a field is complete and cocomplete, meaning it has both small limits and small colimits.*

PROOF.

(1) A direct check: $\Delta(f(I)) = (f \otimes f)\Delta(I) \subset (f \otimes f)(I \otimes C + C \otimes I) \subset f(I) \otimes D + D \otimes f(I)$ and $\epsilon_D(f(I)) = \epsilon_C(I) = 0$.
(2) First, $D_0 \subset f(C_0)$ if f is surjective ([**Rad12**, Proposition 4.1.7]). Second, if g is set-like in C, then $f(g)$ is also set-like and spans a simple subcoalgebra in D. Therefore, $f(C_0) \subset D_0$.
(3) There are multiple approaches to prove this statement. One possible approach is discussed in [**Ago11**] and [**Ago11a**]. □

THEOREM 1.10.

(1) *Every pointed coalgebra is the filtered colimit of its finite-dimensional pointed subcoalgebra.*
(2) *The embedding ι from the category of pointed coalgebras to the category of coalgebras admits a right adjoint functor Φ.*
(3) *The category of pointed coalgebras is bicomplete.*

PROOF.

(1) Any subcoalgebra of a pointed coalgebra is also pointed, and this can be proven using the same proof as the fundamental theorem for coalgebras.
(2) We have an explicit construction given by the formula $\Phi(C) = \sum_\alpha C_\alpha$, where C_α runs over all pointed subcoalgebras of C. Note that $\Phi(C)$ is not only a subcoalgebra of C but also a pointed coalgebra. This is because, by [**Rad12**, Proposition 3.4.3], every simple subcoalgebra in $\Phi(C)$ must be in one of the C_α and thus has dimension one. Moreover, the image of a pointed coalgebra under a coalgebra homomorphism is also a pointed coalgebra. This follows from the fact that the image is a subcoalgebra and the map is surjective onto its image (we can use the previous lemma in this section). A map $f : C \to D$ induces a map $\Phi(f) : \Phi(C) \to \Phi(D)$.

    Now we define the unit $\eta_X : X \to (\Phi\iota)(X)$ to be the identity on a pointed coalgebra X, and the counit $\varepsilon_Y : (\iota\Phi)(Y) \to Y$ as the natural inclusion of the pointed subcoalgebra $\Phi(Y)$ in a coalgebra Y. The maps η and ε are natural transformations. We need to show that the triangular identities hold.

    It is easy to check $(\Phi\iota\Phi)(X) = \Phi(X)$ and $(\iota\Phi\iota)(Y) = \iota(Y)$. In other words, the following triangular identities hold

$$(\Phi(\varepsilon_X) \circ \eta_{\Phi(X)})(\Phi(X)) = \Phi(X),$$
$$(\varepsilon_{\iota(X)} \circ \iota(\eta_X))(\iota(X)) = \iota(X).$$

    It follows from [**Lan10**] that there is an adjunction such that Φ is right adjoint to ι.
(3) (2) shows that the category of pointed coalgebras is a coreflective subcategory of the category of coalgebras. Since the category of coalgebras is complete and cocomplete, it follows from the corollaries in [**HS71**] that the category of pointed coalgebras is also complete and cocomplete. □

2. Algebraic perspective

In this section, we explore the algebraic structure of simply colored coalgebras. We start by examining the general simply colored comonoid in any monoidal abelian category. we show reduced comultiplication is coassociative. Subsequently, we narrow our scope to simply colored coalgebras in the category of R-modules. We prove the equivalence between simply colored coalgebras over a field and pointed coalgebras with a splitting.

2.1. Simply colored comonoid.

In this subsection, we extend the constructions and proofs presented in [**KM22a**, Section 2.3] to a monoidal category. We consider $\mathcal{C}$ to be a small monoidal category $(\mathcal{C}, \otimes, \mathbb{1})$ that also possesses an abelian structure. It is important to note that, at this stage, we do not assume the compatibility of the monoidal structure with the abelian structure.

REMARK 2.1. According to Mac Lane's coherence theorem, the associators and unitors can be assumed to be strict in the calculations. Additionally, the tensor product of multiple objects can be specified without ambiguity. Hence, in this paper, we assume any monoidal category is strict without loss of generality.

2.1.1. *Reduced comultiplication.* Next, let $\delta : \mathbf{C} \to \mathbf{S}$ be a comonoid retraction between two comonoids $\mathbf{C}$ and $\mathbf{S}$ in the category. This means that δ is the left inverse of the monomorphism of comonoids $\iota : \mathbf{S} \to \mathbf{C}$. Since the category $\mathcal{C}$ is abelian, we have $\mathbf{C} \cong \mathbf{S} \oplus \mathbf{I}$, where $\mathbf{I}$ is the kernel of δ. We identify $\mathbf{S}$ with the image of ι, and we consider δ as the composition $\iota \circ \delta$ (which becomes an idempotent). Consequently, $\mathbf{S}$ is a well-defined subcomonoid.

REMARK 2.2. The notion of a subcomonoid is not generally well-defined. While $\Delta(\iota(\mathbf{S})) \subset \iota(\mathbf{S}) \otimes \iota(\mathbf{S})$, it may not be a subobject of $\mathbf{C} \otimes \mathbf{C}$. However, in our case, we require the map $\iota : \mathbf{S} \to \mathbf{C}$ to be a *split* monomorphism of comonoids. The monoidal product of two split monomorphisms is also a split monomorphism, thanks to the interchange law. Hence, $\iota(\mathbf{S})$ is indeed a well-defined subcomonoid of $\mathbf{C}$ that is isomorphic to $\mathbf{S}$.

Consequently, the following equations make sense.

$$\delta \circ \delta = \delta \tag{2.1}$$

$$\epsilon \circ \delta = \epsilon \tag{2.2}$$

$$\Delta \circ \delta = (\delta \otimes \delta) \circ \Delta \tag{2.3}$$

Now we will construct the following $\mathbf{S}$-bicomodule on $\mathbf{C}$ via $\omega_l = (\delta \otimes id) \circ \Delta$ and $\omega_r = (id \otimes \delta) \circ \Delta$.

PROPOSITION 2.1. *$\omega_l = (\delta \otimes id) \otimes \Delta$ and $\omega_r = (id \otimes \delta) \otimes \Delta$ induces an $\mathbf{S}$-bicomodule on $\mathbf{C}$.*

PROOF. We need to check the following equation for the right comodule structure.

$$(\omega_r \otimes id) \circ \omega_r = (id \otimes \Delta) \circ \omega_r. \tag{2.4}$$

Do the computation:

$$
\begin{aligned}
(\omega_r \otimes id) \circ \omega_r &= (((id \otimes \delta) \circ \Delta) \otimes id) \circ (id \otimes \delta) \circ \Delta \\
&= (((id \otimes \delta) \circ \Delta) \otimes \delta) \circ \Delta \\
&= ((id \otimes \delta) \otimes \delta) \circ (\Delta \otimes id) \circ \Delta \\
&= (id \otimes (\delta \otimes \delta)) \circ (id \otimes \Delta) \circ \Delta \\
&= (id \otimes \Delta) \circ (id \otimes \delta) \circ \Delta \\
&= (id \otimes \Delta) \circ \omega_r .
\end{aligned}
$$

Note we have used the coassociativity and the fact δ is a comonoid homomorphism.

$$(id \otimes \epsilon) \circ \omega_r = id. \tag{2.5}$$

Do the computation again:

$$
\begin{aligned}
(id \otimes \epsilon) \circ \omega_r &= (id \otimes \epsilon) \circ (id \otimes \delta) \circ \Delta \\
&= (id \otimes (\epsilon \circ \delta)) \circ \Delta \\
&= (id \otimes \epsilon) \circ \Delta \\
&= id.
\end{aligned}
$$

Here in $(\epsilon \circ \delta) = \epsilon$ we use the fact δ is a comonoid retraction. With similar computation which we ignore, we show the following equations for the left comodule structure.

$$(id \otimes \omega_l) \circ \omega_l = (\Delta \otimes id) \circ \omega_l \tag{2.6}$$

$$(\epsilon \otimes id) \circ \omega_l = id \tag{2.7}$$

Finally, for the bicomodule structure, we need to compute

$$(\omega_l \otimes id) \circ \omega_r = (id \otimes \omega_r) \circ \omega_l \tag{2.8}$$

$$
\begin{aligned}
(\omega_l \otimes id) \circ \omega_r &= (((\delta \otimes id) \circ \Delta) \otimes id) \circ (id \otimes \delta) \circ \Delta \\
&= ((\delta \otimes id) \otimes id) \circ (\Delta \otimes id) \circ (id \otimes \delta) \circ \Delta \\
&= ((\delta \otimes id) \otimes id) \circ (\Delta \otimes \delta) \circ \Delta \\
&= ((\delta \otimes id) \otimes id) \circ ((id \otimes id) \otimes \delta) \circ (\Delta \otimes id) \circ \Delta \\
&= (\delta \otimes (id \otimes id)) \circ (id \otimes (id \otimes \delta)) \circ (id \otimes \Delta) \circ \Delta \\
&= (\delta \otimes (id \otimes id)) \circ (id \otimes ((id \otimes \delta) \circ \Delta)) \circ \Delta \\
&= (\delta \otimes ((id \otimes \delta) \circ \Delta)) \circ \Delta \\
&= (id \otimes ((id \otimes \delta) \circ \Delta)) \circ (\delta \otimes id) \circ \Delta \\
&= (id \otimes \omega_r) \circ \omega_l .
\end{aligned}
$$

□

REMARK 2.3. In the theory of commutative rings, if B is an A-algebra, it follows the multiplication of B is compatible with action of A. Analogously, the retraction-induced bicomodule is compatible with the comonoid structure. The compatibility means the following equations.

$$(\omega_r \otimes id) \circ \Delta = (id \otimes \omega_l) \circ \Delta \tag{2.9}$$

If the category $\mathcal{C}$ is abelian, this equation means the comultiplication factors through "cotensor" over $\mathbf{S}$. We will define what we mean by cotensor later. Now we compute:

$$\begin{aligned}(\omega_r \otimes id) \circ \Delta &= ((id \otimes \delta) \circ \Delta) \otimes id) \circ \Delta \\ &= ((id \otimes \delta) \otimes id) \circ (\Delta \otimes id) \circ \Delta \\ &= (id \otimes (\delta \otimes id)) \circ (id \otimes \Delta) \circ \Delta \\ &= (id \otimes (\delta \otimes id) \circ \Delta) \circ \Delta \\ &= (id \otimes \omega_l) \circ \Delta,\end{aligned}$$

$$(id \otimes \Delta) \circ \omega_l = (\omega_l \otimes id) \circ \Delta, \tag{2.10}$$

$$\begin{aligned}(id \otimes \Delta) \circ \omega_l &= (id \otimes \Delta) \circ (\delta \otimes id) \circ \Delta \\ &= (\delta \otimes \Delta) \circ \Delta \\ &= (\delta \otimes (id \otimes id)) \circ (id \otimes \Delta) \circ \Delta \\ &= ((\delta \otimes id) \otimes id) \circ (\Delta \otimes id) \circ \Delta \\ &= (((\delta \otimes id) \circ \Delta) \otimes id) \circ \Delta \\ &= (\omega_l \otimes id) \circ \Delta.\end{aligned} \tag{2.11}$$

In a similar computation, it follows

$$(id \otimes \omega_r) \circ \Delta = (\Delta \otimes id) \circ \omega_r. \tag{2.12}$$

Moreover, both ω_l and ω_r act on $\mathbf{S}$ as its own comultiplication. Therefore, we have more equations

$$(id \otimes \omega_r) \circ \omega_r = (id \otimes \omega_l) \circ \omega_r = (id \otimes \Delta) \circ \omega_r = (\omega_r \otimes id) \circ \omega_r. \tag{2.13}$$

We will compute some of equations and the rest is left to the readers.

$$\begin{aligned}(id \otimes \omega_r) \circ \omega_r &= (id \otimes (id \otimes \delta) \circ \Delta) \circ (id \otimes \delta) \circ \Delta \\ &= (id \otimes (id \otimes \delta)) \circ (id \otimes \Delta) \circ (id \otimes \delta) \circ \Delta \\ &= (id \otimes (id \otimes \delta)) \circ (id \otimes (\Delta \circ \delta)) \circ \Delta \\ &= (id \otimes (id \otimes \delta)) \circ (id \otimes (\delta \otimes \delta)) \circ (id \otimes \Delta) \circ \Delta \\ &= (id \otimes (\delta \otimes (\delta \circ \delta))) \circ (id \otimes \Delta) \circ \Delta \\ &= (id \otimes (\delta \otimes (\delta))) \circ (id \otimes \Delta) \circ \Delta \\ &= (id \otimes (\delta \otimes (\delta))) \circ (id \otimes \Delta) \circ \Delta \\ &= (id \otimes (\delta^2 \otimes \delta)) \circ (id \otimes \Delta) \circ \Delta \\ &= (id \otimes (\delta \otimes id)) \circ (id \otimes (\delta \otimes \delta)) \circ (id \otimes \Delta) \circ \Delta \\ &= (id \otimes (\delta \otimes id)) \circ (id \otimes \Delta) \circ (id \otimes \delta) \circ \Delta \\ &= (id \otimes ((\delta \otimes id)) \circ \Delta) \circ (id \otimes \delta) \circ \Delta \\ &= (id \otimes \omega_l) \circ \omega_r, \\ (id \otimes \omega_r) \circ \omega_r &= (id \otimes (\delta \otimes (\delta \circ \delta))) \circ (id \otimes \Delta) \circ \Delta \\ &= (id \otimes (\delta \otimes \delta)) \circ (id \otimes \Delta) \circ \Delta \\ &= (id \otimes \Delta) \circ (id \otimes \delta) \circ \Delta \\ &= (id \otimes \Delta) \circ \omega_r.\end{aligned}$$

Also, in a similar computation

$$(\omega_l \otimes id) \circ \omega_l = (\omega_r \otimes id) \circ \omega_l = (\Delta \otimes id) \circ \omega_l = (id \otimes \omega_l) \circ \omega_l. \tag{2.14}$$

DEFINITION 2.4. The *reduced decomposition* associated to the retraction δ is defined as $\overline{\Delta} = \Delta - \omega_r - \omega_l$.

If the category $\mathcal{C}$ is an abelian monodial category, that is, the following is a homomorphism of abelian groups.

$$\mathcal{C}(\mathbf{X}, \mathbf{Y}) \otimes_{\mathbb{Z}} \mathcal{C}(\mathbf{X}', \mathbf{Y}') \to \mathcal{C}(\mathbf{X} \otimes \mathbf{X}', \mathbf{Y} \otimes \mathbf{Y}')$$

In particular, it implies the tensor product is linear in both entries. We call this condition *biadditive*. The reduced decomposition is coassociative in the following sense. It is a generalization of [**KM22a**, Lemma 2.19].

THEOREM 2.5. *If $\otimes$ is biadditive, $(\overline{\Delta} \otimes id)\overline{\Delta} = (id \otimes \overline{\Delta})\overline{\Delta}$.*

PROOF. The left-hand side is

$$\begin{aligned}(\overline{\Delta} \otimes id) \circ \overline{\Delta} = & (\Delta \otimes id) \circ \Delta - (\omega_r \otimes id) \circ \Delta - (\omega_l \otimes id) \circ \Delta \\ & - (\Delta \otimes id) \circ \omega_r + (\omega_r \otimes id) \circ \omega_r + (\omega_l \otimes id) \circ \omega_r \\ & - (\Delta \otimes id) \circ \omega_l + (\omega_r \otimes id) \circ \omega_l + (\omega_l \otimes id) \circ \omega_l.\end{aligned}$$

The right-hand side is

$$\begin{aligned}(id \otimes \overline{\Delta}) \circ \overline{\Delta} = & (id \otimes \Delta) \circ \Delta - (id \otimes \omega_r) \circ \Delta - (id \otimes \omega_l) \circ \Delta \\ & - (id \otimes \Delta) \circ \omega_r + (id \otimes \omega_r) \circ \omega_r + (id \otimes \omega_l) \circ \omega_r \\ & - (\Delta \otimes id) \circ \omega_l + (id \otimes \omega_r) \circ \omega_l + (id \otimes \omega_l) \circ \omega_l.\end{aligned}$$

Note that $(\Delta \otimes id) \circ \Delta = (id \otimes \Delta) \circ \Delta$ by coassociavity. And most of the terms are cancelled by previous computation. We only need to check the following remaining terms.

$$(id \otimes \Delta) \circ \omega_l = (\omega_l \otimes id) \circ \Delta \tag{2.15}$$

$$\begin{aligned}(id \otimes \Delta) \circ \omega_l &= (id \otimes \Delta) \circ (\delta \otimes id) \circ \Delta \\ &= (\delta \otimes \Delta) \circ \Delta \\ &= (\delta \otimes (id \otimes id)) \circ (id \otimes \Delta) \circ \Delta \\ &= ((\delta \otimes id) \otimes id) \circ (\Delta \otimes id) \circ \Delta \\ &= (((\delta \otimes id) \circ \Delta) \otimes id) \circ \Delta \\ &= (\omega_l \otimes id) \circ \Delta\end{aligned}$$

By similar calculation as above,

$$(id \otimes \omega_r) \circ \Delta = (\Delta \otimes id) \circ \omega_r. \tag{2.16}$$

After canceling terms, we obtain $(\bar{\Delta} \otimes id)\bar{\Delta} - (id \otimes \bar{\Delta})\bar{\Delta} = 0$ and the proof is completed. □

In this case, we will refer to $\bar{\Delta}$ as the *reduced comultiplication*. There are two reasons for using the term "comultiplication." Firstly, it is coassociative. Secondly, when considering a coalgebra over a field, the retraction $\delta : C \to S$ leads to a splitting $C = S \oplus I$ for a certain coideal I. An elementary argument (which we will present later) shows that I is closed under the reduced comultiplication induced by I. Consequently, $(I, \bar{\Delta})$ forms a coalgebra without a counit.

2.1.2. *Short discussion on simply colored cooperads.* If $\mathbb{1}$ is a unit object, it naturally possesses a comonoid structure induced by the left and right unitors $\mathbb{1} \cong \mathbb{1} \otimes \mathbb{1}$, and the counit map is simply an isomorphism of comonoids. We can define the set-like comonoid $\mathbf{S}$ over a set S in the category $\mathcal{C}$ as a direct sum of comonoids $\bigoplus_{s \in S} \mathbb{1}_s$, where S is a set and $\mathbb{1} \cong \mathbb{1}_s$ is an isomorphism of comonoids. It is expected that simply colored comonoids can be defined in any monoidal category with an abelian structure that may not necessarily be compatible. The challenge lies in finding a suitable description for the conilpotent condition. The following example of a simply colored cooperad serves as a starting point for the concept of simply colored comonoids.

We use the notations and definitions from [**LV12**]. The ground ring we consider is a field with characteristic zero. Recall that a cooperad is a comonoid in the category of $\mathbb{S}$-modules, which is both an abelian category and a monoidal category with $\bar{\circ}$ as the non-biadditive monoidal product. The construction we employ is similar to that of the conilpotent cooperad in [**LV12**]. We start with a cooperad $\mathbf{S}$ concentrated in arity one, where $\mathbf{S}(0) = C[S]$ is a set-like coalgebra over a set S. Next, we consider a cooperad $\mathbf{C}$ such that $\delta : \mathbf{C} \to \mathbf{S}$ is an operadic retraction. Similar to the coalgebra case, there exists a cooperadic coideal $\mathbf{I}$ such that $\mathbf{C} = \mathbf{S} \oplus \mathbf{I}$. We adopt Loday's method of defining the conilpotent condition, and we define the following concepts in order:

$$\tilde{\Delta}^0 = id,$$
$$\tilde{\Delta}^1 = \Delta + (id\bar{\circ}\delta)\Delta.$$

Then, we can define it iteratively,

$$\tilde{\Delta}^n = (id\bar{\circ}\tilde{\Delta})\tilde{\Delta}^{n-1}$$

and

$$\hat{\Delta}^n = \tilde{\Delta}^n - (id\bar{\circ}((id\bar{\circ}\delta)\Delta))\tilde{\Delta}^{n-1}.$$

This yields a filtration $F_0\mathbf{C} = \mathbf{S}$ and $F_n\mathbf{C} = \ker \hat{\Delta}^n$. Then, we can define the simply colored cooperad is the pair $(\mathbf{C}, \delta)$ s.t. the filtration we just defined is exhaustive (our conilpotent condition). Note that by counit axiom, it can be shown that when $\mathbf{S} \cong \mathbb{1}$ the monoidal unit , the right coaction $\omega_r = (id\bar{\circ}\delta)\Delta$ amounts to "adding a level made up of the unique set-like $|$", which generalizes construction (see [**LV12**, page 169]) using $id\bar{\circ}\eta$.

As a consequence of our generalization to arbitrary monoidal categories, one can extend the discussion for conilpotent cooperad in [**LV12**] to simply colored cooperads, or even simply colored co-properads. This is due to the possibility to define operads and properads as plethysm monoids. So their dual structures are given by a comultiplication. The existence of such a plethysm construction more generally has been given in [**KM22b**] in terms of a plus construction. The details of this connection will be studied in future work.

2.2. Simply colored coalgebras. We begin by introducing the concept of set-like elements.

DEFINITION 2.6.

(1) An element g in a coalgebra C is referred to as *set-like* if $\Delta(g) = g \otimes g \in C \otimes C$, and $\epsilon(g) = 1$.

(2) The *set-like* coalgebra over a set G is a free module with the basis G, where $\Delta(g) = g \otimes g$ and $\epsilon(g) = 1$ for $g \in G$. In this paper, we will denote this coalgebra as $C[G]$.

It is important to note that $C[G] = \bigoplus_{g \in G} Rg$. There is a unique coalgebra structure on R by considering 1 as a set-like element since R is unital. Recall that a retraction $\delta : C \to C[G]$ implies the existence of a well-defined reduced comultiplication $\overline{\Delta} = (id - \delta \otimes id - id \otimes \delta) \circ \Delta$.

DEFINITION 2.7. Let C be a coalgebra, G be a set of set-like elements in C, and $\delta : C \to C[G]$ be a retraction from C to the set-like coalgebra over G. A *simply colored coalgebra* is a triple $(C, G, \delta : C \to C[G])$ such that for every $x \in \ker \delta$, there exists an integer N (which may depend on x) such that $\bar{\Delta}^N(x) = 0$.

REMARK 2.8. In the definition of a simply colored coalgebra (C, G, δ), we do not require G to contain all set-like elements of C. However, the fact G contains all set-like elements of C follows from the conilpotent condition, as we will demonstrate in Corollary 2.5.

EXAMPLE 2.9. A simple example of a simply colored coalgebra is a set-like coalgebra with a zero coideal.

In principle, one can replace $C[G]$ with any coalgebra if the representation of that coalgebra is well understood. Occasionally, we use the notation $C = C[G] \oplus I$ instead of (C, G, δ) (where $I = \ker \delta$) to represent a simply colored coalgebra.

The definition implies the following properties of set-like coalgebras and simply colored coalgebras.

PROPOSITION 2.2.

(1) *Let $C = C[G] \oplus I$ be a simply colored coalgebra, then $\overline{\Delta}(I) \subset I \otimes I \subset C \otimes C$. In fact, I is a $C[G]$-subbicomodule of C.*
(2) *The tensor product of two set-like coalgebras equipped the canonical coalgebra structure*[2] *is also set-like, and thus simply colored*
(3) *The tensor coalgebra of two simply colored coalgebras $C = C[G] \oplus I$ and $D = C[S] \oplus J$ is also a simply colored coalgebra.*

PROOF.

(1) I is a coideal, so $\Delta(I) \subset I \otimes C + C \otimes I = I \otimes C[G] + C[G] \otimes I + I \otimes I$. It is clear that $\bar{\Delta}(I) \subset I \otimes I$ and $(\delta \otimes id)\Delta(I) \subset C[G] \otimes I \subset C[G] \otimes C$ (and $(id \otimes \delta)\Delta(I) \subset I \otimes C[G] \subset C[G] \otimes C$.
(2) The tensor of two free modules is still free. So it suffices to show tensor of set-like elements is also set-like. We compute for g set-like in C and s set-like in D:

$$\begin{aligned}\Delta_{C \otimes D}(g \otimes s) &= (id \otimes \tau_{12} \otimes id) \circ (\Delta_C \otimes \Delta_D)(g \otimes s) \\ &= g \otimes s \otimes g \otimes s.\end{aligned}$$

[2]If C and D are two coalgebras, the tensor product $C \otimes D$ has a natural coalgebra structure:

$$\Delta_{C \otimes D} = (id \otimes \tau_{12} \otimes id) \circ (\Delta_C \otimes \Delta_D)$$

and

$$\epsilon_{C \otimes D} = \epsilon_C \otimes \epsilon_D$$

where $\tau_{12}(c \otimes d) = d \otimes c$ is a linear isomorphism on $C \otimes D$ to $D \otimes C$

(3) $C\otimes D = C[G]\otimes C[S]\oplus C[G]\otimes J\oplus I\otimes C[S]\oplus I\otimes J$. We showed $C[G]\otimes C[S]$ is also a set-like coalgebra. Then we need to show $C[G]\otimes J\oplus I\otimes C[S]\oplus I\otimes J = C\otimes J + I\otimes D$ is an coideal. The counit condition is straightforward. We finish the proof by checking the comultiplication requirement $\Delta(C\otimes J) \subset C\otimes D\otimes C\otimes J + C\otimes J\otimes C\otimes D$ and $\Delta(I\otimes D) \subset I\otimes D\otimes C\otimes D + C\otimes D\otimes I\otimes D$. Therefore, $\Delta(C\otimes J + I\otimes D) \subset (C\otimes J + I\otimes D)\otimes(C\otimes D) + (C\otimes D)\otimes(C\otimes J + I\otimes D)$. □

REMARK 2.10. The statement in the above proposition is also true if $C[G]$ is replaced by any coalgebra S.

EXAMPLE 2.11. Every coaugmented conilpotent coalgebra is simply colored. Recall that an coaugmentation map is an coalgebra homomorphism $u : R \to C$ s.t. $\epsilon \circ u = id$. If ground ring R is a field k, the condition $\epsilon \circ u = id$ is obvious and does not need to specify. This induces a splitting $C = Re \oplus \ker\epsilon$ in which $e = u(1)$. We claim $\ker\epsilon$ is an coideal. It is obvious that $\epsilon(\ker\epsilon) = 0$. It suffices to show $\Delta(\ker\epsilon) \subset \ker\epsilon\otimes C + C\otimes\ker\epsilon$. Let $x \in \ker\epsilon$. In general,

$$\Delta(x) \in (Re\otimes Re)\oplus(Re\otimes\ker\epsilon)\oplus(\ker\epsilon\otimes Re)\oplus(\ker\epsilon\otimes\ker\epsilon).$$

We only need the fact $\Delta(x)$ has no terms in $Re\otimes Re$. By the left counit identity $(\epsilon\otimes id)\circ\Delta(x) = x$, it follows there exists $r \in R$ and $i \in \ker\epsilon$ such that $x = re + i$ and $\Delta(x) - r(e\otimes e) - (e\otimes i) \in (\ker\epsilon\otimes Re)\oplus(\ker\epsilon\otimes\ker\epsilon)$. However, $x \in \ker\epsilon$ implies $r = 0$. Thus, we show $\ker\epsilon$ is an coideal and the projection $\pi : C \to C/\ker\epsilon \cong Re$ is an retraction that agrees with our notion. The reduced comultiplication then becomes $\bar{\Delta}(x) = \Delta(x) - e\otimes x - x\otimes e$, which agrees with the usual notion of reduced comultiplication.

The following proposition provides a major source of simply colored coalgebras.

PROPOSITION 2.3. *Any nonnegatively graded coalgebra $C = \bigoplus_{i\geq 0} C(i)$ whose zeroth homogeneous component $C(0)$ is a set-like coalgebra $C[X]$ over some set X is simply colored with the choice of the coideal $\bigoplus_{i>0} C(i)$. We will call such graded coalgebras space-like.*

PROOF. It suffices to show the conilpotent condition. We prove by induction that if $x \in C(i)$ for $i > 0$, then $\bar{\Delta}^n(x) = 0$ for any $n \geq i$. The base case for $i = 1$ is straightforward. Suppose it is true for $i = k$. Let $x \in C(k+1)$, $\Delta(x) \in \sum_{i=0}^{k+1} C(i)\otimes C(n-i)$. And $\bar{\Delta}(x) \in \sum_{i=1}^{n-1} C(i)\otimes C(n-i)$ because the reduced comultiplication kills the parts $C(0)\otimes C(k+1)$ and $C(k+1)\otimes C(0)$. Each $C(i)$ will be eliminated by at most i-iteration or reduced comultiplication by induction hypothesis. Since reduced comultiplication is billinear and coassociative, it follows x is eliminated by at most $k+1$-iterations of reduced comultiplication. □

EXAMPLE 2.12.

(1) A connected coalgebra is simply colored. Examples of simply colored coalgebras include Connes–Kreimer's Hopf algebra, Goncharov's Hopf algebra for multiple zeta values, and Baues's Hopf algebra for double loop spaces [**GCKT20a**]. These three Hopf algebras can be unified categorically through the framework of Feynman Categories [**GCKT20a**]. Specifically, if a Feynman category is factorization finite and almost connected, the associated Hopf algebra (as shown in [**GCKT20b**, Theorem 1.1]) is

connected and, therefore, simply colored. The three mentioned Hopf algebras can be constructed by selecting appropriate almost connected and factorization finite Feynman categories.

(2) The path coalgebra associated with a quiver is simply colored. This is because it is known to be a (coradically) graded coalgebra whose zero homogeneous components are spanned by set-like elements.

(3) The categorical coalgebra arising from a Feynman category [**KM22a**, Theorem 5.3] is simply colored. A similar notion of categorical coalgebra in [**MRZ23**] is also simply colored.

(4) The component coalgebra [**MP12**] is simply colored.

EXAMPLE 2.13. A nontrivial example of simply colored coalgebra. Let $C = \mathbf{Z} \oplus \mathbf{Z} \oplus \mathbf{Z}/4\mathbf{Z}$ be a $\mathbf{Z}$-module. We give a coalgebra structure on C by setting $g = (1,0,0)$ and $h = (0,1,0)$ two different set-like elements and $x = (0,0,1)$ a g-primitive elements. $\mathbf{Z}g \oplus \mathbf{Z}h$ is a well-defined subcoalgebra of C (because the natural inclusion splits) and there is a natural splitting of coalgebra onto the subcoalgebra just by projection where the coideal is precisely generated by x. The key is that we allow the existence of torsion elements in the coalgebra.

DEFINITION 2.14. Given C a coalgebra and A an algebra, the R-module $\mathrm{Hom}(C, A)$ is a R-algebra with unit $\eta \circ \epsilon$ and multiplication $\star$ given by $(f \star g)(x) = m \circ (f \otimes g) \circ \Delta(x) = \sum_x f(x_{(1)})g(x_{(2)})$. We say $\{\mathrm{Hom}(C, A), \star, \eta \circ \epsilon\}$ is *the convolution algebra.*

If for $f, g \in \mathrm{Hom}(C, A)$ there is $f \star g = g \star f = \eta \circ \epsilon$, we say f is $\star$-inverse (or convolution inverse) to g and vice versa. The antipode of the Hopf algebra H can be equivalently formulated as the unique convolution inverse to the identity id in $\mathrm{Hom}(H, H)$.

The following theorem is a modification of Takeuchi's lemma used in [**KM22a**] and it allows one to simplify the problem of finding convolution inverse.

THEOREM 2.15. *Let F_0 and C be two coalgebras such that C_0 is a direct summand of $C = F_0 \oplus M$ as R-module and C_0 has a unique subcoalgebra structure inherit from C. Suppose for any $x \in C$, there exists N s.t. its image under the composition of canonical projection onto F_0 with (the kernel is M) and iterated comultiplications is zero.*

$$C \xrightarrow{\Delta^N} C^{\otimes N} \xrightarrow{\pi^{\otimes N}} M^{\otimes N}$$

Then the following conditions are equivalent for any element f of $Hom(C, A)$ with an algebra A:

(1) *f is convolution invertible;*

(2) *$f|_{F_0}$ is convolution invertible in $Hom(F_0, A)$.*

When C is a coalgebra over a field and F_0 is the coradical, this is known as Takeuchi's lemma [**Tak71**]. The proof for the modified version is the same as the one in [**KM22a**, Theorem 3.3]. The key is that we translate the condition $\mathrm{Ext}(C/F_C^{QT}, A) = 0$ to the wedge product characterization.

PROPOSITION 2.4. *Let $C = C[G] \oplus I$ be a simply colored coalgebra and $\pi : C \to I$ be the canonical projection induced by splitting. We have $\pi^{\otimes N}(\bar{\Delta}^N(x)) = \pi^{\otimes N}(\Delta^N(x))$ for $x \in I$.*

PROOF. It follows because $\Delta(x)-\overline{\Delta}(x) \in C[G]\otimes C+C\otimes C[G]$ and the difference will be killed when passed to the quotient. $\Delta^N(x) - \bar{\Delta}^N(x)$ will be also eliminated when passed to the quotient by the same reason. □

Corollary 2.5 then follows.

COROLLARY 2.5.

(1) *If $C = C[G] \oplus I$ is simply colored, then G is the set of all set-likes.*
(2) *Given algebra A, any $f \in \mathrm{Hom}(C, A)$ is convolution invertible if $f|_{C[G]} \in \mathrm{Hom}(C|_{C[G]}, A)$ is convolution invertible.*
(3) *We call an bialgebra is simply colored if its coalgebra structure is simply colored. The simply colored bialgebra has an/the antipode if and only if the set of set-likes form a group under the multiplication.*

PROOF. We only need to show (1). Let t be an set-like not in $C[G]$. It cannot be in I as well since $\epsilon(g) \neq 0$. So it has to be the form of $r + i$ for $0 \neq t \in C[G]$ and $0 \neq i \in I$. Since $\Delta^n(t) = t^{\otimes n}$, $\pi^{\otimes n}(\bar{\Delta}^n(t)) = \pi^{\otimes n}(\Delta^n(t)) = i^{\otimes n} \neq 0$ for any $n \geq 0$. However, the conilpotent condition implies that every $i \in I$ can be eliminated by finite iterations of reduced comultiplication. Thus, $\pi^{\otimes N}(\bar{\Delta}^N(t)) = \pi^{\otimes N}(\bar{\Delta}^N(r)) + \pi^{\otimes N}(\bar{\Delta}^N(t)) = \pi^{\otimes N}(\bar{\Delta}^N(t)) = 0$ for N sufficiently large, which is an contradiction. □

2.2.1. *Simply colored coalgebra over a field is equivalent to pointed coalgebra with a splitting.*

THEOREM 2.16. *Let k be a field,*

(1) *Every simply colored coalgebra (C, G, δ) over k can be viewed as a pointed coalgebra with a splitting (C, δ).*
(2) *Conversely, if (C, δ) is a pointed coalgebra with a splitting, it naturally gives rise to a simply colored coalgebra structure (C, G, δ), where G is the set of all set-like elements in C.*

PROOF.

(1) Let $C = C[G] \oplus I$ where $I = \ker \delta$. We show the coalgebra filtration formed by $\{\wedge^n(C[G])\}_{n\geq 0}$ is exhaustive. It follows directly from Proposition 2.4: if $\bar{\Delta}^n(x) = 0$, then $x \in \wedge^n(C[G])$. By [**Rad12**, Exercise 4.1.9], the coradical $C_0 \subset C[G]$. Therefore, every cosimple subcoalgebra is one dimensional and C is pointed.
(2) The first statement follows again from the fact every pointed coalgebra is isomorphic to a subcoalgebra of a path coalgebra associated to a quiver (proposition 1.2) and the fact subcoalgebra of coradically graded coalgebra is coradically graded ([**Rad12**, Exercise 4.4.6]). Now take any splitting $C = C_0 \oplus I$ of pointed coalgebras. The reduced comultiplication is induced by this splitting. By discussion of proof of [**Mon93**, Theorem 5.4.1], it follows $\overline{\Delta}(I_n) \subset I_{n-1} \otimes I_{n-1}$ where $I_n = C_n \cap I$. Since $C_0 \cap I = \emptyset$, every element in I is eliminated through some iterations of reduced comultiplication. □

Therefore, we see simply colored coalgebra and pointed coalgebra with a splitting are the same thing. Since the subcoalgebra of pointed coalgebra over a field is well-defined and also pointed. Corollary 2.6 follows immediately.

COROLLARY 2.6.

(1) *the subcoalgebra of simply colored coalgebra over a field can be realized to be a simply colored coalgebra over a field.*
(2) *A simply colored coalgebra over a field is a sum of finite dimensional simply colored subcoalgebras. Thus, every simply colored coalgebra over a field is a filtered limit of finite dimensional simply colored coalgebras.*

PROOF. Consider a subcoalgebra D of a pointed coalgebra with a splitting $C = C[G] \oplus I$. According to [**Rad12**, Proposition 3.4.3], we have $D \cap C_0 = D \cap C[G] = D_0$. Thus, $D_0 \subseteq C[G]$ is also pointed, and in particular, a set-like coalgebra since $C[G]$ is cosemisimple. The splitting restricts to D as $\delta|_D : D \to D_0$, resulting in a pointed subcoalgebra with a splitting, denoted as $(D, \delta|_D)$.

By the fundamental theorem of coalgebras, every element x in the coalgebra C belongs to a finite-dimensional subcoalgebra. Moreover, any subcoalgebra of a pointed coalgebra is also pointed. Therefore, for any element x, there exists a finite-dimensional pointed subcoalgebra with a splitting where x resides. Consequently, the pointed coalgebra (C, I) with a splitting can be regarded as the sum of all finite-dimensional pointed subcoalgebras with a splitting. □

EXAMPLE 2.17. As a result, the pointed curved coalgebra in [**HL22**] is also a simply colored coalgebra.

2.2.2. *How is simply colored coalgebra "colored"?* The question has already been addressed in [**KM22a**] through the use of orthogonal projectors. This method involves a specific choice of an orthogonal idempotent family, which has been extensively studied in the context of pointed coalgebras (e.g., see [**Rad82**] and [**Rad12**, Chapter 4]). In this section, we provide a summary of the results presented in the aforementioned papers and books.

When $A = R$, the convolution algebra $C^* = \mathrm{Hom}_R(C, R)$ is referred to as the **dual algebra** of C. The multiplication in C^* is inherited from the multiplication in R, thus we omit the star notation. Any right C-comodule M, which is also an R-module, induces a left C^*-module structure on M defined as $f \cdot m = (id \otimes f)\rho_r(m)$. Similarly, any left C-comodule M gives rise to a right C^*-module structure on M defined as $m \cdot f = (f \otimes id)\rho_l(m)$.

DEFINITION 2.18 ([**Rad12**]). Let C be an coalgebra, An **orthogonal family of idempotents** $\mathcal{E}$ of C^* is a family of element $\{e_i\}_{i\in I}$ of C^* s.t. $e_i e_i = e_i$ and $e_i e_j = 0$ for $i \neq j$. Furthermore, a orthogonal family of idempotents $\mathcal{E}$ is called **orthonormal** if the counit $\epsilon = \sum_{i\in I} e_i$.[3]

Denote $L_e(c) = e \cdot c$ and $R_{e^g} = c \cdot e$ for $e \in \mathcal{E}$. The following identites are true (see [**Rad12**, p. 110 and p. 132] for detail). It still works if the ground ring is only assumed to be commutative unital.

LEMMA 2.19 ([**Rad12**]).

(1) $L_e \circ L_{e'} = \delta_{e,e'} L_e$ *and* $R_e \circ R_{e'} = \delta_{e,e'} R_e$*;*
(2) $\sum_{e\in\mathcal{E}} L_e = I_C = \sum_{e\in\mathcal{E}} R_e$ *where* I_C *is the identity action on* C *of* C^**;*

[3]By the fundamental theorem of coalgebras, a coalgebra is a direct limit of finite-dimensional (sub)coalgebras. Thus, its dual becomes an inverse limit of algebras, which possesses a topology that allows for certain infinite sums to make sense. This algebra is sometimes referred to as a *pseudocompact* algebra in the literature.

(3) $(id \otimes L_e) \circ \Delta = \Delta \circ L_e$ *and* $(R_e \otimes id) \circ \Delta = \Delta \circ R_e$;
(4) $\sum_{e \in \mathcal{E}} (L_e \otimes R_e) \circ \Delta = \Delta$;
(5) $L_e \circ R_{e'} = R_{e'} \circ L_e$.

For a simply colored coalgebra $C = C[G] \oplus I$, we can choose the orthogonal family to be $\delta_{g_{g \in G}}$, where $e_g \in C[G]^*$ represents the dual basis to g. This choice is sensible due to the hypothesis that $C[G]$ is free. It can be verified that this family is also orthonormal. As a consequence of the previous lemma, we have the following proposition:

PROPOSITION 2.7. *An $C[G]$-bicomodule M can be decomposed as*

$$M = \bigoplus_{g,g' \in G} {}^{g}M^{g'},$$

where ${}^{g}M^{g'} = e_{g'} \cdot M \cdot e_g$. *Moreover, it has an explicit description* ${}^{g}M^{g'} = m \in {}^{g}M^{g'} : \rho_l(m) = g \otimes m, \rho_r(m) = m \otimes g'$.

The key observation is that every left/right/bi-comodule over $C[G]$ is cosemisimple. In particular, a $C[G]$-bicomodule is equivalent to a G-bigraded module.

REMARK 2.20. Alternatively, one can directly prove the result using the definition of a bicomodule. However, it is important to note that in this case, ${}^{g}M \cap M^{g'} = {}^{g}M^{g'}$ holds due to the orthonormal condition. It is not true that the equality holds in general, and one cannot deduce it from the same argument presented in [**Mon93**, Example 1.6.7]. The conditions of freeness of $C[G]$ and being a bicomodule are essential for the result.

Move back to the main discussion. The splitting $C = C[G] \oplus I$ endows C with a $C[G]$-bicomodule structure and give rise to a reduced comultiplication.

PROPOSITION 2.8. *Given a splitting $C = C[G] \oplus I$, where I is an coideal, it follows*

(1) $\Delta({}^{g'}C^{g}) \subset \sum_{s \in G} {}^{g'}C^{s} \otimes {}^{s}C^{g}$;
(2) $I = \oplus_{g,g' \in G}({}^{g'}C^{g} \cap I)$ *and* $I \cap {}^{g'}C^{g} = {}^{g'}C^{g}$ *for* $g' \neq g$;
(3) *As a result of (1) and (2), for* $x \in {}^{g'}C^{g}$, *we have both* $\Delta(x) - g' \otimes x \in I \otimes C$ *and* $\Delta(x) - x \otimes g \in C \otimes I$.

PROOF.

(1) It follows from Lemma 2.19 and Proposition 2.7.
(2) For $i \in I$, we know $i = \epsilon \cdot i \cdot \epsilon$. Let $i_{gg'} = e'_g \cdot i \cdot e_g$ so $i = \sum_{g,g' \in G} i_{gg'}$ a direct sum where $i_{gg'} = e_g(i_1)e_{g'}(i_2)i_3$. Note if $g' \neq g$, $i_{gg'} \in g \cdot (C[G]) \cdot g' \oplus g \cdot I \cdot g' = g \cdot I \cdot g'$. If $g = g'$, then $i_{gg'} = c_g g \oplus i'_{gg'}$ where $g \in G$ and $i'_{gg'} \in g \cdot I \cdot g' \subset I$. $C[G] \oplus I = C$ implies $c_g = 0$ for all g. We conclude then $I = \oplus_{g,g' \in G}(I \cap {}^{g'}C^{g})$. To show ${}^{g'}C^{g} \subset I \cap {}^{g'}C^{g}$, just apply $\epsilon = \sum_{s \in G} e_s$ to $e_{g'} \cdot x \cdot e_g$ for $x \in C$ and check $\sum_{s \in G} e_{g'}(x_1)e_s(x_2)e_g(x_3) = 0$ by orthogonality.
(3) Observe $\Delta({}^{g'}x^{g}) - g' \otimes {}^{g'}x^{g} \in \sum_{s \in G; s \neq g'} {}^{g'}C^{s} \otimes {}^{s}C^{g}$ and $\epsilon({}^{g'}C^{s}) = 0$. Thus $\Delta(x) - g' \otimes x \in I \otimes C$. The other is similar. □

We show a simply colored coalgebra has the following structure:

- C is a G-bigraded module: $C = \bigoplus_{g,h\in G} {}^gC^h$ and I is a G-bigraded submodule of C that $I = \oplus_{g,h\in G}({}^gC^h \cap I)$. Both are compatible with both comultiplication and the retraction: ${}^gC^g = Rg \oplus {}^gI^g$, if $g \neq h$ then ${}^gC^h = {}^gI^h$ and $\Delta({}^gC^h) \subset \sum_{t\in G} {}^gC^t \otimes {}^tC^h$.
- There is a well-defined coassociative reduced comultiplication $\bar{\Delta}$ induced by δ s.t. For $x \in {}^gC^h$, $\bar{\Delta}(x) = \Delta(x) - x \otimes h - g \otimes x$. Furthermore, for any $i \in I$, there exists a non-negative integer n (that depends on x) s.t. $\bar{\Delta}^n(x) = 0$.

The algebraic structure also determines a simply colored coalgebra. We will take a categorical viewpoint on the algebraic structure of simply colored coalgebras in the next section.

3. Categorical perspective

In this section, we explore the categorical perspective of simply colored coalgebras. The key result we establish is that the category of simply colored coalgebras over a field (equivalently, pointed coalgebras with a splitting) admits all small limits and colimits. The main challenge lies in demonstrating that the category has both products and coequalizers. To overcome this challenge, we introduce the notion of reduced colored coalgebras, which bears resemblance to the approach used in studying conilpotent coaugmented coalgebras by disregarding the coaugmentation.

3.1. Three categories related to simply colored coalgebras. Note for this part we do not assume the ground ring is a field.

DEFINITION 3.1. The category of simply colored coalgebras is defined as follows:

- The objects are precisely simply colored coalgebras (C, G, δ).
- a morphism between (C, G, δ) and (D, H, μ) is precisely a coalgebra homomorphism $f : C \to D$ such that $\mu \circ f = f|_{C[G]} \circ \delta$.

REMARK 3.2. We have shown that the image of a set-like element under a morphism between simply colored coalgebras is also a set-like element. Therefore, the map $f|_{C[G]}$ has the codomain $C[H]$. Additionally, according to the fundamental theorems of coalgebra homomorphism [**Swe69**], a morphism between (C, G, δ) and (D, H, μ) in the category of simply colored coalgebras can be equivalently described as a coalgebra homomorphism f between C and D such that $f(\ker \delta) \subseteq \ker \mu$.

Recall that we have shown how the reduced comultiplication makes a simply colored coalgebra $C = C[G] \oplus I$ to a coalgebra without counit $(I, \overline{\Delta})$. This coalgebra without counit possesses the property that every element in I is eliminated by a finite number of iterations of the reduced comultiplication. Specifically, it is reduced in the following sense.

DEFINITION 3.3. A coalgebra without counit $\overline{C}$ is called *reduced* simply colored coalgebra if

- It is conilpotent: for every $x \in \overline{C}$, there exists a finite natural number n s.t. $\overline{\Delta}^n(x) = 0$.[4]
- There exists a set G and $\overline{C}$ is a $C[G]$-bicomodule. In particular, $\overline{C}$ is a naturally G-bigraded comodule $\overline{C} = \bigoplus_{g,h\in G} {}^g\overline{C}^h$.

[4]Since counit axiom does not hold, the comultiplication is no longer guaranteed to be injective.

- Furthermore, $\overline{C}$ is G-bigraded coalgebra over R: $\overline{C} = \bigoplus_{g,h \in G} {}^g\overline{C}^h$ and $\Delta({}^g\overline{C}^h) \subset \oplus_{s \in S} {}^g\overline{C}^s \otimes {}^s\overline{C}^h$.

The reduced colored coalgebra is then characterized as a pair $(\overline{C}, G)$.

DEFINITION 3.4. The category of reduced (simply) colored coalgebras is defined as follows:

- The objects are precisely reduced colored coalgebras $(\overline{C}, G)$.
- The morphism are precisely pairs $(\overline{f}, i)$ where $\overline{f} : \overline{C} \to \overline{D}$ is a coalgebra homomorphism that only preserves comultiplication and $i : G \to S$ is a morphism of sets.

There is a third category (V, G) formed by forgetting the coalgebra structure of the reduced colored coalgebras.

DEFINITION 3.5. The category of colored modules (over a ring R) has the following:

- The objects are precisely pairs (V, G) where V is a G bigraded R-module with $V = \bigoplus_{g,h \in G} {}^gV^h$
- The morphisms are precisely pairs $(f, i) : (V, G) \to (W, S)$ where $i : G \to S$ is a morphism of sets and $f({}^gV^h) \subset {}^{i(g)}W^{i(h)}$.[5]

What are the relationships between these categories? In the upcoming subsection, we will see in the next part that there exists a cofree functor from the category of colored vector spaces (colored modules over a field) to the category of pointed coalgebras with a splitting (simply colored coalgebras over a field). Unfortunately, this adjunction may not hold for over arbitrary ring R. For the rest of the subsection, we will establish an equivalence between the category of simply colored coalgebras and the category of reduced colored coalgebras. As discussed earlier, we already know how to obtain a reduced colored coalgebra from a simply colored coalgebra. Next, we will show the reverse process by extending the reduced comultiplication $\overline{\Delta}$. The procedure is as follows:

- Join the set-like coalgebra $C[G]$ over the set S as a direct summand to $\overline{C}$, so $C = C[S] \oplus \overline{C}$.
- Extend the counit map ϵ from $C[G]$ by setting $\epsilon(\overline{C}) = 0$
- The comultiplication Δ on $C[G] \oplus \overline{C}$ is defined as follows: its restriction on $C[S]$ is the same as comultiplication on set-like coalgebra. Its restriction on $\overline{C} = \bigoplus_{g,h \in G} {}^g\overline{C}^h$ is constructed as $\Delta(x) = \rho_l(x) + \overline{\Delta}(x) + \rho_r(x)$ for $x \in \overline{C}$. It is helpful to read component-wise that $\Delta(x) = g \otimes x + \overline{\Delta}(x) + x \otimes h$ for $x \in {}^g\overline{C}^h$.

We need to show the resulting construction is indeed a coalgebra, and furthermore, a simply colored coalgebra.

PROPOSITION 3.1. *The construction $C[G] \oplus \overline{C}$ described above produces a simply colored coalgebra $C = C[G] \oplus I$ where $I = \overline{C}$ is an coideal of C.*

[5]In other word, $(f, i) = \bigoplus_{g,h \in G} f|_{{}^gV^h}$ and $\mathrm{Im}(f|_{{}^gV^h}) \subset {}^{i(g)}W^{i(h)}$. On the other hands, if we have a sequence of maps $\phi_{g,h}$ on ${}^gV^h$ s.t. $\mathrm{Im}(\phi_{g,h}) \subset {}^{i(g)}W^{i(h)}$ given a function of sets $i : G \to S$. Then we can construct a morphism of colored modules via gluing $(\phi, i) = \bigoplus_{g,h \in G} \phi_{g,h}$.

PROOF. First, we need to verify both counit and coassociativity axioms are satisfied. The counit axiom is straightforward once the constructed comultiplication is well-defined, whereas the coassociativity requires computations. It is enough to check component-wise. For simplicity, we use the notation ${}^gx^h \in {}^g\overline{C}^h$ and $\overline{\Delta}({}^gx^h) = \sum_{s\in G} {}^gx^s \otimes {}^sx^h$.

$$\begin{aligned}(\Delta \otimes id)\Delta({}^gx^h) &= (\Delta \otimes id)(g \otimes {}^gx^h + \sum_{s\in G} {}^gx^s \otimes {}^sx^h + {}^gx^h \otimes h)\\ &= g \otimes g \otimes {}^gx^h + \sum_{s\in G}(g \otimes {}^gx^s + \sum_{t\in G} {}^gx^t \otimes {}^tx^s + {}^gx^s \otimes s) \otimes {}^sx^h\\ &\quad + (g \otimes {}^gx^h + \sum_{t\in G} {}^gx^t \otimes {}^tx^s + {}^gx^h \otimes h) \otimes h\end{aligned}$$

and

$$\begin{aligned}(id \otimes \Delta)\Delta({}^gx^h) &= (id \otimes \Delta)(g \otimes {}^gx^h + \sum_{s\in G} {}^gx^s \otimes {}^sx^h + {}^gx^h \otimes h)\\ &= g \otimes (g \otimes {}^gx^h) + \sum_{s\in G} {}^gx^s \otimes {}^sx^h + {}^gx^h \otimes h)\\ &\quad + \sum_{s\in G} {}^gx^s \otimes (s \otimes {}^sx^h + \sum_{t'\in G} {}^sx^{t'} \otimes {}^{t'}x^h + {}^sx^h \otimes h)\\ &\quad + {}^gx^h \otimes h \otimes h.\end{aligned}$$

Observe that due to coaasociativity of $\overline{\Delta}$,[6]

$$\sum_{s\in G}\sum_{t\in G} {}^gx^t \otimes {}^tx^s \otimes {}^sx^h = \sum_{s\in G}\sum_{t'\in G} {}^gx^s \otimes {}^sx^{t'} \otimes {}^{t'}x^h$$

Then it follows that $(\Delta \otimes id) \circ \Delta = (id \otimes \Delta) \circ \Delta$. Finally, $I = \overline{C}$ is a coideal can be checked easily. □

To extend the construction of Proposition 3.1 to a functor, we now define how it acts on each morphism.

PROPOSITION 3.2. *A coalgebra homomorphism between simply colored coalgebras induces an homomorphism of coalgebras without counit. On the other hand, A coalgebra homomorphism without counit between reduced colored coalgebras has an extenstion to a coalgebra homomorphism to the simply colored coalgebras.*

PROOF. The first statement is clear from the structure of simply colored coalgebras because coalgebra homomorphism maps set-like elements to set-like elements, $i(g) = f(g)$ for $f : C \to D$ and $\overline{f} = f|_{\overline{C}} : \overline{C} \to \overline{D}$. For the second statement, it is due to the way we construct the reduced comultiplication. If $(\overline{f}, i) : (\overline{C}, G) \to (\overline{D}, S)$ between two reduced colored coalgebras, then there is an extension to the morphism between simply colored coalgebras as $f(g) = i(g)$ for $g \in G$ and $f(x) = \overline{f}(x)$ for $x \in \overline{C}$. We check it is well defined:

- $\Delta(f(x)) = i(g) \otimes \overline{f}(x) + \overline{f}(x') \otimes \overline{f}(x'') + \overline{f}(x) \otimes i(h)$ for $x \in {}^g\overline{C}^h$ where $\overline{f}(\overline{\Delta}(x)) = (\overline{\Delta} \otimes \overline{\Delta})\overline{f}(x) = \overline{f}(x') \otimes \overline{f}(x'')$. In this case, the $x \in {}^g\overline{C}^h$ is sent to $\overline{f}(x) \in {}^{i(g)}\overline{C}^{i(h)}$. Moreover, $\Delta(f(g)) = \Delta(i(g)) = i(g) \otimes i(g)$.

[6]The Sweddler's notation is omitted, but it is easy to see this equation corresponds to $(\overline{\Delta} \otimes id)\overline{\Delta}({}^gx^h) = (id \otimes \overline{\Delta})\overline{\Delta}({}^gx^h)$.

- $\epsilon(f(x)) = 0$ if $x \in \overline{C}$ and $\epsilon(f(g)) = \epsilon(i(g)) = 1$. □

COROLLARY 3.3. *There is a functor F from the category of simply colored coalgebras to the category of reduced colored coalgebras and a functor G that goes in the other direction such that F (and G) gives an isomorphism of categories.*

PROOF. Define $F((C, I)) = (I, G)$ $(C = C[G] \oplus I)$ and $F(f) = (f|_I, f|_{C[G]})$. Also, define G based on Proposition 3.1 and Proposition 3.2. It is straightforward to check both F and G are functors and $F \circ G = \mathrm{Id}$ and $G \circ F = \mathrm{Id}$. □

REMARK 3.6. From now on, we will write a simply colored coalgebra as $C = C[G] \oplus \overline{C}$ where $\overline{C}$ is the associated reduced colored coalgebra.

3.2. Interlude: Cofree construction. From this point forward, we assume that the ground ring is a field. As established by previous theorems, a simply colored coalgebra is equivalent to a pointed coalgebra with a splitting. In his work [**Rad82**], Radford demonstrates that the forgetful functor from the category of pointed coalgebras with a splitting to the category of colored vector spaces (colored bicomodules over k) admits a left adjoint. This left adjoint is known as the cofree pointed coalgebra (with a splitting) over a pair (V, G), and it is constructed using a cotensor coalgebra. Let us recall the notion of the cotensor product.

Given a right C'-comodule M with a right coaction ρ_R and a left C'-comodule N with a left coaction ρ_L, we define the *cotensor product* $M\square_{C'}N$ as the equalizer of $id_M \otimes \rho_L$ and $\rho_R \otimes id_N$:

$$M\square_{C'}N \dashrightarrow M \otimes N \underset{id_M\otimes\rho_l}{\overset{\rho_r\otimes id_N}{\rightrightarrows}} M \otimes C' \otimes N.$$

Since $M \otimes N$ is a vector space, $M\square_{C'}N$ is the kernel of $(\rho_R \otimes id_N - id \otimes \rho_L)$ and subvector space of $M \otimes N$. When the underlying space is a vector space, there is an associativity $(M\square_{C'}N)\square_{C'}Q \cong M\square_{C'}(N\square_{C'}Q)$. Next, we can construct the cotensor coalgebra ([**Nic78**]) over a pair (S, M) of a coalgebra S and S-bicomodule M as follows (It is clear in this case the underlying coring of a comodule, so we omit the subscript S in $\square_S$ as $\square$):

$$CoT_S(M) = \bigoplus_{n\geq 0} M^{\square n} = S \oplus M \oplus (M\square M) \oplus (M\square M\square M) \oplus \dots$$

Since the ground ring is a field, $M^{\square n}$ is a vector subspace in $M^{\otimes n}$. Every element in $M^{\square n}$ could be written as the normal bar notation $[x_1|x_2|\dots x_n]$ without confusion. This cotensor construction can be endowed with a natural coalgebra structure by deconcatenation:

$$\Delta([x_1|\dots|x_n]) = [\rho_l(x_1)|\dots|x_i] + \sum_{i=1}^{n-1}[x_1|\dots|x_i]\square[x_{i+1}|\dots|x_n] + [x_1|x_2|\dots|\rho_r(x_n)]$$

and counit

$$\epsilon|_S = \epsilon_S \text{ and } \epsilon|_{M^{\square n}} = 0 \text{ for all } n.$$

Note the cotensor coalgebra has a clear grading by number of tensor factors of M. Radford shows that $C(V, G) = CoT_{C[G]}(V)$ exhibiting the following universal property.

THEOREM 3.7 (Radford, [**Rad82**]). *Let $C = k[G(C)] \oplus \overline{C}$ be a pointed coalgebra with a splitting, and $C(S,V)$ constructed from pair (S,V). Let f be a linear map from $\overline{C}$ to V. If f is compatible with both bigrading in $\overline{C}$ and V, i.e., there is an set function $\phi : G(C) \to S$ s.t. $f({}^g\overline{C}^h) \subset {}^{\phi(g)} V^{\phi(h)}$. Then there exists a unique coalgebra map $F : C \to C(G,V)$ such that the following diagram of k-linear maps commutes*

$$\begin{array}{ccc} \overline{C} & \xrightarrow{\ f\ } & V \\ & \underset{F|_{\overline{C}}}{\searrow} & \uparrow{\scriptstyle \pi_V} \\ & & C(G,V) \end{array}$$

where π_V is the projection from $C(S,V)$ to V and $F({}^gC^h) \subset {}^{\phi(g)}C(G,V)^{\phi(h)}$ for $x, y \in G(C)$.

$CoT_{C[G]}(V)$ is pointed with the splitting given by $\delta|_{C[G]} = id$ and $\delta|_{\Box^n} = 0$. Now we set $\overline{C(S,V)} = \bigoplus_{n>0} M^{\Box n}$ as the reduced colored coalgebra associated to $C(S,V)$. By the equivalence of reduced colored coalgebras and simply colored colagbras, the universal property becomes

$$\begin{array}{ccc} \overline{C} & \xrightarrow{\ f\ } & V \\ & \underset{\bar{F}}{\searrow} & \uparrow{\scriptstyle \pi_V} \\ & & \overline{C(G,V)} \end{array}.$$

Then, it becomes the cofree construction in the category of reduced colored coalgebras, and we will called it *reduced cofree colored coalgebra* over a field.

EXAMPLE 3.8. This theorem generalizes the known fact that *the tensor coalgebra is cofree in the category of (coaugmented) conilpotent coalgebras* by letting G to be a Singleton.

3.3. Limits and colimits. Finally, we conclude this section by demonstrating that the category of pointed coalgebras with a splitting (equivalently, the category of reduced colored coalgebras over a field) is both complete and cocomplete. Our approach will differ from that of [**HL22**].

THEOREM 3.9. *The category of pointed coalgebras with a splitting has all small coproducts and equalizers*

PROOF. Recall that the coproduct of coalgebras corresponds to the direct sum of coalgebras. Given that the sum of pointed coalgebras is also pointed, and the sum of coideals is also a coideal, the coproduct of pointed coalgebras with a splitting inherits the structure from the coproduct in the category of coalgebras. It is precisely the direct sum of pointed coalgebras with a splitting induced by the direct sum of coideals. The argument for the equalizer follows the same reasoning as presented in [**Ago11**, Theorem 1.1]. □

Now we come to the more challenging aspect. Instead of dealing with pointed coalgebras with a splitting, we will shift our focus to reduced colored coalgebras as they offer a less complex framework.

LEMMA 3.10. *The category of colored vector spaces has all small products.*

PROOF. Given a family Λ of colored vector spaces $(\overline{C_\alpha}, G_\alpha)$ for $\alpha \in \lambda$, we define the colored product $\mathbb{G}(\prod_{\alpha\in\Lambda}\overline{C_\alpha})$ as the vector subspace of $\prod_{\alpha\in\Lambda}\overline{C_\alpha}$ generated by the sequence of homogeneous elements $({}^{g'_\alpha}x_\alpha{}^{g_\alpha})_\alpha$ where each ${}^{g'_\alpha}x_\alpha{}^{g_\alpha} \in {}^{g'_\alpha}\overline{C_\alpha}{}^{g_\alpha}$ for some g'_α and g_α in G_α. $\mathbb{G}(\prod_{\alpha\in\Lambda}\overline{C_\alpha})$ is $(\prod_{\alpha\in\Lambda}G_\alpha)$ bigraded by construction. The natural projection from product of vector spaces p_α from $\prod_{\alpha\in\Lambda}\overline{C_\alpha} \to \overline{C_\beta}$ for a $\lambda \in \Lambda$ restricts to the subspace $\mathbb{G}(\prod_{\alpha\in\Lambda}\overline{C_\alpha})$. The restriction $\tilde{p}_\alpha$ of each projection p_α from the universal property on the subspace $\mathbb{G}(\prod_{\alpha\in\Lambda}\overline{C_\alpha})$ is also surjective. The universal property of product of sets induces a map $\pi_\lambda : \prod_{\alpha'\in\Lambda}G_{\alpha'} \to G_\lambda$. Now we would like to show $((\mathbb{G}(\prod_{\alpha\in\Lambda}\overline{C_\alpha}), \prod_{\alpha\in\Lambda}G_\alpha), \{(\tilde{p}_\alpha, \pi_\alpha)\}_{\alpha\in\lambda})$ satisfies the universal property of product. Let $(\overline{D'}, S)$ be a colored vector space and a family of map $(\psi_\alpha, \xi_\alpha) : (\overline{D'}, S) \to (\overline{C_\alpha}, G_\alpha)$ for each α in a family Λ of a colored vector spaces $(\overline{C_\alpha}, G_\alpha)$. We have $\psi_\alpha({}^g\overline{D'}{}^h) \subset {}^{\xi_\alpha(g)}\overline{C_\alpha}{}^{\xi_\alpha(h)}$ and $\xi_\alpha(S) \subset G_\alpha$. There are three observations:

- The construction of product of sets implies that $\xi_\alpha(S) \subset G_\alpha$ for $\alpha \in \lambda$ induces a morphism of sets $\xi(S) \subset \prod_{\alpha\in\Lambda}G_\alpha$. The map ξ sends $s \in S$ to a sequence of elements $(\xi_\alpha(s))_{\alpha\in\Lambda}$.
- The construction of product of vector spaces implies that $\psi_\alpha(\overline{D'}) \subset \overline{C_\alpha}$ induces a morphism of vector spaces $\phi : \overline{D'} \to \prod_{\alpha\in\lambda}\overline{C_\alpha}$. The map sends $d \in \overline{D'}$ to a sequence of vectors $(\phi_\alpha(d))_{\alpha\in\Lambda}$.
- It is an important observation that the "homogeneous" map $\phi : \overline{D'} \to \prod_{\alpha\in\Lambda}\overline{C_\alpha}$ sends $d \in {}^g\overline{D'}{}^h$ to $(\phi_\alpha(d))_{\alpha\in\Lambda} \in \prod_{\alpha\in\Lambda}{}^{\xi_\alpha(g)}\overline{C_\alpha}{}^{\xi_\alpha(h)}$ because $\psi_\alpha(d) \in {}^{\xi_\alpha(g)}\overline{C_\alpha}{}^{\xi_\alpha(h)}$. Moreover, ${}^{\xi_\alpha(g)}\overline{C_\alpha}{}^{\xi_\alpha(h)}$ is in the image of $\tilde{p}_\alpha$.

Therefore, the image of ϕ is in the colored vector spaces $\mathbb{G}(\prod_{\alpha\in\Lambda}\overline{C_\alpha})$ and preserves the degrees based on the degree map ξ. It follows the family of maps $\{\phi_\alpha, \xi_\alpha\}_{\alpha\in\lambda}$ induces a map $(\phi|_{\mathbb{G}(\prod_{\alpha\in\Lambda}\overline{C_\alpha})}, \xi)$ from $(\overline{D'}, S)$ to $(\mathbb{G}(\prod_{\alpha\in\Lambda}\overline{C_\alpha}), \prod_{\alpha\in\Lambda}G_\alpha)$. The uniqueness of $(\phi|_{\mathbb{G}(\prod_{\alpha\in\Lambda}\overline{C_\alpha})}, \xi)$ follows from the fact that ξ and ϕ are unique up to unique isomorphism because of universal properties of products of sets and vector spaces respectively. □

REMARK 3.11. The infinite product of vector spaces presents several subtleties. One notable instance is the existence of a non-trivial topology: while many vectors in the product can be expressed as an infinite sum of a set of vectors, not all infinite sums of vectors may belong to this product. This issue is reflected in the fact that $\oplus_i \prod_j M_{i,j} \ncong \prod_j \oplus_i M_{i,j}$, and the usual grading cannot be applied. To address this, we are motivated by the concept of the *graded product* of $\mathbb{N}$-graded modules. By leveraging the property of homogeneous maps, $\oplus_i \prod_j M_{i,j}$ becomes the appropriate choice.

THEOREM 3.12. *The category of reduced colored coalgebras over a field has all small coequalizers and small products.*

PROOF. We start with the coeqaulizer first. We would like to coequalizer the morphisms $(f, a), (g, b) : (\overline{C}, G) \to (\overline{D}, S)$. It begins with the three observations:

- In the category of sets, there exists the coequalizer (up to unique isomorphism) of morphisms of sets $a, b : G \to S$, which is a quotient set $\frac{S}{\sim}$ and the equivalence relation $\sim$ is generated by $a(x) \sim b(x)$ for all $x \in G$. Let the natural map from S to $\frac{S}{\sim}$ be p.

- $\overline{D}$ can be endowed with a new bigrading based on $\frac{S}{\sim}$. We set ${}^{[g]}\overline{D}^{[h]}$ equal
$$\bigoplus_{s,t\in S;p(s)=p(g),p(t)=p(h)} {}^{g}\overline{D}^{h} \text{ for } [g]=p(g), g\in S.$$
The comultiplication is compatible with the new bigrading: if $\Delta({}^{g}\overline{D}^{h}) \subset \sum_{t\in S}^{g} \overline{D}^{t} \otimes {}^{t}\overline{D}^{h}$, then $\Delta({}^{[g]}\overline{D}^{[h]}) \subset \sum_{t\in S} {}^{[g]}\overline{D}^{[t]} \otimes {}^{[t]}\overline{D}^{[h]} = \sum_{[t]\in\frac{S}{\sim}} {}^{[g]}\overline{D}^{[t]} \otimes {}^{[t]}\overline{D}^{[h]}$ because ${}^{s}\overline{D}^{t} \subset {}^{[s]}\overline{D}^{[t]}$ for any $s,t\in S$ by construction.
- $f(x)-g(x) \in {}^{[a(g)]}\overline{D}^{[a(h)]}$ for $x\in {}^{g}\overline{C}^{h}$ and $p(a(g)) = p(b(g)) = [a(g)]$ $(p(a(h)) = p(b(h)) = [a(h)])$. It implies that
$$\operatorname{Im}(f-g) = \bigoplus_{[s],[t]\in\frac{S}{\sim}} ({}^{[s]}\overline{D}^{[t]} \cap \operatorname{Im}(f-g)),$$
i.e., it is generated by homogeneous elements. f and g that restrict to ${}^{g}\overline{C}^{h}$ is coequalized in $\frac{{}^{[a(g)]}\overline{D}^{[a(h)]}}{{}^{[a(g)]}\overline{D}^{[a(h)]}\cap \operatorname{Im}(f-g)}$. In addition, $\operatorname{Im}(f-g)$ is a graded coideal $\bigoplus_{\text{cond}_2} {}^{s'}\overline{C}^{t'}$. The check is the same as in the case of normal coalgebra, but we do not need to check counit condition.

We claim $(\bigoplus_{[s],[t]\in\frac{S}{\sim}} \frac{{}^{[s]}\overline{D}^{[t]}}{{}^{[s]}\overline{D}^{[t]}\cap\operatorname{Im}(f-g)}, (\pi,p))$ is the coequalizer (up to unique isomorphism) of (f,a) and (g,b) where π is the natural projection which is homogeneous according to p. Suppose there is another $((\overline{B},Q),(h,c))$ that coequalizes the two morphisms. On the level of set, there exists a unique morphism $\tilde{c}$ from $\frac{S}{\sim} \to Q$ s.t. $\tilde{c}\circ p = c$.

$$\begin{array}{ccccc} G & \overset{a}{\underset{b}{\rightrightarrows}} & S & \xrightarrow{p} & \frac{S}{\sim} \\ & & & \searrow^{c} & \downarrow{\scriptstyle \exists!\tilde{c}} \\ & & & & Q \end{array}$$

This induces the following diagram of homogeneous maps on the level of vector spaces. Fixing two element $[s]$, $[t]$ in $\frac{S}{\sim}$. Name **condition 1** to be $p(a(m)) = p(b(m)) = [s]$, $p(a(n)) = p(b(n)) = [t]$ and $m,n\in G$. Likewise, name the **condition 2** to be $p(s') = [s]$, $p(t') = [t]$ and $s',t'\in S$. The goal is to simplify the writing under the plus sign. There exists a coequalizer on the level of vector spaces.

$$\begin{array}{ccccc} \bigoplus_{\text{cond}_1} {}^{m}\overline{C}^{n} & \overset{(f,a)_{\text{res}}}{\underset{(g,b)_{\text{res}}}{\rightrightarrows}} & \bigoplus_{\text{cond}_2} {}^{s'}\overline{D}^{t'} & \xrightarrow{\tilde{\pi}} & \frac{{}^{[s]}\overline{D}^{[t]}}{\operatorname{Im}(f-g)_{\text{res}}} \\ & & & \searrow^{(h,c)} & \downarrow{\scriptstyle \tilde{h}_{[s],[t]}} \\ & & & & {}^{c(s)}\overline{B}^{c(t)} \end{array}$$

Note $\bigoplus_{\text{cond}_2} {}^{s'}\overline{D}^{t'} = {}^{[s]}\overline{D}^{[t]}$ as vector spaces (but not colored ones). Thus the natural projection $\tilde{\pi}$ is in fact the restriction of (π,p). The dash arrow is unique due to isomorphism theorems of vector spaces; it is defined as $\tilde{h}_{[s],[t]}(\tilde{\pi}(s)) = h(s)$ (remember $\tilde{\pi}$ is surjective). $\operatorname{Im}(f-g)_{\text{res}}$ is the image of $f-g$ that restricts on $\bigoplus_{\text{cond}_1} {}^{m}\overline{C}^{n}$. Because of previous observations, $\operatorname{Im}(f-g)_{\text{res}}$ is Furthermore,
$$\frac{{}^{[s]}\overline{D}^{[t]}}{{}^{[s]}\overline{D}^{[t]}\cap\operatorname{Im}(f-g)} = \frac{{}^{[s]}\overline{D}^{[t]}}{\operatorname{Im}(f-g)_{\text{res}}}.$$
Then glue $\tilde{h}_{[s],[t]}$ for each element $[s],[t]\in\frac{S}{\sim}$; it yields $(\tilde{h},\tilde{c}) = \bigoplus_{[s],[t]\in\frac{S}{\sim}} \tilde{h}_{[s],[t]}$ as a morphism between colored vector spaces. Since $\operatorname{Im}(f-g)$ is an coideal, this

morphism is furthermore a morphism between reduced colored coalgebras, which is unique because $\tilde{c}$ is unique on the level of sets and $\tilde{c}$ is unqieu on the level of vector spaces (and respect the bigrading). It shows $(\bigoplus_{[s],[t]\in\frac{S}{\sim}}\frac{{}^{[s]}\overline{D}^{[t]}}{{}^{[s]}\overline{D}^{[t]}\cap \mathrm{Im}(f-g)},(\pi,p))$ is the coequalizer between (f,a) and (g,b).

Next, the idea of constructing the product is the same as the one in [**Ago11**, Theorem 1.1], but there are some modification to the argument: Let $(\overline{C_i},G_i)_{i\in I}$ be a a family of reduced colored coalgebras and (Lemma 3.10) the associated reduced cofree colored coalgebra $\overline{C(\mathbb{G}(\prod_{i\in I}\overline{C_i}),\prod_{i\in I}G_i)}$ over the colored vector spaces $\prod_{i\in I}\overline{C_i}$ where we forget the coalgebra structure of $\overline{C_i}$. The rest follows [**Ago11**]. Let $(D,\prod_{i\in I}G_i)$ be the sum of all subcoalgebras $(E,\prod_{i\in I}G_i)$ of $\overline{C(\mathbb{G}(\prod_{i\in I}\overline{C_i}),\prod_{i\in I}G_i)}$ such that $(\tilde{p_i},\pi_i)\circ(\pi_{\mathbb{G}(\prod_{i\in I}\overline{C_i})},id)\circ(j_E,id)$ where (j_E,id) is the canonical inclusion from E to $\mathbb{G}(\prod_{i\in I}\overline{C_i})$. Then one can complete the proof by following the same method in [**Ago11**] to check

$$((D,\prod_{i\in I}G_i),((\tilde{p_i},\pi_i)\circ(\pi_{\mathbb{G}(\prod_{i\in I}\overline{C_i})},id)\circ(j_E,id))_{i\in I})$$

satisfies the universal property and thus the product of the family I in the category of reduced colored coalgebras. □

Lastly, we get Corollary 3.4.

COROLLARY 3.4. *The category of pointed coalgebras with a splitting (or equivalently, the category of reduced colored coalgebras) has all small limits and colimits.*

REMARKS 3.5. Our argument also works if the underlying R-module is either $\mathbb{N}$-graded or differential $\mathbb{N}$-graded. First, we need to endow $C[G]$ with a trivial graded structure that it lies only in zero degree and trivial differentials. Second, every morphisms considered in the paper should be also homogeneous and commutes with differentials.

Acknowledgments

The author would like to express gratitude to R. Kaufmann, M. Rivera, and L. Guo for their valuable suggestions and insightful comments, which greatly contributed to the improvement of the original ideas. I would like to extend my special thanks to the organizers of the AMS special session on higher structures in topology, geometry, and physics, where I had the opportunity to present a preliminary version of this work.

References

[Ago11a] A. L. Agore, *Categorical constructions for Hopf algebras*, Comm. Algebra **39** (2011), no. 4, 1476–1481, DOI 10.1080/00927871003705583. MR2804687

[Ago11] A. L. Agore, *Limits of coalgebras, bialgebras and Hopf algebras*, Proc. Amer. Math. Soc. **139** (2011), no. 3, 855–863, DOI 10.1090/S0002-9939-2010-10542-7. MR2745638

[Chi03] William Chin, *A brief introduction to coalgebra representation theory*, Hopf algebras, Lecture Notes in Pure and Appl. Math., vol. 237, Dekker, New York, 2004, pp. 109–131. MR2051734

[D97] D. Woodcock, *Some categorical remarks on the representation theory of coalgebras*, Comm. Algebra **25** (1997), no. 9, 2775–2794, DOI 10.1080/00927879708826022. MR1458729

[GCKT20a] Imma Gálvez-Carrillo, Ralph M. Kaufmann, and Andrew Tonks, *Three Hopf algebras from number theory, physics & topology, and their common background I: operadic & simplicial aspects*, Commun. Number Theory Phys. **14** (2020), no. 1, 1–90, DOI 10.4310/CNTP.2020.v14.n1.a1. MR4060970

[GCKT20b] Imma Gálvez-Carrillo, Ralph M. Kaufmann, and Andrew Tonks, *Three Hopf algebras from number theory, physics & topology, and their common background II: general categorical formulation*, Commun. Number Theory Phys. **14** (2020), no. 1, 91–169, DOI 10.4310/CNTP.2020.v14.n1.a2. MR4060971

[HL22] J. Holstein and A. Lazarev, *Categorical Koszul duality.* part B, Adv. Math. **409** (2022), no. part B, Paper No. 108644, 52, DOI 10.1016/j.aim.2022.108644. MR4477015

[HS71] Horst Herrlich and George E. Strecker, *Coreflective subcategories*, Trans. Amer. Math. Soc. **157** (1971), 205–226, DOI 10.2307/1995843. MR280561

[KM22a] Ralph M. Kaufmann and Yang Mo, *Pathlike co/bialgebras and their antipodes with applications to bi- and Hopf algebras appearing in topology, number theory and physics*, SIGMA Symmetry Integrability Geom. Methods Appl. **18** (2022), Paper No. 053, 42, DOI 10.3842/SIGMA.2022.053. MR4450299

[KM22b] Ralph M. Kaufmann and Michael Monaco, *Plethysm products, element and plus constructions*, 2022, Preprint: arXiv:2211.11946.

[Lan10] Saunders Mac Lane, *Categories for the working mathematician*, 2nd ed., Graduate Texts in Mathematics, vol. 5, Springer-Verlag, New York, 1998. MR1712872

[LV12] Jean-Louis Loday and Bruno Vallette, *Algebraic operads*, Grundlehren der mathematischen Wissenschaften [Fundamental Principles of Mathematical Sciences], vol. 346, Springer, Heidelberg, 2012, DOI 10.1007/978-3-642-30362-3. MR2954392

[Mon93] Susan Montgomery, *Hopf algebras and their actions on rings*, CBMS Regional Conference Series in Mathematics, vol. 82, Published for the Conference Board of the Mathematical Sciences, Washington, DC; by the American Mathematical Society, Providence, RI, 1993, DOI 10.1090/cbms/082. MR1243637

[MP12] J. P. May and K. Ponto, *More concise algebraic topology*, Chicago Lectures in Mathematics, University of Chicago Press, Chicago, IL, 2012. Localization, completion, and model categories. MR2884233

[MRZ23] Emilio Minichiello, Manuel Rivera, and Mahmoud Zeinalian, *Categorical models for path spaces*, Adv. Math. **415** (2023), Paper No. 108898, 48, DOI 10.1016/j.aim.2023.108898. MR4544567

[Nic78] Warren D. Nichols, *Bialgebras of type one*, Comm. Algebra **6** (1978), no. 15, 1521–1552, DOI 10.1080/00927877808822306. MR506406

[Rad82] David E. Radford, *On the structure of pointed coalgebras*, J. Algebra **77** (1982), no. 1, 1–14, DOI 10.1016/0021-8693(82)90274-5. MR665161

[Rad12] David E. Radford, *Hopf algebras*, Series on Knots and Everything, vol. 49, World Scientific Publishing Co. Pte. Ltd., Hackensack, NJ, 2012. MR2894855

[Swe69] Moss E. Sweedler, *Hopf algebras*, Mathematics Lecture Note Series, W. A. Benjamin, Inc., New York, 1969. MR252485

[Tak71] Mitsuhiro Takeuchi, *Free Hopf algebras generated by coalgebras*, J. Math. Soc. Japan **23** (1971), 561–582, DOI 10.2969/jmsj/02340561. MR292876

Department of Mathematics, Purdue University, West Lafayette, Indiana 47907
Email address: mo12@purdue.edu

Contemporary Mathematics
Volume **802**, 2024
https://doi.org/10.1090/conm/802/16073

On examples and classification of Frobenius objects in Rel

Ivan Contreras, Adele Long, Sophia Marx, and Rajan Amit Mehta

ABSTRACT. We give some new examples of Frobenius objects in the category of sets and relations **Rel**. One example is a groupoid with a twisted counit. Another example is the set of conjugacy classes of a group. We also classify Frobenius objects in **Rel** with two or three elements, and we compute the associated surface invariants using the partition functions of the corresponding topological quantum field theories (TQFT).

1. Introduction

A basic result in topological quantum field theory (TQFT) is the correspondence between 2-dimensional (oriented) TQFTs and commutative Frobenius algebras [**1**, **6**]. This result can be placed in a more general framework by defining a (commutative) Frobenius object in a (symmetric) monoidal category $\mathcal{C}$ (see, e.g., [**9**]). Then a proof of the above correspondence can be reinterpreted as a proof that the 2-dimensional oriented cobordism category is isomorphic to the free symmetric monoidal category generated by one commutative Frobenius object. The upshot of this is that the study of Frobenius objects in any symmetric monoidal category $\mathcal{C}$ has topological significance.

In this paper, we study Frobenius objects in the category **Rel**, whose objects are sets and whose morphisms are relations of sets. Such structures appeared in [**7**] (also see [**8**]), where it was shown that special dagger Frobenius objects in **Rel** are in correspondence with groupoids. This result was extended in [**12**], where it was shown that a Frobenius object in **Rel** can be encoded in a simplicial set equipped with an automorphism of the set of 1-simplices, satisfying certain properties.

Frobenius objects in the category of spans (which is closely related to **Rel**) have also been recently considered. In [**13**], it was shown that symmetric Frobenius objects in **Span** that are coherent (in the 2-categorical sense) are in correspondence with cyclic 2-Segal sets. An analogous correspondence at the 1-categorical level was given in [**3**]. There is a symmetric monoidal functor **Span** $\to$ **Rel**, so any Frobenius object in **Span** gives rise to one in **Rel**. It turns out [**3**, **11**] that **Span** can be viewed as a set-theoretic model for the Wehrheim–Woodward symplectic category

2020 *Mathematics Subject Classification.* Primary 18B10, 18B40, 18C40, 18N50, 20L05, 57R56.

Key words and phrases. Category of relations, Frobenius algebra, groupoid, simplicial set, topological quantum field theory.

The first author thanks the Amherst College Provost and Dean of the Faculty's Research Fellowship (2021–2022).

©2024 American Mathematical Society

[14, 15]. Thus the study of Frobenius objects in **Rel** provides a relatively simple setting to better understand TQFT with values in the symplectic category. Since symplectic groupoids are examples of Frobenius objects in the symplectic category, we expect this direction of research to shed light on the role of symplectic groupoids as reduced phase spaces of the Poisson sigma model, a 2-dimensional topological field theory [2]. There may also be connections to higher-dimensional TQFT, such as Dijkgraaf–Witten theory [5].

The main results of this paper are as follows:

- We describe new examples of Frobenius objects in **Rel**. One is a generalization of the groupoid example which allows for a twist that can spoil the special and dagger properties. As we will see, the twists are necessary to obtain nontrivial topological invariants. Another example is the set of conjugacy classes of a group. These examples can have multivalued multiplication relations.
- We completely classify Frobenius objects in **Rel** with two or three elements. We find that, up to isomorphism, there are 5 Frobenius objects in **Rel** with two elements, and there are 25 Frobenius objects in **Rel** with three elements. This suggests that most examples are not groupoids, since only 5 of the three-element examples are groupoids.
- For all of the two- and three-element Frobenius objects in **Rel**, we compute the associated topological invariants for closed oriented surfaces. These invariants can be viewed as boolean-valued functions of the genus of the surface.

Our low-cardinality classification was done completely by hand, but it involves some simplifications and systematic calculations that should allow for an extension to higher cardinalities with some computer assistance.

2. Frobenius objects and TQFT

In this section, we review the definition of a Frobenius object in a (symmetric) monoidal category $\mathcal{C}$, and we briefly explain the relationship to 2-dimensional TQFT.

2.1. Frobenius objects in a monoidal category.

Let $\mathcal{C}$ be a monoidal category with monoidal unit 1 and monoidal product $\otimes$. For simplicity, we will assume that $\mathcal{C}$ is strict monoidal, though everything in this section works more generally, with appropriate modification.

DEFINITION 2.1. A *Frobenius object* in $\mathcal{C}$ is an object $X \in \mathrm{Ob}(\mathcal{C})$ equipped with morphisms

- $\eta : 1 \to X$ (Unit)
- $\mu : X \otimes X \to X$ (Multiplication)
- $\varepsilon : X \to 1$ (Counit)

satisfying the following axioms.

(1) Unitality: $\mu \circ (\mathbf{1} \otimes \eta) = \mu \circ (\eta \otimes \mathbf{1}) = \mathbf{1}$
(2) Associativity: $\mu \circ (\mathbf{1} \otimes \mu) = \mu \circ (\mu \otimes \mathbf{1})$
(3) Nondegeneracy: There exists $\beta : 1 \to X \otimes X$ such that $(\varepsilon \otimes \mathbf{1}) \circ (\mu \otimes \mathbf{1}) \circ (\mathbf{1} \otimes \beta) = (\mathbf{1} \otimes \varepsilon) \circ (\mathbf{1} \otimes \mu) \circ (\beta \otimes \mathbf{1}) = \mathbf{1}$.

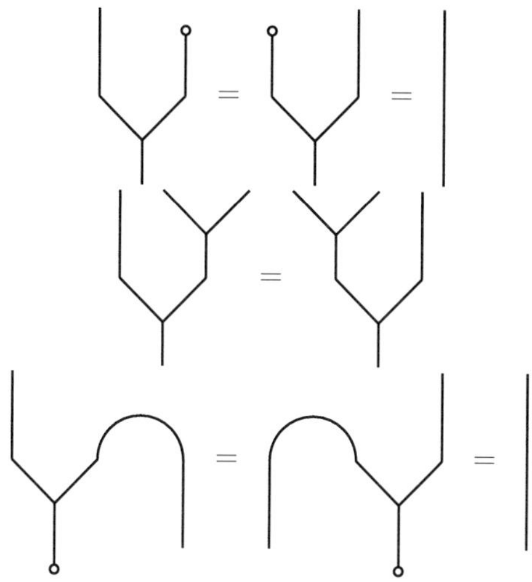

FIGURE 1. The unitality, associativity, and nondegeneracy axioms via string diagrams.

In the case where $\mathcal{C}$ is the monoidal category of vector spaces with the tensor product, one recovers the notion of *Frobenius algebra.*

It can be helpful to use string diagrams to describe morphisms built out of the structure morphisms for a Frobenius object. We denote the unit, multiplication, and counit by the following diagrams, read from top to bottom.

It can be proven that β in the nondegeneracy condition is unique. It is denoted as follows.

The equations in the axioms can be rewritten using string diagrams; see Figure 1.

Given a Frobenius object, we may define a comultiplication $\delta : X \to X \otimes X$ as follows.

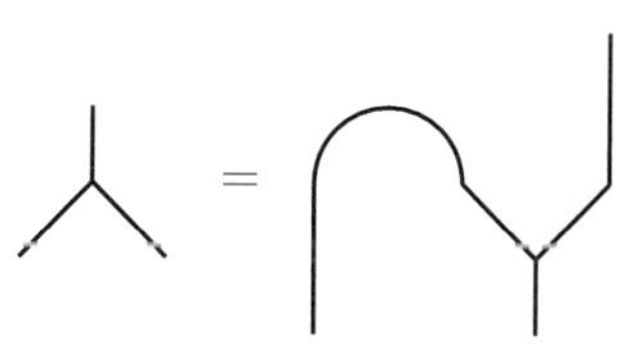

One can then use the axioms to show that the comultiplication is counital and coassociative. Another nice exercise for the reader is to show that the *Frobenius condition* in Figure 2 is satisfied.

We note that the definition of Frobenius object can be stated in different equivalent ways; see, for example, [**9**, **12**].

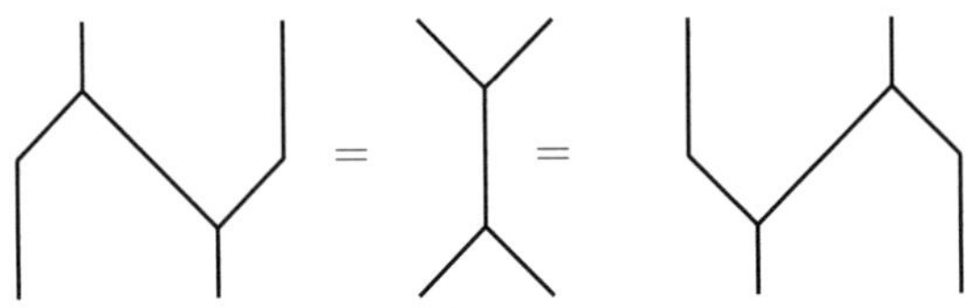

FIGURE 2. The Frobenius condition.

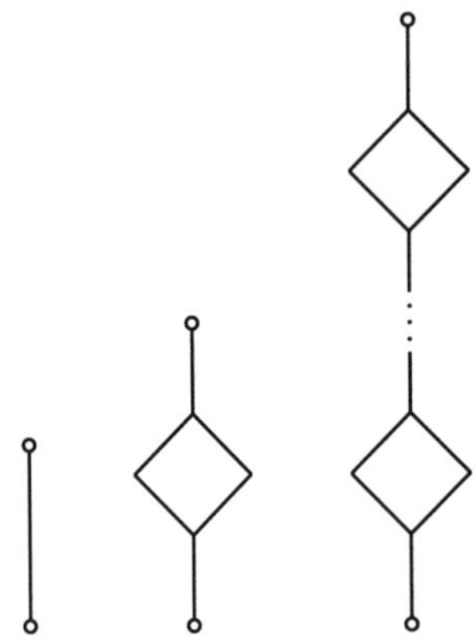

FIGURE 3. Invariants associated to the sphere, torus, and higher genus surfaces.

2.2. Commutativity and relation to TQFT. If $\mathcal{C}$ is a symmetric monoidal category, then we can define commutative Frobenius objects. Specifically, a Frobenius object X in $\mathcal{C}$ is commutative if $\mu \circ \tau = \mu$, where τ is the braiding isomorphism from $X \otimes X$ to itself.

A well-known result **[1, 6]** is that commutative Frobenius objects in $\mathcal{C}$ correspond to $\mathcal{C}$-valued 2-dimensional topological field theories, i.e. symmetric monoidal functors from the 2D oriented cobordism category to $\mathcal{C}$. We refer the reader to **[9]** for details about cobordism categories and a concise proof of this result.

In particular, a commutative Frobenius object gives invariants of closed orientable surfaces. The invariants appear as the output of the *partition function* Z of the theory, which takes closed orientable surfaces as input and takes values in the commutative monoid $\mathrm{Mor}_{\mathcal{C}}(1,1)$. The partition function is explicitly given by the formula

$$Z(\Sigma_g) = \varepsilon \circ (\mu \circ \delta)^g \circ \eta, \tag{2.1}$$

where Σ_g is the closed orientable surface of genus g. The corresponding string diagrams are in Figure 3.

REMARK 2.2. The reason commutativity appears in the context of TQFT is that the relation $\mu \circ \tau = \mu$ holds in the 2D oriented cobordism category; a pair of pants with twisted legs is topologically equivalent to an untwisted pair of pants. Nonetheless, (2.1) makes sense without the commutativity requirement (or even without requiring $\mathcal{C}$ to have a symmetric structure), and one could ask whether (2.1) has a topological interpretation in this case. Such an interpretation is given in **[10]**, where it is shown that Frobenius objects correspond to planar open string topological field theories.

3. Frobenius objects in Rel

In this section, we briefly review the category of relations **Rel**, and we then describe Frobenius objects in **Rel** associated to groupoids and conjugacy classes.

3.1. The category of relations. The objects of **Rel** are sets. A morphism from a set X to a set Y is a relation, i.e., a subset of $X \times Y$. Given relations $R \subseteq X \times Y$ and $S \subseteq Y \times Z$, the composition $S \circ R \subseteq X \times Z$ is defined as

$$S \circ R = \{(x, z) \mid (x, y) \in R \text{ and } (y, z) \in S \text{ for some } y \in Y\}.$$

The identity morphism from X to X is the diagonal relation:

$$\mathbf{1} = \{(x, x) \mid x \in X\}.$$

The Cartesian product gives **Rel** the structure of a symmetric monoidal category. The monoidal unit is the one-point set, which we denote as $\{\bullet\}$.

It can sometimes be convenient to think of a relation $R \subseteq X \times Y$ as a generalized map that can take any number of values. To make this idea precise, we define a map $\widetilde{R} : X \to \mathcal{P}(Y)$, where $\mathcal{P}(Y)$ is the power set of Y, given by

$$\widetilde{R}(x) = \{y \in Y \mid (x, y) \in R\}.$$

This gives a one-to-one correspondence between relations $R \subseteq X \times Y$ and maps $\widetilde{R} : X \to \mathcal{P}(Y)$. In terms of the latter, composition is given by

$$\widetilde{S \circ R}(x) = \bigcup_{y \in \widetilde{R}(x)} \widetilde{S}(y). \tag{3.1}$$

The data of a Frobenius object in **Rel** consists of a set X, together with subsets $\eta \subseteq \{\bullet\} \times X \cong X$, $\varepsilon \subseteq X \times \{\bullet\} \cong X$, and $\mu \subseteq X \times X \times X$. In the remainder of the paper, we will view η and ε as subsets of X, and we will usually consider μ via the corresponding map $\widetilde{\mu} : X \times X \to \mathcal{P}(X)$.

When checking that the data $(X, \eta, \varepsilon, \widetilde{\mu})$ satisfies the axioms of a Frobenius object, unitality and associativity can be checked directly, but the nondegeneracy axiom is a bit clumsy as given. In **[12]**, a useful characterization of the nondegeneracy axiom is given. The condition says that, for every $x \in X$, there exists a unique $y \in X$ such that $\widetilde{\mu}(x, y) \cap \varepsilon$ is nonempty. If this condition is satisfied, the correspondence $x \mapsto y$ determines a bijection $\hat{\alpha} : X \to X$.

From **[12]**, we also have that, if $(X, \eta, \varepsilon, \widetilde{\mu})$ is a Frobenius object in **Rel**, then the comultiplication, viewed as a map $\widetilde{\delta} : X \to \mathcal{P}(X \times X)$, can be expressed in terms of $\widetilde{\mu}$ and the bijection $\hat{\alpha}$, as follows:

$$\widetilde{\delta}(x) = \{(\hat{\alpha}(y), z) \mid z \in \widetilde{\mu}(y, x)\}. \tag{3.2}$$

Finally, we note that commutativity can be simply expressed by the condition that $\widetilde{\mu}(x, y) = \widetilde{\mu}(y, x)$ for all $x, y \in X$.

3.2. Disjoint unions. Let $(X, \eta_X, \varepsilon_X, \widetilde{\mu}_X)$ and $(Y, \eta_Y, \varepsilon_Y, \widetilde{\mu}_Y)$ be Frobenius objects in **Rel**. Then we may define a Frobenius structure on the disjoint union $X \sqcup Y$ as follows:

- $\eta_{X \sqcup Y} = \eta_X \sqcup \eta_Y$,
- $\varepsilon_{X \sqcup Y} = \varepsilon_X \sqcup \eta_Y$, and
- the multiplication is in block form, i.e. for $x, x' \in X$ and $y, y' \in Y$, we set

$$\widetilde{\mu}_{X \sqcup Y}(x, x') = \widetilde{\mu}_X(x, x'),$$
$$\widetilde{\mu}_{X \sqcup Y}(y, y') = \widetilde{\mu}_Y(y, y'),$$
$$\widetilde{\mu}_{X \sqcup Y}(x, y) = \widetilde{\mu}_{X \sqcup Y}(y, x) = \varnothing.$$

It is fairly straightforward to check that the axioms of a Frobenius object hold for $X \sqcup Y$ as a consequence of the fact that they hold for X and Y.

3.3. Example: Groupoids. In **[7]** it was shown that groupoids can be seen as Frobenius objects in **Rel**. Conversely, they showed that any Frobenius object in **Rel** satisfying the extra properties of being *special* and *dagger* is associated to a groupoid. Here, we briefly review this correspondence before introducing the additional possibility of a twisted counit.

Recall that a *groupoid* is a small category where all morphisms are invertible. In more concrete terms, a groupoid G consists of sets G_0, G_1, equipped with *source* and *target* maps $s, t : G_1 \to G_0$ and a *multiplication* operation $(g, h) \mapsto g \cdot h$, defined when $s(g) = t(h)$ for $g, h \in G_1$, such that

(1) $s(g \cdot h) = s(h)$ and $t(g \cdot h) = t(g)$ for $g, h \in G_1$ such that $s(g) = t(h)$,
(2) $(g \cdot h) \cdot k = g \cdot (h \cdot k)$ for all $g, h, k \in G_1$ such that $s(g) = t(h)$ and $s(h) = t(k)$,
(3) there exists an *identity* map $e : G_0 \to G_1$ such that $s \circ e = t \circ e = \mathrm{id}$ and $g \cdot e(s(g)) = e(t(g)) \cdot g = g$ for all $g \in G_1$,
(4) there exists an *inverse map* $G_1 \to G_1$, $g \mapsto g^{-1}$, such that $s(g^{-1}) = t(g)$, $t(g^{-1}) = s(g)$, $g \cdot g^{-1} = e(t(g))$, and $g^{-1} \cdot g = e(s(g))$ for all $g \in G_1$.

As with groups, the identity and inverse maps for a groupoid are unique.

Given a groupoid G, the corresponding Frobenius object in **Rel** given in **[7]** is as follows:

- $X = G_1$,
- $\eta = e(G_0)$,
- $\widetilde{\mu}(g, h) = \{g \cdot h\}$,
- $\varepsilon = e(G_0)$.

Associativity and unitality hold as a result of conditions (2) and (3) above. Nondegeneracy holds as a result of condition (4); in particular, the associated bijection $\hat{\alpha}$ is given by $\hat{\alpha}(g) = g^{-1}$.

This example can be generalized by introducing the possibility of "twisting" the counit as follows. Let $\sigma : G_0 \to G_1$ be a bisection, i.e. a map such that $t \circ \sigma = 1$ and $s \circ \sigma$ is an automorphism, and set $\varepsilon = \sigma(G_0)$. With this choice of counit, one can see that the nondegeneracy condition still holds, with $\hat{\alpha}(g) = g^{-1} \cdot \sigma(t(g))$. We will see in Section 4 that this generalization is necessary to obtain nontrivial topological invariants.

3.4. Example: Conjugacy classes. Let G be a group, and let X be the set of conjugacy classes of G. Set $\eta = \varepsilon = \{[e]\}$, and let $\widetilde{\mu} : X \times X \to \mathcal{P}(X)$ be given by

$$\widetilde{\mu}(C_1, C_2) = \{C_3 \mid g_1 g_2 = g_3 \text{ for some } g_i \in C_i\}.$$

Then X is a Frobenius object in **Rel**. The associated bijection $\hat{\alpha}$ is given by $\hat{\alpha}([g]) = [g^{-1}]$. We note that nondegeneracy holds due to the not-completely-obvious fact that, for any $g \in G$, the conjugacy class of g^{-1} is the set of inverses of elements of the conjugacy class of g.

This is in fact an example of a commutative Frobenius object in **Rel**, since gh and hg are conjugate to each other for all $g, h \in G$:

$$g^{-1}(gh)g = hg.$$

It is also an example in which the multiplication can be multi-valued. This is notable, because all examples of Frobenius objects in **Rel** given in **[7]** and **[12]** have (at most) single-valued multiplication.

In Section 6, we will see the set of conjugacy classes of S_3 appearing as an explicit example.

4. Topological invariants

Any commutative Frobenius object in **Rel** gives rise to surface invariants (see Section 2.2). These invariants take values in $\mathrm{Mor}_{\mathbf{Rel}}(\{\bullet\}, \{\bullet\}) \cong \mathcal{P}(\{\bullet\} \times \{\bullet\}) \cong \mathcal{P}(\{\bullet\})$. It can be useful to identify the values with Booleans, where False is identified with $\varnothing$ and True is identified with $\{\bullet\}$. Under this identification, the monoidal operation, arising from composition of relations, corresponds to the AND operator.

Thus, the partition function associated to a commutative Frobenius object in **Rel**, given by (2.1), can be viewed as a Boolean-valued function of the genus g.

Let $(X, \eta, \varepsilon, \widetilde{\mu})$ be a commutative Frobenius object in **Rel**, with comultiplication and associated bijection $\hat{\alpha}$ as described in Section 3.1. Let $S = \delta \circ \mu$. Using (3.1) and (3.2), we have

$$\begin{aligned} \widetilde{S}(x) &= \widetilde{\mu \circ \delta}(x) \\ &= \bigcup_{\phi \in \widetilde{\delta}(x)} \widetilde{\mu}(\phi) \\ &= \bigcup_{y \in X} \bigcup_{z \in \widetilde{\mu}(y,x)} \widetilde{\mu}(\hat{\alpha}(y), z). \end{aligned} \tag{4.1}$$

This general formula may not seem very enlightening, but in practice it often leads to a relatively nice description of S. From this, we can determine $S^g \circ \eta$, which, as a relation from $\{\bullet\}$ to X, can be viewed as a subset of X. Then the partition function $Z(\Sigma_g) = \varepsilon \circ S^g \circ \eta$ is True if $(S^g \circ \eta) \cap \varepsilon$ is nonempty and False if $(S^g \circ \eta) \cap \varepsilon = \varnothing$.

PROPOSITION 4.1. *Let G be a finite abelian group. For fixed $\omega \in G$, let $X = G$ be the Frobenius object in* **Rel** *associated to G with counit $\varepsilon = \{\omega\}$ (see Section 3.3). Then the associated partition function is given by*

$$Z(\Sigma_g) = \begin{cases} T \text{ if } (g-1)\omega = 0 \\ F \text{ otherwise.} \end{cases} \tag{4.2}$$

PROOF. In this situation, we have the single-valued multiplication $\widetilde{\mu}(x,y) = \{x+y\}$ and $\hat{\alpha}(x) = \omega - x$. Then, from (4.1) we have

$$\begin{aligned} \widetilde{S}(x) &= \bigcup_{y\in G} \{\hat{\alpha}(y) + y + x\} \\ &= \bigcup_{y\in G} \{\omega + x\} \qquad (4.3) \\ &= \{\omega + x\}. \end{aligned}$$

We then obtain

$$\begin{aligned} \widetilde{S^g}(x) &= \{g\omega + x\}, \\ S^g \circ \eta &= \{g\omega\}, \end{aligned}$$

so $Z(\Sigma_g) = \varepsilon \circ S^g \circ \eta = T$ if and only if $g\omega = \omega$. □

The following result will be useful later in computing partition functions for the low-cardinality cases.

LEMMA 4.2. *Let $(X, \eta, \varepsilon, \widetilde{\mu})$ be a commutative Frobenius object in* **Rel**. *If $\eta \cap \varepsilon \neq \varnothing$, then the associated partition function is given by $Z(\Sigma_g) = T$ for all g.*

PROOF. Let $e \in X$ be such that $e \in \eta \cap \varepsilon$. From the unitality axiom, we have that $\widetilde{\mu}(e,e) = \{e\}$. It follows that $\hat{\alpha}(e) = e$, so we see from (4.1) that $e \in \widetilde{S}(e)$. We then have that $e \in (S^g \circ \eta) \cap \varepsilon$ (including the case $g = 0$), and hence $Z(\Sigma_g) = T$. □

REMARK 4.3. Because the partition function associated to a Frobenius object in **Rel** is Boolean-valued, it can be viewed as a logical proposition depending on the variable g. This point of view can lead to a more intuitively clear description of the partition function. For example, in the situation of Proposition 4.1 the partition function is given by the proposition "$g \equiv 1 \pmod{|\omega|}$". In the situation of Lemma 4.2, the partition function is given by the proposition "True".

5. The two element case

5.1. The classification. In this section, we classify the Frobenius objects in **Rel** where $X = \{a, b\}$ is a two element set. Recall that a Frobenius structure (in **Rel**) on X is given by three pieces of data: the unit $\eta \subseteq X$, the counit $\varepsilon \subseteq X$, and the multiplication $\mu \subseteq X \times X \times X$. When X has two elements there are *a priori* $2^2 = 4$ possibilities for each of η and ε and $2^8 = 256$ possibilities for μ, giving a total of $2^{12} = 4096$ possibilities for the data. As we will see, the axioms for a Frobenius object impose strong restrictions, leaving only five possibilities up to isomorphism.

As in Section 3, we describe μ via associated map $\widetilde{\mu} : X \times X \to \mathcal{P}(X)$, which can be represented by a multiplication table.

$\widetilde{\mu}(a,a)$	$\widetilde{\mu}(a,b)$
$\widetilde{\mu}(b,a)$	$\widetilde{\mu}(b,b)$

This is similar to the multiplication tables that students encounter when learning group theory, except that, because $\widetilde{\mu}$ takes values in the power set $\mathcal{P}(X)$, the entries in the table are subsets of X, rather than elements of X.

We begin by considering the different possibilities for the unit relation η. Up to isomorphism, there are three different cases: $\eta = \varnothing, \{a\}, \{a,b\}$. In each case, we can then see what restrictions on μ arise from unitality.

One side of the unitality axiom requires that, for each x, there exists $e \in \eta$ such that $x \in \widetilde{\mu}(x, e)$, and there does not exist any $e \in \eta$ such that $y \in \widetilde{\mu}(x, e)$ for $y \neq x$. The other side of the unitality axiom requires a similar condition with the inputs of $\widetilde{\mu}$ reversed. This condition immediately rules out the case where $\eta = \varnothing$.

In the case where $\eta = \{a\}$, unitality determines three of the four entries in the multiplication table.

$\{a\}$	$\{b\}$
$\{b\}$	

Unitality doesn't impose any constraints on the remaining entry $\widetilde{\mu}(b, b)$.

In the case where $\eta = \{a, b\}$, unitality completely determines the multiplication table.

$\{a\}$	$\varnothing$
$\varnothing$	$\{b\}$

Next, we consider the different possibilities for the counit ε. From [**12**], we know that the unit and counit relations must have the same number of elements. This gives us a small number of cases, and in each case we can see what restrictions on μ arise from nondegeneracy. The nondegeneracy condition (as described in Section 3.1) requires that, in each row or column, there exists a unique entry containing an element of ε.

- $\eta = \{a\}$, $\varepsilon = \{a\}$.
  Here, nondegeneracy requires that $\widetilde{\mu}(b, b)$ contains a, leaving two possibilities for the multiplication table.

$\{a\}$	$\{b\}$
$\{b\}$	$\{a\}$

$\{a\}$	$\{b\}$
$\{b\}$	$\{a, b\}$

- $\eta = \{a\}$, $\varepsilon = \{b\}$. Here, nondegeneracy requires that $\widetilde{\mu}(b, b)$ does not contain b, leaving two possibilities for the multiplication table.

$\{a\}$	$\{b\}$
$\{b\}$	$\{a\}$

$\{a\}$	$\{b\}$
$\{b\}$	$\varnothing$

- $\eta = \{a, b\}$, $\varepsilon = \{a, b\}$. As we saw above, the multiplication table is already determined by the unit axiom, but we observe that nondegeneracy is also satisfied in this situation.

We are now left with only five possibilities (two of which have the same multiplication but different counits), and it remains to check associativity in each case. In three of the five cases (Cases 1, 3, and 5 in Figure 1), we can recognize the multiplication table as being that of a groupoid. In the remaining two cases, we can systematically check associativity in the following way. For each element of $X \times X \times X$, we compute the image in $\mathcal{P}(X)$ under both sides of the associativity equation, and see whether the two images are equal. Here we use the rule for compositions in (3.1).

We find that that associativity holds in both remaining cases, giving us the following result.

THEOREM 5.1. *Up to isomorphism, there are five Frobenius objects in* **Rel** *with two elements. They are listed in Table* 1.

TABLE 1. Frobenius objects in **Rel** with two elements and their partition functions, given as logical propositions.

<table>
<tr><th>Case</th><th>Unit</th><th>Counit</th><th>Multiplication</th><th>Partition function</th></tr>
<tr><td>1</td><td>$\{a\}$</td><td>$\{a\}$</td><td><table><tr><td>$\{a\}$</td><td>$\{b\}$</td></tr><tr><td>$\{b\}$</td><td>$\{a\}$</td></tr></table></td><td>True</td></tr>
<tr><td>2</td><td>$\{a\}$</td><td>$\{a\}$</td><td><table><tr><td>$\{a\}$</td><td>$\{b\}$</td></tr><tr><td>$\{b\}$</td><td>$\{a,b\}$</td></tr></table></td><td>True</td></tr>
<tr><td>3</td><td>$\{a\}$</td><td>$\{b\}$</td><td><table><tr><td>$\{a\}$</td><td>$\{b\}$</td></tr><tr><td>$\{b\}$</td><td>$\{a\}$</td></tr></table></td><td>g is odd</td></tr>
<tr><td>4</td><td>$\{a\}$</td><td>$\{b\}$</td><td><table><tr><td>$\{a\}$</td><td>$\{b\}$</td></tr><tr><td>$\{b\}$</td><td>$\varnothing$</td></tr></table></td><td>$g = 1$</td></tr>
<tr><td>5</td><td>$\{a,b\}$</td><td>$\{a,b\}$</td><td><table><tr><td>$\{a\}$</td><td>$\varnothing$</td></tr><tr><td>$\varnothing$</td><td>$\{b\}$</td></tr></table></td><td>True</td></tr>
</table>

REMARK 5.2. As mentioned above, cases 1, 3, and 5 in Figure 1 are Frobenius objects corresponding to groupoids. Cases 1 and 3 correspond to the group $\mathbb{Z}_2$ with the two possible choices of counits, and case 5 corresponds to the trivial groupoid with two objects. Case 4 also has an interesting interpretation as the cohomology of S^2 (see [**12**]). At the moment, we do not have a conceptual interpretation of case 2.

5.2. Calculation of partition functions. One can immediately see from the multiplication tables in Figure 1 that all five two-element Frobenius objects in **Rel** are commutative, so they correspond to **Rel**-valued 2D oriented TQFTs, and their partition functions give invariants of oriented surfaces.

From Lemma 4.2, we have that the partition function for cases 1, 2, and 5 (when given as a logical proposition, c.f. Remark 4.3) is "True".

For case 3, Proposition 4.1 applies. Since the counit element b is of order 2, the partition function is "g is odd".

For the remaining case 4, we can calculate the partition function directly, following the procedure outlined in Section 4. In this case, we have $\hat{\alpha}(a) = b$ and $\hat{\alpha}(b) = a$. From (4.1), we compute that $\widetilde{S}(a) = \{b\}$ and $\widetilde{S}(b) = \varnothing$. From this we see that $Z(\Sigma_0) = F$, $Z(\Sigma_1) = T$, and $Z(\Sigma_g) = F$ for $g \geq 2$. Thus, as a logical proposition, the partition function is "$g = 1$".

These results are listed in the final column of Table 1.

6. The three element case

In this section, we classify the Frobenius objects in **Rel** where $X = \{a, b, c\}$ is a three element set. Although there is more work involved, our approach is essentially the same as in Section 5, where we first impose unitality, then nondegeneracy, and then associativity. Note that we often omit calculations that are similar to others that have been already been presented.

Up to isomorphism, we have the following cases for the unit: $\eta = \{a\}, \{a, b\}$, and $\{a, b, c\}$. As in Section 5, unitality rules out the case $\eta = \varnothing$.

6.1. The case $\eta = \{a\}$.

In this case, unitality partially determines the multiplication table as follows.

$\{a\}$	$\{b\}$	$\{c\}$
$\{b\}$		
$\{c\}$		

Up to isomorphism, there are two possible counits: $\varepsilon = \{a\}, \{b\}$.

6.1.1. *The case* $\varepsilon = \{a\}$. In this case, nondegeneracy restricts the remaining entries to the two following scenarios.

$\{a\}$	$\{b\}$	$\{c\}$
$\{b\}$	includes a	does not include a
$\{c\}$	does not include a	includes a

$\{a\}$	$\{b\}$	$\{c\}$
$\{b\}$	does not include a	includes a
$\{c\}$	includes a	does not include a

This still leaves many possibilities, but we can make the situation more tractable by considering each possibility for $\widetilde{\mu}(b, b)$ and seeing what restrictions on the other entries arise from associativity.

- $\widetilde{\mu}(b, b) = a$. In this case,

$$\widetilde{\mu \circ (\mu \times \mathbf{1})}(b, b, c) = \bigcup_{x \in \widetilde{\mu}(b,b)} \widetilde{\mu}(x, c) = \widetilde{\mu}(a, c) = \{c\},$$

whereas

$$\widetilde{\mu \circ (\mathbf{1} \times \mu)}(b, b, c) = \bigcup_{x \in \widetilde{\mu}(b,c)} \widetilde{\mu}(b, x),$$

so associativity requires that $\widetilde{\mu}(b, c) = \{c\}$. Similarly,

$$\widetilde{\mu \circ (\mu \times \mathbf{1})}(c, b, b) = \bigcup_{x \in \widetilde{\mu}(c,b)} \widetilde{\mu}(x, b),$$

whereas

$$\widetilde{\mu \circ (\mathbf{1} \times \mu)}(c, b, b) = \{c\},$$

so associativity requires that $\widetilde{\mu}(c, b) = \{c\}$.

Next, we consider

$$\widetilde{\mu \circ (\mu \times \mathbf{1})}(b, c, c) = \bigcup_{x \in \widetilde{\mu}(b,c)} \widetilde{\mu}(x, c) = \widetilde{\mu}(c, c)$$

and

$$\widetilde{\mu \circ (\mathbf{1} \times \mu)}(b, c, c) = \bigcup_{x \in \widetilde{\mu}(c,c)} \widetilde{\mu}(b, x).$$

Since $\widetilde{\mu}(c, c)$ includes a in this scenario, we see that the latter expression contains $\widetilde{\mu}(b, a) = \{b\}$. So associativity requires $\{a, b\} \subseteq \widetilde{\mu}(c, c)$. This leaves us with cases 1 and 2 in Table 2, where we can check that associativity holds in full.

- $\widetilde{\mu}(b,b) = \{a,b\}$. In this case, a similar series of calculations lead to cases 3 and 4 in Table 2.
- $\widetilde{\mu}(b,b) = \{a,c\}$. Here, there are four cases, two of which are isomorphic (under exchange of b and c) to cases 1 and 3 in Table 2, and two of which are new, cases 5 and 6.
- $\widetilde{\mu}(b,b) = \{a,b,c\}$. Here, there are four cases, three of which are isomorphic (under exchange of b and c) to cases 2, 4, and 6 in Table 2, and one of which is new, case 7.
- $\widetilde{\mu}(b,b) = \varnothing$. In this case,

$$\widetilde{\mu \circ (\mu \times \mathbf{1})}(b,b,c) = \varnothing,$$

whereas

$$\widetilde{\mu \circ (\mathbf{1} \times \mu)}(b,b,c) = \bigcup_{x \in \widetilde{\mu}(b,c)} \widetilde{\mu}(b,x),$$

which contains b because $a \in \widetilde{\mu}(b,c)$ and $b \in \widetilde{\mu}(b,a)$. Thus associativity is contradicted and this case is ruled out.
- $\widetilde{\mu}(b,b) = \{b\}$. This gives us cases 8 and 9 in Table 2.
- $\widetilde{\mu}(b,b) = \{c\}$. This gives us cases 10 and 11 in Table 3.
- $\widetilde{\mu}(b,b) = \{b,c\}$. Here, there are three cases, two of which are isomorphic (under exchange of b and c to cases 9 and 11, and one of which is new, case 12.

6.1.2. *The case* $\varepsilon = \{b\}$. In this case nondegeneracy restricts the remaining entries to the following scenario.

$\{a\}$	$\{b\}$	$\{c\}$
$\{b\}$	does not include b	does not include b
$\{c\}$	does not include b	includes b

As above, we can consider each possibility for $\widetilde{\mu}(b,b)$ and see what restrictions on the other entries arise from associativity. This leads to cases 13–20 in Tables 3 and 4.

6.2. The case $\eta = \{a,b\}$. In this case, unitality partially determines the multiplication table as follows.

$\{a\}$	$\varnothing$	
$\varnothing$	$\{b\}$	

Additionally, unitality requires that $\widetilde{\mu}(a,c) \cup \widetilde{\mu}(b,c) = \{c\}$ and $\widetilde{\mu}(c,a) \cup \widetilde{\mu}(c,b) = \{c\}$. In particular, $\widetilde{\mu}(a,c)$, $\widetilde{\mu}(b,c)$, $\widetilde{\mu}(c,a)$, and $\widetilde{\mu}(c,b)$ do not include a or b.

Up to isomorphism, there are two possible counits: $\varepsilon = \{a,b\}$ and $\varepsilon = \{a,c\}$. We consider each case separately.

6.2.1. $\varepsilon = \{a,b\}$. In this case, nondegeneracy requires that $\widetilde{\mu}(c,c)$ contain a or b. Up to isomorphism, we can assume $b \in \widetilde{\mu}(c,c)$.

We know that $\widetilde{\mu}(b,c)$ is either $\{c\}$ or $\varnothing$. If $\widetilde{\mu}(b,c) = \varnothing$, then

$$\widetilde{\mu \circ (\mu \times \mathbf{1})}(b,c,c) = \varnothing,$$

whereas

$$\widetilde{\mu \circ (\mathbf{1} \times \mu)}(b, c, c) = \bigcup_{x \in \widetilde{\mu}(c,c)} \widetilde{\mu}(b, x),$$

which would include b because $b \in \widetilde{\mu}(c, c)$ and $b \in \widetilde{\mu}(b, b)$. This contradicts associativity, so we conclude that $\widetilde{\mu}(b, c) = \{c\}$. Similar calculations show that $\widetilde{\mu}(c, b) = \{c\}$ and $\widetilde{\mu}(a, c) = \widetilde{\mu}(c, a) = \varnothing$.

Suppose $a \in \widetilde{\mu}(c, c)$. Then

$$\widetilde{\mu \circ (\mu \times \mathbf{1})}(a, c, c) = \varnothing,$$

whereas

$$\widetilde{\mu \circ (\mathbf{1} \times \mu)}(a, c, c) = \bigcup_{x \in \widetilde{\mu}(c,c)} \widetilde{\mu}(a, x),$$

which would include a. We conclude that $a \notin \widetilde{\mu}(c, c)$.

This leaves two possibilities, cases 21 and 22 in Table 4. We can see that associativity does hold in these cases by recognizing that they are disjoint unions of the trivial 1-element Frobenius object $\{a\}$ with cases 1 and 2 in Table 4.

6.2.2. $\varepsilon = \{a, c\}$. In this case, nondegeneracy implies that $\widetilde{\mu}(a, c) = \widetilde{\mu}(c, a) = \varnothing$, that $\widetilde{\mu}(b, c) = \widetilde{\mu}(c, b) = \{c\}$, and that $\widetilde{\mu}(c, c)$ does not have a or c. This leaves two possibilities, cases 23 and 24 in Table 4. We note that they are disjoint unions of the trivial 1-element Frobenius object $\{a\}$ with cases 3 and 4 in Table 4.

6.3. The case $\eta = \{a, b, c\}$. In this case, the unit axiom completely determines the multiplication table, and there is only one possible counit. This is case 25 in Table 4.

THEOREM 6.1. *Up to isomorphism, there are 25 Frobenius objects in* **Rel** *with three elements. They are listed in Tables* 2–4.

REMARK 6.2. For some of the three-element Frobenius objects in **Rel**, we can give conceptual explanations. Cases 10 and 17 correspond to the group $\mathbb{Z}_3$ with the two (up to isomorphism) possible choices of counit. Cases 21, 23, and 25 correspond to groupoids. Case 3 corresponds to the conjugacy classes of S_3, as discussed in Section 3.4. But the majority of the cases do not arise from known general constructions, and the exhaustive classification that we have done was needed in order to find them all.

6.4. Calculation of partition functions. All of the three-element Frobenius objects in **Rel** are commutative, so their partition functions give oriented surface invariants. By Lemma 4.2, we have that the partition function, as a logical proposition, is "True" for cases 1–12 and 21–25. Case 17 is the group $\mathbb{Z}_3$ with nontrivial counit, so from Proposition 4.1 we have that the partition function is "$g \equiv 1 \pmod 3$".

This leaves seven cases to check directly, but since they have similarities there are some calculations that can be done for all of them simultaneously. First, since they all have $\eta = \{a\}$ and $\varepsilon = \{b\}$, it follows that $\hat{\alpha}(a) = b$, $\hat{\alpha}(b) = a$, and $\hat{\alpha}(c) = c$. Second, for all of them we have from (4.1) that

$$\begin{aligned}\widetilde{S}(a) &= \bigcup_{y \in X} \widetilde{\mu}(\hat{\alpha}(y), y) \\ &= \widetilde{\mu}(b, a) \cup \widetilde{\mu}(a, b) \cup \widetilde{\mu}(c, c) \\ &= \widetilde{\mu}(c, c),\end{aligned}$$

where in the last step we have used the fact that $b \in \widetilde{\mu}(c,c)$ in the cases we are considering.

In cases 15, 16, 18, 19, and 20, where $a \in \widetilde{\mu}(c,c)$, it follows that $\widetilde{\mu}(c,c) \subseteq S^g \circ \eta$ for all $g \geq 1$, so $Z(\Sigma_g) = T$ for $g \geq 1$. Thus, as a logical proposition, the partition function is "$g \geq 1$".

For case 13, we have $\widetilde{S}(a) = \widetilde{\mu}(c,c) = \{b\}$, and we also calculate

$$\widetilde{S}(b) = \bigcup_{y \in X} \bigcup_{z \in \widetilde{\mu}(y,b)} \widetilde{\mu}(\hat{\alpha}(y), z) = \widetilde{\mu}(\hat{\alpha}(a), b) = \widetilde{\mu}(b,b) = \varnothing.$$

From this we have $S \circ \eta = \{b\}$ and $S^g \circ \eta = \varnothing$ for $g \geq 2$, so $Z(\Sigma_1) = T$ and $S(\Sigma_g) = F$ for $g \geq 2$. As a logical proposition, the partition function is "$g = 1$".

For case 14, we have $\widetilde{S}(a) = \widetilde{\mu}(c,c) = \{b,c\}$, $\widetilde{S}(b) = \varnothing$ as in case 13, and we also calculate

$$\widetilde{S}(c) = \bigcup_{y \in X} \bigcup_{z \in \widetilde{\mu}(y,c)} \widetilde{\mu}(\hat{\alpha}(y), z) = \widetilde{\mu}(b,b) \cup \widetilde{\mu}(c,b) \cup \widetilde{\mu}(c,c) = \{b,c\}.$$

From this we have $S^g \circ \eta = \{b,c\}$ for $g \geq 1$, so $S(\Sigma_g) = T$ for $g \geq 1$. As a logical proposition, the partition function is "$g \geq 1$".

These results are listed in the final columns of Tables 2–4.

7. Future directions

A natural generalization is the full classification of Frobenius objects in **Rel** of higher cardinality, and the computation of their corresponding topological invariants. As suggested in the previous section, the computations become more involved as cardinality increases, so developing code to solve the "equations" associated to the axioms would help to accomplish this objective. It would also be interesting to see if there exists analogue in **Rel** of the Artin–Wedderburn classification theorem of semisimple algebras.

An ongoing work in progress [**4**] intends to extend the characterization of Frobenius objects in **Rel** and **Span** to higher categories. In particular, we plan to connect the simplicial description of various flavors of Frobenius objects with 2-Segal spaces [**13**].

Since we can view Frobenius objects in **Rel** as a generalization of groupoids, it could be interesting to see whether certain concepts in the theory of groupoids (such as Morita equivalence) can be extended.

As stated in the introduction, this project arose with the aim of better understanding TQFTs with values in the symplectic category. Our long-term goal is to characterize such structures and connect them to topological field theories such as the Poisson sigma model and Dijkgraaf–Witten theory.

Acknowledgments

We would like to thank Pavel Mnev and Walker Stern for stimulating discussions on topics related to this paper, and the referee for useful suggestions.

TABLE 2. Frobenius objects in **Rel** with three elements and their partition functions, part 1.

Case	Unit	Counit	Multiplication	Partition function
1	$\{a\}$	$\{a\}$	$\begin{array}{\|c\|c\|c\|}\hline \{a\} & \{b\} & \{c\} \\ \hline \{b\} & \{a\} & \{c\} \\ \hline \{c\} & \{c\} & \{a,b\} \\ \hline \end{array}$	True
2	$\{a\}$	$\{a\}$	$\begin{array}{\|c\|c\|c\|}\hline \{a\} & \{b\} & \{c\} \\ \hline \{b\} & \{a\} & \{c\} \\ \hline \{c\} & \{c\} & \{a,b,c\} \\ \hline \end{array}$	True
3	$\{a\}$	$\{a\}$	$\begin{array}{\|c\|c\|c\|}\hline \{a\} & \{b\} & \{c\} \\ \hline \{b\} & \{a,b\} & \{c\} \\ \hline \{c\} & \{c\} & \{a,b\} \\ \hline \end{array}$	True
4	$\{a\}$	$\{a\}$	$\begin{array}{\|c\|c\|c\|}\hline \{a\} & \{a\} & \{c\} \\ \hline \{b\} & \{a,b\} & \{c\} \\ \hline \{c\} & \{c\} & \{a,b,c\} \\ \hline \end{array}$	True
5	$\{a\}$	$\{a\}$	$\begin{array}{\|c\|c\|c\|}\hline \{a\} & \{b\} & \{c\} \\ \hline \{b\} & \{a,c\} & \{b,c\} \\ \hline \{c\} & \{b,c\} & \{a,b\} \\ \hline \end{array}$	True
6	$\{a\}$	$\{a\}$	$\begin{array}{\|c\|c\|c\|}\hline \{a\} & \{b\} & \{c\} \\ \hline \{b\} & \{a,c\} & \{b,c\} \\ \hline \{c\} & \{b,c\} & \{a,b,c\} \\ \hline \end{array}$	True
7	$\{a\}$	$\{a\}$	$\begin{array}{\|c\|c\|c\|}\hline \{a\} & \{b\} & \{c\} \\ \hline \{b\} & \{a,b,c\} & \{b,c\} \\ \hline \{c\} & \{b,c\} & \{a,b,c\} \\ \hline \end{array}$	True
8	$\{a\}$	$\{a\}$	$\begin{array}{\|c\|c\|c\|}\hline \{a\} & \{b\} & \{c\} \\ \hline \{b\} & \{b\} & \{a,b,c\} \\ \hline \{c\} & \{a,b,c\} & \{b,c\} \\ \hline \end{array}$	True
9	$\{a\}$	$\{a\}$	$\begin{array}{\|c\|c\|c\|}\hline \{a\} & \{b\} & \{c\} \\ \hline \{b\} & \{b\} & \{a,b,c\} \\ \hline \{c\} & \{a,b,c\} & \{c\} \\ \hline \end{array}$	True

TABLE 3. Frobenius objects in **Rel** with three elements and their partition functions, part 2.

<table>
<tr><th>Case</th><th>Unit</th><th>Counit</th><th>Multiplication</th><th>Partition function</th></tr>
<tr><td>10</td><td>$\{a\}$</td><td>$\{a\}$</td><td><table><tr><td>$\{a\}$</td><td>$\{b\}$</td><td>$\{c\}$</td></tr><tr><td>$\{b\}$</td><td>$\{c\}$</td><td>$\{a\}$</td></tr><tr><td>$\{c\}$</td><td>$\{a\}$</td><td>$\{b\}$</td></tr></table></td><td>True</td></tr>
<tr><td>11</td><td>$\{a\}$</td><td>$\{a\}$</td><td><table><tr><td>$\{a\}$</td><td>$\{b\}$</td><td>$\{c\}$</td></tr><tr><td>$\{b\}$</td><td>$\{c\}$</td><td>$\{a,b\}$</td></tr><tr><td>$\{c\}$</td><td>$\{a,b\}$</td><td>$\{b,c\}$</td></tr></table></td><td>True</td></tr>
<tr><td>12</td><td>$\{a\}$</td><td>$\{a\}$</td><td><table><tr><td>$\{a\}$</td><td>$\{b\}$</td><td>$\{c\}$</td></tr><tr><td>$\{b\}$</td><td>$\{b,c\}$</td><td>$\{a,b,c\}$</td></tr><tr><td>$\{c\}$</td><td>$\{a,b,c\}$</td><td>$\{b,c\}$</td></tr></table></td><td>True</td></tr>
<tr><td>13</td><td>$\{a\}$</td><td>$\{b\}$</td><td><table><tr><td>$\{a\}$</td><td>$\{b\}$</td><td>$\{c\}$</td></tr><tr><td>$\{b\}$</td><td>$\varnothing$</td><td>$\varnothing$</td></tr><tr><td>$\{c\}$</td><td>$\varnothing$</td><td>$\{b\}$</td></tr></table></td><td>$g=1$</td></tr>
<tr><td>14</td><td>$\{a\}$</td><td>$\{b\}$</td><td><table><tr><td>$\{a\}$</td><td>$\{b\}$</td><td>$\{c\}$</td></tr><tr><td>$\{b\}$</td><td>$\varnothing$</td><td>$\varnothing$</td></tr><tr><td>$\{c\}$</td><td>$\varnothing$</td><td>$\{b,c\}$</td></tr></table></td><td>$g\geq 1$</td></tr>
<tr><td>15</td><td>$\{a\}$</td><td>$\{b\}$</td><td><table><tr><td>$\{a\}$</td><td>$\{b\}$</td><td>$\{c\}$</td></tr><tr><td>$\{b\}$</td><td>$\{a\}$</td><td>$\{c\}$</td></tr><tr><td>$\{c\}$</td><td>$\{c\}$</td><td>$\{a,b\}$</td></tr></table></td><td>$g\geq 1$</td></tr>
<tr><td>16</td><td>$\{a\}$</td><td>$\{b\}$</td><td><table><tr><td>$\{a\}$</td><td>$\{b\}$</td><td>$\{c\}$</td></tr><tr><td>$\{b\}$</td><td>$\{a\}$</td><td>$\{c\}$</td></tr><tr><td>$\{c\}$</td><td>$\{c\}$</td><td>$\{a,b,c\}$</td></tr></table></td><td>$g\geq 1$</td></tr>
<tr><td>17</td><td>$\{a\}$</td><td>$\{b\}$</td><td><table><tr><td>$\{a\}$</td><td>$\{b\}$</td><td>$\{c\}$</td></tr><tr><td>$\{b\}$</td><td>$\{c\}$</td><td>$\{a\}$</td></tr><tr><td>$\{c\}$</td><td>$\{a\}$</td><td>$\{b\}$</td></tr></table></td><td>$g\equiv 1 \pmod 3$</td></tr>
<tr><td>18</td><td>$\{a\}$</td><td>$\{b\}$</td><td><table><tr><td>$\{a\}$</td><td>$\{b\}$</td><td>$\{c\}$</td></tr><tr><td>$\{b\}$</td><td>$\{c\}$</td><td>$\{a,c\}$</td></tr><tr><td>$\{c\}$</td><td>$\{a,c\}$</td><td>$\{a,b,c\}$</td></tr></table></td><td>$g\geq 1$</td></tr>
</table>

TABLE 4. Frobenius objects in **Rel** with three elements and their partition functions, part 3.

<table>
<tr><th>Case</th><th>Unit</th><th>Counit</th><th>Multiplication</th><th>Partition function</th></tr>
<tr><td>19</td><td>$\{a\}$</td><td>$\{b\}$</td><td><table><tr><td>$\{a\}$</td><td>$\{b\}$</td><td>$\{c\}$</td></tr><tr><td>$\{b\}$</td><td>$\{a,c\}$</td><td>$\{a,c\}$</td></tr><tr><td>$\{c\}$</td><td>$\{a,c\}$</td><td>$\{a,b\}$</td></tr></table></td><td>$g \geq 1$</td></tr>
<tr><td>20</td><td>$\{a\}$</td><td>$\{b\}$</td><td><table><tr><td>$\{a\}$</td><td>$\{b\}$</td><td>$\{c\}$</td></tr><tr><td>$\{b\}$</td><td>$\{a,c\}$</td><td>$\{a,c\}$</td></tr><tr><td>$\{c\}$</td><td>$\{a,c\}$</td><td>$\{a,b,c\}$</td></tr></table></td><td>$g \geq 1$</td></tr>
<tr><td>21</td><td>$\{a,b\}$</td><td>$\{a,b\}$</td><td><table><tr><td>$\{a\}$</td><td>$\varnothing$</td><td>$\varnothing$</td></tr><tr><td>$\varnothing$</td><td>$\{b\}$</td><td>$\{c\}$</td></tr><tr><td>$\varnothing$</td><td>$\{c\}$</td><td>$\{b\}$</td></tr></table></td><td>True</td></tr>
<tr><td>22</td><td>$\{a,b\}$</td><td>$\{a,b\}$</td><td><table><tr><td>$\{a\}$</td><td>$\varnothing$</td><td>$\varnothing$</td></tr><tr><td>$\varnothing$</td><td>$\{b\}$</td><td>$\{c\}$</td></tr><tr><td>$\varnothing$</td><td>$\{c\}$</td><td>$\{b,c\}$</td></tr></table></td><td>True</td></tr>
<tr><td>23</td><td>$\{a,b\}$</td><td>$\{b,c\}$</td><td><table><tr><td>$\{a\}$</td><td>$\varnothing$</td><td>$\varnothing$</td></tr><tr><td>$\varnothing$</td><td>$\{b\}$</td><td>$\{c\}$</td></tr><tr><td>$\varnothing$</td><td>$\{c\}$</td><td>$\{b\}$</td></tr></table></td><td>True</td></tr>
<tr><td>24</td><td>$\{a,b\}$</td><td>$\{b,c\}$</td><td><table><tr><td>$\{a\}$</td><td>$\varnothing$</td><td>$\varnothing$</td></tr><tr><td>$\varnothing$</td><td>$\{b\}$</td><td>$\{c\}$</td></tr><tr><td>$\varnothing$</td><td>$\{c\}$</td><td>$\varnothing$</td></tr></table></td><td>True</td></tr>
<tr><td>25</td><td>$\{a,b,c\}$</td><td>$\{a,b,c\}$</td><td><table><tr><td>$\{a\}$</td><td>$\varnothing$</td><td>$\varnothing$</td></tr><tr><td>$\varnothing$</td><td>$\{b\}$</td><td>$\varnothing$</td></tr><tr><td>$\varnothing$</td><td>$\varnothing$</td><td>$\{c\}$</td></tr></table></td><td>True</td></tr>
</table>

References

[1] Lowell Abrams, *Two-dimensional topological quantum field theories and Frobenius algebras*, J. Knot Theory Ramifications **5** (1996), no. 5, 569–587, DOI 10.1142/S0218216596000333. MR1414088

[2] Ivan Contreras, *Groupoids, Frobenius algebras and Poisson sigma models*, Mathematical aspects of quantum field theories, Math. Phys. Stud., Springer, Cham, 2015, pp. 413–426. MR3330248

[3] Ivan Contreras, Molly Keller, and Rajan Amit Mehta, *Frobenius objects in the category of spans*, Rev. Math. Phys. **34** (2022), no. 10, Paper No. 2250036, 34, DOI 10.1142/S0129055X22500362. MR4514384

[4] Ivan Contreras, Rajan Mehta, and Walker Stern, *Frobenius and commutative pseudomonoids in the bicategory of spans*, Preprint, `arXiv:2311.15342`, 2023.

[5] Robbert Dijkgraaf and Edward Witten, *Topological gauge theories and group cohomology*, Comm. Math. Phys. **129** (1990), no. 2, 393–429. MR1048699

[6] Robertus Henricus Dijkgraaf, *A geometrical approach to two-dimensional conformal field theory*, PhD Thesis (1989).

[7] Chris Heunen, Ivan Contreras, and Alberto S. Cattaneo, *Relative Frobenius algebras are groupoids*, J. Pure Appl. Algebra **217** (2013), no. 1, 114–124, DOI 10.1016/j.jpaa.2012.04.002. MR2965909

[8] Chris Heunen and Jamie Vicary, *Categories for quantum theory*, Oxford Graduate Texts in Mathematics, vol. 28, Oxford University Press, Oxford, 2019. An introduction, DOI 10.1093/oso/9780198739623.001.0001. MR3971584

[9] Joachim Kock, *Frobenius algebras and 2D topological quantum field theories*, London Mathematical Society Student Texts, vol. 59, Cambridge University Press, Cambridge, 2004. MR2037238

[10] Aaron D. Lauda, *Frobenius algebras and ambidextrous adjunctions*, Theory Appl. Categ. **16** (2006), No. 4, 84–122. MR2210667

[11] David Li-Bland and Alan Weinstein, *Selective categories and linear canonical relations*, SIGMA Symmetry Integrability Geom. Methods Appl. **10** (2014), Paper 100, 31, DOI 10.3842/SIGMA.2014.100. MR3299000

[12] Rajan Amit Mehta and Ruoqi Zhang, *Frobenius objects in the category of relations*, Lett. Math. Phys. **110** (2020), no. 7, 1941–1959, DOI 10.1007/s11005-020-01281-1. MR4114394

[13] Walker H. Stern, *2-Segal objects and algebras in spans*, J. Homotopy Relat. Struct. **16** (2021), no. 2, 297–361, DOI 10.1007/s40062-021-00282-8. MR4266208

[14] Katrin Wehrheim and Chris T. Woodward, *Functoriality for Lagrangian correspondences in Floer theory*, Quantum Topol. **1** (2010), no. 2, 129–170, DOI 10.4171/QT/4. MR2657646

[15] Alan Weinstein, *A note on the Wehrheim-Woodward category*, J. Geom. Mech. **3** (2011), no. 4, 507–515, DOI 10.3934/jgm.2011.3.507. MR2888017

Department of Mathematics, Amherst College, 31 Quadrangle Drive, Amherst, Massachusetts 01002

Email address: `icontreraspalacios@amherst.edu`

Department of Mathematics & Statistics, Smith College, 44 College Lane, Northampton, Massachusetts 01063

Email address: `AdeleLRLong@gmail.com`

Department of Mathematics & Statistics, University of Massachusetts, Amherst, 710 N Pleasant St, Amherst, Massachusetts 01003

Email address: `semarx@umass.edu`

Department of Mathematics & Statistics, Smith College, 44 College Lane, Northampton, Massachusetts 01063

Email address: `rmehta@smith.edu`

Contemporary Mathematics
Volume **802**, 2024
https://doi.org/10.1090/conm/802/16074

Stirling decomposition of graph homology in genus 1

Benjamin C. Ward

ABSTRACT. We prove that commutative graph homology in genus $g = 1$ with $n \geq 3$ markings has a direct sum decomposition whose summands have rank given by Stirling numbers of the first kind. These summands are computed as the homology of complexes of certain decorated trees which have an elementary combinatorial description.

1. Introduction

The purpose of this paper is to compute the homology of a family of chain complexes which arose in my study of higher operations on graph homology. We will denote these chain complexes as $\mathcal{S}_{n,k}$. Although they arose from the study of some rather technical topics (eg ∞-modular operads and the Feynman transform), it is possible to give a very elementary description of them in terms of the combinatorics of decorated trees, and that is the approach we take herein.

The computation of the homology $H_*(\mathcal{S}_{n,k})$ is a new result, given for the first time in this article. Specifically, we prove (Theorem 4.1) that the Betti numbers of $\mathcal{S}_{n,k}$ are

$$\beta_i(\mathcal{S}_{n,k}) = \begin{cases} |s_{n,k}| & \text{if } i = n \\ 0 & \text{else,} \end{cases} \tag{1.1}$$

where $s_{n,k}$ denotes the signed Stirling number of the first kind, whose absolute value counts the number of permutations of the set $\{1, \dots, n\}$ which can be written as a product of k disjoint cycles.

Let me briefly explain, in basic terms, the context in which this result can be applied. By (commutative) graph homology, we refer to a chain complex spanned by isomorphism classes of graphs. The degree of a graph is its number of edges, and the differential is given by a signed sum of edge contractions. This complex splits over the genus g of the graph. A variant of this construction allows n labeled markings of the vertices. In this way, for each pair (g, n) we get a chain complex of such graphs, and we endeavor to compute its rational homology. The expert reader has surely noticed several important technical details which we've glossed over in this informal description.

This article concerns the calculation in the case $g = 1$ and $n \geq 3$. In this case, the rational homology may be determined from [**CGP22**, Theorem 1.2], where the authors determine the homotopy type of a cell complex $\Delta_{1,n}$ whose reduced cellular

---

2020 *Mathematics Subject Classification.* Primary 55U15, 18M70.

©2024 American Mathematical Society

chain complex coincides (up to a shift in degree) with the graph complex described above. In particular they show

$$\beta_i(\Delta_{1,n}) = \begin{cases} \dfrac{(n-1)!}{2} & \text{if } i = n-1 \\ 0 & \text{else,} \end{cases} \tag{1.2}$$

and moreover prove that as an S_n-module, $H_{n-1}(\Delta_{1,n})$ is the representation induced from a particular 1-dimensional representation of the dihedral group.

Our homology calculation above (equation (1.1)) gives a new proof of equation (1.2), and it gives a different description of the S_n-module structure which adds some computational facility. Namely we prove (Corollary 5.5) that

$$H_*(\Delta_{1,n+1}) \cong \bigoplus_{i=1}^{\lfloor n/2 \rfloor} H_*(\mathcal{S}_{n,2i}), \tag{1.3}$$

and hence the rank of $H_{n-1}(\Delta_{1,n})$ is equal to the number of permutations in S_{n-1} which can be written using an even number of disjoint cycles (counting fixed points as cycles of length 1). Depending on the parity of n, these coincide with either the even or the odd permutations and in either case, it's exactly half. Whence our confirmation of equation (1.2).

In addition to giving a new proof of equation (1.2), the decomposition in equation (1.3) tells us something new about the S_{n+1}-module structure. The isomorphism in equation (1.3) holds S_{n+1} equivariantly, meaning that the irreducible decomposition of a given $H_n(\Delta_{1,n+1})$ also splits accordingly. Moreover, we conjecture (Section 5) that each of the sequences $H_n(\mathcal{S}_{n,n-i})$ is representation stable (after tensoring with the sign representation). This means that the irreducible decomposition of a given $H_n(\Delta_{1,n+1})$ can be broken into pieces, all but one of which is determined by lower n. It is not too difficult to compute the first few nontrivial stable sequences by hand, see Section 5.5, although the representation stability aspects fall somewhat outside of our parameters here.

Indeed the goal of this paper is to give a broadly accessible description of the complexes $\mathcal{S}_{n,k}$ and to compute their homology. We call these complexes "Stirling complexes" and they are defined in Section 3. Section 2 gives a brief recollection of Stirling numbers and proves an identity (Proposition 2.2) which is the shadow of our main theorem when interpreted in terms of the Euler characteristic. The homology calculation is given in Section 4, and we conclude with a discussion of the decomposition of equation (1.3) in Section 5.

Finally, it should be pointed out that in attempting to present this problem in a hands-on and elementary way, we have necessarily but perhaps unfairly circumvented the use of more sophisticated tools which could also be used to attack this problem. This includes the theory of Grobner bases [**DK10**], since our main calculation could be recast as the Koszulity of Lie graph homology in genus 1, viewed as an infinitesmal bimodule over its underlying genus 0 operad. The novice reader should also be advised that this article has largely bypassed a discussion of the influential and sophisticated literature on graph complexes for which the works [**Kon94**], [**Wil15**], [**KWZ17**], [**Mer21**] may serve as a possible starting point for further reading.

Conventions. Symmetric groups are denoted by S_n, or S_X for a finite set X. The irreducible representation of the symmetric group associated to a partition λ

will be denoted V_λ. We denote shift operators for graded vector spaces by $\Sigma^{\pm}$. We work over the field of rational numbers throughout.

2. Stirling numbers

Let S_n denote the symmetric group of permutations of the set $\{1, \ldots, n\}$. Let $S_{n,k} \subset S_n$ denote the subset consisting of permutations having k disjoint cycles, where fixed elements are considered cycles of length 1. For example, the magnitude of the sets $S_{n,k}$ for $n \leq 7$ is given in the following table.

(2.1)

n \ k	1	2	3	4	5	6	7
1	1						
2	1	1					
3	2	3	1				
4	6	11	6	1			
5	24	50	35	10	1		
6	120	274	225	85	15	1	
7	720	1764	1624	735	175	21	1

By convention we consider $S_{n,k}$ to be the empty set, so the blanks should be considered zeros.

This triangle of numbers has been long studied under the name of the (unsigned) Stirling numbers of the first kind. We define the signed Stirling numbers of the first kind to be

$$s_{n,k} := (-1)^{n-k}|S_{n,k}|.$$

From now on we will just call these the "Stirling numbers" for short.

We refer to [**Cha02**, Chapter 8] for a comprehensive overview of the Stirling numbers and their history. In this article we will restrict attention to the small bits and pieces of this theory which we need. We first mention a few basic facts that we will use below. Their verification is immediate.

LEMMA 2.1. *Stirling numbers of the first kind satisfy:*

(1) $\sum_k |s_{n,k}| = n!$

(2) $s_{n,n-1} = -\binom{n}{2}$

(3) $s_{n,n-2} = \frac{1}{4}\binom{n}{3}(3n-1)$

(4) *If* $n \geq 2$ *then* $\sum_k s_{n,k} = 0$.

The main theorem of this paper will be a categorification of the following slightly less trivial fact about Stirling numbers.

PROPOSITION 2.2.

$$s_{n,k} = \sum_{m=k+1}^{n+1} \binom{m-1}{k} s_{n+1,m}. \tag{2.2}$$

PROOF. We emphasize that these are signed Stirling numbers, so we're considering an alternating sum and this result arises as a corollary of our main theorem (Theorem 4.1) by taking the Euler characteristic. However, let us give the combinatorial proof which is the shadow of the proof of our main theorem below.

To unify the two, let us choose to regard S_{n+1} as the permutation group of the set $\{0,1,\ldots,n\}$ via the isomorphism $\{0,1,\ldots,n\} \cong \{1,\ldots,n,n+1\}$ which fixes $1,\ldots,n$.

Fix n and k with $k \leq n$. Given a permutation in S_{n+1} with m cycles, there are $\binom{m-1}{k}$ ways to choose k cycles which do not contain 0. We may therefore view each term in the right hand side of equation (2.2) as counting the number of permutations in S_{n+1} with m cycles, k of which are distinguished, subject to the rule that 0 is not in a distinguished cycle. Let X_m be the set of such choices. In particular $|X_m| = \binom{m-1}{k}|S_{n+1,m}|$.

The set X_m can be written as a disjoint union indexed by the choice of a non-empty set $D \subset \{1,\ldots,n\}$ and a permutation $\sigma \in S_D$ corresponding to the product of the k distinguished cycles in the given element. Let $X_{m,D,\sigma} \subset X_m$ be the subset appearing in index (D,σ). Counting the number of elements in each index, we may write:

$$\sum_{m=k+1}^{n+1} \binom{m-1}{k} s_{n+1,m} = \sum_{m=k+1}^{n+1} \sum_{D\subset\{1,\ldots,n\}} \sum_{\sigma\in S_{D,k}} (-1)^{m+n+1}|X_{m,D,\sigma}| \tag{2.3}$$

where $S_{D,k}$ is the set of permutations of D having k cycles.

Note that $X_{m,D,\sigma} \subset X_m$ is nonempty only if $m-k \leq n+1-|D|$. Therefore, reindexing the previous equation we have

$$\sum_{m=k+1}^{n+1} \binom{m-1}{k} s_{n+1,m} = \sum_{D\subset\{1,\ldots,n\}} \sum_{\sigma\in S_{D,k}} \sum_{m=k+1}^{n+1-|D|+k} (-1)^{m+n+1}|X_{m,D,\sigma}| \tag{2.4}$$

By considering the product of the nondistinguished cycles, each element in $X_{m,D,\sigma}$ specifies a unique permutation of $\{0\} \cup (\{1,\ldots,n\} \setminus D)$ having $m-k$ cycles, and conversely, so $|X_{m,D,\sigma}| = (-1)^{m-k+n+1-|D|} s_{n+1-|D|,m-k}$. Therefore, for each fixed D and σ we have:

$$\begin{aligned}\sum_{m=k+1}^{n+1-|D|+k} (-1)^{m+n+1}|X_{m,D,\sigma}| &= (-1)^{k+|D|} \sum_{m=k+1}^{n+1-|D|+k} s_{n+1-|D|,m-k} \\ &= (-1)^{k+|D|} \sum_{i=1}^{n+1-|D|} s_{n+1-|D|,i}.\end{aligned}$$

Applying the final statement of Lemma 2.1, we see that this sum is zero unless $n+1-|D| = 1$. This happens if and only if $D = \{1,\ldots,n\}$, i.e., every nonzero element appears in a distinguished cycle. Thus equation (2.4) becomes

$$\sum_{m=k+1}^{n+1} \binom{m-1}{k} s_{n+1,m} = (-1)^{n+k} \sum_{\sigma\in S_{n,k}} |X_{k+1,\{1,\ldots,n\},\sigma}|.$$

Finally, the set $|X_{k+1,\{1,\ldots,n\},\sigma}|$ counts the number permutations of S_{n+1} which fix 0 and have $k+1$ cycles, such that the product of the k cycles not containing 0 is equal to σ. Clearly there is exactly one such permutation, namely $(0)\cdot\sigma$, from which we conclude

$$\sum_{m=k+1}^{n+1} \binom{m-1}{k} s_{n+1,m} = (-1)^{n+k}|S_{n,k}| = s_{n,k}.$$

□

3. Stirling complexes

In this section we will define a chain complex $\mathcal{S}_{n,k}$ whose Euler characteristic may be calculated via the alternating sum in Proposition 2.2 to satisfy $\chi(\mathcal{S}_{n,k}) = s_{n,k}$. For lack of better terminology we call $\mathcal{S}_{n,k}$ "Stirling complexes".

In Section 4 below we will calculate the homology of the Stirling complexes, and in Section 5 we will give an application of this computation to graph homology. The purpose of this section is simply to give their definition. The Stirling complexes arose in [**War22**] as certain subcomplexes of the Feynman transform of Lie graph homology. However, it is possible to give a much more down to earth definition of them, and that is the approach we take here.

Finally, we remark that the reader should be sure to not confuse the chain complex $\mathcal{S}_{n,k}$, defined in this section, with the set $S_{n,k}$ defined in the previous section. These objects live in different categories, and while both have symmetric group actions, the two actions are not comparable. One relationship between these two objects will be given later on, in Theorem 4.1.

3.1. Decorated trees. We will define the Stirling complexes as a span of certain decorated trees. For a formal definition of a tree we refer to the appendix, namely Section A.3.

Informally, a tree t is a 1-dimensional CW complex with is connected and simply connected. The 0-cells are called vertices, the set of which is denoted $V(\mathsf{t})$. The 1-cells are called edges, the set of which is denoted $E(\mathsf{t})$. An n-tree is a tree along with a function $\ell\colon \{0, \dots, n\} \to V(\mathsf{t})$ called the leg labeling. The leg labeling is depicted graphically by $|\ell^{-1}(v)|$ line segments extending from v, labeled bijectively by the elements of the set $\ell^{-1}(v)$. These line segments are called the legs of the tree. From this perspective what we call "edges" might also be called "internal edges" in that they run between two vertices where-as legs do not.

Our formal definition of a graph (Section A.1) views edges as consisting of two halves, called flags, and views legs as consisting of a single flag. With this convention, every vertex has a set of adjacent flags, which is canonically identified with the union of the sets of adjacent edges and adjacent legs. The valence of a vertex is defined to be the order of this set. An n-tree may be viewed as a directed graph, with edges directed toward the leg labeled by 0. This in turn specifies a unique output flag at each vertex, the remainder of the flags are called inputs. A tree is called stable if it has at least 2 input flags.

For the remainder of this article, including in the following definition, all trees are assumed to be stable n-trees for some n. The Stirling complexes will be spans of such trees along with additional decorations.

DEFINITION 3.1. A Stirling tree (t, v, A) is a tree t along with the choice of the following additional data:

(1) a vertex v of t, which we call the distinguished vertex (or DV for short).
(2) a distinguished subset A of the input flags adjacent to the distinguished vertex, whose elements we call alternating flags, such that $|A| \geq 2$.

A Stirling tree having $n+1$ legs and k alternating flags is said to be of type (n, k). See Figure 1. We will often refer to Stirling trees simply as t, leaving the decorations v and A implicit.

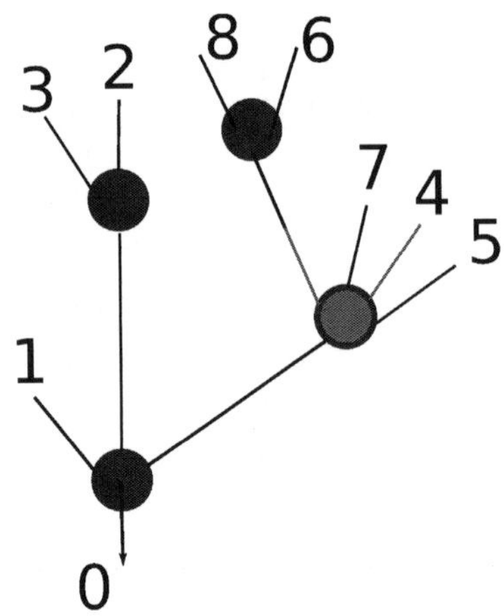

FIGURE 1. A Stirling tree with four vertices, three edges and nine legs. The distinguished vertex and the alternating flags are depicted in red. This Stirling tree is said to be of type $(8, 2)$.

We remark that we could loosen the requirement that $|A| \geq 2$ if we allowed trees whose distinguished vertex was not necessarily stable, but our intended application only requires this case, so we make this assumption for simplicity.

3.2. The underlying spaces. In this article we consider Stirling trees up to structure preserving isomorphisms. In particular, isomorphisms must preserve the leg labels, the distinguished vertex and the set of alternating flags. Stirling complexes will be constructed as the sum over isomorphism classes of Stirling trees.

Each summand will in turn be spanned by a choice of additional sign data, which will be encoded as an order on the set of edges and the set of alternating flags, modulo even permutations. To formalize this, given a finite set $X = \{x_1, \dots, x_n\}$, we define $det(X)$ to be the top exterior power of the vector space spanned by X. Explicitly, $det(X)$ is a 1-dimensional vector space spanned by $x_1 \wedge \cdots \wedge x_n$ and subject to the rule that $x_1 \wedge \cdots \wedge x_n = sgn(\sigma) x_{\sigma(1)} \wedge \cdots \wedge x_{\sigma(n)}$ for a permutation $\sigma \in S_X$.

We are now prepared to define the vector spaces underlying the Stirling complexes.

DEFINITION 3.2. Define $\mathcal{S}_{n,k,i}$ to be

$$\mathcal{S}_{n,k,i} = \bigoplus_{(\mathrm{t},v,A)} det(E(\mathrm{t})) \otimes det(A)$$

with direct sum taken over all isomorphism classes of Stirling trees of type (n, k) having i edges.

For example, the Stirling tree pictured in Figure 1 specifies a one dimensional subspace of $\mathcal{S}_{8,2,3}$. The choice of an order on the set of edges and the set of alternating flags would further specify a nonzero element in this subspace.

LEMMA 3.3. *$\mathcal{S}_{n,k,i}$ is nonzero if and only if $0 \leq i \leq n - k$.*

PROOF. Suppose $\mathcal{S}_{n,k,i} \neq 0$. Then clearly $i \geq 0$. Such a tree has $2i + n + 1$ flags, but by stability must also have at least $3(|V| - 1) + k + 1$ flags, where $|V|$

denotes the number of vertices of the tree. Using the fact that $|V| - 1 = i$ we see $n \geq i + k$ as desired. Conversely, if we construct any tree with $n - k$ edges (for example by arranging all the vertices and edges along a vertical line), we may contract the edges one at a time to show that each such $\mathcal{S}_{n,k,i}$ is not zero. $\square$

3.3. The symmetric group action. The vector space $\mathcal{S}_{n,k,i}$ is naturally an S_n-module via the permutation action on leaf labels. For example, $\mathcal{S}_{n,n,0}$ is the alternating representation and $\mathcal{S}_{n,n-1,0}$ is the tensor product of the alternating representation with the permutation representation, i.e., $V_{1^n} \oplus V_{2,1^{n-2}}$. However, the description of $\mathcal{S}_{n,k,i}$ given in terms of the Feynman transform (Theorem 5.4 below) makes it clear that this "obvious" S_n action is the restriction of an S_{n+1}-module structure. Although not needed for the homology calculation below, let us describe this action in this subsection.

First, we let S_{n+1} act on the set of isomorphism classes of n-tress by permutation of the leg labels, using the unique isomorphism $\{0, \dots, n\} \cong \{1, \dots, n+1\}$ which fixes $1, \dots, n$ and exchanges 0 and $n+1$. Let $\sigma \in S_{n+1}$ and $\mathtt{t}$ be an n-tree. When comparing $\mathtt{t}$ and $\sigma\mathtt{t}$, observe that there is a natural bijective correspondence between their edges, their vertices, and the flags at each vertex, although which flags are inputs and outputs need not be preserved. Given a vertex v of $\mathtt{t}$, we use the notation σv to denoted the corresponding vertex of $\sigma\mathtt{t}$. Similarly for edges and flags, as well as sets of flags.

Now suppose we have a Stirling tree $(\mathtt{t}, v, A)$ of type (n, k) and $\sigma \in S_{n+1}$. The tree $\sigma\mathtt{t}$ comes with a distinguished vertex σv, as well as a distinguished subset of the flags adjacent to σv, namely σA. Since the edges and alternating flags are naturally identified, there is a map

$$det(E(\mathtt{t})) \otimes det(A) \to det(E(\sigma\mathtt{t})) \otimes det(\sigma A) \tag{3.1}$$

which simply preserves the order of each wedge product. Note however since σ needn't fix $n+1$, there is no guarantee that each flag in σA is an input flag.

With this in mind, we define the action of S_{n+1} on $\mathcal{S}_{n,k,i}$ in two cases. Fix $\sigma \in S_{n+1}$. Given a Stirling tree $(\mathtt{t}, v, A)$, if σA does not contain the output flag of σv, then we define the action of σ on the summand of $\mathcal{S}_{n,k,i}$ indexed by $(\mathtt{t}, v, A)$ to be given by the map in equation (3.1).

If, on the other hand, the output flag of σv, call it z, is contained in the set σA, we proceed as follows. Let B be the set of flags adjacent to σv which are not contained in σA. Note that B is not empty, since under these conditions the output of v corresponds to an input of σv which is not in σA, hence is in B.

For each $b \in B$, let $(\sigma A)_b := (\sigma A \setminus z) \cup \{b\}$. The permutation σ induces a natural bijective correspondence between the elements of A and the elements of $(\sigma A)_b$ given by the intermediary σA, and exchanging z with b. Therefore, there is a canonical map

$$\sigma_b \colon det(E(\mathtt{t})) \otimes det(A) \to det(E(\sigma\mathtt{t})) \otimes det((\sigma A)_b)$$

given by canonically identifying the edges (resp. flags) on the left bijectively with those on the right. In this case we define the action of σ on the summand of $\mathcal{S}_{n,k,i}$ indexed by $(\mathtt{t}, v, A)$ to be given by the map

$$\sum_{b \in B} -\sigma_b. \tag{3.2}$$

We remark that in light of Theorem 5.4, this rule is merely a translation of [**War23**, Lemma 3.2], which in this terminology tells us how to rewrite a term containing a distinguished output flag as a linear combination of those having only distinguished input flags.

3.4. The differential. We are now prepared to define the chain complex $\mathcal{S}_{n,k}$. It will have the following form,

$$0 \to \mathcal{S}_{n,k,n-k} \to \cdots \to \mathcal{S}_{n,k,i+1} \to \mathcal{S}_{n,k,i} \to \cdots \to \mathcal{S}_{n,k,0} \to 0.$$

By convention $\mathcal{S}_{n,k,i}$ is considered to be of total degree $i+k$, so that $\mathcal{S}_{n,k}$ is concentrated between degrees k and n.

The differential d will be defined as a sum over edge contractions, as in the case of the bar construction of an operad [**GK94**]. Specifically, this means d applied to the summand of $\mathcal{S}_{n,k,i}$ corresponding to a Stirling tree (t, v, A) will be of the form $d = \sum_{e \in \mathsf{t}} d_e$, and we proceed to define d_e.

Fix such a Stirling tree (t, v, A) and choose an edge e of t. We may contract the edge to produce a new tree t/e. If this edge does not contain an alternating flag, then the edge collapse specifies a Stirling tree whose distinguished vertex and alternating flags are canonically identified with those of the source. Thus we have a map

$$det(E(\mathsf{t})) \otimes det(A_{\mathsf{t}}) \xrightarrow{d_e} det(E(\mathsf{t}/e)) \otimes det(A_{\mathsf{t}/\mathsf{e}}) \tag{3.3}$$

which by convention removes the edge e in the last position of the wedge product of the edges of t.

Now suppose, on the other hand, that the edge $e = \{a, b_0\}$ consists of input flag $a \in A$ and output flag b_0. When e is contracted, the alternating flag a is lost and we account for this by summing over the newly adjacent flags as replacements, see Figure 2. To make this precise, define B to be the set of input flags at the vertex whose output is b_0, and define $A_b := (A \setminus a) \cup \{b\}$, for each $b \in B$. Observe that the elements of A and A_b are in canonical bijective correspondence, with bijection exchanging a and b and preserving all other elements. This in turn induces an isomorphism $det(A) \to det(A_b)$. We then define

$$d_{e,b} \colon det(E(\mathsf{t})) \otimes det(A) \to det(E(\mathsf{t}/e)) \otimes det(A_b) \tag{3.4}$$

by taking this isomorphism on the right hand tensor factor and, as above, taking the map which removes e in the last position of the wedge product on the left hand tensor factor. Finally, we define $d_e := \sum_{b \in B} d_{e,b}$ in this case.

LEMMA 3.4. $d^2 = 0$.

PROOF. Start with a Stirling tree (t, v, A). We will show that $d^2 = 0$ when applied to the summand of $\mathcal{S}_{n,k,i}$ indexed by this Stirling tree, from which the claim follows. Note that if t has fewer than two edges, then $d^2 = 0$ just by degree considerations. Else, choose two edges e_1, e_2 of t, and we will argue that

$$d_{e_1} \circ d_{e_2} = -d_{e_2} \circ d_{e_1}$$

from which the statement follows. We divide this verification into four cases.

CASE I. Suppose that neither e_1 nor e_2 contains an alternating flag. In this case, we apply equation (3.3) twice and $d_{e_1} \circ d_{e_2}$ and $d_{e_2} \circ d_{e_1}$ are related by applying the transposition swapping e_1 and e_2, producing a factor of -1 on the $det(E(\mathsf{t}))$ tensor factor, while the alternating flags are preserved at each stage.

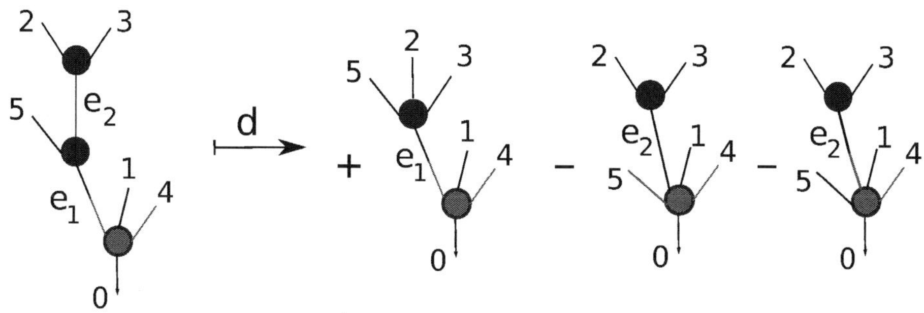

FIGURE 2. The Stirling tree t on the left specifies an element of $\mathcal{S}_{5,2,2}$, taking $e_1 \wedge e_2 \in det(E(\mathrm{t}))$ and taking the alternating flags in the order indicated by the choice of planar embedding. The differential applied to this element is indicated on the right. The first differential term comes from contracting the edge which does not contain an alternating flag, the latter two come from contracting the edge which does contain an alternating flag. One may verify that $d^2 = 0$ in this case by contracting the remaining edge in each of the three terms on the right. The result is three pairs of identical terms appearing with opposite sign.

For the remaining cases, for $i \in \{1,2\}$, define v_i to be the vertex adjacent and above e_i and define B_i to be the set of input flags adjacent to v_i.

CASE II. Suppose that both e_1 and e_2 contain an alternating flag. In this case, twice applying equation (3.4) we have

$$d_{e_1} \circ d_{e_2} = \sum_{(b_1,b_2)\in B_1\times B_2} d_{e_1,b_1} \circ d_{e_2,b_2}$$

and similarly in the opposite order. Here it suffices to observe that for each pair (b_1, b_2) we have $d_{e_1,b_1} \circ d_{e_2,b_2} = -d_{e_2,b_2} \circ d_{e_1,b_1}$. This follow as in the previous case, with the -1 arising from the $det(E(\mathrm{t}))$ tensor factor from the permutation of the two edges. Since the wedge product of the alternating flags does not depend on the order of the edge contractions, no sign arises on the $det(A)$ factor.

CASE III. Suppose (without loss of generality) that e_1 contains an alternating flag and e_2 does not, and suppose that e_2 is not adjacent to v_1. In this case, applying equation (3.4) for d_{e_1} and applying equation (3.3) for d_{e_2} we have:

$$d_{e_1} \circ d_{e_2} = \sum_{b\in B_1} d_{e_1,b} \circ d_{e_2}$$

and similarly in the opposite order. It then suffices to verify $d_{e_1,b_1} \circ d_{e_2} = -d_{e_2} \circ d_{e_1,b_1}$, with the sign coming from the permutation of the two edges, as in the previous cases.

CASE IV. We are thus left with the most interesting case, namely e_1 contains an alternating flag and e_2 is adjacent to v_1. This case is depicted in Figure 2. Let $b_0 \in B_1$ be the input flag of the edge e_2. If we contract e_2 first, the flags adjacent to v_1 are altered since b_0 is removed, but the flags in B_2 become adjacent to v_1.

Hence:

$$d_{e_1} \circ d_{e_2} = \sum_{b \in (B_1 \setminus b_0) \cup B_2} d_{e_1,b} \circ d_{e_2} \tag{3.5}$$

On the other hand, if we contract e_1 first we may single out the term d_{e_1,b_0}. Since application of this map results in b_0 being an alternating flag, we must apply equation (3.4) when composing with d_{e_2}. For each other $b_1 \in B_1$, we apply equation (3.3). Hence:

$$d_{e_2} \circ d_{e_1} = \sum_{b_1 \in (B_1 \setminus b_0)} d_{e_2} \circ d_{e_1,b_1} + \sum_{b_2 \in B_2} d_{e_2,b_2} \circ d_{e_1,b_0} \tag{3.6}$$

To conclude, it suffices to observe that the terms in equations (3.5) and (3.6) are negatives of each other. This follows from the fact that for $b \in (B_1 \setminus b_0) \cup B_2$, we have:

$$-d_{e_1,b} \circ d_{e_2} = \begin{cases} d_{e_2} \circ d_{e_1,b} & \text{if } b \in B_1 \setminus b_0 \\ d_{e_2,b} \circ d_{e_1,b_0} & \text{if } b \in B_2 \end{cases}$$

with the -1 factor again arising on the $det(E(\mathtt{t}))$ tensor factor by transposition of the edges. □

The differential on each $\mathcal{S}_{n,k}$ is compatible with the S_{n+1}-action defined in the previous subsection. This follow from the identification of $\mathcal{S}_{n,k}$ with a subcomplex of the Feynman transform, see Theorem 5.4, but it can also be verified directly from the above descriptions as we now indicate.

PROPOSITION 3.5. *For each $\sigma \in S_{n+1}$, we have $\sigma d = d\sigma$.*

PROOF. Start with a Stirling tree $(\mathtt{t}, v, A)$. We will show that $\sigma d = d\sigma$ when applied to the summand of $\mathcal{S}_{n,k,i}$ indexed by this Stirling tree, from which the claim follows. If $\mathtt{t}$ has no edges the claim is vacuous, so assume $\mathtt{t}$ has at least one edge. Since the edges of $\mathtt{t}$ and $\sigma\mathtt{t}$ coincide (i.e., are in canonical bijective correspondence) it suffices to show that $\sigma d_e = d_e \sigma$ for each edge e of $\mathtt{t}$.

Let us first observe that if $\sigma \in S_n$, i.e., if σ fixes 0, then the claim is immediate, since the permutation simply relabels the input legs of the tree. Which flags are alternating does not change upon such a permutation, only their labels change, and this operation clearly commutes with edge contraction. So it suffices to restrict attention to the case that $\sigma = (0\ i)$ for some $1 \leq i \leq n$ from which the general case follows.

Fix $\sigma = (0\ i)$ for such an i and fix an edge e. Let a be the input flag adjacent to v which is on the unique shortest path from the leg labeled by i to v. Let $e = \{x, y\}$ where x is an input flag and y is an output flag. To show $\sigma d_e = d_e \sigma$, one separates the verification into several cases based on the possibilities for a, x and y. These cases are as follows:

(1) $a = x$ and $a \in A$,
(2) $a \in A$ and y is adjacent to v,
(3) $a \in A$, x is adjacent to v and $x \notin A$,
(4) $a, x \in A$, and $x \neq a$,
(5) none of the above, i.e., $a \notin A$ or e is not adjacent to v.

Case 5 is the simplest, since the terms coincide on the nose with no cancellation. The first four cases are very similar to each other, so we carefully work through Case 1 and leave the verification of the others to the reader.

For this, assume that $a = x$ and $a \in A$. In this case the edge e is adjacent to v. Let w be the vertex whose output is y and let B be the set of inputs of w. The unique shortest path connecting v to the leg labeled by i necessarily contains one input of w, call it $b_0 \in B$. Finally, define C to be the set of flags adjacent to v.

By equation (3.2) we have:

$$d_e\sigma = d_e \circ (-\sum_{C\setminus a} \sigma_c) = -\sum_{C\setminus a} d_e \circ \sigma_c \tag{3.7}$$

On the other hand,

$$\sigma d_e = \sum_{b\in B} \sigma d_{e,b} = \sigma d_{e,b_0} + \sum_{b\neq b_0} \sigma d_{e,b}$$

Since the image of $\sigma d_{e,b_0}$ has a distinguished output, we apply equation (3.2) to it to rewrite this term as $\sigma d_{e,b_0} = \sum -\sigma_c d_{e,b_0}$, where here c ranges over the input flags of the vertex of $\sigma\mathrm{t}$ formed by contracting e. This set is (canonically identified with) $(B \setminus b_0) \cup (C \setminus a)$. Therefore

$$\sigma d_e = \left(\sum_{c\in B\setminus b_0} -\sigma_c d_{e,b_0} + \sum_{c\in C\setminus a} -\sigma_c d_{e,b_0} \right) + \sum_{b\in B\setminus b_0} \sigma d_{e,b}.$$

The first and third sums cancel. Finally, observe that for $c \in C \setminus a$ we have $\sigma_c d_{e,b_0} = d_e\sigma_c$, since both simply contract the edge e and replace a with c as the new alternating flag. Hence the middle sum in the previous equation is exactly $d_e\sigma$ by equation (3.7), as desired. □

4. Homology of Stirling complexes

THEOREM 4.1. *The Betti numbers of the complex $\mathcal{S}_{n,k}$ are:*

$$\beta_i(\mathcal{S}_{n,k}) = \begin{cases} |s_{n,k}| & \text{if } i = n \\ 0 & \text{else} \end{cases}.$$

PROOF. Fix n, k. If $n = k$ the statement is trivial, so assume $n > k$.

Define $\mathcal{A}$ to be the span of those trees such that one or both of the following properties hold:

(i) the distinguished vertex has valence $> k + 1$ (so not all inputs are alternating)
(ii) the distinguished vertex is not the root vertex.

Observe that $\mathcal{A} \subset \mathcal{S}_{n,k}$ is a subcomplex. Indeed if a tree satisfies property (i) then differential terms arising from it will also satisfy property (i), while if a tree satisfies property (ii) then differential terms arising from it will satisfy property (i) when edges adjacent to the distinguished vertex are contracted, and will satisfy property (ii) when edges not adjacent to the distinguished vertex are contracted. The heart of the proof is the following claim.

CLAIM. $\mathcal{A}$ has no homology.

To prove this claim, we will filter the complex $\mathcal{A}$. For a Stirling tree $\mathrm{t} \in \mathcal{A}$ there is a unique shortest path connecting the distinguished vertex and the root vertex. We'll refer to this path as the "0-to-DV path". Let $P(\mathrm{t}) \subset E(\mathrm{t})$ be the set of edges on this path. Let $p(\mathsf{t})$ and $e(\mathsf{t})$ be the number of (path) edges, taking values $0 \leq p(\mathsf{t}) \leq e(\mathsf{t})$.

Define the reach of a tree to be $r(\mathsf{t}) = 2e(\mathsf{t}) - p(\mathsf{t}) - \nu(\mathsf{t})$, where $\nu(\mathsf{t})$ is the "vcd check" defined by

$$\nu(\mathsf{t}) := \begin{cases} 1 & \text{if the distinguished vertex has valence equal to } k+1, \\ 0 & \text{if the distinguished vertex has valence not equal to } k+1. \end{cases}$$

We adopt the terminology "vcd vertex" to refer to the distinguished vertex of a Stirling tree t for which $\nu(\mathsf{t}) = 1$.

Since we've assumed $n > k$, the reach takes values in $0 \leq r(\mathsf{t}) \leq 2(n-k) - 2$. It achieves the upper bound if $e(\mathsf{t}) = n - k - 1$, $\nu(\mathsf{t}) = 0$ and $P(\mathsf{t})$ is empty (i.e., the root vertex and distinguished vertex coincide) or if $e(\mathsf{t}) = n - k$ and $p(\mathsf{t}) = 1$. Note that if $e(\mathsf{t}) = n - k$ then $\nu(\mathsf{t}) = 1$, since this is the maximum possible number of edges (by Lemma 3.3).

Reach defines a filtration because contracting an edge which is not adjacent to a vcd vertex lowers the reach by at least one, while contracting an edge which is adjacent to a vcd vertex lowers the reach by 1 if said edge is not on the 0-to-DV path and it preserves the filtration degree if the edge is on the 0-to-DV path. Thus in all cases the reach is not increased by the differential and so we have an increasing filtration

$$\mathcal{A}_0 \subset \mathcal{A}_1 \subset \cdots \subset \mathcal{A}_r \subset \mathcal{A}_{r+1} \subset \cdots \subset \mathcal{A}_{2n-2k-2}$$

where $\mathcal{A}_r$ denotes the span of those Stirling trees in $\mathcal{A}$ of reach $\leq r$.

Consider the associated graded. Each $\mathcal{A}_{r+1}/\mathcal{A}_r$ may be described as the span of trees of reach equal to r, with differential given by projection of the original differential to the quotient. By the above analysis, all differential terms vanish in this quotient, except in the case where we have a vcd vertex and we contract the unique edge of maximum height on the 0-to-DV path (i.e., the edge immediately below the distinguished vertex). Let us call this differential δ_0.

We therefore conclude that each Stirling tree of type (n,k) satisfying the properties (i) and (ii) above corresponds to a 1-dimensional subspace of the complex $\mathcal{A}$ which is either the source or the target of exactly one nontrivial differential δ_0 in the associated graded. Indeed, each quotient complex $\mathcal{A}_{r+1}/\mathcal{A}_r$ splits as a direct sum of complexes of the form

$$0 \to det(E(\mathsf{t})) \otimes det(A_{\mathsf{t}}) \xrightarrow{\delta_0} det(E(\mathsf{t}/\mathsf{e})) \otimes det(A_{\mathsf{t}/\mathsf{e}}) \to 0 \tag{4.1}$$

where t is a Stirling tree of type (n,k) and reach r whose distinguished vertex is a vcd vertex and where e is the unique edge on the 0-to-DV path of maximum height. See Figure 3. Since the differential in equation (4.1) is nonzero, it must be an isomorphism. Thus the associated graded has no homology, and so neither does $\mathcal{A}$.

Having shown that $\mathcal{A}$ has no homology, to compute the homology of $\mathcal{S}_{n,k}$, it suffices to compute the homology of the quotient $\mathcal{S}_{n,k}/\mathcal{A}$. The quotient complex is spanned by those Stirling trees which satisfied neither condition (i) nor condition (ii) above. This means such a tree must have the root vertex equal to the distinguished vertex and be of vcd type. In the quotient, the nonzero differential terms can only contract edges that are not adjacent to the distinguished vertex. To each such tree, associate a partition of the set $\{1, \dots, n\}$ by declaring two numbers to belong to the same block if and only if the leaves they label share the same nearest alternating

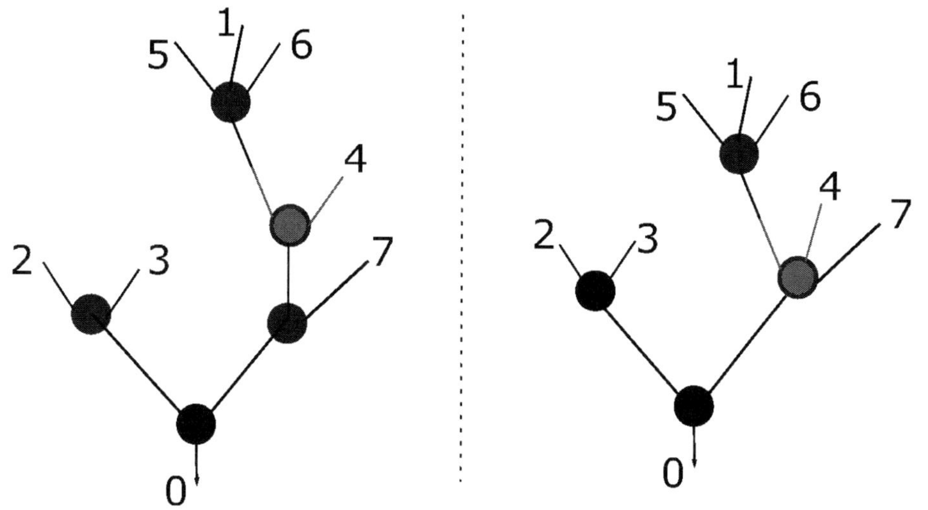

FIGURE 3. The Stirling tree of type $(7, 2)$ on the left corresponds to a 1-dimensional subspace of $\mathcal{A}$ having reach $r = 8-2-1 = 5$. In the associated graded $\mathcal{A}_5/\mathcal{A}_4$, its unique nonzero differential term lands in the summand corresponding the Stirling tree on the right, which we confirm has reach $r = 6 - 1 - 0 = 5$.

flag. By definition, the nearest alternating flag is the input flag adjacent to the root vertex that we first encounter when we move along the unique shortest path connecting a leaf to the root. In particular, the blocks in the partition must be nonempty, and the number of blocks is k.

Such a partition is unchanged by the differential in the quotient $\mathcal{S}_{n,k}/\mathcal{A}$, and so this complex splits over the choice of such partitions. Therefore to compute the homology of the quotient, it suffices to compute the homology of each summand. For this, we appeal to Koszulity of the commutative operad. Indeed each summand may be viewed as a tensor product over the blocks of the partition of a complex isomorphic to the bar construction of the commutative operad in arity equal to the size of the block. This is simply because each branch emanating from an alternating flag has only standard edge contractions, with no distinguished vertices. Appealing to Koszul duality of the commutative and Lie operads, denoted Com and Lie, the homology of a summand corresponding to a partition $b_1, \ldots, b_k$ is isomorphic to $\otimes_i \mathsf{Lie}(|b_i|)$, concentrated in the maximal degree, shown above to be n.

Finally we recall that the dimension of the constituent arities of the Lie operad satisfy $\mathsf{Lie}(|b_i|) = (b_i - 1)!$, which is also the number of cycles of length b_i that can be formed from a set of size b_i. Therefore, the dimension of the homology in degree n is equal to the number of ways to partition $\{1, \ldots, n\}$ into k nonempty blocks, and to subsequently associate a cycle of length $|b_i|$ with the numbers appearing in the block b_i. In other words, $\beta_n(\mathcal{S}_{n,k})$ is the number of permutations on n letters having k cycles, or $|s_{n,k}|$. □

REMARK 4.2. As mentioned above, Proposition 2.2 can be seen as a corollary of this Theorem. For this, compute the Euler characteristic $\chi(\Sigma^{-k}\mathcal{S}_{n,k})$ in two ways. As the alternating sum of Betti numbers we find $s_{n,k}$. On the other hand, if we compute the alternating sum of the dimensions of the complex itself, we can use Koszulity of the commutative operad to collapse subtrees not containing a distinguished vertex, and label the vertices resulting from the collapse with the space of Lie words of appropriate arity. The degree of the result is determined by the valence of the distinguished vertex, call it m, and is equal to $n+1-m$. The number of all such trees, after this collapse, is counted by the ways to partition the set $\{0,\dots,n\}$ times the dimension of the space of Lie words on each block of the partition which, as the proof above indicates, is $|s_{n+1,m}|$, times the possible choices of alternating flags, which is $\binom{m-1}{k}$. We therefore see the alternating sum arising on the right hand side of equation (2.2).

5. Application to commutative graph homology

In this final section we would like to say how the Stirling complexes arose as a natural object of study, and what Theorem 4.1 tells us in this context. Specifically, the complexes $\mathcal{S}_{n,k}$ are isomorphic to subcomplexes of the Feynman transform of Lie graph homology in genus 1. We will give a recollection of these prerequisites, and we refer to [**War22**] and [**War23**] for more details.

5.1. Modular operads. Modular operads were introduced by Getzler and Kapranov in [**GK98**] as a generalization of the notion of an operad. Modular operads permit composition along all graphs, not just along trees. Here we consider graphs whose vertices carry a labeling by a nonnegative integer, called the genus of a vertex, which allows us to keep track of loop contractions. We call these modular graphs (see Section A.2). Similar to an operad, compositions are generated by single edge contractions, hence the following definition.

DEFINITION 5.1. A (dg) modular operad M is a collection of S_n modules $\mathsf{M}(g,n)$ indexed over pairs of natural numbers g and n with $2g+n\geq 3$, along with two families of linear maps called loop contractions and bridge contractions:

- Loop contractions: for all g, all $n\geq 2$ and all pairs $1\leq i<j\leq n$, there exists a linear map
$$\circ_{ij}\colon \mathsf{M}(g,n)\to\mathsf{M}(g+1,n-2).$$
- Bridge contractions: for all g_1,g_2,n_1,n_2 and each $1\leq i\leq n_1$ and $1\leq j\leq n_2$ there exists a linear map
$${}_i\circ_j\colon \mathsf{M}(g_1,n_1)\otimes\mathsf{M}(g_2,n_2)\to\mathsf{M}(g_1+g_2,n_1+n_2-2).$$

The terminology "bridge" and "loop" contractions comes from the classification of edges in a graph as either bridges or loops (Section A.1). Collectively we call these compositions "edge contractions".

Edge contractions are posited to satisfy a series of axioms analogous to the associativity and S_n-equivariance axioms for an operad. These axioms should be intuitively understood via the graphical intuition which views each $\mathsf{M}(g,n)$ as a vertex with label g and valence n. The loop contraction $\circ_{ij}$ is viewed as gluing the legs labeled i and j to form a "loop" before contracting the result. The bridge contraction ${}_i\circ_j$ glues together the corresponding legs on two vertices to form a "bridge" before contracting the result. The associativity axioms assert that contracting edges of a graph in any order produces the same result, regardless of the order of the contractions.

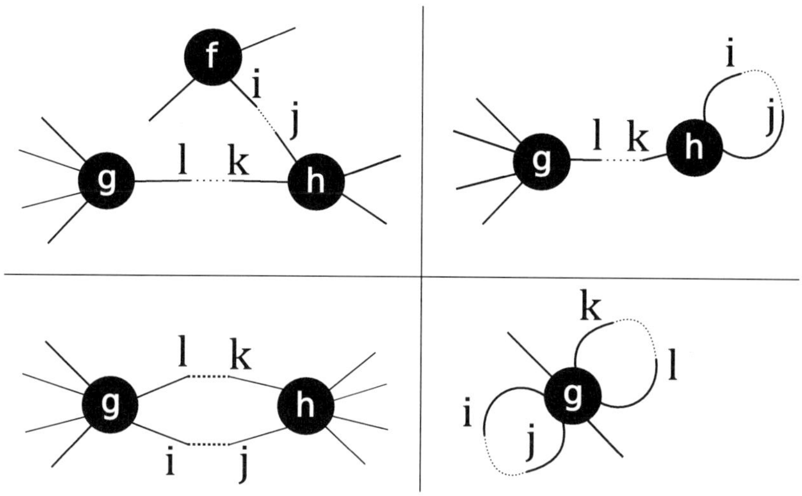

FIGURE 4. The associativity relations in a modular operad are indexed by graphs with two edges. These relations are divided into four types, corresponding to the four pictured two-edged graphs. The leg labels are suppressed except for those which were used in the formation of the two edges. The formed edges are indicated with dashed lines. The target of the composition of operations is determined by contracting the edges, while keeping track of the total genus via the genus label.

With this intuition the associativity axioms for a modular operad are indexed by graphs with two edges and are divided into the following four families. See also Figure 4.

- Contraction of two bridges ${}_{i'}\circ_{j'} \cdot {}_{l}\circ_{k} = {}_{l'}\circ_{k'} \cdot {}_{i}\circ_{j}$
- Contraction of a loop and a bridge $\circ_{i'j'} \cdot {}_{l}\circ_{k} = {}_{l'}\circ_{k'} \cdot \circ_{ij}$
- Contraction of parallel bridges $\circ_{i'j'} \cdot {}_{l}\circ_{k} = \circ_{l'k'} \cdot {}_{i}\circ_{j}$
- Contraction of two loops $\circ_{i'j'} \cdot \circ_{lk} = \circ_{l'k'} \cdot \circ_{ij}$

The notation i', j', l', k' means the label of the leg formerly labeled by i, j, k, l. The precise label will depend on a chosen convention for relabeling legs after edge contraction. See [**War22**, Page 33] for one such convention.

Finally, both loop contraction and bridge contraction satisfy compatibility with the symmetric group actions. These compatibility conditions are of the form $\circ_{ij} \cdot \sigma = \circ_{ij}(\sigma)\circ_{\sigma^{-1}(i)\sigma^{-1}(j)}$ and ${}_{i}\circ_{j} \cdot (\sigma, \tau) = \sigma_{i}\circ_{j}\tau \cdot {}_{\sigma^{-1}(i)}\circ_{\tau^{-1}(j)}$ for certain permutations $\circ_{ij}(\sigma)$ and $\sigma_{i}\circ_{j}\tau$ determined by permutations σ and τ and leg labels i and j. As in the case of an operad, these equivariance axioms may be understood by forming one-edged graphs as specified by the edge contraction and the permutation, and asserting that if the labeled one-edged graphs are the same then the operations are the same.

5.2. The Feynman transform. The Feynman transform [**GK98**] is the modular operadic generalization of the bar construction. Viewed as a functor, its input is a modular operad, call it M, and its output may be understood as the direct sum over graphs whose vertices are labeled by M, (or by its linear dual depending on conventions). The result is a family of chain complexes $\mathsf{FT}(\mathsf{M})(g,n)$ for each total genus g and number of legs n, with differential given by a sum over edge contractions (or dually edge expansions). When an edge is contracted, the modular operad structure of M is used to determine the new vertex label.

To make the above description precise, we proceed as follows. Let M be a modular operad. Let γ be a modular graph (see Section A.2). Let $V(\gamma)$ and $E(\gamma)$ denote the sets of vertices and (internal) edges of γ respectively. If $v \in V(\gamma)$ has genus $g(v)$, valence $n(v)$ and set of adjacent flags $A(v)$, we define a vector space $\mathsf{M}(v) := \mathsf{M}(g, A(v))$ where $\mathsf{M}(g,-)$ is viewed as a functor from the category of all finite sets and bijections via left Kan extension. In particular $\mathsf{M}(v) \cong \mathsf{M}(g(v), n(v))$, but noncanonically so. We then define

$$\mathsf{M}(\gamma) := \otimes_{v\in V(\gamma)} \mathsf{M}(v).$$

Additionally, to γ we associate the one-dimensional graded vector space $\mathfrak{K}(\gamma) := det(E(\gamma))$, where det was defined in Section 3.2. With these pre-requisites, we define the chain complex

$$\mathsf{FT}(\mathsf{M})(g,n) := \bigoplus_{\substack{\text{iso. classes of}\\ (g,n)-\text{graphs } \gamma}} \mathsf{M}(\gamma) \otimes_{Aut(\gamma)} \mathfrak{K}(\gamma), \tag{5.1}$$

with differential given by the sum over edge contractions. Here $Aut(\gamma)$ denotes the group of leg fixing automorphisms of γ.

Each chain complex $\mathsf{FT}(\mathsf{M})(g,n)$ is bigraded by the number of edges of the underlying graph and the sum of the degrees of the vertex labels, which we call the internal degree. We write $\mathsf{FT}(\mathsf{M})(g,n)^{r,s}$ for the span of those graphs having r edges and internal degree s. The differential in the Feynman transform contracts one edge, using the degree 0 modular operad structure maps, so it has the form

$$\mathsf{FT}(\mathsf{M})(g,n)^{r,s} \stackrel{d}{\longrightarrow} \mathsf{FT}(\mathsf{M})(g,n)^{r-1,s}, \tag{5.2}$$

and so the complex splits into a direct sum of subcomplexes indexed over internal degree. The Stirling complexes, studied above, are examples of such subcomplexes for a particular choice of M as we shall subsequently explain.

We remark that the Feynman transform as originally considered by Getzler and Kapranov [**GK98**] is the linear dual of the definition given here, so it may be more appropriate to call this the coFeynman transform.

5.3. Lie graph homology. Given a modular operad M, the family of chain complexes $\mathsf{FT}(\mathsf{M})(g,n)$ can themselves be composed along graphs by freely grafting the legs of the graph together. Thus the Feynman transform forms something akin to a modular operad, but the degree of the compositions is nonzero. The notion of $\mathfrak{K}$-modular operads, in which composition formally has degree 1, was also introduced in [**GK98**] as a target for the Feynman transform. Conversely, an analogous construction produces a modular operad from a $\mathfrak{K}$-modular operad, and so we view the Feynman transform as a pair of functors

$$\{\text{modular operads}\} \overset{\mathsf{FT}}{\leftrightarrows} \{\mathfrak{K}\text{-modular operads}\}. \tag{5.3}$$

As above, we write Lie for the Lie operad. We may choose to view Lie as a modular operad by declaring the structure map corresponding to any graph of positive genus to be 0. Alternatively, we can take the shifted operadic suspension of the Lie operad and then extend the result by 0 to higher genus, and the result will be a $\mathfrak{K}$-modular operad. Denote this $\mathfrak{K}$-modular operad $\Sigma\mathfrak{s}^{-1}\mathsf{Lie}$. By "Lie graph homology" in this article we refer to the modular operad $H_*(\mathsf{FT}(\Sigma\mathfrak{s}^{-1}\mathsf{Lie}))$.

Lie graph homology was studied in [**CHKV16**], where it was shown to be calculable as the group homology of a family of groups $\Gamma_{g,n}$, and so we use notation

$$H_*(\Gamma)(g,n) := H_*(\Gamma_{g,n}) = H_*(\mathsf{FT}(\Sigma\mathfrak{s}^{-1}\mathsf{Lie}))(g,n),$$

and the notation $H_*(\Gamma)$ for the associated modular operad.

Below, we will connect the Stirling complexes to the genus 1 component of the Feynman transform of the modular operad $H_*(\Gamma)$. This in turn depends only on the graded vector spaces $H_*(\Gamma_{g,n})$ for $g \in \{0,1\}$ and the edge contractions between them. This structure can be completely characterized as follows.

First the homology of these spaces is given by Lemma 5.2

LEMMA 5.2 ([**CHKV16**]). *When $g \leq 1$, the homology of $\Gamma_{g,n}$ is:*

$$H_k(\Gamma_{0,n}) = \begin{cases} V_n & \text{if } k = 0 \\ 0 & \text{else} \end{cases}$$

and

$$H_k(\Gamma_{1,n}) = \begin{cases} V_{n-k,1^k} & \text{if } k \text{ is even and } 0 \leq k \leq n-1 \\ 0 & \text{else.} \end{cases}$$

Second, to characterize the modular operadic structure maps between these spaces, we first observe that edge contractions between degree 0 vertices will be isomorphisms. Thus the only nontrivial structure maps landing in genus ≤ 1 are contractions along bridge edges adjacent to a genus 0 and genus 1 vertex. To characterize these operations, we describe the vector spaces $H_k(\Gamma_{1,k+l})$ in terms of such compositions as follows:

LEMMA 5.3 ([**War23**]). *Fix any nonzero vectors $\mu_l \in H_0(\Gamma_{0,l+1})$ and $\alpha_k \in H_k(\Gamma_{1,k+1})$. The vector space $H_k(\Gamma_{1,k+l})$ has a basis consisting of edge contractions of the form $\sigma(\mu_l \circ_1 \alpha_k)$ over all $\sigma \in S_{k+l-1}$.*

To be specific, this result is a corollary of [**War23**, Lemma 3.2] which gives the precise form of the relations as well.

Here we choose to view our graphs as labeled by $\{0,\dots,n-1\} \cong \{1,\dots,n\}$ via the isomorphism sending 0 to n and fixing those i with $0 < i < n$. Then $\circ_1 := {}_1\circ_0$ denotes the operadic composition which glues the root (by convention the leaf labeled by 0) of α on to the leaf labeled by 1 of μ. The permutations σ fix 0, hence the lemma says that has a basis given by those compositions in-which 0 is adjacent to the root (see Figure 5).

5.4. Stirling complexes and the Feynman transform of Lie graph homology. With an understanding of the modular operad $H_*(\Gamma)$ in genus ≤ 1, we now turn to consideration of the chain complex $\mathsf{FT}(H_*(\Gamma))(1,n)$. By equation (5.2), we know that this complex splits over internal degree. By Lemma 5.2 we know that it is nonzero only when the internal degree is even. The following theorem relates the Stirling complexes to the summands corresponding to nonzero internal degree.

THEOREM 5.4. *Let k be a positive, even integer. There is an isomorphism of chain complexes*

$$\mathsf{FT}(H_*(\Gamma))(1,n+1)^{*,k} \cong \mathcal{S}_{n,k}.$$

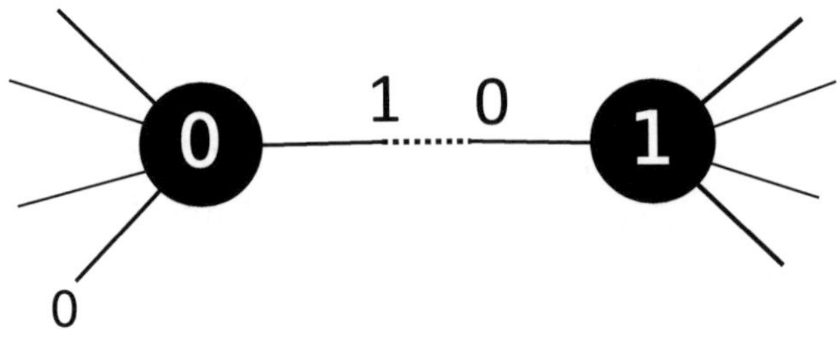

FIGURE 5. Compose the two indicated vertices via the composition $\circ_1$. The genus 1 vertex carries a label in $H_k(\Gamma_{1,k+1})$ and the genus 0 vertex carries a label in $H_0(\Gamma_{0,l+1})$. After choices of generators of these one dimensional vector spaces, call them α_k and μ_l respectively, each way to label the legs in the pictured tree results in a class in $H_k(\Gamma_{1,k+l})$. Lemma 5.3 says that the $\binom{k+l-1}{k}$ such classes form a basis.

PROOF. Fix such a k, and assume $k \leq n$, else the statement is vacuous. Consider the chain complex $\mathsf{FT}(H_*(\Gamma))(1, n+1)^{*,k}$. By equation (5.1), this chain complex is a direct sum indexed over graphs of type $(1, n+1)$, i.e., total genus 1 and with $n+1$ labeled legs.

By definition, the total genus of a modular graph is the sum of the genera of the vertex labels with the first Betti number of the given graph. If the vertex labels all have genus 0, then the internal degree would be 0 (by Lemma 5.2). Since this is not the case, there must be a vertex of positive genus. Since the total genus is 1, this can only happen if the graph is a tree and there is a unique vertex of genus 1, with all other vertices of genus 0. We thus conclude that the homogeneous elements in $\mathsf{FT}(H_*(\Gamma))(1, n+1)^{i,k}$ correspond to trees with i edges and $n+1$ labeled legs, along with a distinguished vertex (namely the unique vertex of genus 1).

Given such an homogeneous element, the degree 0 vertices of valence $l+1$ are labeled with homology classes from $H_*(\Gamma_{0,l+1})$ which is simply the ground field concentrated in degree 0. The genus 1, or distinguished, vertex is therefore labeled with a homology class in $H_k(\Gamma_{1,m})$ for some m.

Using the basis of Lemma 5.3, we may thus view an homogeneous element in $\mathsf{FT}(H_*(\Gamma))(1, n+1)^{i,k}$ as a tree with a distinguished vertex which is itself labeled by a tree with one edge and two vertices. Among these two "inner" vertices, the upper one is labeled by the genus 1 alternating class α_k and the lower one is labeled by the genus 0 class μ_l. In order to specify a Stirling tree from this data, we color all legs adjacent to the genus 1 vertex red, and all the others black. See Figure 6.

This gives an isomorphism of graded vector spaces. In hindsight, the formula for the differential of the Stirling complexes given above (Section 3.4) was chosen such that this isomorphism of graded vector spaces is also an isomorphism of chain complexes. To see that this is the case, observe that both differentials are given by sums over edge contraction, and it's immediate that the terms coincide when contracting an edge which is not adjacent to the distinguished vertex, or when contracting an edge which is adjacent to the distinguished vertex but is nonalternating/adjacent to the lower inner vertex. The only nontrivial case, then, is to show that contracting an edge adjacent to the upper/alternating inner vertex corresponds to the description given above for contracting an edge in a Stirling tree

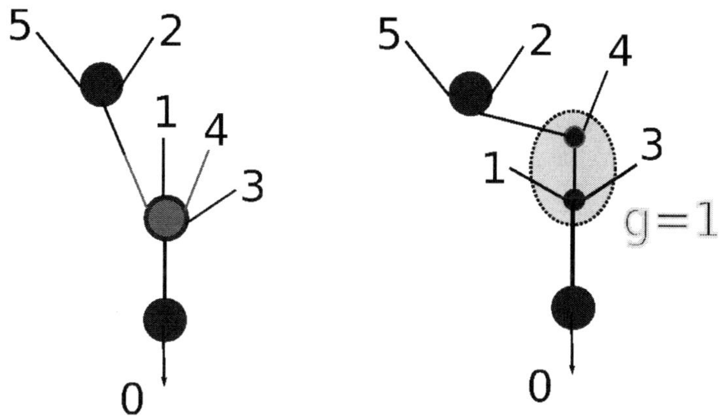

FIGURE 6. The Stirling tree on the left corresponds to a modular graph of total genus 1 on the right. The unique genus 1 vertex (with dotted boundary) is itself labeled by a composition of the commutative product μ_l with the alternating class α_k. (Pictured is the case $k = 2$.) This composition specifies the label of the genus 1 vertex.

when said edge contains an alternating flag. This in turn requires rewriting a one-edge composition landing in $H_k(\Gamma_{1,m})$ in terms of the chosen basis. A formula for this rewriting rule is given in [**War23**, Lemma 3.2], and the description in terms of Stirling trees is simply a reformulation of the statement of that result. □

Combining this result with Theorem 4.1, we have the following corollary pertaining to commutative graph homology. Let $\overline{\mathsf{Com}}$ be the modular operad defined by $\overline{\mathsf{Com}}(g, n) = \mathbb{Q}$, all of whose structure maps are the canonical isomorphisms.

COROLLARY 5.5. *There is an isomorphism of S_{n+1}-modules*

$$H_{*+1}(\mathsf{FT}(\overline{\mathsf{Com}}))(1, n+1) \cong \bigoplus_{i=1}^{\lfloor n/2 \rfloor} H_*(\mathcal{S}_{n,2i}).$$

PROOF. Consider the chain complex

$$\mathsf{FT}(H_*(\Gamma))(1, n+1) = \mathsf{FT}(H_*(\mathsf{FT}(\Sigma\mathfrak{s}^{-1}\mathsf{Lie})))(1, n+1).$$

If it were the case that the modular operad $\mathsf{FT}(\Sigma\mathfrak{s}^{-1}\mathsf{Lie})$ were formal, then this complex would be acyclic, because $\mathsf{FT}^2(\Sigma\mathfrak{s}^{-1}\mathsf{Lie}(1, n+1))$ has the same homology as $\Sigma\mathfrak{s}^{-1}\mathsf{Lie}(1, n+1)$, which is zero by definition. Alas, this modular operad is not formal (indeed we've computed a portion of its homology above). However, in this setting, the bicomplex $\mathsf{FT}(H_*(\Gamma))(1, n+1)^{g,k}$ is the E_0 page of a spectral sequence converging to 0 (called the genus label filtration spectral sequence, [**War22**, Section 4.5.1]).

The higher pages of this spectral sequence have differentials inherited from the homotopy transfer theorem applied to the modular operad $H_*(\Gamma)$. These differentials necessarily increase the internal degree. However, above the row $k = 0$, the E_1-page is concentrated in total degree n (combining Theorems 4.1 and 5.4). Therefore, the $k = 0$ row of the E_1 page must be concentrated in degree $n + 1$, and moreover must be isomorphic to the direct sum of the homologies of $\mathcal{S}_{n,k}$ over all even $k > 0$. In particular, from Theorem 4.1 we conclude that the homology of $\mathsf{FT}(H_*(\Gamma))(1,n+1)^{*,0}$ is concentrated in a single degree and has rank equal to the number of even (or odd) permutations of S_n, namely $n!/2$.

Finally we observe that the modular operads $\overline{\mathsf{Com}}$ and $H_0(\Gamma)$ (taking only the degree 0 summands in each biarity) coincide, hence

$$\mathsf{FT}(H_*(\Gamma))(1,n+1)^{*,0} = \mathsf{FT}(H_0(\Gamma))(1,n+1) = \mathsf{FT}(\overline{\mathsf{Com}})(1,n+1),$$

which finishes the proof. □

REMARK 5.6. Before proving Theorem 4.1, I had calculated the result by hand for low values of n. For example the case of $n+1=5$ is pictured in [**War22**, Figure 14], but Theorem 4.1 was needed to ensure that the analogous picture holds for all n.

The chain complexes which comprise the $\mathfrak{K}$-modular operad $\mathsf{FT}(\overline{\mathsf{Com}})$ coincide (up to a shift in degree) with the reduced cellular chains of the spaces $\Delta_{g,n}$, and so this corollary establishes equation (1.3) given in the introduction.

5.5. Representation stability. As mentioned in Section 1, Corollary 5.5 agrees with [**CGP22**] on rank, but gives a different perspective on the S_{n+1}-module structure. We end this article with a modest conjecture.

CONJECTURE 5.7. *For each i the sequence of S_{n+1}-modules $H_n(\mathcal{S}_{n,n-i})$ is representation stable, after tensoring with the alternating representation.*

Although filling in all the details is a bit beyond the scope of this article, let me give some idea of why this seems likely. The first step would be to show that for each r, the sequence $\mathcal{S}_{n,n-i,r}\otimes V_{1^n}$ is a finitely generated FI-module. Here r is the number of edges. Recall [**CEF15**] that an FI-module is simply a functor from the category of finite sets and injections. For this, given an injection $\iota\colon \{0,\dots,n\} \hookrightarrow \{0,\dots,n+l\}$, and a tree whose leaves are labeled $\{0,\dots,n\}$, we can simply relabel leg j with $\iota(j)$, and then add alternating flags to the distinguished vertex which are labeled by the elements not in the image of ι. The factor of V_{1^n} ensures that the order of the added flags plays no role.

This putative FI-module would then be finitely generated by those trees whose distinguished vertex has valence $k+1$, and so the general theory of finitely generated FI-modules [**CEF15**] would apply to show that each sequence $\mathcal{S}_{n,n-i,r}$ is representation stable. Once this is established, it would be straight-forward to argue that the S_{n+1}-equivariant Euler characteristic, and hence the homology, is representation stable as well. Having established representation stability, a next step would then be to determine the stable range and calculate some stable multiplicities.

Let us give some confirmation of this conjecture in the case of small i. When $i=0$, we have already seen that $H_n(\mathcal{S}_{n,n})$ is the alternating representation of S_{n+1}, which is representation stable after tensoring with the sign representation.

When $i=1$, each complex $\mathcal{S}_{n,n-1}$ is only nonzero in two degrees corresponding to number of edges $r=0$ and $r=1$. From [**CHKV16**] we know $\mathcal{S}_{n,n-1,0}\cong V_{2,1^{n-1}}$ and we may compute $\mathcal{S}_{n,n-1,1}$ as an induced representation using the Littlewood–Richardson rule [**FH91**] to be $V_{2,1^{n-1}}\oplus V_{3,1^{n-2}}$. Thus, by Theorem 4.1 the homology is $H_n(\mathcal{S}_{n,n-1})\cong V_{3,1^{n-2}}$, which again is representation stable after tensoring with the alternating (aka sign) representation.

Finally, when $i = 2$ we can determine $H_n(\mathcal{S}_{n,n-2})$ via Theorem 4.1 by calculating the S_{n+1}-equivariant Euler characteristic. This requires $n \geq 4$ and just by dimension considerations, after Lemma 2.1(3), $n = 4$ must be outside the stable range. However, in the case $n = 5$ we find, by a slightly tedious tabulation of characters, that

$$H_n(\mathcal{S}_{n,n-2}) \cong V_{3,3} \oplus V_{2,2,1^2} \oplus V_{3,2,1} \oplus V_{5,1},$$

and using the hook length formula it is a straight forward computation to show that the stable sequence of S_n-modules which continues this S_6-representation (after tensoring with the sign representation) has dimension equal to

$$\frac{1}{4}\binom{n}{3}(3n-1),$$

as predicted by Lemma 2.1.

To conclude, let me remark that it would be interesting to use the semi-classical modular operadic character [**Get98**] to compute the character polynomials of each of these putative representation stable sequences. Indeed, if we restrict attention to the modular operad whose only nonzero terms are $H_0(\Gamma_{0,n})$ and $H_k(\Gamma_{1,n})$, ranging over all n, the character of the Feynman transform can be read off from the results of *op. cit.*, and Conjecture 5.7 implies, after [**CEF15**, Theorem 1.5], that the characters of these sequences will eventually be polynomials.

Appendix A. Trees and graphs

In an attempt to keep this article reasonably self-contained, we spell out our conventions regarding graphs and trees in this appendix. We refer to, e.g., [**BM08**], [**KW17**] and [**War22**] for more detail about this viewpoint.

A.1. Graphs: Definition and basic terminology.

DEFINITION A.1. A graph $\gamma = (V, F, a, \iota)$ is the following data:

- A finite nonempty set V called the vertices of the graph.
- A finite set F called the flags of the graph.
- A function $a\colon F \to V$.
- A function $\iota\colon F \to F$ with the property that ι^2 is the identity function.

Fix a graph $\gamma = (V, F, a, \iota)$ and let $v \in V$ and $f \in F$. When discussing graphs we use the following terminology. If $a(f) = v$ we say that f is **adjacent** to v. The **valence** of v is $|a^{-1}(v)|$. If $\iota(f) = f$ we say f is a **leg** of γ. If $\iota(f) \neq f$ we say that the unordered pair $\{f, \iota(f)\}$ is an **edge** of γ. The set of edges of γ is denoted $E(\gamma)$ or just E if the context is clear. The set of legs of γ will be denoted $\mathrm{leg}(\gamma)$. Observe $|F| = 2|E| + |\mathrm{leg}(\gamma)|$. An edge $\{f, \iota(f)\}$ will be called a **loop** if $a(f) = a(\iota(f))$ and will be called a **bridge** if $a(f) \neq a(\iota(f))$.

A **path** in a graph γ is a finite sequence of flags $\{f_1, \dots, f_r\}$ such that for each pair f_i and f_{i+1} either $a(f_i) = a(f_{i+1})$ or $\iota(f_i) = f_{i+1}$. The length of a path is the number of flags in the sequence. Given such a path, if $a(f_1) = v$ and $a(f_r) = w$ then the path is said to run from v to w. We specifically allow the empty path from any vertex to itself, considered to be of length 0.

A **subgraph** of γ is a pair of subsets $V' \subset V$ and $F' \subset F$ which are closed under a and ι. In a graph γ, define an equivalence relation on the vertices V via $v \sim w \iff \exists$ a path running from v to w. Let $V_1, \dots, V_c$ be the equivalence classes of this relation. Partition F by defining $F_i := \{f \in F : a(f) \in V_i\}$. It is straightforward to show that each (V_i, F_i) is a subgraph of γ, we call these subgraphs the **connected components** of γ. The **genus** of a graph having c components is defined to be $\beta_1(\gamma) = c - |V| + |E|$. If $c = 1$ we say that γ is **connected.** It is an elementary exercise to show that given a connected graph of genus 0 along with vertices v and w, there is a unique shortest path running from v to w.

A.2. Modular graphs. A **genus labeled** graph is a graph along with a function $g\colon V \to \mathbb{Z}_{\geq 0}$. We call the nonnegative integer $g(v)$ the genus label of the vertex. If γ is a genus labeled graph, we define its **total genus** to be

$$g(\gamma) := \beta_1(\gamma) + \sum_{v \in V} g(v).$$

A genus labeled graph is said to be **stable** if $2g(v) + |a^{-1}(v)| \geq 3$ for each $v \in V$.

A **modular graph** of type (g, n) refers to a stable, connected graph of total genus g along with a bijection $\text{leg}(\gamma) \cong \{1, \dots, n\}$. We call the number associated to a given leg its labeling. Morphisms, in particular isomorphisms, of modular graphs are presumed to preserve the genus labels and leg labels.

A.3. Trees. A **tree** is a connected graph of genus 0. Trees will often be denoted t. A tree is stable if $|a^{-1}(v)| \geq 3$ for each $v \in V$. An n-tree is a tree with $n+1$ legs, along with a bijection between the sets $\{0, \dots, n\}$ and $\text{leg}(\mathsf{t})$, called the leg labeling. The leg labeled by 0 is called the root of an n-tree. The vertex adjacent to the root is called the root vertex.

In an n-tree, each vertex has a distinguished adjacent flag called the output. The output of the root vertex is defined to be the root. For any other vertex, call it v, the output is defined to be the first flag on the unique shortest path running from v to the root vertex. All flags which are not output flags are called input flags. Morphisms, in particular isomorphisms, of n-trees are presumed to preserve the leg labels. In particular, relabeling the legs can produce a nonisomorphic n-tree.

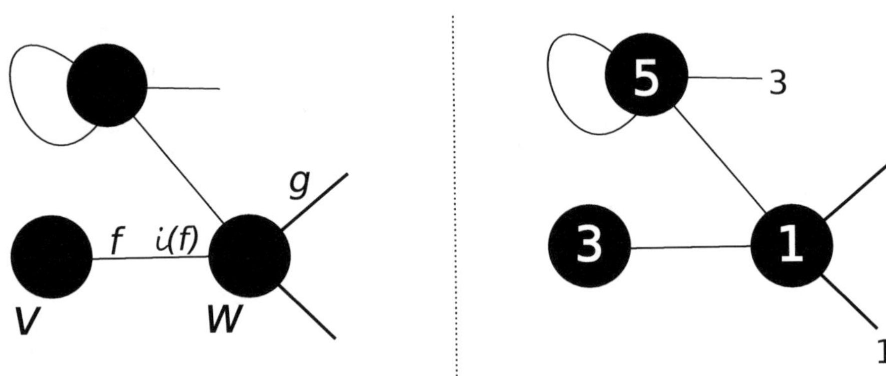

FIGURE 7. Left: a diagramatic depiction of a graph with three vertices and 9 flags having two bridges, one loop and three legs. The labeled vertices v and w and labeled flags f and g satisfy $a(f) = v$, $a(g) = w$ and $\iota(g) = g$. Right: adding labels to indicate a modular graph of type (10,3).

By convention, a stable n-tree determines a modular graph of type $(0, n+1)$ by declaring that each vertex is of genus 0 and by relabeling the legs according to the isomorphism $\{0, \ldots, n\} \cong \{1, \ldots, n+1\}$ which sends 0 to $n+1$ and fixes the set $\{1, \ldots, n\}$.

A.4. Graphical depiction of a graph. To a graph we may associate a diagram in which the vertices are depicted as nodes and the flags are depicted as arcs or line segments. If $a(f) = v$, the line segment corresponding to f is drawn adjacent to v at one end, call it the vertex end. If $f \neq \iota(f)$ then these two flags are joined at their nonvertex ends to form a graphical depiction of the edge. The legs of the graph are depicted as line segments with loose ends. Leg and genus labels may be added to the diagram to depict modular graphs and/or n-trees (see Figure 7). A graph can in turn be extracted from such a diagram, although the underlying graph depends only on the combinatorics of the diagram and not on any embedding in Euclidean space.

Acknowledgments

It is my sincere pleasure to thank Ralph Kaufmann, Martin Markl, and Sasha Voronov for their invitation to participate in the Special Session on Higher Structures in Topology, Geometry and Physics at the AMS Spring Central Meeting, March 2022. I would also like to thank an anonymous referee for comments which improved an earlier version of this article.

References

[BM08] Dennis V. Borisov and Yuri I. Manin, *Generalized operads and their inner cohomomorphisms*, Geometry and dynamics of groups and spaces, Progr. Math., vol. 265, Birkhäuser, Basel, 2008, pp. 247–308, DOI 10.1007/978-3-7643-8608-5_4. MR2402406

[CEF15] Thomas Church, Jordan S. Ellenberg, and Benson Farb, *FI-modules and stability for representations of symmetric groups*, Duke Math. J. **164** (2015), no. 9, 1833–1910, DOI 10.1215/00127094-3120274. MR3357185

[CGP22] Melody Chan, Søren Galatius, and Sam Payne, *Topology of moduli spaces of tropical curves with marked points*, Facets of algebraic geometry. Vol. I, London Math. Soc. Lecture Note Ser., vol. 472, Cambridge Univ. Press, Cambridge, 2022, pp. 77–131. MR4381898

[Cha02] Charalambos A. Charalambides, *Enumerative combinatorics*, CRC Press Series on Discrete Mathematics and its Applications, Chapman & Hall/CRC, Boca Raton, FL, 2002. MR1937238

[CHKV16] James Conant, Allen Hatcher, Martin Kassabov, and Karen Vogtmann, *Assembling homology classes in automorphism groups of free groups*, Comment. Math. Helv. **91** (2016), no. 4, 751–806, DOI 10.4171/CMH/402. MR3566523

[DK10] Vladimir Dotsenko and Anton Khoroshkin, *Gröbner bases for operads*, Duke Math. J. **153** (2010), no. 2, 363–396, DOI 10.1215/00127094-2010-026. MR2667136

[FH91] William Fulton and Joe Harris, *Representation theory*, Graduate Texts in Mathematics, vol. 129, Springer-Verlag, New York, 1991. A first course; Readings in Mathematics, DOI 10.1007/978-1-4612-0979-9. MR1153249

[Get98] E. Getzler, *The semi-classical approximation for modular operads*, Comm. Math. Phys. **194** (1998), no. 2, 481–492, DOI 10.1007/s002200050365. MR1627677

[GK94] Victor Ginzburg and Mikhail Kapranov, *Koszul duality for operads*, Duke Math. J. **76** (1994), no. 1, 203–272, DOI 10.1215/S0012-7094-94-07608-4. MR1301191

[GK98] E. Getzler and M. M. Kapranov, *Modular operads*, Compositio Math. **110** (1998), no. 1, 65–126, DOI 10.1023/A:1000245600345. MR1601666

[Kon94] Maxim Kontsevich, *Feynman diagrams and low-dimensional topology*, First European Congress of Mathematics, Vol. II (Paris, 1992), Progr. Math., vol. 120, Birkhäuser, Basel, 1994, pp. 97–121. MR1341841

[KW17] Ralph M. Kaufmann and Benjamin C. Ward, *Feynman categories* (English, with English and French summaries), Astérisque **387** (2017), vii+161. MR3636409

[KWZ17] Anton Khoroshkin, Thomas Willwacher, and Marko Živković, *Differentials on graph complexes*, Adv. Math. **307** (2017), 1184–1214, DOI 10.1016/j.aim.2016.05.029. MR3590540

[Mer21] Sergei Merkulov, *Grothendieck-Teichmüller group, operads and graph complexes: a survey*, Integrability, quantization, and geometry II. Quantum theories and algebraic geometry, Proc. Sympos. Pure Math., vol. 103, Amer. Math. Soc., Providence, RI, [2021] ©2021, pp. 383–445, DOI 10.1090/pspum/103.2/01863. MR4285704

[War22] Benjamin C. Ward, *Massey products for graph homology*, Int. Math. Res. Not. IMRN **11** (2022), 8086–8161, DOI 10.1093/imrn/rnaa346. MR4425832

[War23] Benjamin C. Ward, *Wheel graph homology classes via Lie graph homology*, J. Noncommut. Geom. **17** (2023), no. 2, 693–717, DOI 10.4171/jncg/508. MR4592884

[Wil15] Thomas Willwacher, *M. Kontsevich's graph complex and the Grothendieck-Teichmüller Lie algebra*, Invent. Math. **200** (2015), no. 3, 671–760, DOI 10.1007/s00222-014-0528-x. MR3348138

Department of Mathematics and Statistics, Bowling Green State University, Bowling Green, Ohio 43403

Email address: `benward@bgsu.edu`

Contemporary Mathematics
Volume **802**, 2024
https://doi.org/10.1090/conm/802/16075

A combinatorial model for the moduli of bordered Riemann surfaces and a compactification

Ralph M. Kaufmann and Javier Zúñiga

ABSTRACT. We construct a combinatorial moduli space closely related to the KSV-compactification of the moduli space of bordered marked Riemann surfaces. The open part arises from symmetric metric ribbon graphs. The compactification is obtained by considering sequences of noncontractible subgraphs. This leads to a partial real blow-up of rational cells that together form a compact orbi-cell space. For genus zero the constructed space gives an orbi-cell decomposition of the corresponding analytic moduli space decorated by real numbers and a compactification of this space. In higher genus the relation is more involved, as we briefly explain. The spaces we construct are of interest in their own right as they are constructed directly from an interesting class of graphs.

1. Introduction

Let $\mathcal{M}_{g,n}^{dec}$ denote the (decorated) moduli space of Riemann surfaces of genus g and n marked points together with decorations, i.e., a choice of a positive real number for each marked point so that they add up to one. Let $\mathcal{M}_{g,n}^{comb}$ denote the (combinatorial) moduli space of metric ribbon graphs (or fat graphs) of genus g with n boundary cycles so that the sum of lengths of all edges adds up to one half. The combinatorial moduli space has a natural orbicell decomposition. A known result of Harer, Mumford, and Thurston [**Har86**], Penner [**Pen87**], and Bowdich and Epstein [**BE88**], states that the decorated and combinatorial moduli spaces are homeomorphic. This homeomorphism can be realized by quadratic differentials with quadratic residues (the ribbon graph model) or by hyperbolic metrics with horocycles (the fat graph model). As a consequence, the space $\mathcal{M}_{g,n}^{dec}$ has an orbicell decomposition given by $\mathcal{M}_{g,n}^{comb}$.

Konsevich's proof of Witten's conjecture in [**Kon92**] requires a compactification of the combinatorial moduli space emulating the Deligne-Mumford compactification of the moduli space of Riemann surfaces. This compactification losses a lot of information (it is more coarse) when the surface degenerates in certain ways, which is codified topologically by the so called *genus defect*. The details of this compactification together with a finer compactification recovering the missing data was

2020 *Mathematics Subject Classification*. Primary 32G15, 57Q15; Secondary 57R18.
Key words and phrases. Bordered Riemann surfaces, ribbon graphs, moduli spaces.

©2024 American Mathematical Society

given by Looijenga in [**Loo95**]. A further extension of Looijenga's compactification was given in [**Zúñ15**] using the real compactification proposed by [**KSV95**].

For bordered Riemann surfaces a few combinatorial models have been proposed including [**God07**] and [**Cos07**] using BW graphs, see [**Ega15**] for a summary. These models are homotopical in nature. In this article, we propose a combinatorial model based on ribbon graphs and the double construction. The advantage is that our model identifies the homeomorphism type of a concrete space, but it can be related directly to the moduli of bordered surfaces only for genus zero with one boundary component and some special cases in genus one. We also discuss some relations between out model and the model in [**Cos07**], and also a connection with the KSV compactification of the moduli space. This type of space is also of interest in terms of open/closed field theory and actions [**KP06, Kau10**].

This article is structured as follows. In Section 2, we review the construction of polytopes and their truncation. In Section 3, we introduce ribbon graphs and in Section 4, we define their moduli spaces. In Section 5, we define the combinatorial blow-up of the moduli spaces in the previous section. In Section 6, we review the theory of Strebel–Jenkins differentials on Riemann surfaces. In Section 7, we define several moduli spaces of interest: decorated moduli spaces, moduli spaces of symmetric surfaces and different compactifications. We also define the space of interest in this article: the moduli space of symmetric ribbon graphs. In Section 8, we give a few examples of these new spaces. In Section 9, we establish some connections with known moduli spaces and BW graphs.

2. Polytopes

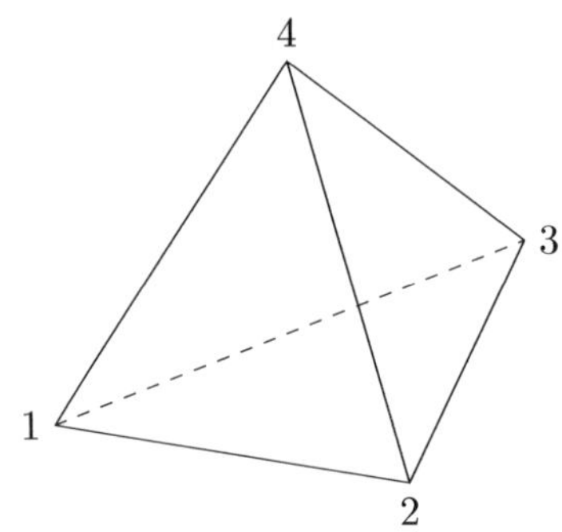

FIGURE 2.1. The simplex Δ_S where $S = \{1,2,3,4\}$.

Given a finite set S we denote by Δ_S the standard simplex on S. Figure 2.1 shows the tetrahedron. In this case if Δ_S can be given coordinates so that $x_1 + x_2 + x_3 + x_4 = 1$ where $x_1 \geq 0$. Notice that as $x_1, x_3 \to 0$, we obtain the interval $\Delta_{3,4}$. For a set S we denote by P_S the permutohedron on S which can be defined as a *truncation* or *blow-up* of Δ_S along all faces in increasing dimension. When $S = \{1,2,3,4\}$, Figure 2.2 shows P_S. The faces are obtained by considering proper, nonempty sequences of subsets of S. For example, the sequence $\{1,2,3,4\} \supset \{1,3\}$ gives the rectangle $P_{\{2,4\}} \times P_{\{1,3\}}$. In the previous example, as $x_1, x_3 \to 0$, the rectangle is obtained by blowing-up $\{1,3\}^c = \{2,4\}$.

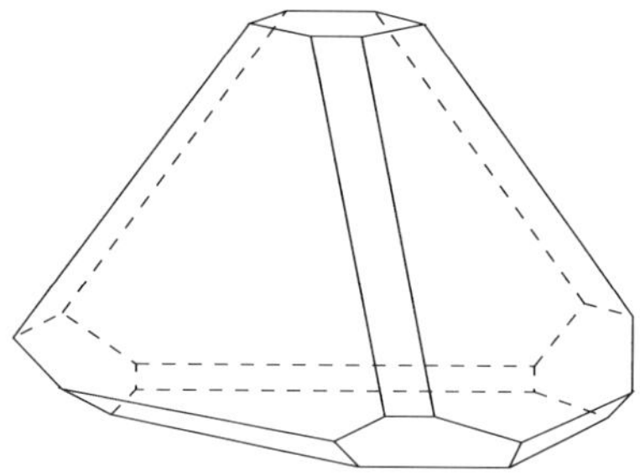

FIGURE 2.2. The permutahedron P_S where $S = \{1,2,3,4\}$.

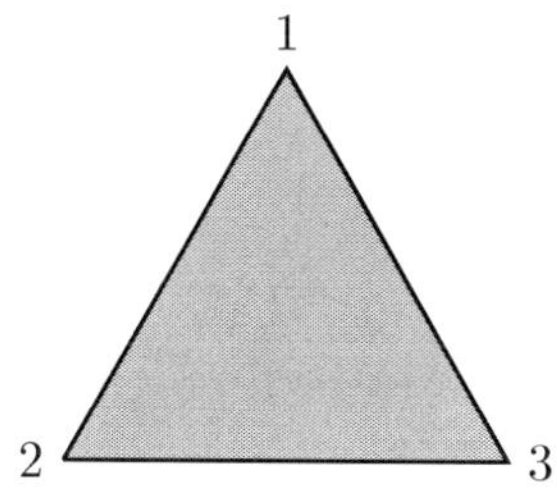

FIGURE 2.3. The simplex Δ_S where $S = \{1,2,3\}$.

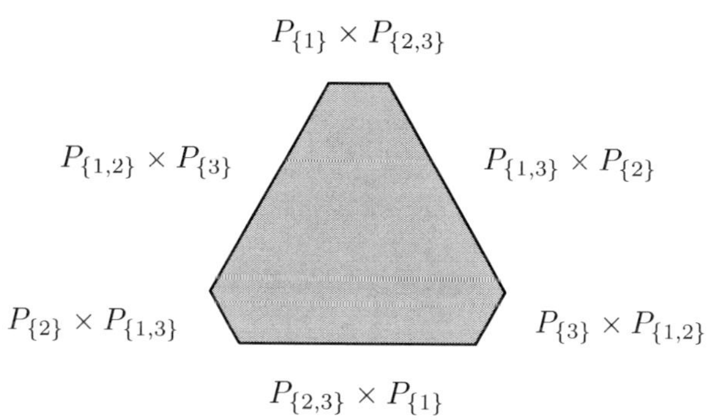

FIGURE 2.4. The permutahedron P_S where $S = \{1,2,3\}$.

From the simplex in Figure 2.3, and using that labeling, we can obtain the permutohedron P_S in Figure 2.4 together with all faces.

It is possible to obtain other polytopes (nestohedra) by blowing-up (truncating) only certain lower dimensional faces of a simplex. One way to indicate which faces to blow-up in Δ_S is by providing a subset $B \subset 2^S \setminus \{\varnothing\}$.

DEFINITION 2.1. For a given set S, let N_B be the result of blowing-up Δ_S along the faces determined by B and increasing by dimension.

With the definition above, $N_\varnothing = \Delta_S$ and $N_{2^S\setminus\{\varnothing\}} = P_S$. For an intermediate example, if we consider $S = \{1,2,3,4\}$ and

$$B = \{\{4\},\{1,4\},\{2,4\},\{3,4\}\}$$

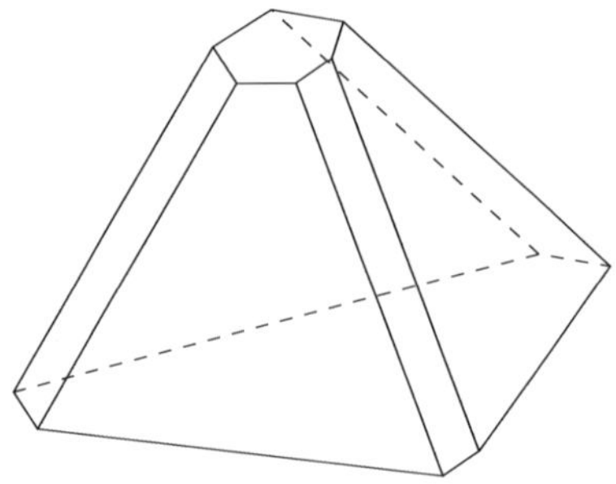

FIGURE 2.5. An intermediate polytope.

we obtain the polytope in Figure 2.5 by blowing up first the top vertex of the tetrahedron in Figure 2.1 and then the corresponding three edges.

REMARK 2.2. It is possible to use the language of hypegraph polytopes as in [**COI19**] to deal with the polytopes in this sections.

3. Ribbon graphs

DEFINITION 3.1. A *ribbon graph* Γ is a connected graph with vertices of valence at least three together with a subdivision of edges into half-edges, also called *edge refinement*, and a cyclic order on the set of incident half-edges to a vertex.

If H is the set of half-edges and v is a vertex of Γ, let H_v be the set of half-edges adjacent to this vertex. A cyclic ordering at a vertex v is an ordering of H_v up to cyclic permutation. Once a cyclic ordering of H_v is chosen, a cyclic permutation of H_v is defined (an element of $\mathfrak{S}_{H_v}$): it moves a half-edge to the next in the cyclic order. Define by σ_0 the element of $\mathfrak{S}_H$ (the group of permutations acting on H) which is the product of all the cyclic permutations at every vertex. Also, let σ_1 be the involution that interchanges the two half-edges on each edge of Γ. This combinatorial data completely defines the ribbon graph.

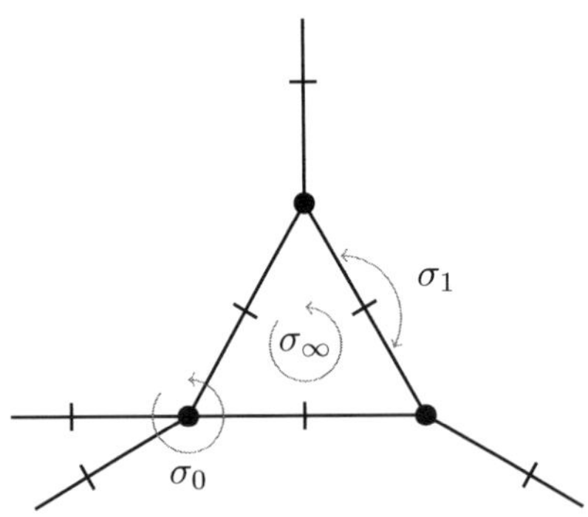

FIGURE 3.1. An example of the permutations that define a ribbon graph around a boundary cycle.

To be more precise, a ribbon graph Γ can be build out of combinatorial data in the following way. Let H be a finite set of even cardinality. Let $\sigma_1 \in \mathfrak{S}_H$ be an involution without fixed points and $\sigma_0 \in \mathfrak{S}_H$ be such that σ_0 is a product of cyclic permutations with disjoint support. A vertex of Γ is then given as an orbit of σ_0, while an edge is an orbit of σ_1. The set of vertices can be identified with $V(\Gamma) = H/\sigma_0$ and the set of edges with $E(\Gamma) = H/\sigma_1$. The graph thus obtained is required to be connected.

Let $\sigma_\infty = \sigma_0^{-1}\sigma_1 \in \mathfrak{S}_H$. The set of *cusps* is defined as $C(\Gamma) = H/\sigma_\infty$. The half-edges in the orbit of a cusp form a cyclically ordered set of half-edges called a *boundary cycle.* Given a boundary cycle, all edges and vertices that are associated to those half-edges via σ_0 or σ_1 form a graph called a *boundary subgraph.* For example, in Figure 3.1 the boundary cycle represented in the middle (and corresponding to three half edges) forms a boundary subgraph with three edges and three vertices (the middle triangle as a subgraph). Notice also that knowing σ_1 and σ_∞ completely determines the ribbon graph structure since $\sigma_0 = \sigma_1\sigma_\infty^{-1}$.

An *orientation of an edge* can be defined as an order on its corresponding half edges and we can use the notation $\vec{e} = \overrightarrow{h\sigma_1(h)}$ where h is a half-edge. The involution σ_1 switches the orientation of an edge.

A *loop* is an edge incident to only one vertex. A *tree* is a connected graph T with trivial $\bar{H}_*(T)$.

DEFINITION 3.2. An *isomorphism* of ribbon graphs is a graph isomorphism of the edge refinement preserving the cyclic orders on each vertex.

Therefore, two graphs Γ, Γ' are isomorphic when there is a bijection $\eta : H \to H'$ between the set of half-edges of these two graphs that commutes with σ_0, σ_0' and σ_1, σ_1'. In particular, the boundary cycles are preserved, *i.e.*, η also commutes with σ_∞, σ_∞'.

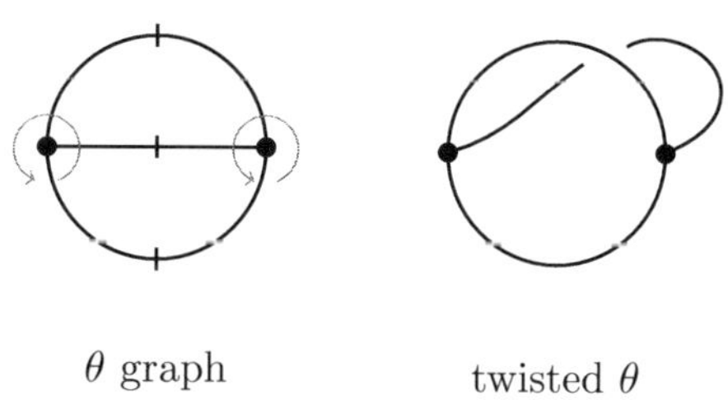

FIGURE 3.2. Two nonisomorphic ribbon graphs of different topological types.

Let $\mathrm{Aut}(\Gamma)$ denote the automorphism group of the edge refinement as a usual graph. The automorphism group of a ribbon graph, denoted by $\mathrm{Aut}^{rg}(\Gamma)$, is the group of automorphisms of the edge refinement that also preserve the cyclic order on every vertex. Figure 3.2 shows two isomorphic graphs that are not isomorphic as ribbon graphs.

When replacing the edges by *ribbons* and gluing them according to the cyclic ordering, one obtains an orientable surface of genus g with n holes as in Figure 3.3. Each hole can be retracted to a boundary subgraph. Since the surface retracts to the graph, its Euler characteristic is $\chi = 2 - 2g - n$. Figure 3.4 shows three nonisomorphic graphs of type $(0, 3)$: the theta graph, the double noose and the figure eight respectively. All ribbon graphs of type $(1, 1)$ are shown in Figure 3.5: The twisted theta graph and the twisted figure eight.

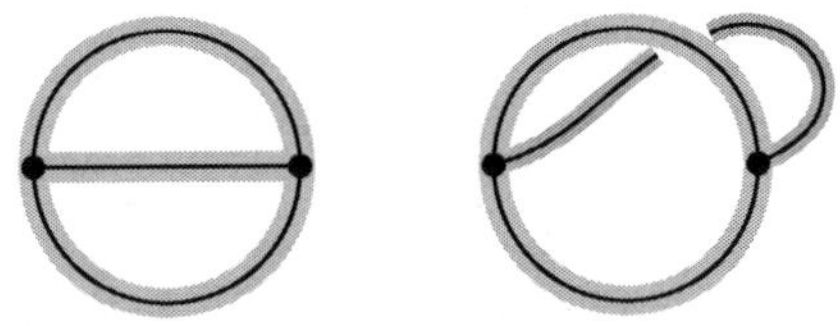

FIGURE 3.3. Two surfaces obtained from thickening ribbon graphs.

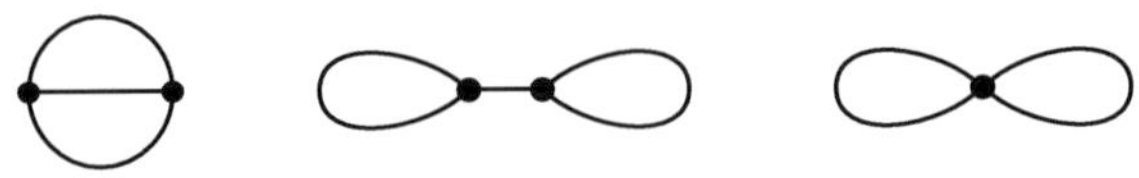

FIGURE 3.4. Nonisomorphic ribbon graphs of type $(0,3)$.

FIGURE 3.5. Nonisomorphic ribbon graphs of type $(1,1)$.

Analogously to a labeling of the marked points on a surface, the boundary cycles or cusps of a ribbon graph can be *labeled*.

DEFINITION 3.3. For a labeled ribbon graph, let $\mathrm{Aut}_{\partial}^{rg}(\Gamma)$ denote the ribbon graph automorphism group preserving the labels.

It can be shown that up to isomorphism, there is exactly one labeled theta graph and three labeled double nooses. Since the twisted theta graph has only one boundary cycle, there is only one twisted labeled theta graph up to isomorphism. From now on, we assume that all ribbon graphs have labeled boundary cycles.

A metric on a ribbon graph is a function $l : E(\Gamma) \to \mathbb{R}_+$. A metric can be represented by a point in $\mathbb{R}_+^{|E(\Gamma)|}$. A metric is *unital* when the sum of the lengths of the boundary cycles is equal to one. This is equivalent to requiring that the sum of all edges equals one half since each edge is counted twice in the sum of the lengths of boundary cycles. A unital metric can be then represented by a point in $int(\Delta_{E(\Gamma)})$.

To every metric ribbon graph Γ we associate an oriented punctured surface $\mathrm{Surf}(\Gamma)$ as follows. To each oriented edge $\vec{e} = \overrightarrow{h\sigma_1(h)}$ we attach an oriented triangle $K_{\vec{e}} = C|\vec{e}|$, where $|\vec{e}|$ is a segment of length $l(e)$ and C refers to the cone construction. The base of $K_{\vec{e}}$ is identified with the base of $K_{\sigma_1(\vec{e})}$. Next we paste the right-hand edge of $K_{\vec{e}}$ with the left-hand edge of $K_{\sigma_\infty(\vec{e})}$. All vertices opposite to the bases of all triangles in the same boundary cycle are identified. Such points are in one to one correspondence with the orbits of σ_∞; which is why we call them cusps. Gluing triangles is done in a way compatible with the orientation of each triangle, and thus, this punctured surface is triangularized. Let $v' = |V(\Gamma)|, e' = |E(\Gamma)|$ and n' denote the number of vertices, edges and cusps, respectively. It is easy to shows that there are $2e'$ faces, $3e'$ edges and $v' + n'$ points. The surface $\mathrm{Surf}(\Gamma)$ has genus $g = (2 - v' + e' - n')/2$. This surface comes with a natural orientation given by the tiles since they are naturally oriented and their orientations match each other

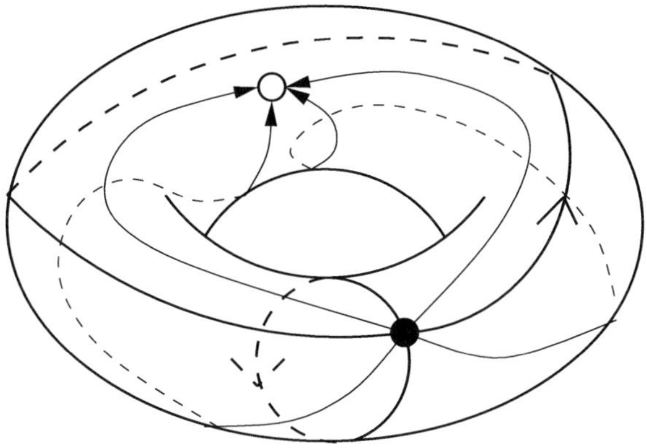

FIGURE 3.6. Surface obtained from a metric twisted figure eight.

because of the way we have glued them. Figure 3.6 shows the surface obtained from a metric twisted figure eight. The meridian and longitude joint at the dark point in the front of the torus and form the metric twisted figure eight. There are four cones based at the ribbon graph, their walls run from the dark point in the front of the torus to the hollow point at the back.

4. Combinatorial moduli space

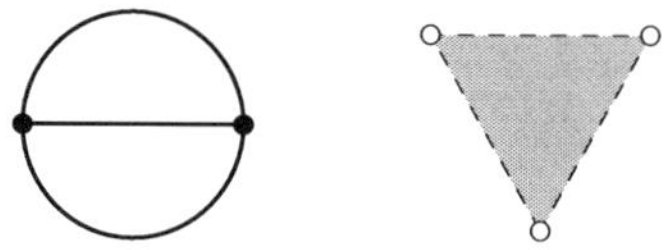

FIGURE 4.1. Rational cell of the theta graph.

DEFINITION 4.1. For a given ribbon graph Γ, its associated *rational cell* is the quotient space $int\,(\Delta_{E(\Gamma)})/\operatorname{Aut}_{\partial}^{rg}(\Gamma)$ where $int\,(\Delta_{E(\Gamma)})$ is the space of unital metrics of Γ and the action is by given by permutation of edges.

Figure 4.1 shows the theta graph and its corresponding rational cell. The automorphism group is trivial in this case.

The contraction of an edge that is not a cycle results in a new ribbon graph (with induced cyclic orders on each vertex) of the same topological type. This process is called *edge collapse.* The process can be generalized to the collapse of any disjoint union of subtrees. The edge collapse can be used to glue rational cells. For example, the collapse of one edge of the theta graph or the middle edge in the double noose results in the figure eight graph. This induces a gluing of a higher dimensional rational cell to a lower dimensional one. The orbispace of metric ribbon graphs of a fixed topological type (g, n) up to isomorphism is the combinatorial moduli space $\mathcal{M}_{g,n}^{comb}$ obtained by gluing all rational cells of type (g, n). The orbicell structure is studied in detail in [**MP98**].

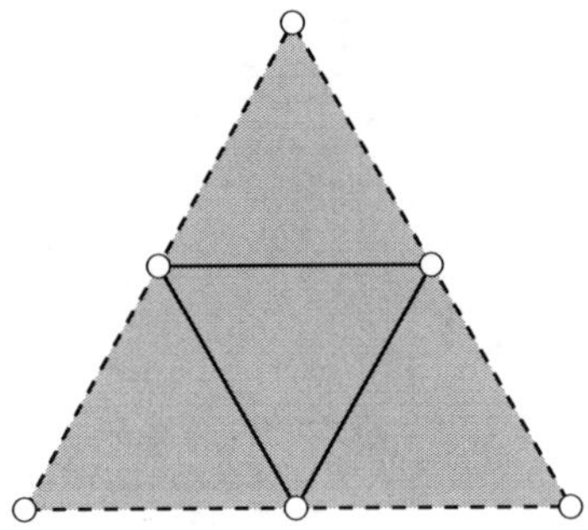

FIGURE 4.2. The orbispace $\mathcal{M}_{0,3}^{comb}$.

Figure 4.2 shows the combinatorial moduli space $\mathcal{M}_{0,3}^{comb}$. The 2-dimensional rational cell in the middle corresponds with the theta graph. The other three 2-dimensional rational cells correspond with the three nonisomorphic ways to label the boundary cycles of the double noose graph. The three one-dimensional rational cells correspond with the three nonisomorphic ways to label the boundary cycles of the figure eight graph.

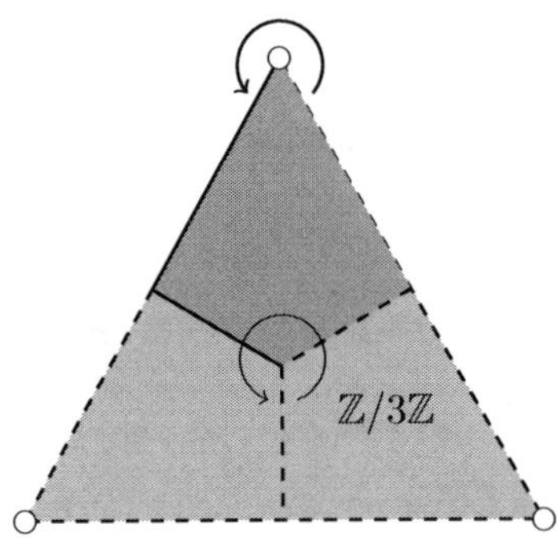

FIGURE 4.3. The orbispace $\mathcal{M}_{1,1}^{comb}$.

Figure 4.3 shows the combinatorial moduli space $\mathcal{M}_{1,1}^{comb}$. There is only one 2-dimensional rational cell corresponding with the twisted theta graph and depicted as the whole triangle. The automorphism group of the twisted theta graph is $\mathbb{Z}/3\mathbb{Z}$. The action of the automorphism group on the rational cell is by rotations, thus the shaded region represents the fundamental domain. This actions glues the lower left edge of the fundamental domain to the lower right edge. The top half of the left edge of the fundamental domain corresponds with the twisted figure eight. The automorphism group of the twisted figure eight is $\mathbb{Z}/2\mathbb{Z}$ and its action glues the top half of the left edge to top half of the right edge. As a result, $\mathcal{M}_{1,1}^{comb}$ is homeomorphic to a sphere minus a point.

5. Combinatorial blow-up

Given a ribbon graph Γ, define $A \subset 2^{E(\Gamma)} \setminus \{\varnothing\}$ as the family of subsets of edges corresponding with subgraphs of Γ that are not a union of trees. Let B be the family defined by taking the complement of every element in A.

DEFINITION 5.1. Since the action of $\mathrm{Aut}_\partial^{rg}(\Gamma)$ can be extended to the boundary of N_B, we define the *compact rational cell* associated to Γ as $N_B/\mathrm{Aut}_\partial^{rg}(\Gamma)$. Gluing compact rational cells along the edge collapse of the union of trees, we obtain the compact combinatorial moduli space

$$\underline{\mathcal{M}}_{g,n}^{comb} = \coprod_{\Gamma \text{ of type } (g,n)} N_B/\mathrm{Aut}_\partial^{rg}(\Gamma).$$

The subspace $\underline{\mathcal{M}}_{g,n}^{comb} \setminus \mathcal{M}_{g,n}^{comb}$ is formed by new cells created by the combinatorial blow-up. It is possible to identified these new cells by combinatorial objects that are generalizations of metric ribbon graphs. These new graphs are in fact more general than the semistable ribbon graphs defined in [**Zúñ15**], but we will not try to define them here in full generality. This will be addressed in future work, here we will be content with a description in terms of truncations given by chains of inclusions.

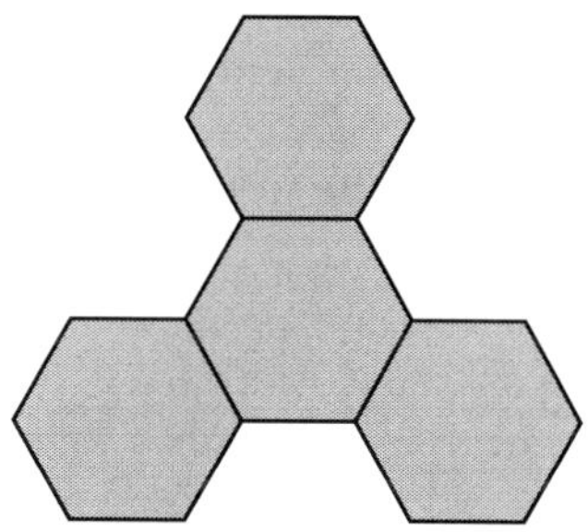

FIGURE 5.1. The orbispace $\underline{\mathcal{M}}_{0,3}^{comb}$.

Figure 5.1 shows $\underline{\mathcal{M}}_{0,3}^{comb}$. In this case all vertices are truncated because the only subgraphs that are unions of trees are single edges. In this case all automorphism groups are trivial.

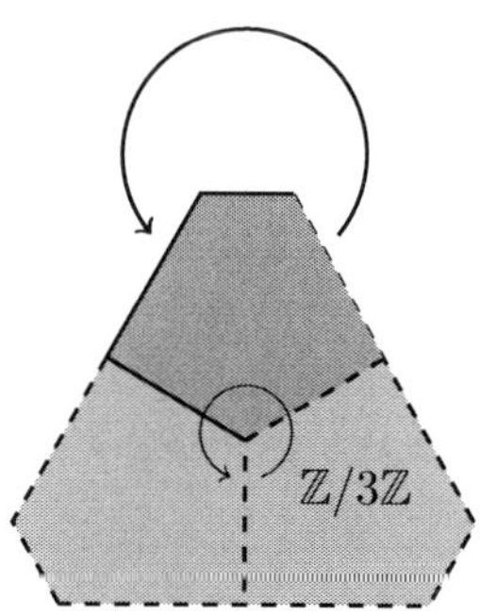

FIGURE 5.2. The orbispace $\underline{\mathcal{M}}_{1,1}^{comb}$.

The space $\underline{\mathcal{M}}_{1,1}^{comb}$ can be seen Figure 5.2. The vertex of the single 2-dimensional compact rational cell is truncated. The gluings due to the group actions of $\mathbb{Z}/3\mathbb{Z}$ and $\mathbb{Z}/2\mathbb{Z}$ work as before, but in this case a boundary isomorphic to a circle is created due to the truncation. The resulting compact moduli space is homeomorphic to a surface minus an open disc.

6. Quadratic differentials

DEFINITION 6.1. A *meromorphic quadratic differential* on a Riemann surface Σ is a meromorphic section of $(T^*\Sigma)^{\odot 2}$, the second symmetric power of the cotangent bundle.

The notions of zero and order of a zero do not depend on the local representation of a quadratic differential. In the same way, the notion of pole and order of a pole are invariant by change of coordinates. Zeros and poles are called *critical points.* A pole of order two of a quadratic differential is called a *double pole* and a pole of order one is called a *simple pole.* Given a representation in local coordinates $f(z)dz^2$ around a double pole q, we can express f as

$$f(z) = \frac{a_{-2}}{z^2} + \frac{a_{-1}}{z} + a_0 + \cdots$$

and call the term a_{-2} its *quadratic residue.* It can be shown that this number does not depend on the choice of local coordinates.

Quadratic differentials define certain curves on the Riemann surface. If $q = f(z)dz^2$ is a meromorphic quadratic differential, then the parametric curve $\vec{r} : (a,b) \to C$ is called a *horizontal trajectory* or *leaf* of q if

$$f(\vec{r}(t)) \left(\frac{d\vec{r}(t)}{dt}\right)^2 > 0$$

and a *vertical trajectory* if

$$f(\vec{r}(t)) \left(\frac{d\vec{r}(t)}{dt}\right)^2 < 0.$$

DEFINITION 6.2. A *Strebel–Jenkins differential* is a meromorphic quadratic differential with only double poles with negative quadratic residues.

Strebel–Jenkins differentials have only two kinds of leaves: closed ones (surrounding a double pole) and critical ones (connecting zeroes). The union of critical leaves and zeroes forms the *critical graph.* The vertical trajectories connect the double poles to the critical graph and are orthogonal to the closed leaves under the metric induced by $\sqrt{q}$. The following existence and uniqueness theorem follows from the work of Jenkins and Strebel (see [**Str84**] and [**Loo95**, Theorem 7.6]).

THEOREM 6.3 (Strebel). *Given a Riemann Surface of genus g with n marked points and decorations in $int(\Delta^{n-1})$, there exists a unique quadratic differential with the following properties. It is holomorphic in the puncture surface. The union of closed leaves form semi-infinite cylinders around the marked points. The quadratic residues coincide with the decorations.*

7. Moduli spaces and decorations

Let $\mathcal{M}_{g,n}$ denote the moduli space of Riemann surfaces of genus g and n marked points. The topological type of these surfaces is defined as (g,n). The interior of a Riemann surface of type (g,n) is the result of taking away the marked points, thus obtaining a *punctured* surface.

Following [**Liu20**], let $\mathbb{R}\mathcal{M}_{(g,b,k),(n,m)}$ denote the moduli of smooth symmetric compact Riemann surfaces with $2n+m$ marked points where a symmetric surface is a pair (Σ,σ) of a surface Σ and σ is an antiholomorphic involution. The index

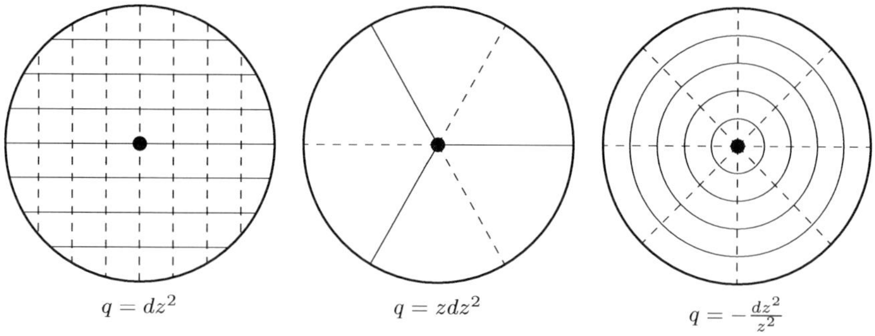

FIGURE 6.1. Different behaviors of Strebel–Jenkins differentials. The solid lines represent horizontal trajectories and the dotted one vertical trajectories.

g fixes the genus of Σ and b is the number of components of the fixed points of σ that is, $\Sigma^\sigma \cong \sqcup_b \mathbb{S}^1$. The index k fixes the orientability, it is 0 if $\Sigma/\langle\sigma\rangle$ is orientable and it is 1 if $\Sigma/\langle\sigma\rangle$ is not orientable. The symmetry interchanges n pairs of marked points away from Σ^σ and fixes m marked points in Σ^σ.

Denote by $\mathbb{R}\mathcal{M}_{g,n,m}$ the moduli space of symmetric surfaces of genus g with (n, m) marked points. This space is a disjoint union of the spaces $\mathbb{R}\mathcal{M}_{(g,b,k),(n,m)}$ ranging over all the different topological types depending on b and k. Forgetting the symmetry induces a map $f : \mathbb{R}\mathcal{M}_{g,n,m} \to \mathcal{M}_{g,2n+m}$ that is generically injective, but it fails to be injective when the automorphism group of the marked surface Σ is bigger than the automorphism group of the marked symmetric surface (Σ, σ).

Following [**Liu20**] again, let $\mathcal{M}_{g,b,n,\vec{m}}$ be the moduli space of bordered Riemann surfaces of genus g with b boundary components isomorphic to $\mathbb{S}^1$, n marked points away from the boundary and $\vec{m}$ marked points on the boundary. Here the boundary components are also marked.

There is an alternative theory in terms of hyperbolic geometry which leads to the same combinatorics, cf. [**Pen04, KP06**].

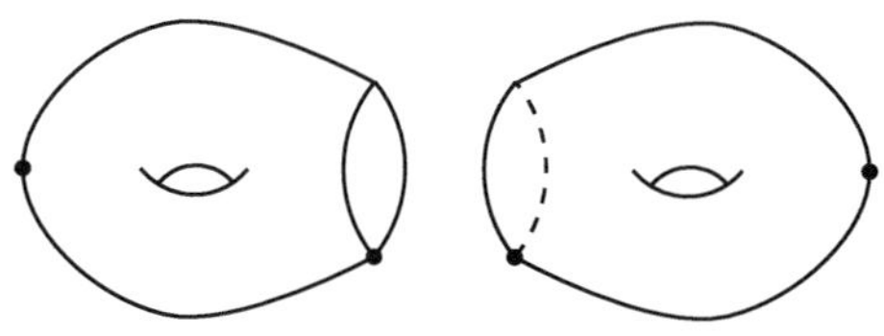

FIGURE 7.1. The double of a surface of type $(1, 1, 1, 1)$.

The *complex double* or simply *double* of a surface of topological type $(g, b, n, \vec{m})$ is a surface of topological type $(2g + b - 1, 2n + m)$. The doubled surface has a complex structure and a real involution σ whose fixed points lie in the boundary of the bordered surface as in Figure 7.1. By definition, the Euler characteristic of the bordered surface is

$$\chi = 2 - 2g - b - n - m/2,$$

this is half the Euler characteristic of the doubled surface. A doubled surface is in fact symmetric. This observations induces a finite to one map $d : \mathcal{M}_{g,b,n,\vec{m}} \to \mathbb{R}\mathcal{M}_{2g+b-1,n,m}$.

By considering stable Riemann surfaces with possible nodal singularities one can define the Deligne–Mumford compactifications of these spaces ([**Abi80**], [**Sep91**], [**Sil92**], [**Liu20**]). Moreover, the composition of the double and forgetful map

$$\overline{\mathcal{M}}_{g,b,n,\vec{m}} \xrightarrow{d} \overline{\mathbb{R}\mathcal{M}}_{(2g+b-1,b,0),(n,m)} \xrightarrow{f} \overline{\mathcal{M}}_{2g+b-1,2n+m}$$

has similar characteristics as before, in particular the double map is generically two to one.

The KSV compactification $\underline{\mathcal{M}}_{g,n}$ is obtained by decorating the nodes of a stable surface with relative phase parameters [**KSV95**]. Geometrically, this is the same as a real tangent direction: a ray in the complex tensor product of the tangent spaces on each side of a node.

The decorated moduli space of Riemann surfaces of genus g and n marked points is by definition

$$\mathcal{M}_{g,n}^{dec} = \mathcal{M}_{g,n} \times int(\Delta^{n-1}).$$

Each marked point is decorated by a positive real number so that the sum of decorations is one. In [**MP98**, Section 5], a homeomorphism $\mathcal{M}_{g,n} \times \mathbb{R}_+^n \to RGB_{g,n}^{met}$ is described. By restricting the domain to the standard simplex, we obtain the map $\varphi : \mathcal{M}_{g,n}^{dec} \to \mathcal{M}_{g,n}^{comb}$.

This map is realized by Strebel's theory of quadratic differentials. Given a Riemann surface Σ with decorations, Strebel's theorem produces a Strebel–Jenkins differential whose critical graph is a metric ribbon graph Γ. For the inverse, given a metric ribbon graph Γ, it is possible to give complex charts to Surf(Γ) to obtain a Riemann surface Σ. The decorations of each marked point is the length of the corresponding boundary subgraph.

Each rational cell of the moduli space $\mathcal{M}_{0,3}^{comb}$ depicted in Figure 4.2 can be represented by the critical graph of a Strebel–Jenkins differential on the Riemann surface with three marked points. This is shown in Figure 7.2 and illustrates how the map φ works. This last figure also illustrates how the metric ribbon graphs deform as we move around the moduli space.

Let us define $\underline{\varphi} : \mathcal{M}_{g,n}^{dec} \to \underline{\mathcal{M}}_{g,n}^{comb}$ by simply changing the codomain of φ and notice that the closure of the image of this new map is $\underline{\mathcal{M}}_{g,n}^{comb}$. The seven compact rational cells numbered in green in Figure 7.3 can be represented via $\underline{\varphi}$ by the seven critical graphs numbered in Figure 7.4. This also illustrates the deformations of metric ribbon graphs. The critical trajectories of the sufaces labeled as 4, 5, 6, 7 and 8 correspond with generalizations of the concept of metric ribbon graph. For example, let $a, b \in E(\Gamma)$ be the loops of the double noose in 3. As these loops contract, that is, their lengths $l(a)$ and $l(b)$ go to zero, the ribbon graph degenerates to an interval (no longer a proper metric ribbon graph). However, due to the blow-up, those two lengths are not lost: they form a one dimensional family represented by the two smaller surfaces with two marked points that are *attached* to the degenerated ribbon graph. The one parameter famliy is $\Delta_{a,b}$.

For a bordered Riemann surface of type $(g, b, n, \vec{m})$, let $c_i \geq 0$ with $1 \leq i \leq n$ denote decorations for the marked points in the interior of the surface and let $o_j \geq 0$ with $1 \leq j \leq m$ denote decorations for the marked points in the boundary of the

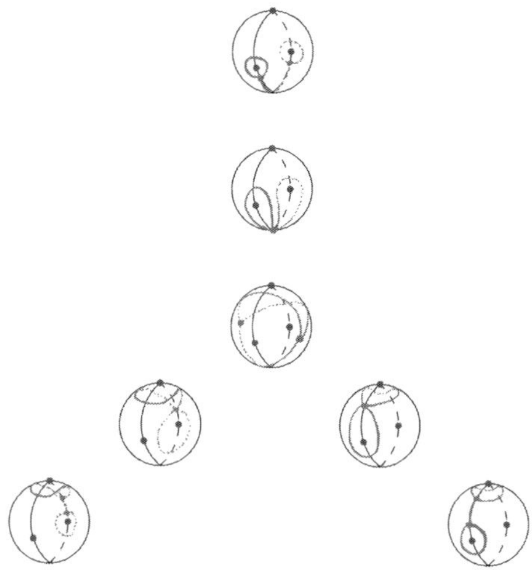

FIGURE 7.2. Critical graphs on a surface of type $(0,3)$ corresponding with the seven rational cells of $\mathcal{M}_{0,3}^{comb}$. The critical graphs appear in a lighter shade of grey on the surface of Riemann spheres with three marked points.

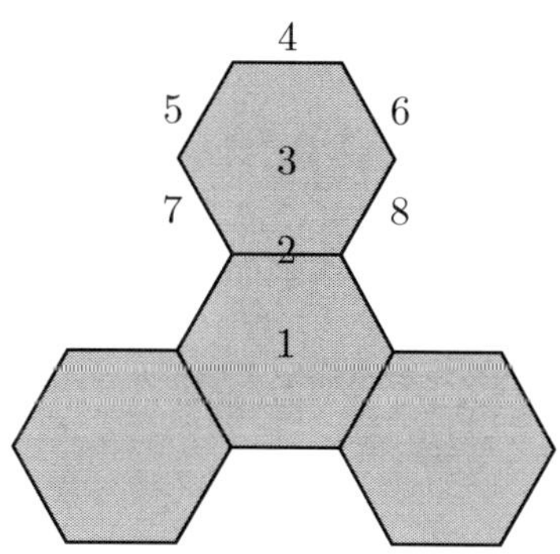

FIGURE 7.3. Seven compact rational cells in $\underline{\mathcal{M}}_{0,3}^{comb}$ labeled by numbers.

surface. Suppose that $2\sum c_i + \sum o_j = 1$. All these decorations can be considered in the interior of the $(n+m-1)$-dimensional simplex $int(\Delta^{n+m-1})$.

DEFINITION 7.1. The decorated moduli space of bordered surfaces of type $(g,b,n,\vec{m})$ is defined as

$$\mathcal{M}_{g,b,n,\vec{m}}^{dec} = \mathcal{M}_{g,b,n,\vec{m}} \times int(\Delta^{n+m-1}).$$

The decorated moduli space of symmetric surfaces of type (g,b,k) and (n,m) marked points is defined as

$$\mathbb{R}\mathcal{M}_{(g,b,k),(n,m)}^{dec} = \mathbb{R}\mathcal{M}_{(g,b,k),(n,m)} \times int(\Delta^{n+m-1})$$

where we require that the decorations of conjugate marked points be equal.

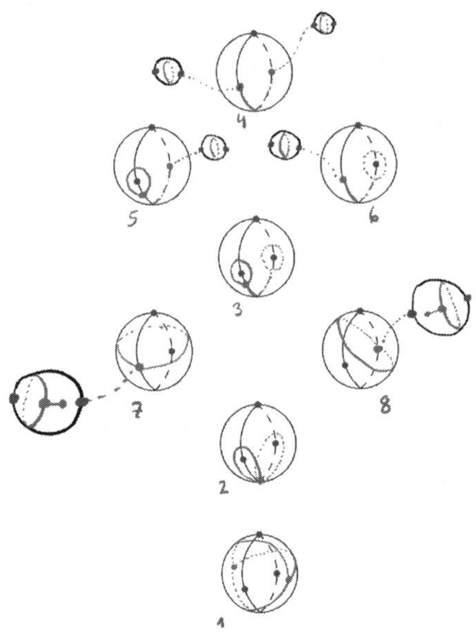

FIGURE 7.4. The corresponding seven critical graphs appear inside Riemann spheres with three marked points in a lighter shade of grey.

By doubling the decorations of the n interior marked points, we obtain a composition

$$\mathcal{M}^{dec}_{g,b,n,\vec{m}} \xrightarrow{d} \mathbb{R}\mathcal{M}^{dec}_{(2g+b-1,b,0),(n,m)} \xrightarrow{f} \mathcal{M}^{dec}_{2g+b-1,2n+m}$$

with similar characteristics as before.

DEFINITION 7.2. Let $\mathcal{S}^{dec}_{g,b,n,\vec{m}} = f \circ d(\mathcal{M}^{dec}_{g,b,n,\vec{m}})$. The moduli space of symmetric metric ribbon graphs is defined by

$$\mathcal{S}^{comb}_{g,b,n,\vec{m}} = \varphi(\mathcal{S}^{dec}_{g,b,n,\vec{m}})$$

and

$$\underline{\mathcal{S}}^{comb}_{g,b,n,\vec{m}} = cl(\underline{\varphi}(\mathcal{S}^{dec}_{g,b,n,\vec{m}}))$$

is the moduli space of stable symmetric metric ribbon graphs.

THEOREM 7.3. *The combinatorial moduli space $\underline{\mathcal{S}}^{comb}_{g,b,n,\vec{m}}$ is a compact Hausdorff space and it is a compactification of the decorated moduli space $\mathcal{S}^{dec}_{g,b,n,\vec{m}}$.*

PROOF. The combinatorial moduli space $\underline{\mathcal{S}}^{comb}_{2g+b-1,2n+m}$ is a compact, Hausdorff space since it is a finite union of compact rational cells. This implies that $\underline{\mathcal{S}}^{comb}_{g,b,n,\vec{m}}$ is a compact Hausdorff space. Moreover, since

$$\underline{\mathcal{S}}^{comb}_{2g+b-1,2n+m} \setminus \mathcal{S}^{comb}_{2g+b-1,2n+m}$$

is made of lower dimensional rational faces and $\underline{\varphi}$ is a homeomorphism, then the decorated space $\mathcal{S}^{dec}_{g,b,n,\vec{m}}$ compactifies to $\underline{\mathcal{S}}^{comb}_{g,b,n,\vec{m}}$ □

The real dimension of this moduli space is $6g + 3b + 3n + 2m - 7$. This follow by counting dimensions of the bordered moduli space and the simplex defining the decorations. The lowest dimensional examples with boundary and negative Euler characteristic are shown in Table 7.1.

TABLE 7.1. Euler characteristic of a bordered surfaces of the corresponding topological type and dimension of the corresponding combinatorial moduli space.

g	b	n	m	χ	dim
0	1	0	3	-1/2	2
0	1	1	1	-1/2	1
0	1	1	2	-1	3
0	1	2	0	-1	2
0	2	0	1	-1/2	1
0	2	0	2	-1	3
0	2	1	0	-1	2

It would be interesting to define $\underline{\mathcal{S}}^{comb}_{g,b,n,\vec{m}}$ in a purely combinatorial way. That would involve a combinatorial notion of bordered ribbon graphs and their semistable versions.

8. A few examples

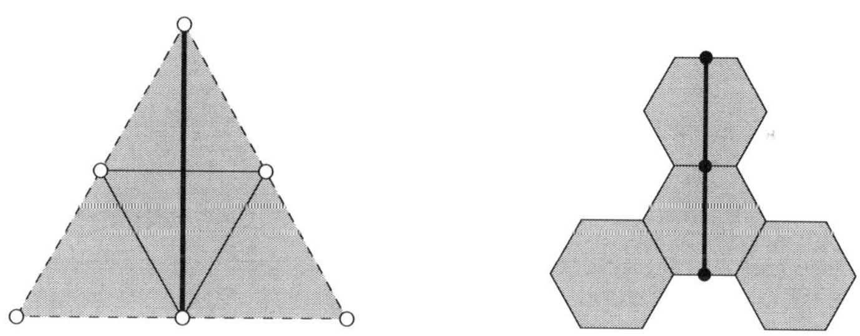

FIGURE 8.1. The spaces $\mathcal{S}^{comb}_{0,1,1,1}$ and $\underline{\mathcal{S}}^{comb}_{0,1,1,1}$ are depicted in the middle with thicker lines.

The moduli space $\mathcal{S}^{comb}_{0,1,1,1}$ sits inside $\mathcal{M}^{comb}_{0,3}$ and must be one dimensional. Considering symmetric versions of the theta graph, figure eight and double noose, Figure 8.1 shows on the left $\mathcal{S}^{comb}_{0,1,1,1}$ represented by the thicker subspace in the middle of the triangle. On the right of Figure 8.1 the space $\underline{\mathcal{S}}^{comb}_{0,1,1,1}$ is depicted as the thicker subspace of $\underline{\mathcal{M}}^{comb}_{0,3}$. The compact combinatorial moduli space consists of three 0-dimensional cells and two 1-dimensional cells.

Figure 8.2 shows the five critical graphs corresponding to the five compact rational cells of the space $\underline{\mathcal{S}}^{dec}_{0,1,1,1}$.

The space $\underline{\mathcal{S}}^{comb}_{0,2,0,1}$ is one dimensional and sits inside $\underline{\mathcal{M}}^{comb}_{1,1}$, this is depicted by the thicker line in Figure 8.3. The moduli space $\underline{\mathcal{M}}^{comb}_{0,2,0,1}$ consists of two 0-dimensional compact rational cells and one 1-dimensional compact rational cell. A

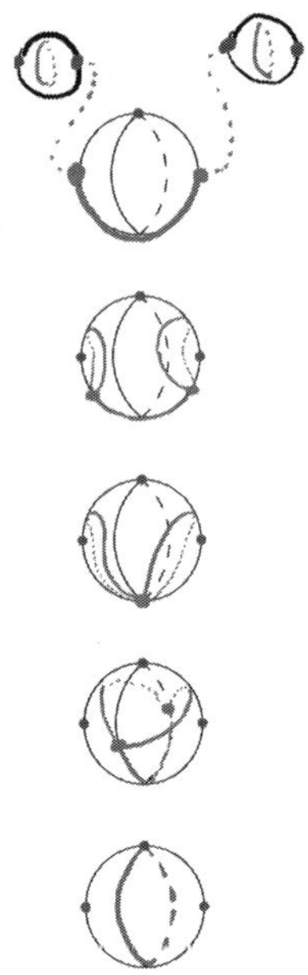

FIGURE 8.2. Critical graphs corresponding with the compact rational cells of $\underline{\mathcal{S}}^{comb}_{0,1,1,1}$. The critical graphs appear inside the Riemann with three marked points spheres in a lighter shade of grey.

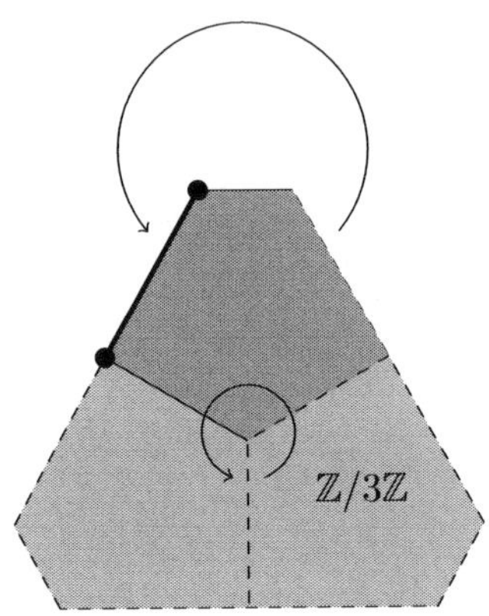

FIGURE 8.3. The space $\underline{\mathcal{S}}^{comb}_{0,2,0,1}$ is the thicker edge with two univalent vertices.

typical symmetric critical graph in $\underline{\mathcal{M}}^{comb}_{1,1}$ describing a point in $\underline{\mathcal{S}}^{comb}_{0,2,0,1}$ is shown in Figure 8.4.

Figure 8.5 shows a theta graph as a symmetric critical trajectory in $\underline{\mathcal{M}}^{comb}_{0,3}$. In fact, all graphs of type $(0,3)$ can be placed in the double of the disc with three marked points on the boundary as symmetric graphs. This implies that $\underline{\mathcal{M}}^{comb}_{0,1,0,3} \cong \underline{\mathcal{M}}^{comb}_{0,3}$ whose picture can be found in Figure 5.1.

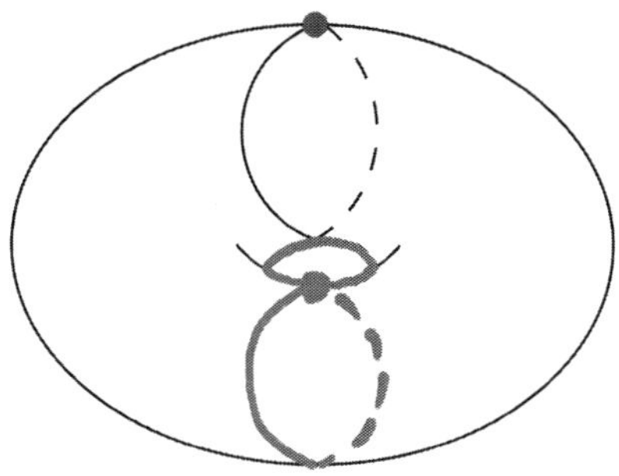

FIGURE 8.4. Critical graph in $\underline{\mathcal{S}}^{comb}_{0,2,0,1}$. The critical graph is the twisted figure eight on the bottom-center of the torus.

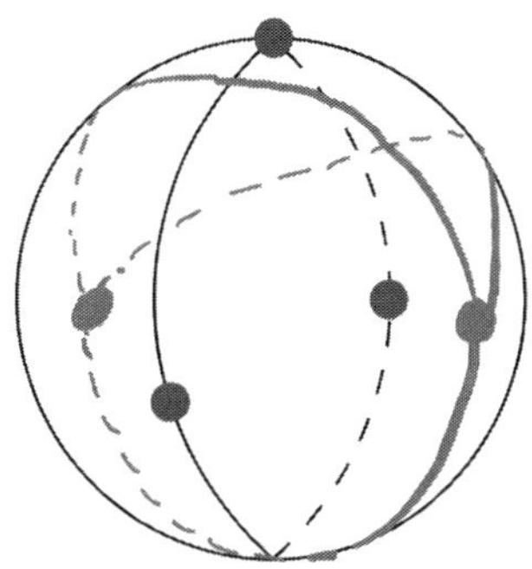

FIGURE 8.5. Critical graph representing a point in $\underline{\mathcal{S}}^{comb}_{0,1,0,3}$. The critical graph is the theta graph inside the sphere represented with a lighter shade of grey.

$\underline{\mathcal{S}}^{comb}_{0,2,1,0}$ consists of only one 2-dimensional cell as shown in Figure 8.6. That figure also shows the double of annulus with one marked interior point, that is, the symmetric torus with two marked points. In the torus we see the critical trajectory right in the middle which is a ribbon graph of type $(1,2)$.

The space $\underline{\mathcal{S}}^{comb}_{0,1,2,0}$ is formed by three 2-dimensional cells as depicted in Figure 8.7. In that figure the 2-dimensional cells are labeled together with two additional 1-dimensional cells. They correspond with the critical graphs in Figure 8.8. The double of the disc with two interior marked points is the symmetric Riemann sphere with four marked points.

9. Relations to known spaces

Ideally, we would like to construct orbicell decompositions for the moduli spaces $\mathbb{R}\mathcal{M}^{dec}_{(2g+b-1,b,0),(n,m)}$. However since the forgetful map f is not injective in general, it is not possible to do so in a straightforward manner. For $g=0$ and $b=1$, the forgetful map is injective since the moduli spaces for Riemann spheres are fine (see

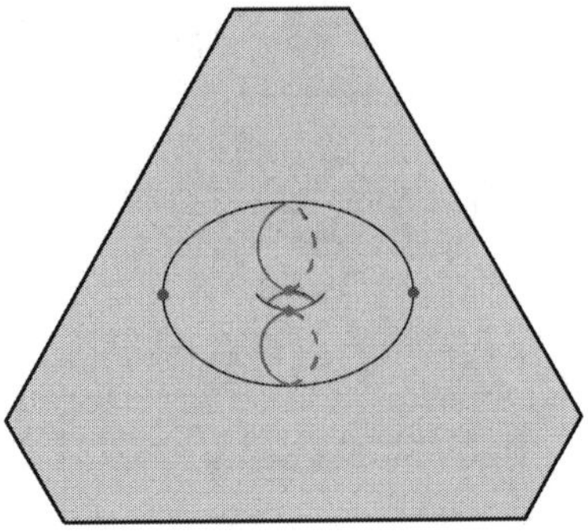

FIGURE 8.6. The space $\underline{\mathcal{S}}^{comb}_{0,2,1,0}$ and the critical graph representing the interior. The critical graph is represented by three loops on the middle of the torus in a lighter shade of grey.

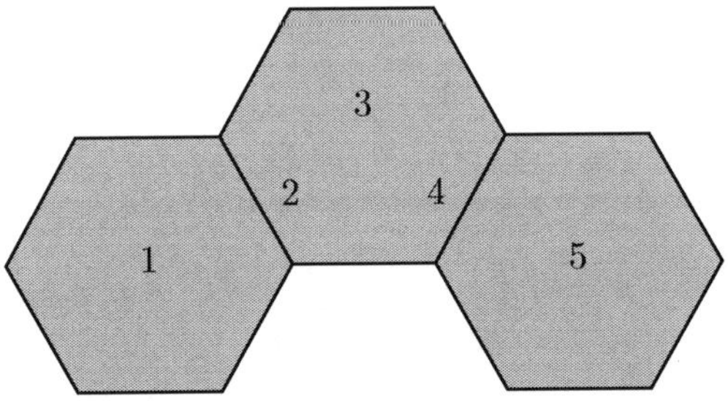

FIGURE 8.7. The space $\underline{\mathcal{S}}^{comb}_{0,1,2,0}$ The numbers label different racional cells.

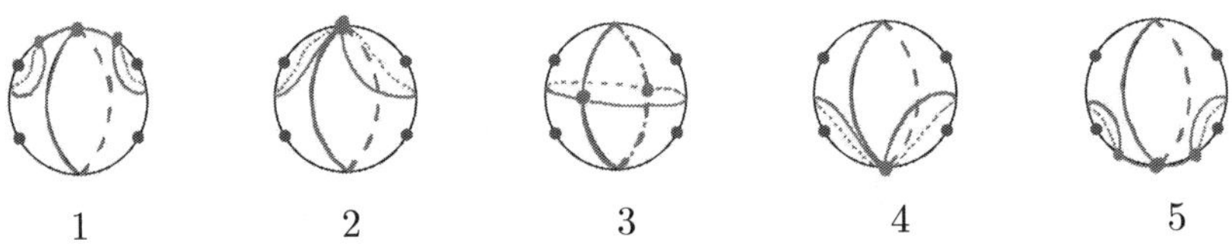

FIGURE 8.8. Critical trajectories in a symmetric sphere with four marked points. The numbers correspond with the rational cells in Figure 8.7.

[**Cey06**]). The genus one examples in the previous section are also decompositions for the corresponding real moduli spaces because the real and complex automorphism group of an elliptic curve are equal, thus the injectivity of the forgetful map follows (see [**Sil92**]).

A possible way forward would be to study the difference between complex and real automorphisms of a real curve. As suggested by the referee, one can consider the cone of the morphism

$$\Psi_* : C_\bullet(\mathcal{M}_{g,b,n,\vec{m}}, \mathbb{Q}) \to C_\bullet(\mathcal{M}_{2g+b-1,2n+m}, \mathbb{Q}) \cong C_\bullet(\mathcal{M}^{comb}_{2g+b-1,2n+m}, \mathbb{Q})$$

The complex on the lefthand side admits a combinatorial model in terms of BW ribbon graphs. The complex on the right admits a model in terms of classical ribbon graphs. The corresponding cone can be considered as the combinatorial tool to detect the differences between complex and real isomorphisms of the double of a bordered surface.

There is also a connection between $\underline{S}^{comb}_{g,b,n,\vec{m}}$ and $\underline{\mathcal{M}}_{2g+b-1,2n+m}$. It should be possible to construct a nonsurjective map

$$\pi : \underline{S}^{comb}_{g,b,n,\vec{m}} \to \underline{\mathcal{M}}_{2g+b-1,2n+m}$$

which in the interior corresponds with φ^{-1} and forgetting the decorations by real numbers. On the boundary, the construction of the stable surface can be achieved by considering sequences of nested subgraphs as in [**Zúñ15**]. The decorations by real tangent directions at the nodes can be obtained by remembering how the subgraphs degenerate. The exact nature of this map and its fibers needs further investigation.

In [**Cos07**] a combinatorial model $D_{g,b,m,n}$ is introduced formed by BW graphs as defined in [**WW16**]. This complex is weakly homotopic to a partial compactification of the moduli space $\mathcal{M}_{g,b,n,\vec{m}}$. It is interesting to note a relation between BW graphs and the graphs presented here. The first three examples in the previous section have marked points in the boundary and one can obtain a BW graph choosing half of one of the critical trajectories.

FIGURE 9.1. BW graphs in $D_{0,1,1,1}$, $D_{0,2,1,0}$, and $D_{0,1,3,0}$, respectively.

At the top of Figure 8.2 there is a graph consisting of an edge with two vertices and two extra circles. One half of the single edge (say the left half) gives the first BW graph in Figure 9.1 with only one white vertex corresponding with the interior marked point and one half-edge corresponding to the boundary marked point. Notice that this graph only has one boundary cycle since $b = 1$. Figure 8.4 shows the twisted figure eight as a critical graph. One half of this graph (the left half) gives the second BW graph on Figure 9.1 with two boundary cycles since $b = 2$, one black trivalent vertex, and one half-edge since $m = 1$. There are no white vertices since $n = 0$. Finally, Figure 8.5 shows the theta graph as a critical graph. Again, one half of this graph (the left half again) gives the third BW graph in Figure 9.1 with only one boundary cycle since $b = 1$, three half-edges since $m = 1$, and no white vertex since $n = 0$. For the last two examples $m = 0$ and the same idea can not be applied. However, for the critical trajectory in Figure 8.6, after contracting the middle loop, there is a dual graph in the left half of the torus: a loop with one edge and one white vertex at the interior marked points. This BW graph has two boundary cycles since $b = 2$. For the critical graph marked as "3" in Figure 8.8, after contracting the middle horizontal loop, there is a dual BW graph in the left half with two white vertices at the interior marked points joined by a single edge. This seems to indicate a way to extract BW graphs in $D_{g,b,m,n}$ from $\underline{S}^{comb}_{g,b,n,\vec{m}}$ in a similar way in which one chooses a bordered Riemann surface out of a symmetric one.

Acknowledgment

We would like to thank the referee for pointing out several shortcomings of the initial version of this article and also for suggesting the way for improving it. His contributions made this work more precise, organized and readable.

References

[Abi80] William Abikoff, *The real analytic theory of Teichmüller space*, Lecture Notes in Mathematics, vol. 820, Springer, Berlin, 1980. MR590044

[BE88] B. H. Bowditch and D. B. A. Epstein, *Natural triangulations associated to a surface*, Topology **27** (1988), no. 1, 91–117, DOI 10.1016/0040-9383(88)90008-0. MR935529

[Cey06] Özgür Ceyhan, *On moduli of pointed real curves of genus zero*, Proceedings of Gökova Geometry-Topology Conference 2006, Gökova Geometry/Topology Conference (GGT), Gökova, 2007, pp. 1–38. MR2404948

[COI19] Pierre-Louis Curien, Jovana Obradović, and Jelena Ivanović, *Syntactic aspects of hypergraph polytopes*, J. Homotopy Relat. Struct. **14** (2019), no. 1, 235–279, DOI 10.1007/s40062-018-0211-9. MR3913975

[Cos07] Kevin Costello, *A dual version of the ribbon graph decomposition of moduli space*, Geom. Topol. **11** (2007), 1637–1652, DOI 10.2140/gt.2007.11.1637. MR2350462

[Ega15] Daniela Egas, *Comparing fat graph models of moduli space*, 2015.

[God07] Véronique Godin, *The unstable integral homology of the mapping class groups of a surface with boundary*, Math. Ann. **337** (2007), no. 1, 15–60, DOI 10.1007/s00208-006-0025-7. MR2262776

[Har86] John L. Harer, *The virtual cohomological dimension of the mapping class group of an orientable surface*, Invent. Math. **84** (1986), no. 1, 157–176, DOI 10.1007/BF01388737. MR830043

[Kau10] Ralph M. Kaufmann, *Open/closed string topology and moduli space actions via open/closed Hochschild actions*, SIGMA Symmetry Integrability Geom. Methods Appl. **6** (2010), Paper 036, 33, DOI 10.3842/SIGMA.2010.036. MR2647315

[Kon92] Maxim Kontsevich, *Intersection theory on the moduli space of curves and the matrix Airy function*, Comm. Math. Phys. **147** (1992), no. 1, 1–23. MR1171758

[KP06] Ralph M. Kaufmann and R. C. Penner, *Closed/open string diagrammatics*, Nuclear Phys. B **748** (2006), no. 3, 335–379, DOI 10.1016/j.nuclphysb.2006.03.036. MR2242677

[KSV95] Takashi Kimura, Jim Stasheff, and Alexander A. Voronov, *On operad structures of moduli spaces and string theory*, Comm. Math. Phys. **171** (1995), no. 1, 1–25. MR1341693

[Liu20] C.-C. Melissa Liu, *Moduli of J-holomorphic curves with Lagrangian boundary conditions and open Gromov-Witten invariants for an S^1-equivariant pair*, J. Iran. Math. Soc. **1** (2020), no. 1, 5–95. MR4528678

[Loo95] Eduard Looijenga, *Cellular decompositions of compactified moduli spaces of pointed curves*, The moduli space of curves (Texel Island, 1994), Progr. Math., vol. 129, Birkhäuser Boston, Boston, MA, 1995, pp. 369–400, DOI 10.1007/978-1-4612-4264-2_13. MR1363063

[MP98] M. Mulase and M. Penkava, *Ribbon graphs, quadratic differentials on Riemann surfaces, and algebraic curves defined over $\overline{\mathbf{Q}}$*, Asian J. Math. **2** (1998), no. 4, 875–919, DOI 10.4310/AJM.1998.v2.n4.a11. Mikio Sato: a great Japanese mathematician of the twentieth century. MR1734132

[Pen87] R. C. Penner, *The decorated Teichmüller space of punctured surfaces*, Comm. Math. Phys. **113** (1987), no. 2, 299–339. MR919235

[Pen04] R. C. Penner, *Decorated Teichmüller theory of bordered surfaces*, Comm. Anal. Geom. **12** (2004), no. 4, 793–820. MR2104076

[Sep91] M. Seppälä, *Moduli spaces of stable real algebraic curves*, Ann. Sci. École Norm. Sup. (4) **24** (1991), no. 5, 519–544. MR1132756

[Sil92] R. Silhol, *Compactifications of moduli spaces in real algebraic geometry*, Invent. Math. **107** (1992), no. 1, 151–202, DOI 10.1007/BF01231886. MR1135469

[Str84] Kurt Strebel, *Quadratic differentials*, Ergebnisse der Mathematik und ihrer Grenzgebiete (3) [Results in Mathematics and Related Areas (3)], vol. 5, Springer-Verlag, Berlin, 1984, DOI 10.1007/978-3-662-02414-0. MR743423

[WW16] Nathalie Wahl and Craig Westerland, *Hochschild homology of structured algebras*, Adv. Math. **288** (2016), 240–307, DOI 10.1016/j.aim.2015.10.017. MR3436386

[Zúñ15] Javier Zúñiga, *Compactifications of moduli spaces and cellular decompositions*, Algebr. Geom. Topol. **15** (2015), no. 1, 1–41, DOI 10.2140/agt.2015.15.1. MR3325730

Purdue University, Department of Mathematics, 150 N. University Street West Lafayette, Indiana 47907-2067

Email address: `rkaufman@math.purdue.edu`

Departmento de Economía, Universidad del Pacífico, Av. Salaverry 2020, Jesús María, Lima 11 - Perú

Email address: `zuniga_jj@up.edu.pe`

Contemporary Mathematics
Volume **802**, 2024
https://doi.org/10.1090/conm/802/16076

Calculations for plus constructions

Michael Monaco

ABSTRACT. In recent work by Kaufmann and the author, we define a plus construction for monoidal categories and showed that if the monoidal category is a unique factorization category, then the plus construction yields a Feynman category. In this paper, we will use different methods to construct UFCs and demonstrate how the plus construction reproduces and clarifies existing constructions through explicit computations.

Introduction

In the framework of Feynman categories, the plus construction plays a fundamental role in describing algebras, twists, and more recently monoid definitions. In [**KM22**], they extend the plus construction to a type of monoidal category which they call a unique factorization category. We will use techniques from different areas to give explicit examples of these unique factorization categories and give applications of the plus construction. This paper will focus on four fundamental examples: the trivial Feynman category, the category of finite sets, the category of cospans, and the category of spans.

As described by Kaufmann and Ward in [**KW17**], the plus constructions $(\mathfrak{F}^{triv})^+$ and $(\mathcal{F}in\mathcal{S}et)^+$ are related to monoids and operads respectively. We will add to this by considering the nc-plus construction introduced in [**KM22**] and operads with multiplication.

The category of cospans has finite sets as objects and equivalence classes of diagrams $S \to V \leftarrow T$ as morphisms. In algebraic topology, cospans have proven to be a useful way to work with cobordisms and related structures. Examples of this sort can be found in the work of Grandis [**Gra07b, Gra07a, Gra08**], Feshbach and Voronov [**FV11**], and Steinebrunner [**Ste21**], just to name a few. Following this line of thought, we will show how Frobenius algebras naturally arise from a combinatorial version of the plus construction. We will also use the notion of a structured cospan due to Baez and Courser [**BC19, Cou20**] to give colored versions of properads. The nc-plus construction of cospans and its relation to mergers will also be discussed.

In [**KM22**], they show that the category of spans is a hereditary unique factorization category. This implies that $\mathcal{S}pan^+$ is a Feynman category. Despite having some interesting properties, the gadgets corepresented by $\mathcal{S}pan^+$ have not been

2020 *Mathematics Subject Classification.* Primary 18M60.

©2024 American Mathematical Society

explored to the same extent as the gadgets corepresented by $\mathcal{C}ospan^+$. To better understand the algebraic significance of this structure, we will briefly survey some of the combinatorial and categorical properties of $\mathcal{S}pan$. We will then end by describing the relation of $\mathcal{S}pan$ and $\mathcal{S}pan^+$ to bialgebras.

1. Plus constructions, Feynman categories, and UFCs

In [**Law86**], Lawvere advocates taking descriptions of mathematical objects as categories "seriously". For example, a group G canonically determines a category ΣG with one object and G as a set of morphisms. Applying this attitude to ΣG, one recovers many classical ideas such as representations, intertwining operators, induced representations, and Frobenius reciprocity as special cases of different categorical constructions. Because of this, it is common to blur the distinction between these concepts. However, it is useful to keep this distinction to separate the datum describing the structure (the set G with a binary operation and certain properties) from the structure itself (the category ΣG).

These sorts of distinctions are especially important in the study of operad-like structures. In [**KW17**], Kaufmann and Ward introduced a special type of monoidal category called a Feynman category to encode different "types" of operad-like structure. In their formalism, a strong monoidal functor $\mathcal{O} : \mathcal{F} \to \mathcal{C}$ encodes an operad-like structure of "type $\mathcal{F}$" in a category $\mathcal{C}$. For Feynman categories that come from a plus construction, a strong monoidal functor $\mathcal{O} : \mathcal{F}^+ \to \mathcal{C}$ determines a category $\mathcal{F}_\mathcal{O}$ by a so-called indexed enrichment. This process of going from the datum $\mathcal{O}$ to the category $\mathcal{F}_\mathcal{O}$ plays a key role in describing algebras and the theory of twists. In [**KM22**], they extend this by defining the plus construction for a larger class of monoidal categories and show that $\mathcal{M}^+$ is a Feynman category when $\mathcal{M}$ is a unique factorization category. In this section, we will briefly recall the definitions of unique factorization categories and the plus construction.

1.1. Unique factorization categories.

NOTATION 1.1. Let $\mathcal{C}^\boxtimes$ denote the free symmetric monoidal category generated by the category $\mathcal{C}$.

DEFINITION 1.2. A *unique factorization category* (UFC) is a triple $(\mathcal{M}, \mathcal{P}, \jmath)$ where $\mathcal{M}$ is a symmetric monoidal category, $\mathcal{P}$ is a groupoid, and $\jmath$ is a functor $\jmath : \mathcal{P} \to \mathrm{Iso}(\mathcal{M} \downarrow \mathcal{M})$ such that:

(1) The functor $\jmath^\boxtimes : \mathcal{P}^\boxtimes \to \mathrm{Iso}(\mathcal{M} \downarrow \mathcal{M})$ induced from the monoidal structure of $\mathcal{M}$ induces an equivalence of categories.
(2) The slice categories of $\mathcal{M}$ are essentially small.

We will often refer to a unique factorization category by its underlying monoidal category $\mathcal{M}$. The pair $(\mathcal{P}, \jmath)$ will be called a *basis of morphisms* and its elements will be called *irreducibles* or *basic morphisms*.

DEFINITION 1.3. [**KM22**] A *Feynman category* $\mathcal{F}$ is a unique factorization category with additional data $(\mathcal{V}, \imath)$ such that:

(1) The functor $\imath^\boxtimes : \mathcal{V}^\boxtimes \xrightarrow{\sim} \mathrm{Iso}(\mathcal{M})$ induces an equivalence.
(2) The irreducible morphisms are given by $\mathcal{P} = \mathrm{Iso}(\mathcal{F} \downarrow \mathcal{V})$ with $\jmath = (id_\mathcal{F}, id_\mathcal{F}, \imath)$ as a compatible choice of basic morphisms $\mathcal{P}$ making $\mathcal{F}$ into a unique factorization category.

A choice of such a pair $(\mathcal{V}, \imath)$ will be called a *basis of objects* and its elements will be called *irreducibles* or *basic objects.*

DEFINITION-PROPOSITION 1.4. [**KM22**] *A basis for morphisms* $(\mathcal{P}, \jmath)$ *is* hereditary *if for every pair of composable morphisms* (ϕ_0, ϕ_1), *with* $\phi_1 \circ \phi_0 = \phi$, *and decomposition into irreducible morphisms*

$$\phi_0 \simeq \bigotimes_{v \in V} \phi_{0,v}, \quad \phi_1 \simeq \bigotimes_{w \in W} \phi_{1,w}, \text{ and } \phi = \bigotimes_{u \in U} \phi_u \tag{1.1}$$

there exists a partition of $V \amalg W = \amalg_{u \in U} P_u$ *indexed by* U, *such that for each* $u \in U$ *there is a decomposition pair* $(\phi_{0,u}, \phi_{1,u})$ *of the* ϕ_u, *viz.* $\phi_{1,u}\phi_{0,u} = \phi_u$, *such that*

$$\phi_{0,u} \simeq \bigotimes_{v \in P_u \cap V} \phi_{0,v} \text{ and } \phi_{1,u} \simeq \bigotimes_{w \in P_u \cap W} \phi_{1,w} \tag{1.2}$$

A unique factorization category is a hereditary UFC *if its basis is hereditary.*

REMARK 1.5. Unique factorization categories are versatile structures which admit many descriptions. In [**KM22**], these conditions were equivalently formulated as right Ore conditions. In [**KM22**, Proposition 6.31], they also show that a hereditary UFC $\mathcal{M}$ naturally determines an indexing functor $\mathcal{M} \to \mathcal{C}ospan$. Categories equipped with indexing functors that satisfy the appropriate conditions play an important role in the work of Steinebrunner [**Ste21**] where they go by the name *labeled cospan categories.* Moreover, Hackney and Beardsley [**BH22**] describe a connection between these labeled cospan categories and Segal presheaves of a category of level graphs.

1.2. Plus constructions of categories.

DEFINITION 1.6. [**KM22**] Given a category $\mathcal{C}$, define $\mathcal{C}^{\boxtimes+}$ so that

(1) The objects are words $\phi_1 \boxtimes \ldots \boxtimes \phi_n$ of morphisms $\phi_i \in \mathcal{C}$.
(2) The morphisms are generated by two types of basic morphisms:
ISOMORPISMS: are words $(\sigma_1 \Downarrow \sigma_1') \boxtimes \ldots \boxtimes (\sigma_n \Downarrow \sigma_n')$ where the σ_i and σ_i' are isomorphisms of $\mathcal{C}$.
γ-MORPHISMS: for every composable pair (ϕ_1, ϕ_0) there is a generator.
$$\gamma_{\phi_1,\phi_0} : \phi_1 \boxtimes \phi_0 \to \phi_1 \circ \phi_0$$
(3) There are several relations including the typical ones like associativity, identities, and interchange as well as some new ones like equivariance with respect to isomorphisms. See [**KM22**] for details.

DEFINITION 1.7. [**KM22**] In the case where $\mathcal{C} = \mathcal{M}$ is a monoidal category, there is a refinement:

(1) The *nc-plus construction* $\mathcal{M}^{nc+}$ is obtained from $\mathcal{M}^{\boxtimes+}$ by adjoining new generators $\mu_{\phi_0,\phi_1} : \phi_1 \boxtimes \phi_2 \to \phi_1 \otimes \phi_2$ and imposing new relations.
(2) The *(localized) plus construction* $\mathcal{M}^{+loc}$ is defined to be the localization of μ. That is, we add the morphism μ^{-1} and mod out by $\mu \circ \mu^{-1} = \mu^{-1} \circ \mu =$ id.

In principle, these localizations can be difficult to compute. However when $\mathcal{M}$ is a hereditary unique factorization category, [**KM22**] defines a plus construction $\mathcal{M}^+$ which is a monoidally equivalent to $\mathcal{M}^{+loc}$. Moreover, the hereditary property is characterized as a right Ore condition, so there is a right roof calculus making the computations tractable. Since we only work with hereditary UFCs, we will not make a distinction between $\mathcal{M}^{+loc}$ and $\mathcal{M}^{+}$ and we will refer to both of them as "the plus construction".

PROPOSITION 1.8. [**KM22**] *The plus construction of a hereditary unique factorization category is a Feynman category.*

2. Trivial Feynman category

Define the *trivial category* 1 to be the category with a single object $*$ and a single identity morphism. Define the *trivial Feynman category* $\mathfrak{F}^{triv}$ so that $\mathcal{F} = 1^{\boxtimes}$ and $\mathcal{V} = 1$. Explicitly, the objects of $\mathfrak{F}^{triv}$ are strings $*^{\boxtimes n}$ with the empty string as the unit and the morphisms are the commutativity constraints.

2.1. Monoids. For a Feynman category $\mathfrak{F} = (\mathcal{F}, \mathcal{V}, \imath)$, a strong monoidal functor $\mathcal{O} : \mathcal{F} \to \mathcal{C}$ is called an $\mathcal{F}$-op in $\mathcal{C}$. The name is supposed to evoke the idea that $\mathcal{O}$ is an operad-like structure of "$\mathcal{F}$-type". In our case, an op of the trivial Feynman category $\mathfrak{F}^{triv}$ is essentially a choice of an object in $\mathcal{C}$. Interestingly, a lot of theory can be "boot-strapped" from this simple example. For a full explanation of this idea, we refer the reader to [**Kau21**, Section 3.4]. For our purposes, we single out the following fact.

PROPOSITION 2.1 (Kaufmann [**Kau21**]). *As a combinatorial object, $(\mathfrak{F}^{triv})^{+}$ is equivalent to the monoidal category $\mathcal{S}urj^{<}$ whose objects are finite sets and whose morphisms are surjections with ordered fibers. Moreover $(\mathfrak{F}^{triv})^{+}$ corepresents monoids as a Feynman category.*

EXAMPLE 2.2. To understand $(\mathfrak{F}^{triv})^{+}$ as a combinatorial category, it is helpful to think of it in terms of the diagrams of [**Kau21**]. In their depiction, an object of $(\mathfrak{F}^{triv})^{+}$ is represented as a string consisting of the basic object id_* and a morphism is represented as a way of stacking these objects.

(2.1)

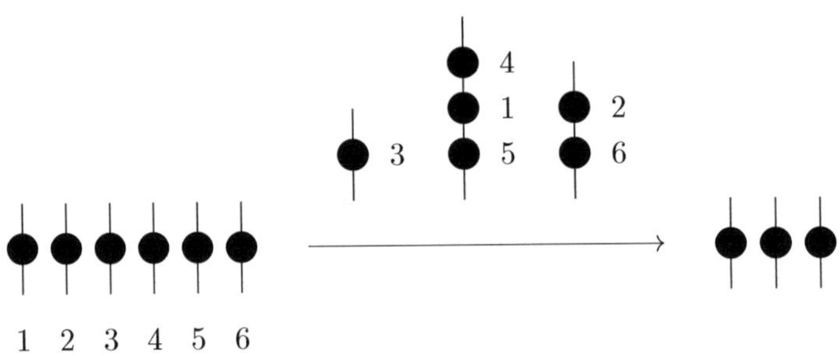

2.2. Graded monoids. If we acknowledge the importance of the trivial Feynman category for the plus construction, then it is natural to consider the nc-plus construction of this category as well. First, observe that the category $(\mathfrak{F}^{triv})^{nc+}$ has a basic object $\mathrm{id}_{*^{\boxtimes n}}$ for each natural number n, hence any $(\mathfrak{F}^{triv})^{nc+}$-op will naturally involve a grading.

To fix notation, we denote a graded object by $A = \{A_i\}_{i \in \mathbb{N}}$. When finite coproducts exist, there is a canonical tensor product of graded objects in $\mathcal{C}$.

$$(A \bullet B)_n = \coprod_{n=p+q} A_p \otimes B_q \tag{2.2}$$

The monoidal unit for this product is then

$$(1_{Gr})_n = \begin{cases} 1_{\mathcal{C}}, & n = 0 \\ 0_{\mathcal{C}}, & n > 0 \end{cases} \tag{2.3}$$

Note that a pointing $\eta : 1_{Gr} \to A$ in graded $\mathcal{C}$-objects amounts to a pointing $1_{\mathcal{C}} \to A_0$ in $\mathcal{C}$. Hence the unit conditions for η say that the following is an isomorphism.

$$A_n \to 1_{\mathcal{C}} \otimes A_n \to A_0 \otimes A_n \to A_n$$

PROPOSITION 2.3. *The category $(\mathcal{F}^{triv})^{nc+}$ corepresents monoid objects in the category of symmetric sequences of $\mathcal{C}$-monoids.*

PROOF. Fix a strong monoidal functor $\mathcal{O} : (\mathcal{F}^{triv})^{nc+} \to \mathcal{C}$. Define an $\mathbb{S}_n$-module $M_{\mathcal{O}}(n) = \mathcal{O}(\mathrm{id}_{*^{\boxtimes n}})$ with the action $\mathbb{S}_n \to \mathrm{Aut}(M_{\mathcal{O}})$ coming from commutativity constraint. All together, this forms an $\mathbb{S}$-module $M_{\mathcal{O}} = \{M_{\mathcal{O}}(n)\}_{n\in\mathbb{N}}$. The image of $\mathcal{O}$ on the morphism $\gamma : \mathrm{id}_{*^{\boxtimes n}} \boxtimes \mathrm{id}_{*^{\boxtimes n}} \to \mathrm{id}_{*^{\boxtimes n}}$ is a morphism $M_{\mathcal{O}}(n) \otimes M_{\mathcal{O}}(n) \to M_{\mathcal{O}}(n)$ which makes each $M_{\mathcal{O}}(n)$ into a monoid. The image of $\mathcal{O}$ on the μ-morphism $\mu : \mathrm{id}_{*^{\boxtimes n}} \boxtimes \mathrm{id}_{*^{\boxtimes m}} \to \mathrm{id}_{*^{\boxtimes (n+m)}}$ is a morphism $M_{\mathcal{O}}(n) \otimes M_{\mathcal{O}}(m) \to M_{\mathcal{O}}(n+m)$. These assemble into a morphism $M_{\mathcal{O}} \bullet M_{\mathcal{O}} \to M_{\mathcal{O}}$. Hence we know that $M_{\mathcal{O}} = \{M_{\mathcal{O}}(n)\}_{n\in\mathbb{N}}$ is at least a monoid object in the category of $\mathbb{N}$-graded sets.

The interchange relation for the plus construction implies that the following diagram commutes.

$$\begin{array}{ccc} M_{\mathcal{O}}(n) \otimes M_{\mathcal{O}}(m) \otimes M_{\mathcal{O}}(n) \otimes M_{\mathcal{O}}(m) & \longrightarrow & M_{\mathcal{O}}(n+m) \otimes M_{\mathcal{O}}(n+m) \\ \downarrow & & \downarrow \\ M_{\mathcal{O}}(n) \otimes M_{\mathcal{O}}(m) & \longrightarrow & M_{\mathcal{O}}(n+m) \end{array} \tag{2.4}$$

Therefore $M_{\mathcal{O}} \bullet M_{\mathcal{O}} \to M_{\mathcal{O}}$ respects the monoid structure of $M_{\mathcal{O}}$ making it into a monoid objects in the category of symmetric sequences of $\mathcal{C}$-monoids. □

COROLLARY 2.4. *Any monoid (an $(\mathfrak{F}^{triv})^{+}$-op) pulls-back to a monoid object in the category of symmetric monoids (an $(\mathfrak{F}^{triv})^{nc+}$-op).*

PROOF. Abstractly, this is just a pullback of the quotient functor $(\mathfrak{F}^{triv})^{nc+} \to (\mathfrak{F}^{triv})^{+}$. Concretely, given a monoid M, we define the graded object $A = \{A_i\}_{i\in\mathbb{N}}$ by $A_i = M^{\otimes i}$. Define $\mu : A_p \otimes A_q \to A_{p+q}$ to be concatenation. To define $\gamma : A_n \otimes A_n \to A_n$, write the multiplication of M as $m : M \otimes M \to M$ and let C be the commutativity constraint associated to the following permutation.

$$\begin{pmatrix} 1 & 1+n & \dots & n & n+n \\ 1 & 2 & \dots & 2n-1 & 2n \end{pmatrix} \tag{2.5}$$

Then $\gamma : A_n \otimes A_n \to A_n$ is the following composition:

$$A_n \otimes A_n \overset{C}{\to} (A \otimes A)^{\otimes n} \overset{m^{\otimes n}}{\to} A^{\otimes n} = A_n.$$ □

3. Finite sets

The category $\mathcal{F}inSet$ of finite sets is a Feynman category where the singleton sets are the basic objects and any map $f : X \to Y$ can be factored into a collection of maps $\{f^{-1}(y) \to \{y\}\}_{y\in Y}$. We will see a connection to operads and make a new connection to operads with multiplication. We will also point out a structural similarity to Young tableaux which we think is interesting.

3.1. Operads. We can think of the plus construction as ascending upwards in some algebraic hierarchy. For example, we have seen that "above" objects ($\mathfrak{F}^{triv}$-ops), there are monoids ($(\mathfrak{F}^{triv})^+$-ops). Similarly, "above" commutative monoids ($\mathcal{F}in\mathcal{S}et$-ops), there are operads ($\mathcal{F}in\mathcal{S}et^+$-ops).

PROPOSITION 3.1 (Kaufmann [**Kau21**]). *The category $\mathcal{F}in\mathcal{S}et$ of finite sets corepresents commutative monoids.*

PROPOSITION 3.2 (Kaufmann [**Kau21**]). *Combinatorially, $\mathcal{F}in\mathcal{S}et^+$ is equivalent to a Borisov–Manin category of graphs with rooted corollas as objects and the generating morphisms have level trees as ghost graphs. Moreover, they corepresent operads as Feynman categories.*

EXAMPLE 3.3. Let ϕ_n denote some n-to-1 map in $\mathcal{F}in\mathcal{S}et$. Then the morphisms $\gamma_{\phi_2,\phi_1 \amalg \phi_3} \to \phi_4$ in $\mathcal{F}in\mathcal{S}et^+$ can be identified with a morphism in a Borisov–Manin category.

(3.1) 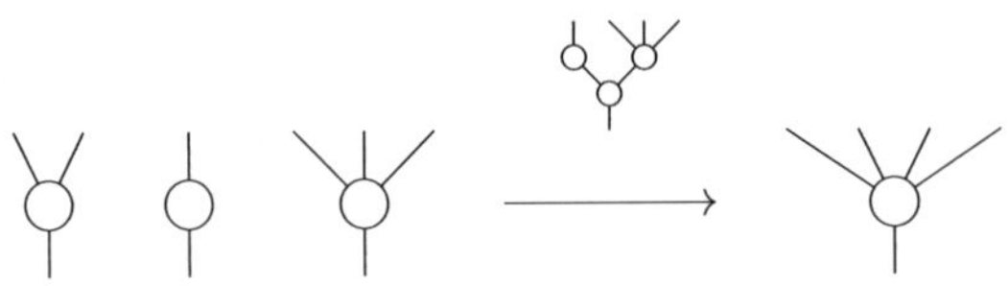

The image of $\gamma_{\phi_2,\phi_1 \amalg \phi_3} \to \phi_4$ under a strong monoidal functor $\mathcal{O} : \mathcal{F}in\mathcal{S}et^+ \to \mathcal{C}$ is the same thing as an operadic composition $\mathcal{O}(2) \otimes \mathcal{O}(1) \otimes \mathcal{O}(3) \to \mathcal{O}(4)$.

PROPOSITION 3.4 (Kaufmann [**Kau21**]). *Combinatorially, $(\mathcal{F}in\mathcal{S}et^<)^+$ is equivalent to a decorated Borisov–Manin category of graphs with planar rooted corollas as objects and the generating morphisms have level trees as ghost graphs. Moreover, they corepresent nonsymmetric operads as Feynman categories.*

3.2. Operads with multiplication. We will show how to obtain a Feynman category which corepresents operads with multiplication by using a slight modification of the plus construction. Rather than shift the focus from a particular example to a general construction, we will simply introduce this modification as an ad hoc construction.

PROPOSITION 3.5. *Define $\mathfrak{O}_\pi$ by starting with $\mathcal{F}in\mathcal{S}et^+$ and allowing words of morphisms of the form $(\sigma_1 \Downarrow \pi_1) \boxtimes \ldots \boxtimes (\sigma_n \Downarrow \pi_n)$ where σ_i are isomorphisms and π_i are surjections. Then $\mathfrak{O}_\pi$ is a Feynman category that corepresents operads with multiplication.*

PROOF. The added actions $(\sigma \Downarrow \pi)$ can be factored as $(\sigma_1 \Downarrow t_1)\boxtimes \ldots \boxtimes(\sigma_n \Downarrow t_n)$ where t_i are surjections with singletons as codomains. The source of $(\sigma_n \Downarrow t_n)$ is an aggregate of corollas and the target is a single corolla. Hence the new morphisms meet the necessary conditions, so $\mathfrak{O}_\pi$ is indeed a Feynman category.

Let $\mathcal{O} : \mathfrak{O}_\pi \to \mathcal{C}$ be a strong monoidal functor. Since $\mathfrak{O} \simeq \mathcal{F}in\mathcal{S}et^+$ is present as a subcategory, we still have the $\mathbb{S}_n$-actions and operadic compositions.

$$\mathcal{O}(\phi_{n_1}) \otimes \ldots \otimes \mathcal{O}(\phi_{n_k}) \otimes \mathcal{O}(\phi_k) \xrightarrow{\gamma} \mathcal{O}\left(\phi_{\sum_i n_i}\right) \tag{3.2}$$

On the other hand, the new actions introduce a multiplication.

$$\mathcal{O}(\phi_n) \otimes \mathcal{O}(\phi_m) \xrightarrow{\sim} \mathcal{O}(\phi_n \amalg \phi_m) \xrightarrow{\mathcal{O}(Id \Downarrow \pi)} \mathcal{O}(\phi_{n+m}) \tag{3.3}$$

Therefore $\mathcal{O}$ is the same data as an operad with multiplication. $\square$

COROLLARY 3.6. *Pulling back along the inclusion* $\iota : \mathcal{F}in\mathcal{S}et^+ \to \mathfrak{O}_\pi$ *forgets the multiplication structure.* □

EXAMPLE 3.7. Let FinSet denote the skeletal category with the sets $\underline{n} = \{1, \dots, n\}$ as objects and functions as morphism. Then define $\mathfrak{O}_\pi^{skel}$ in a similar manner by starting with FinSet^+ and allowing words of the form $(\sigma_1 \Downarrow \pi_1) \boxtimes \dots \boxtimes (\sigma_n \Downarrow \pi_n)$. Fix an operad $\mathcal{O} : \mathsf{FinSet}^+ \to \mathcal{V}ect$ and write $\mathcal{O}(n) = \mathcal{O}(\underline{n} \to \underline{1})$ as short hand. Now define a strong monoidal functor $\mathcal{O}^{nc} : \mathfrak{O}_\pi^{skel} \to \mathcal{V}ect$ on basic objects by

$$\mathcal{O}^{nc}(n) = \bigoplus_{n=\sum_{i=1}^k n_i} \mathcal{O}(n_1) \otimes \dots \otimes \mathcal{O}(n_k). \tag{3.4}$$

(1) The operadic composition in $\mathcal{O}^{nc}$ is given by summing over all possible operadic compositions in $\mathcal{O}$.
(2) The multiplication is defined by commutativity of finite colimits and tensors followed by inclusion.

(3.5)

$$\begin{array}{c}
\left(\bigoplus_{n=\sum_{i=1}^N n_i} \mathcal{O}(n_1) \otimes \dots \otimes \mathcal{O}(n_N) \right) \otimes \left(\bigoplus_{m=\sum_{j=1}^M m_j} \mathcal{O}(m_1) \otimes \dots \otimes \mathcal{O}(m_M) \right) \\
\downarrow \sim \\
\bigoplus_{n=\sum_{i=1}^N n_i} \bigoplus_{m=\sum_{j=1}^M m_j} \mathcal{O}(n_1) \otimes \dots \otimes \mathcal{O}(n_N) \otimes \mathcal{O}(m_1) \otimes \dots \otimes \mathcal{O}(m_M) \\
\rightarrowtail \\
\bigoplus_{m+n=\sum_{k=1}^P p_k} \mathcal{O}(p_1) \otimes \dots \otimes \mathcal{O}(p_P)
\end{array}$$

3.3. Young tableaux. We can think of a surjection $p : E \twoheadrightarrow B$ as a B-indexed partition of a set E by taking fibers $\{p^{-1}(b) : b \in B\}$. In the representation theory of symmetric groups, partitions are encoded by Young tableaux/tabloids. In this section, we will look at this structure from the point of view of the plus construction.

3.3.1. *Tabloids and tableaux as nc-plus constructions.*

DEFINITION 3.8. To match existing conventions, we will work with the following skeletal categories:

(1) Let Surj be the subcategory of $\mathcal{F}in\mathcal{S}et$ whose objects are sets $\underline{n} = \{1, \dots, n\}$ and whose morphisms are surjections.
(2) Let $\mathsf{Surj}^<$ be the category with sets $\underline{n} = \{1, \dots, n\}$ as the objects and surjections with an ordering on each fiber as the morphisms.

EXAMPLE 3.9 (Young tabloids). Consider (3.6), we will think of an object f of Surj^{nc+} on the left as the diagram on the right where the rows are the fibers of f. Since the fibers are not ordered, the corresponding rows are not ordered either which gives us a Young tabloid.

(3.6)

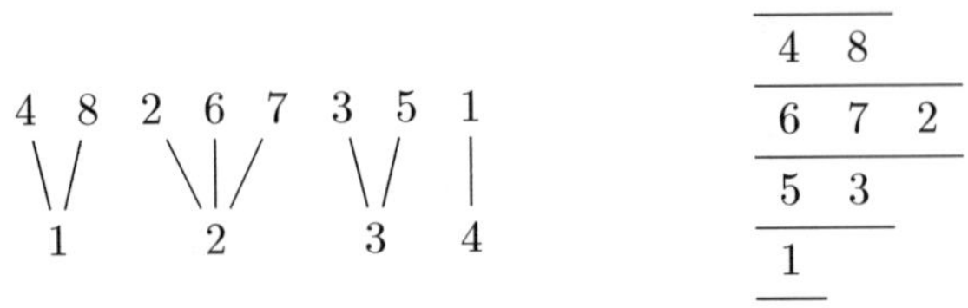

Under this interpretation, precomposing by a surjection lengthens the rows and postcomposing by an injection adds rows. One place where adding columns and rows appears naturally is the branching rules for representations of the symmetric group.

EXAMPLE 3.10 (Young tableaux). Like before, we will think of an object of $(\mathsf{Surj}^{<})^{nc+}$ as a diagram like the one in (3.7). In this case, the fibers are ordered, so the diagram is a Young tableau.

(3.7)

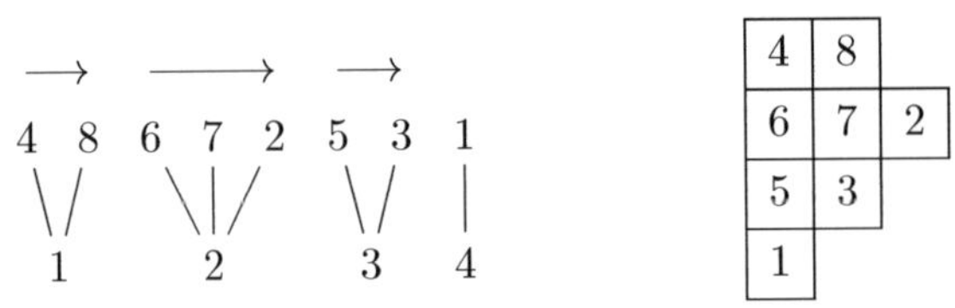

3.3.2. *Descending row condition.* Typically, Young tableaux are written so that the length of the rows decreases from top to bottom. Consequently, the following tableaux are considered distinct.

(3.8)

We will formulate this condition in terms of the underlying groupoid of $(\mathsf{Surj}^{<})^{nc+}$ and $(\mathsf{Surj})^{nc+}$.

DEFINITION 3.11. Given a function $f : X \to Y$ in the category of finite sets, we will say the automorphism $\sigma : Y \to Y$ is *rigid* if $|f^{-1}(y)| = |(\sigma \circ f)^{-1}(y)|$ implies that $\sigma(y) = y$.

(1) Let $\mathcal{YT}ableaux$ be the subcategory of $(\mathsf{Surj}^{<})^{nc+}$ whose morphisms are isomorphism $(\sigma \Downarrow \sigma')$ such that σ' is rigid.
(2) Let $\mathcal{YT}abloids$ be the subcategory of $(\mathsf{Surj})^{nc+}$ whose morphisms are isomorphism $(\sigma \Downarrow \sigma')$ such that σ' is rigid.

EXAMPLE 3.12. The descending row condition can be understood as restricting to these subgroupoids. For example, the left morphism of (3.9) which exchanges the first and second row is in $\mathcal{YT}ableaux$. However, the right morphism of (3.9) which exchanges the first and third row is not in $\mathcal{YT}ableaux$.

(3.9)

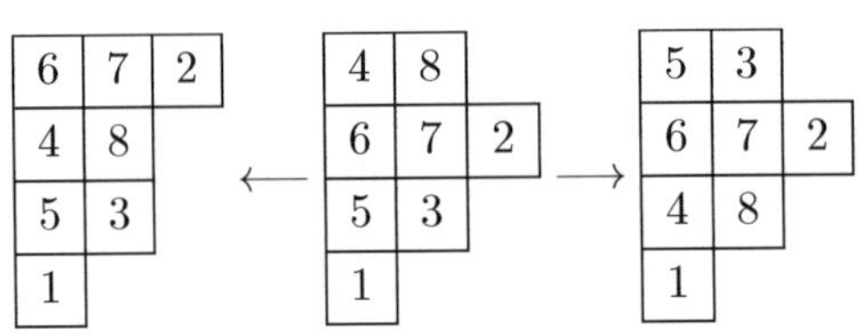

PROPOSITION 3.13. *The automorphism group of a particular Young tableau in $\mathcal{YT}ableaux$ is trivial.*

PROOF. Let $(\sigma \Downarrow \sigma') : f \to f$ be an automorphism in $\mathcal{YT}ableaux$. This implies that the following diagram commutes.

$$\begin{array}{ccc} X & \xrightarrow{f} & Y \\ \big\downarrow{\scriptstyle \sigma} & & \big\downarrow{\scriptstyle \sigma'} \\ X & \xrightarrow{f} & Y \end{array} \tag{3.10}$$

For each $y \in Y$, the isomorphism σ maps the fiber $(\sigma' \circ f)^{-1}(y)$ bijectively onto the fiber $f^{-1}(y)$. Hence $\sigma' = \mathrm{id}$ since σ' is assumed to be rigid. Moreover, σ must be an identity to preserve the order on the fibers. □

COROLLARY 3.14. *Each tableau in $\mathcal{YT}ableaux$ is uniquely isomorphic to a tableau where the width of the rows decreases from top to bottom.*

PROOF. By repeatedly transposing a wider row below with a narrower row above, we obtain a morphism from a particular tableau to a tableau with rows of decreasing length. By the previous proposition, this is necessarily unique. □

4. Cospans

We will show how cospans are connected to graphs, properads, and Frobenius algebras.

DEFINITION 4.1. Define the category $\mathcal{C}ospan$ so that:

(1) The objects are finite sets.
(2) The morphisms are diagrams $S \to V \leftarrow T$ called *cospans* modulo isomorphisms in the middle.

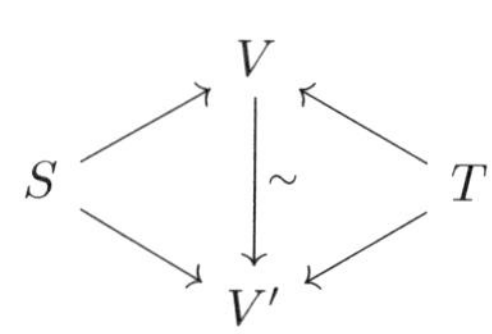

(3) Define composition by taking a pushout in the middle of a pair of cospans.

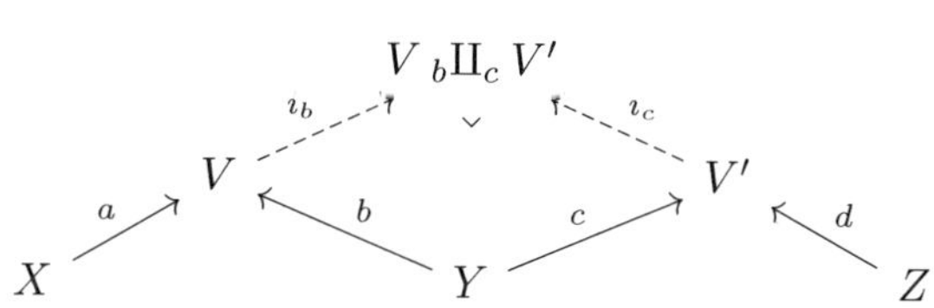

4.1. Zero-to-zero morphisms.

One subtlety of cospans is the existence of nontrivial "zero-to-zero morphisms" such as $\varnothing \to pt \leftarrow \varnothing$. We will discuss how these morphisms differ qualitatively from the others and survey different ways of working with them.

4.1.1. *Subcategories.* In some contexts, the nontrivial zero-to-zero morphisms may not be appropriate. In these cases, we can safely disregard them by considering different subcategories.

DEFINITION 4.2. For a cospan $S \to V \leftarrow T$, we refer to $S \to V$ as the left leg and $V \leftarrow T$ the right leg.

(1) Define the subcategory $'\mathcal{C}ospan$ of *non–unital cospans* to be the subcategory where the left legs $S \to V$ are surjective.
(2) Define the subcategory $\mathcal{C}ospan'$ of *non–co–unital cospans* to be the subcategory where the right legs $V \leftarrow T$ are surjective.
(3) Define $'\mathcal{C}ospan'$ to be the subcategory where both legs $S \to V$ and $V \leftarrow T$ are surjective.

Because pushouts preserve epimorphisms, these all form categories.

EXAMPLE 4.3. Cospans of the form $S \to \{*\} \leftarrow T$ fail to form a subcategory of $\mathcal{C}ospan$. For example, the composition of $S \to \{*\} \leftarrow \varnothing$ and $\varnothing \to \{*\} \leftarrow T$ yields $S \to \{*\} \amalg \{*\} \leftarrow T$. However, in the cospan categories of Definition 4.2, the composition of $S \to \{*\} \leftarrow T$ and $T \to \{*\} \leftarrow U$ always yields $S \to \{*\} \leftarrow U$ because of the surjectivity requirements on one or both legs.

4.1.2. *Factorization and lone morphisms.* When the zero-to-zero morphisms are present, the notion of factorization needs a slight modification. For example, the cospan $\varnothing \to V \leftarrow \varnothing$ factors as $\coprod_{v\in V}(\varnothing \to \{v\} \leftarrow \varnothing)$ with $\varnothing \to \{*\} \leftarrow \varnothing$ as the generating morphism. However, if we consider this cospan as an object of $\mathrm{Iso}(\mathcal{C}ospan \downarrow \mathcal{C}ospan)$, the automorphism group of $\varnothing \to V \leftarrow \varnothing$ is trivial rather than $\mathrm{Aut}(V)$. This means the first condition of a UFC does not quite capture the factorization, but it can with a simple modification.

NOTATION 4.4. Consider $(\mathbb{N}, +)$ as a discrete monoidal category and let $\mathbb{N}[I]$ denote the monoidal subcategory of $\mathbb{N}^I$ where the objects have finite support.

DEFINITION 4.5. Augment the data of a unique factorization category with a set L and an injection $L \rightarrowtail \mathrm{Hom}(1_{\mathcal{M}}, 1_{\mathcal{M}})$. Using the monoidal structure of $\mathcal{M}$ and the inclusion $\mathrm{Iso}(1_{\mathcal{M}} \downarrow 1_{\mathcal{M}}) \hookrightarrow \mathrm{Iso}(\mathcal{M} \downarrow \mathcal{M})$, these induce the following functor.

$$\jmath^{\boxtimes} \times l : \mathcal{P}^{\boxtimes} \times \mathbb{N}[L] \to \mathrm{Iso}(\mathcal{M} \downarrow \mathcal{M}) \tag{4.1}$$

We can think of the image of $L \rightarrowtail \mathrm{Hom}(1_{\mathcal{M}}, 1_{\mathcal{M}})$ as the irreducible *lone morphisms.*

EXAMPLE 4.6. For $\mathcal{C}ospan$, take $L = \{*\}$ and define $L \rightarrowtail \mathrm{Hom}(\varnothing, \varnothing)$ so that $*$ is assigned to $\varnothing \to \{*\} \leftarrow \varnothing$. By the Eckmann–Hilton argument, the only way these lone morphisms can compose is by accumulating. For example in $\mathcal{C}ospan$, the composition of $\varnothing \to \{1\} \leftarrow \varnothing$ and $\varnothing \to \{2\} \leftarrow \varnothing$ is the cospan $\varnothing \to \{1\} \amalg \{2\} \leftarrow \varnothing$.

4.1.3. *Corelations.* In applied situations, a morphism in a cospan category is thought of as a process where the source represents the inputs and the target represents outputs. Under this interpretation, a process with zero inputs and zero outputs should be disregarded whenever they occur. This is formalized by corelations.

DEFINITION 4.7 (Fong–Zanasi [**FZ17**]). Define the *(first) category of corelations* $\mathcal{C}orel_I$ as follows:

(1) The objects are sets.

(2) The morphisms are isomorphism classes of jointly surjective cospans. In other words, cospans $S \to V \leftarrow T$ such that the induced map $S \amalg T \to V$ is a surjection.

(3) In general, a composition of two jointly surjective cospans might compose to some $S \amalg T \to V$ that is not jointly surjective. However, we can restrict the codomain to get a surjection $S \amalg T \twoheadrightarrow V'$. We take this to be the composition in $\mathcal{C}orel_I$.

DEFINITION 4.8 (Fong–Zanasi [**FZ17**]). Let $\mathcal{M}$ denote the collection of injections. Define the *(second) category of corelations* $\mathcal{C}orel_{II}$ so that the morphisms are equivalence classes of cospans where two cospans are considered equivalent if there is a zig-zag of morphisms in $\mathcal{M}$ connecting them.

$$
\begin{array}{ccccccccc}
S & = & S & = & \cdots & = & S & = & S \\
\downarrow & & \downarrow & & & & \downarrow & & \downarrow \\
V_1 & \xrightarrow{\in \mathcal{M}} & V_2 & \xleftarrow{\in \mathcal{M}} & \cdots & \xrightarrow{\in \mathcal{M}} & V_{n-1} & \xleftarrow{\in \mathcal{M}} & V_n \\
\uparrow & & \uparrow & & & & \uparrow & & \uparrow \\
T & = & T & = & \cdots & = & T & = & T
\end{array}
\tag{4.2}
$$

EXAMPLE 4.9. The span $S \xrightarrow{l} V \xleftarrow{r} T$ is equivalent to a span $S \to im(l) \cup im(r) \leftarrow T$ without any "lone vertices" by way of the injection $im(l) \cup im(r) \hookrightarrow V$.

PROPOSITION 4.10 (Fong–Zanasi [**FZ17**]). *The categories $\mathcal{C}orel_I$ and $\mathcal{C}orel_{II}$ defined above are equivalent.*

REMARK 4.11. Fong and Zanasi [**FZ17**] prove a more general result which shows that the definitions of $\mathcal{C}orel_I$ and $\mathcal{C}orel_{II}$ are valid in any category $\mathcal{C}$ with pushouts and a factorization system $(\mathcal{E}, \mathcal{M})$ such that $\mathcal{M}$ is stable under pushouts. Moreover, they show that the equivalence of Proposition 4.10 still holds.

4.2. Properads and Frobenius algebras.

4.2.1. *Cospan corepresents special Frobenius algebras.* A strong monoidal functor $A : \mathcal{C}ospan \to \mathcal{C}$ determines an object $A = A(pt)$, a multiplication $\mu = A(\{1,2\} \to \{1\} \leftarrow \{1\})$, and a comultiplication $\Delta = A(\{1\} \to \{1\} \leftarrow \{1,2\})$. Since the legs of a cospan are set maps, μ and Δ must be commutative. The Frobenius N-condition is a consequence of composing cospans by pushouts.

(4.3) 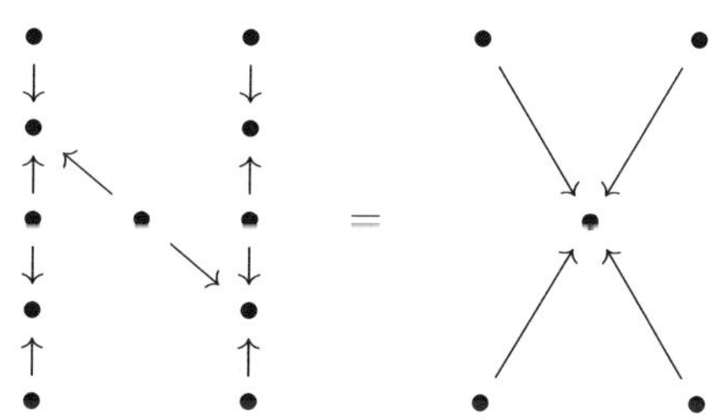

However, composition by pushout automatically implies that $A \xrightarrow{\Delta} A \otimes A \xrightarrow{\mu} A$ is the identity, see Diagram (4.4). Hence $\mathcal{C}ospan$-ops have an additional property that is not guaranteed by the usual axioms for a Frobenius algebra. Instead, $\mathcal{C}ospan$ corepresents what are called *special Frobenius algebras* which were first identified by Carboni and Walters in [**CW87**]. If one uses corelations, one gets the *extra special Frobenius algebras* as described by Coya and Fong in [**CF17**].

(4.4)

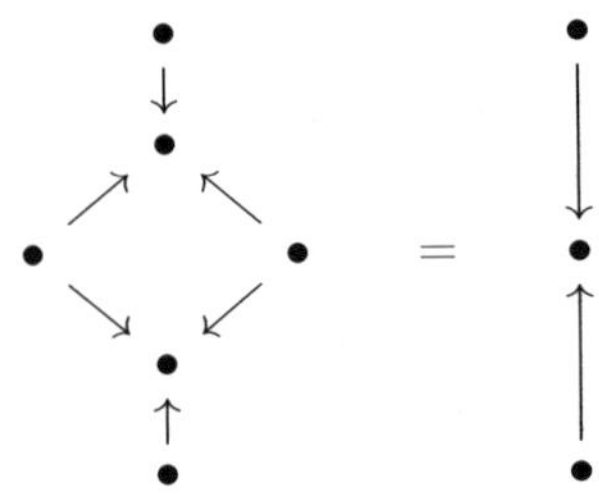

4.2.2. *Genus data.* Considering the equivalence of commutative Frobenius algebras and 2D topological quantum field theories proved by Abrams [**Abr96**], we see that $\mathcal{C}ospan$ would need to be equipped with an extra "genus datum" in order to corepresent commutative Frobenius algebras. To understand the nature of this genus datum, we will consider the following connection between cospans and the properads of Vallette [**Val07**].

PROPOSITION 4.12. [**KM22**] *Combinatorially, $\mathcal{C}ospan^+$ is equivalent to a Borisov–Manin category of graphs such that:*

(1) *The objects are directed aggregates of corollas.*
(2) *The morphisms are generated by monoidal products and compositions of morphisms with connected two-level ghost graphs.*

Moreover, $\mathcal{C}ospan^+$ corepresents properads as a Feynman category.

SKETCH. The objects of $\mathcal{C}ospan^+$ are cospans $S \to V \leftarrow T$. Let $F = S \amalg T$ be the flags with the elements of S as the in-flags and the elements of T as the out-flags. Let V be the vertices. The universal property of coproducts gives a map $\partial : S \amalg T \to V$. This determines a Borisov–Manin graph with no edges.

A typical generating morphism in $\mathcal{C}ospan^+$ is depicted in Diagram (4.5). Under the graphical interpretation of cospans, the γ-morphisms of the plus construction separate the vertices into two levels and match the out-flags of the top vertices to the in-flags of the bottom vertices.

(4.5)

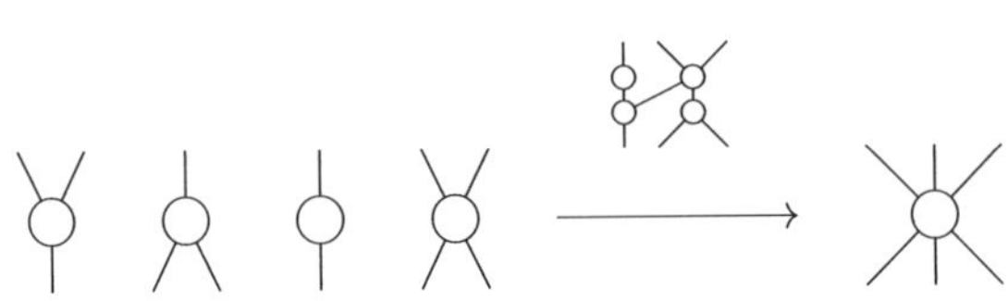

The assumption that the generating morphisms are connected excludes graph mergers. □

DEFINITION 4.13 (Genus properad). This graphical description is convenient and nicely complements the work of Berger and Kaufmann in [**BK22**] where they give a categorical and combinatorial account of different structures and operations coming from string topology.

In particular, they demonstrate that the genus datum can be understood as a strong monoidal functor $\mathcal{O}_{\text{genus}} : (\mathcal{G}raphs, \amalg) \to (\mathcal{S}et, \times)$ which assigns a genus labeling to each vertex. Specializing to our situation, define the *genus properad*

$\mathcal{O}^*_{\text{genus}} : \mathcal{C}ospan^+ \to (\mathcal{S}et, \times)$ to be the strong monoidal functor obtained by the following composition.

$$\mathcal{C}ospan^+ \hookrightarrow (\mathcal{G}raphs^{\text{dir}}, \amalg) \twoheadrightarrow (\mathcal{G}raphs, \amalg) \stackrel{\mathcal{O}_{\text{genus}}}{\to} (\mathcal{S}et, \times) \tag{4.6}$$

4.2.3. *Indexed enrichments.* With the extended notion of a plus construction described in [**KM22**], the indexed enrichments of Kaufmann and Ward [**KW17**] can be adapted to unique factorization categories. We will describe this briefly here and use it to establish the desired connection to Frobenius algebras.

DEFINITION 4.14 (Kaufmann–Ward [**KW17**]). Given a strong monoidal functor $D : \mathcal{M}^+ \to \mathcal{C}$, define the (enriched) category $\mathcal{M}_{\mathcal{O}}$ as follows:

(1) The objects are the same as the original $\mathcal{M}$.
(2) The hom $\mathcal{C}$-objects are $\text{Hom}_{\mathcal{M}_{\mathcal{O}}}(X, Y) = \coprod_{\phi \in \text{Hom}_{\mathcal{M}}(X,Y)} D(\phi)$.
(3) The composition is induced by the γ-morphisms $D(\phi) \otimes D(\psi) \to D(\phi \circ \psi)$ of the plus construction.

EXAMPLE 4.15. The indexed enrichment of an operad $\mathcal{O} : \mathcal{F}in\mathcal{S}et^+ \to \mathcal{C}$ produces a new category $\mathcal{F}in\mathcal{S}et_{\mathcal{O}}$ which corepresents $\mathcal{O}$-algebras. For more examples of this sort, we refer the reader to [**Kau21**].

COROLLARY 4.16. *The category $\mathcal{C}ospan_{\mathcal{O}^*_{\text{genus}}}$ corepresents commutative Frobenius monoids.*

PROOF. With the extra genus datum, the morphisms of $\mathcal{C}ospan_{\mathcal{O}^*_{\text{genus}}}$ agree with the Abrams description of Frobenius algebras as 2D topological quantum field theories. □

4.3. Structured cospans. There are several methods for constructing cospan categories equipped with extra structure such as the decorated cospans of Fong [**Fon15**] and the structured cospans of Baez and Courser [**BC19, Cou20**]. In this section, we will briefly survey structured cospans and use it to describe colored versions of cospans and properads.

DEFINITION 4.17 (Courser [**Cou20**]). Given a functor $F : \mathcal{G} \to \mathcal{C}$ where $\mathcal{C}$ has pushouts, there is a double category $\mathcal{C}ospan(F)$ of *structured cospans*:

(1) The objects are the same as $\mathcal{G}$ and the vertical morphisms are the morphisms of $\mathcal{G}$.
(2) The horizontal morphisms are cospans with F applied to both feet, see Diagram (4.7). Horizontal composition is given by taking pushouts.

$$\begin{array}{ccc} X & \longrightarrow\!\!\!\!\!\!\!\!\!+\;\; & Y \\ \sigma\big\downarrow & \Downarrow\Phi & \big\downarrow\tau \\ X' & \longrightarrow\!\!\!\!\!\!\!\!\!+\;\; & Y' \end{array} \quad := \quad \begin{array}{ccccc} F(X) & \longrightarrow & V & \longleftarrow & F(Y) \\ \sigma\big\downarrow & & \big\downarrow\Phi & & \big\downarrow\tau \\ F(X') & \longrightarrow & V' & \longleftarrow & F(Y') \end{array} \tag{4.7}$$

REMARK 4.18. The functor F is denoted by L in Baez and Courser [**BC19**] which stands for "left adjoint" since the functor often is indeed a left adjoint in their work. For us, this is generally not the case, so we drop that convention.

PROPOSITION 4.19 ([**Cou20**, Section 3.4]). *If $\mathcal{G}$ and $\mathcal{C}$ are symmetric monoidal categories and $F : \mathcal{G} \to \mathcal{C}$ is a strong symmetric monoidal functor, then the double category $\mathcal{C}ospan(F)$ becomes symmetric monoidal in a canonical way.*

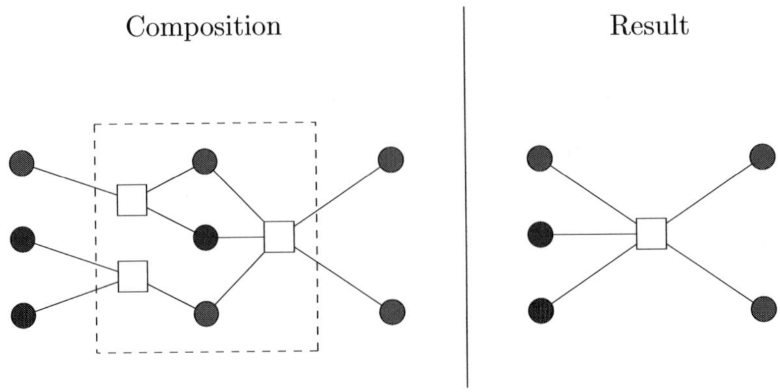

FIGURE 1. A composition of two morphisms in the colored cospan category.

EXAMPLE 4.20 (Cospans and properads with colors). Let $\mathcal{V}$ be a discrete category with two objects R ("red") and B ("blue"). There is a canonical strong monoidal functor $clr : \mathcal{V}^{\otimes} \to \mathcal{F}in\mathcal{S}et$ which sends an object of length n to the set $\underline{n}$. By the previous result, this defines a symmetric monoidal category $\mathcal{C}ospan(clr)$. Composition is illustrated in Figure 1. Moreover, $\mathcal{C}ospan(clr)^+$ corepresents 2-colored properads.

4.4. Props and mergers. We always have a localization functor

$$L_{\mathcal{M}} : \mathcal{M}^{nc+} \to \mathcal{M}^+$$

for any monoidal category $\mathcal{M}$ which sends a basic object $f_1 \boxtimes \ldots \boxtimes f_n$ to $f_1 \otimes \ldots \otimes f_n$. If there is another functor $J : \mathcal{M}^{nc+} \to \mathcal{M}^+$, we can incorporate these as an extra structure on a plus construction which allows "mergers" of morphisms.

DEFINITION 4.21. Given a functor $J : \mathcal{M}^{nc+} \to \mathcal{M}^+$, define a new category by starting with $\mathcal{M}^+$ then formally add a generating morphism $B_\phi : L_{\mathcal{M}}(\phi) \to J(\phi)$ for each basic object ϕ in $\mathcal{M}^{nc+}$. We then add relations making B_ϕ natural in the sense that the following diagram commutes for each morphism Φ of $\mathcal{M}^{nc+}$.

$$\begin{array}{ccc} L_{\mathcal{M}}(\phi) & \xrightarrow{L_{\mathcal{M}}(\Phi)} & L_{\mathcal{M}}(\psi) \\ {\scriptstyle B_\phi}\downarrow & & \downarrow{\scriptstyle B_\psi} \\ J(\phi) & \xrightarrow[J(\Phi)]{} & J(\psi) \end{array} \tag{4.8}$$

EXAMPLE 4.22. In the category of cospans, there is a functor $J : Cospan^{nc+} \to Cospan^+$ which sends a basic object $S \to V \leftarrow T$ to the basic object $S \to pt \leftarrow T$. The morphisms B_ϕ are mergers in the ordinary sense. Naturality for isomorphisms is just equivariance. Naturality for $\gamma : \phi_0 \boxtimes \phi_1 \to \phi_0 \circ \phi_1$ is a sort of interchange, see Figure 2.

5. Spans

Similar to cospans, the category $\mathcal{S}pan$ is defined so that:

(1) The objects are finite sets.

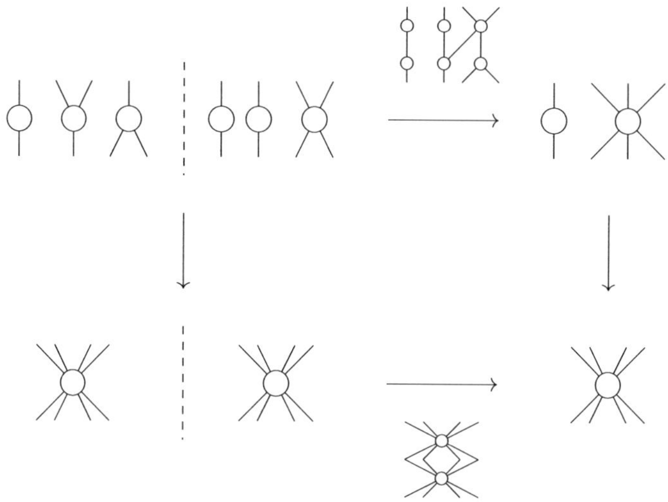

FIGURE 2. The naturality condition of B on the γ-morphisms. The dashed line indicates a $\boxtimes$ that became an $\otimes$ after applying either the functor L or J.

(2) The morphisms are diagrams $S \leftarrow V \rightarrow T$ called *spans* modulo isomorphisms in the middle.

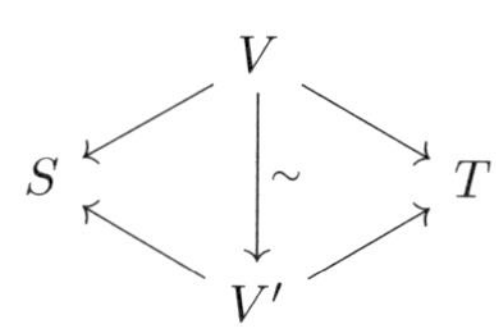

(3) Define composition by taking a pullback in the middle of a pair of spans.

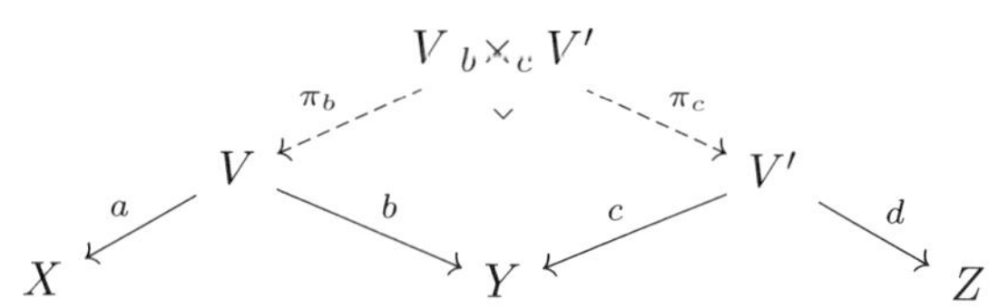

PROPOSITION 5.1. [**KM22**] *Span is a hereditary unique factorization category.*

5.1. Relations. Classically, a relation between two sets X and Y is a subset $R \subseteq X \times Y$. From a categorical point of view, an injection $r : R \rightarrowtail X \times Y$ determines a span $X \leftarrow R \rightarrow Y$, so relations can be understood as a subcategory of $\mathcal{S}pan$.

PROPOSITION 5.2 (well known). *Let $r_0 : R_0 \rightarrowtail X \times Y$ and $r_1 : R_1 \rightarrowtail Y \times Z$ be relations with the following associated spans:*

$$X \overset{a_0}{\leftarrow} R_0 \overset{b_0}{\rightarrow} Y,$$

$$Y \overset{a_1}{\leftarrow} R_1 \overset{b_1}{\rightarrow} Z.$$

The composition $(a_1, b_1) \circ (a_0, b_0)$ of spans is the same as the span associated to the composition $r_1 \circ r_0$ of relations.

PROOF. Consider the composition of $X \overset{a_0}{\leftarrow} R_0 \overset{b_0}{\rightarrow} Y$ and $Y \overset{a_1}{\leftarrow} R_1 \overset{b_1}{\rightarrow} Z$. Let $R_2 = \{(x_0, x_1) \in R_0 \times R_1 : b_0(x_0) = a_1(x_1)\}$ so that $R_0 \overset{\pi_0}{\leftarrow} R_2 \overset{\pi_1}{\rightarrow} R_1$ forms the pullback of $R_0 \overset{b_0}{\rightarrow} Y \overset{a_1}{\leftarrow} R_1$. Hence the following span represents the composition.

$$X \overset{a_0}{\leftarrow} R_0 \overset{\pi_0}{\leftarrow} R_2 \overset{\pi_1}{\rightarrow} R_1 \overset{b_1}{\rightarrow} Z \tag{5.1}$$

This gives $r_2 : R_2 \rightarrowtail R_0 \times R_1 \rightarrowtail X \times Z$. The relation r_2 is precisely the composition $r_1 \circ r_0$. □

PROPOSITION 5.3. *The category of relations is a hereditary unique factorization category.*

PROOF. Since relations form a monoidal subcategory of *Span*, this result follows from the fact that *Span* is a unique factorization category. □

5.2. Matrices. We can also think of *Span* as a categorified version of a matrix. To see this, note that each span $S \leftarrow V \rightarrow T$ corresponds to a map $M : V \rightarrow S \times T$ by the universal property of the product. The map M is determined by its fibers $M_{s,t} = M^{-1}(s,t)$, so M can be understood as a sort of matrix where the coordinates are sets rather than numbers. This analogy is strengthened by the following fact.

PROPOSITION 5.4 (well known). *Given the maps $M : V \rightarrow A \times B$ and $N : U \rightarrow B \times C$, let $M \circ N$ denote their composition as a pair of spans. Then the coordinates of $M \circ N$ are given by a categorified matrix multiplication:*

$$(M \circ N)_{a,c} \cong \coprod_{b \in B} M_{a,b} \times N_{b,c}. \tag{5.2}$$

PROOF. Let $A \leftarrow V \rightarrow B$ be the span for M and let $B \leftarrow U \rightarrow C$ be the span for N. Write the pullback as $V \leftarrow W \rightarrow U$. As shorthand, we will use subscripts for preimages. For example, V_a is the preimage of $a \in A$ under the map $V \rightarrow A$. We will also use double subscripts for intersections, so $V_{ab} = V_a \cap V_b$.

First, a straightforward pullback argument shows that the 2-by-2 square below is a pullback.

$$\begin{array}{ccccccc} W_{ac} & \longrightarrow & W_a & \longrightarrow & V_a & \longrightarrow & \{a\} \\ \downarrow & & \downarrow & & \downarrow & & \downarrow \\ W_c & \longrightarrow & W & \longrightarrow & V & \longrightarrow & A \\ \downarrow & & \downarrow & & \downarrow & & \\ U_c & \longrightarrow & U & \longrightarrow & B & & \\ \downarrow & & \downarrow & & & & \\ \{c\} & \longrightarrow & C & & & & \end{array} \tag{5.3}$$

Then for each point $b \in B$, the dashed arrow is uniquely determined by the universal property of the pullback.

(5.4)

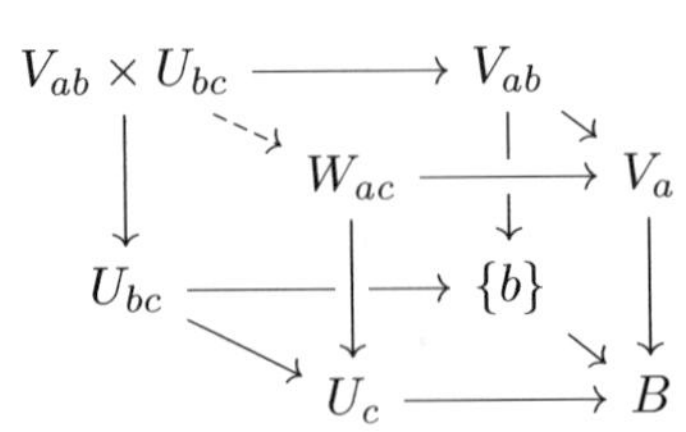

Now consider the diagrams below. Since (1) and (2) are pullbacks, the rectangle (1,2) is also a pullback. By commutativity of (5.4), rectangle (1,2) is the same as (3,4). Since (3,4) and (4) are pullbacks, (3) is also a pullback.

$$\begin{array}{ccccc} V_{ab} \times U_{bc} & \longrightarrow & V_{ab} & \longrightarrow & V_a \\ \downarrow & (1) & \downarrow & (2) & \downarrow \\ U_{bc} & \longrightarrow & \{b\} & \longrightarrow & B \end{array} \qquad \begin{array}{ccccc} V_{ab} \times U_{bc} & \dashrightarrow & W_{ac} & \longrightarrow & V_a \\ \downarrow & (3) & \downarrow & (4) & \downarrow \\ U_{bc} & \longrightarrow & U_c & \longrightarrow & B \end{array} \tag{5.5}$$

Diagram (5) below is a pullback, hence the diagram (3,5) is a pullback for all $b \in B$.

$$\begin{array}{ccccc} V_{ab} \times U_{bc} & \longrightarrow & U_{bc} & \longrightarrow & \{b\} \\ \downarrow & (3) & \downarrow & (5) & \downarrow \\ W_{ac} & \longrightarrow & U_c & \longrightarrow & B \end{array} \tag{5.6}$$

By pullback stability, the following diagram is also a pullback.

$$\begin{array}{ccc} \coprod_{b\in B} V_{ab} \times U_{bc} & \longrightarrow & \coprod_{b\in B}\{b\} \\ \downarrow & & \downarrow \\ W_{ac} & \longrightarrow & B \end{array} \tag{5.7}$$

Since pullbacks preserve isomorphisms, we get $W_{ac} \cong \coprod_{b\in B} V_{ab} \times U_{bc}$, as desired. $\square$

5.3. Black and white graphs. We can also think of the span $V_w \overset{w}{\leftarrow} E \overset{b}{\rightarrow} V_b$ as a black and white graph where the edge $e \in E$ is connected to a white vertex $w(e) \in V_w$ and a black vertex $b(e) \in V_b$. To define a composition of these black and white graphs, suppose we have the following black and white graphs such that $V_b = V'_w$:

$$\Gamma = (V_w \leftarrow E \rightarrow V_b) \text{ and } \Gamma' = (V'_w \leftarrow E' \rightarrow V'_b).$$

Define $\Gamma \wedge \Gamma'$ to be the following pullback.

$$\begin{array}{ccccccc} & & & E^\wedge = E_{\partial_b} \times_{\partial'_w} E' & & & \\ & & \swarrow & & \searrow & & \\ & E & & & & E' & \\ \swarrow & & \searrow & & \swarrow & & \searrow \\ V_w & & & V_b = V'_w & & & V'_b \end{array} \tag{5.8}$$

This new graph has V_w as its white vertices and V'_b as its black vertices. There is an edge between vertices $w \in V_w$ and $b' \in V'_b$ in the new graph whenever there is a vertex $x \in V_b = V'_w$ with an edge between x and w in the first graph and an edge between x and b' in the second graph.

EXAMPLE 5.5. Consider the $\wedge$-operation depicted in (5.9) between a simplified version of the constellation Orion with edges labeled by letters and another black and white graph with edges labeled by numbers. The black and white vertices that are matched are depicted as squares.

(5.9)

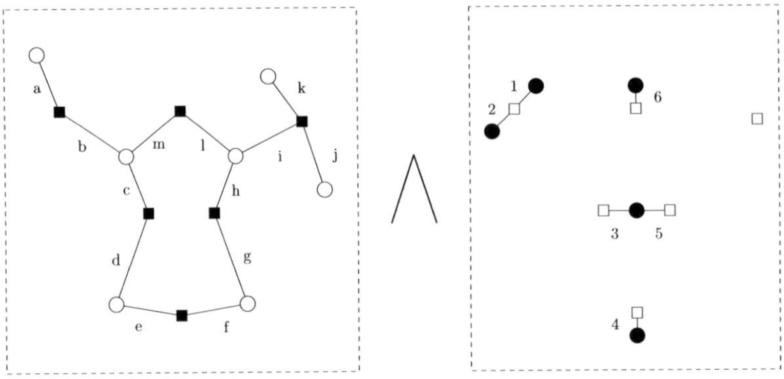

To evaluate this, one can imagine superimposing the two graphs so that the squares are matched and then connecting the black and white vertices accordingly to obtain (5.10). The elbow of the arm holding the club gets "cloned", Orion's belt gets tightened, and the black vertex that is part of the shield is removed.

(5.10)

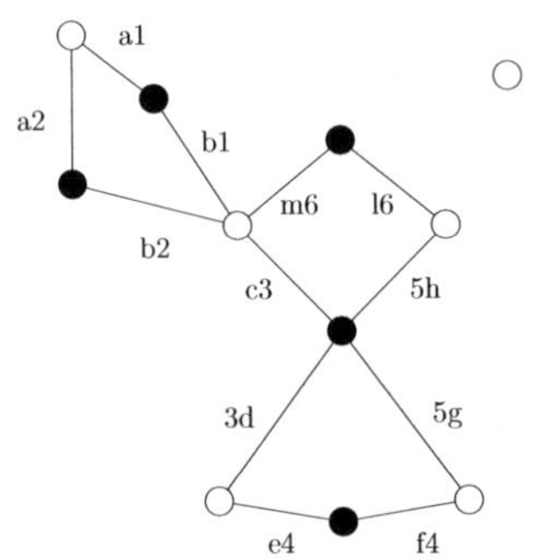

5.4. Bialgebras. The two middle arrows in a composition of two spans forms a cospan. We know that cospans factor as $\{V_y \to \{y\} \leftarrow V'_y\}_{y \in Y}$. Hence it suffices to look at these basic cospans to understand the composition of spans. For instance, the pullback of $\{a, b, c\} \to pt \leftarrow \{1, 2\}$ is the span $\{a, b, c\} \leftarrow \{a, b, c\} \times \{1, 2\} \to \{1, 2\}$. Writing out the elements gives the following picture.

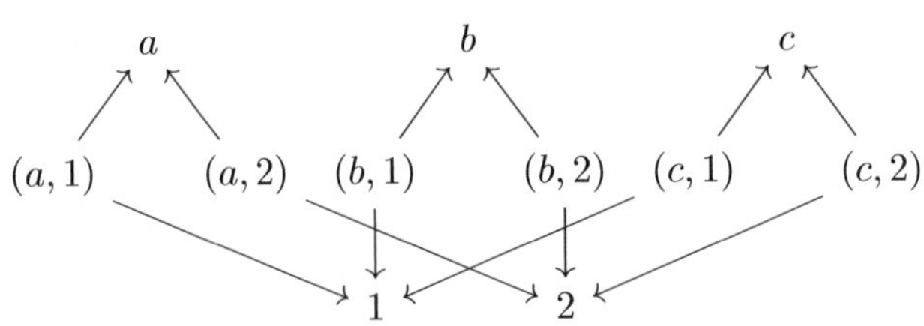

One can think of this picture as the complete graph between vertices $\{a, b, c\}$ and vertices $\{1, 2\}$ with each pair connected by an "edge" in $\{a, b, c\} \times \{1, 2\}$ exactly once.

This complete graph picture is a natural way to describe how multiplication and comultiplication interact in a bialgebra. For example, the relation $\Delta \circ \mu = (\mu \otimes \mu) \circ (\mathrm{id} \otimes C \otimes \mathrm{id}) \circ (\Delta \otimes \Delta)$ for bialgebras can be depicted graphically by

Diagram (5.11) where μ is represented as a circle and Δ is represented as a square.

(5.11)

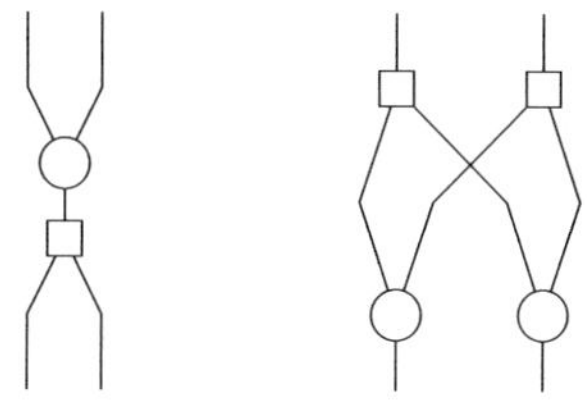

Conversely, we can recover the complete graph picture from this relation by using associativity.

(5.12)

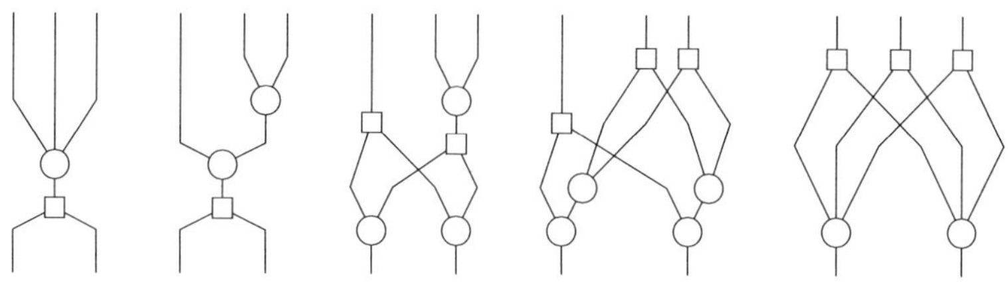

PROPOSITION 5.6 (Lack [**Lac04**]). *Span corepresents commutative bimonoids, that is there is a one-to-one correspondence between bicommutative bimonoids in $\mathcal{C}$ and strong monoidal functors $M : \mathcal{S}pan \to \mathcal{C}$.*

EXAMPLE 5.7. Let A be a commutative unital monoid in $\mathcal{S}et$. For a finite set I, let $T_A(I)$ denote the set of maps $I \to A$. We will think of $T_A(I)$ as the set of I-tuples of A. Define a strong monoidal functor $M_A : \mathcal{S}pan \to \mathcal{S}et$ by sending the span $X \overset{l}{\leftarrow} V \overset{r}{\rightarrow} Y$ to the map

$$T_A(X) \overset{l^*}{\to} T_A(V) \overset{\mu_r}{\to} T_A(Y) \tag{5.13}$$

where μ_r and l^* are defined below.

PRODUCT AND UNIT: Given a map $I \overset{r}{\to} J$ of finite sets, define a map $\mu_r : T_A(I) \to T_A(J)$ so that the tuple $t \in T_A(I)$ gets sent to tuple $t' \in T_A(J)$ defined by

$$t'(j) = \sum_{i \in r^{-1}(j)} t(i). \tag{5.14}$$

We will also use the convention that the empty sum is the unit of A. This is well-defined because A is commutative and I is finite.

COPRODUCT AND COUNIT: Given a map $I \overset{l}{\leftarrow} J$, define $l^* : T_A(I) \to T_A(J)$ by precomposition. In words, the coproduct is duplication and the counit is deletion.

5.5. Plus construction of Span. Building on the combinatorics described in Section 5.3, we will draw the objects of $\mathcal{S}pan^+$ as black and white graphs with the white vertices on top and the black vertices on the bottom.

(5.15)

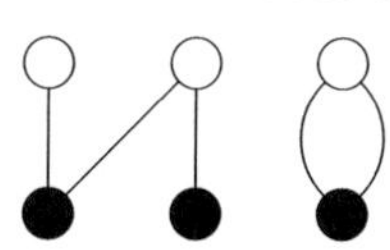

Under this interpretation, the γ-morphisms of the plus construction can be understood as a process which matches black and white vertices and then resolves the connections according to the rule described in Section 5.3.

(5.16)

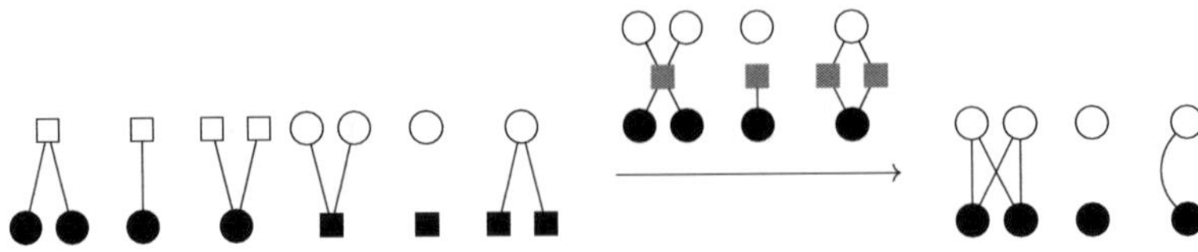

In a rough sense, this picture is dual to the one we described for $\mathcal{C}ospan^+$ in Proposition 4.12. For $\mathcal{C}ospan^+$, flags were grouped by vertices. The γ-morphisms combine multiple groupings into a single grouping of flags. For $\mathcal{S}pan^+$, vertices are connected by edges. The γ-morphisms resolve a series of connections into a single connection.

To understand the algebraic significance of $\mathcal{S}pan^+$, consider some of the results for the other combinatorial categories.

TRIVIAL FEYNMAN CATEGORY: Strong monoidal functors $\mathfrak{F}^{triv} \to \mathcal{C}$ are objects in $\mathcal{C}$. Strong monoidal functors $(\mathfrak{F}^{triv})^+ \to \mathcal{C}$ are monoids in $\mathcal{C}$. (Proposition 2.1).

FINITE SETS: Strong monoidal functors $\mathcal{F}in\mathcal{S}et \to \mathcal{C}$ are commutative algebras in $\mathcal{C}$. (Proposition 3.1) Strong monoidal functors $\mathcal{F}in\mathcal{S}et^+ \to \mathcal{C}$ are operads in $\mathcal{C}$. (Proposition 3.2).

COSPANS: Strong monoidal functors $\mathcal{C}ospan \to \mathcal{C}$ are special bicommutative Frobenius algebras in $\mathcal{C}$. (Section 4.2.1) Strong monoidal functors $\mathcal{C}ospan^+ \to \mathcal{C}$ are properads in $\mathcal{C}$. (Proposition 4.12).

Strong monoidal functors $\mathcal{S}pan \to \mathcal{C}$ are bicommutative bialgebras (Section 5.4). Hence, by analogy, a strong monoidal functor $\mathcal{S}pan^+ \to \mathcal{C}$ encodes a kind of bialgebra just like operads encode a kind of algebra.

EXAMPLE 5.8 (Associative $\mathcal{S}pan^+$-op). Define a strong monoidal functor $A : \mathcal{S}pan^+ \to \mathcal{S}et$ so that $A(X \stackrel{l}{\leftarrow} R \stackrel{r}{\to} Y) = OF(l) \times OF(r)$ where $OF(f)$ is the set of orders on the fibers of a function f. This datum composes in the canonical fashion. For example, suppose we have ordered fibers $\{a < b < c\} \to pt \leftarrow \{1 < 2\}$, then the composition would produce the following ordered fibers.

$$\begin{array}{ll} \{a\} \leftarrow \{(a,1) < (a,2)\} & \\ & \{(a,1) < (b,1) < (c,1)\} \to \{1\} \\ \{b\} \leftarrow \{(b,1) < (b,2)\} & \\ & \{(a,2) < (b,2) < (c,2)\} \to \{2\} \\ \{c\} \leftarrow \{(c,1) < (c,2)\} & \end{array}$$

If we think of an object x of $\mathcal{S}pan^+$ as a black and white graph, then an element of $A(x)$ is an ordering of the edges at each vertex of x.

(5.17)

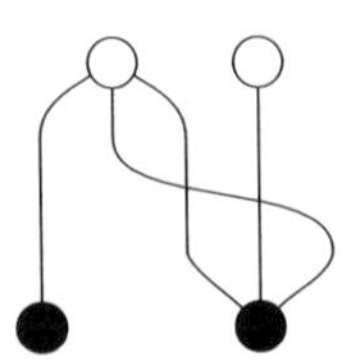

The γ-morphisms compose this data so that the order of the black and white vertices incorporates the ordering of the intermediate vertices.

(5.18)

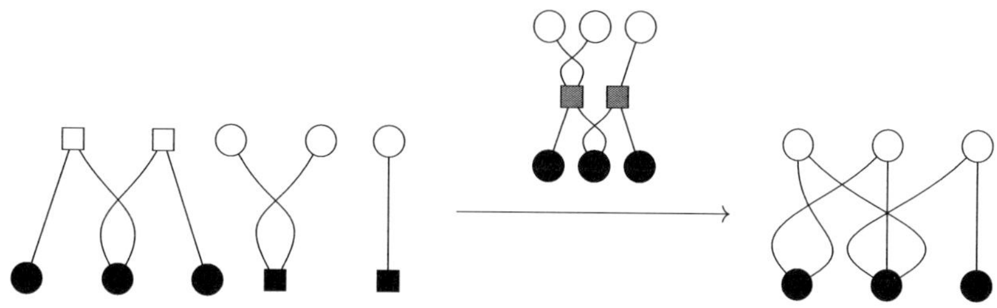

PROPOSITION 5.9. *The indexed enrichment $\mathcal{S}pan_A$ corepresents associative bimonoids.*

PROOF. The only difference between $\mathcal{S}pan_A$ and $\mathcal{S}pan$ is that permuting fibers changes the morphisms of $\mathcal{S}pan_A$ but keeps the morphisms of $\mathcal{S}pan$ the same. Hence a strong monoidal functor $B : \mathcal{S}pan_A \to \mathcal{C}$ is the same data as a bimonoid which is not necessarily commutative or cocommutative. $\square$

Acknowledgments

We would like to thank Ralph Kaufmann for his continued support and thank Philip Hackney and Jan Steinebrunner for helpful discussions on this topic. We also appreciate the work of the referee and the helpful feedback they gave.

References

[Abr96] Lowell Abrams, *Two-dimensional topological quantum field theories and Frobenius algebras*, J. Knot Theory Ramifications **5** (1996), no. 5, 569–587, DOI 10.1142/S0218216596000333. MR1414088

[BC19] John C. Baez and Kenny Courser, *Structured cospans*, Theory Appl. Categ. **35** (2020), Paper No. 48, 1771–1822. MR4170469

[BH22] Jonathan Beardsley and Philip Hackney, *Labelled cospan categories and properads*, J. Pure Appl. Algebra **228** (2024), no. 2, Paper No. 107471, 62, DOI 10.1016/j.jpaa.2023.107471. MR4609536

[BK22] Clemens Berger and Ralph M. Kaufmann, *Trees, graphs and aggregates: a categorical perspective on combinatorial surface topology, geometry, and algebra*, arXiv:2201.10537, 2022.

[CF17] Brandon Coya and Brendan Fong, *Corelations are the prop for extraspecial commutative Frobenius monoids*, Theory Appl. Categ. **32** (2017), Paper No. 11, 380–395. MR3633708

[Cou20] Kenny Allen Courser, *Open Systems: A Double Categorical Perspective*, ProQuest LLC, Ann Arbor, MI, 2020. Thesis (Ph.D.)–University of California, Riverside. MR4106273

[CW87] A. Carboni and R. F. C. Walters, *Cartesian bicategories. I*, J. Pure Appl. Algebra **49** (1987), no. 1-2, 11–32, DOI 10.1016/0022-4049(87)90121-6. MR920513

[Fon15] Brendan Fong, *Decorated cospans*, Theory Appl. Categ. **30** (2015), Paper No. 33, 1096–1120. MR3384097

[FV11] Zheyan Wan and Juven Wang, *Higher anomalies, higher symmetries, and cobordisms I: classification of higher-symmetry-protected topological states and their boundary fermionic/bosonic anomalies via a generalized cobordism theory*, Ann. Math. Sci. Appl. **4** (2019), no. 2, 107–311, DOI 10.4310/AMSA.2019.v4.n2.a2. MR4020366

[FZ17] Brendan Fong and Fabio Zanasi, *A universal construction for (co)relations*, 7th Conference on Algebra and Coalgebra in Computer Science, LIPIcs. Leibniz Int. Proc. Inform., vol. 72, Schloss Dagstuhl. Leibniz-Zent. Inform., Wadern, 2017, pp. Art. No. 12, 16. MR3746663

[Gra07a] Marco Grandis, *Collared cospans, cohomotopy and TQFT (cospans in algebraic topology. II)*, Theory Appl. Categ. **18** (2007), No. 19, 602–630. MR2369111

[Gra07b] Marco Grandis, *Higher cospans and weak cubical categories (cospans in algebraic topology. I)*, Theory Appl. Categ. **18** (2007), No. 12, 321–347. MR2326435

[Gra08] Marco Grandis, *Cubical cospans and higher cobordisms (cospans in algebraic topology. III)*, J. Homotopy Relat. Struct. **3** (2008), no. 1, 273–308. MR2426182

[Kau21] Ralph M. Kaufmann, *Feynman categories and representation theory*, Representations of algebras, geometry and physics, Contemp. Math., vol. 769, Amer. Math. Soc., [Providence], RI, [2021] ©2021, pp. 11–84, DOI 10.1090/conm/769/15419. MR4254096

[KM22] Ralph M. Kaufmann and Michael Monaco, *Plus constructions, plethysm, and unique factorization categories with applications to graphs and operad-like theories*, 2022.

[KW17] Ralph M. Kaufmann and Benjamin C. Ward, *Feynman categories* (English, with English and French summaries), Astérisque **387** (2017), vii+161. MR3636409

[Lac04] Stephen Lack, *Composing PROPS*, Theory Appl. Categ. **13** (2004), No. 9, 147–163. MR2116328

[Law86] F. W. Lawvere, *Taking categories seriously*, Rev. Colombiana Mat. **20** (1986), no. 3-4, 147–178. Seminar and workshop on category theory and applications (Bogotá, 1983). MR948965

[Ste21] Jan Steinebrunner, *The surface category and tropical curves*, Preprint, `arXiv:2111.14757` (2021).

[Val07] Bruno Vallette, *A Koszul duality for PROPs*, Trans. Amer. Math. Soc. **359** (2007), no. 10, 4865–4943, DOI 10.1090/S0002-9947-07-04182-7. MR2320654

Deprtment of Mathematics, Purdue University, West Lafayette, Indiana 47907
Email address: `monacom@purdue.edu`

Contemporary Mathematics
Volume **802**, 2024
https://doi.org/10.1090/conm/802/16077

Segal conditions for generalized operads

Philip Hackney

ABSTRACT. This note is an introduction to several generalizations of the *dendroidal sets* of Moerdijk–Weiss. Dendroidal sets are presheaves on a category of rooted trees, and here we consider indexing categories whose objects are other kinds of graphs with loose ends. We examine the *Segal condition* for presheaves on these graph categories, which is one way to identify those presheaves that are a certain kind of generalized operad (for instance wheeled properad or modular operad). Several free/forgetful adjunctions between different kinds of generalized operads can be realized at the presheaf level using only the left Kan extension/restriction adjunction along a functor of graph categories. These considerations also have bearing on homotopy-coherent versions of generalized operads, and we include some questions along these lines.

1. Introduction

A simplicial set X is said to *satisfy the Segal condition* if the maps

$$\begin{array}{lcl} X_2 & \xrightarrow{d_2, d_0} & X_1 \times_{X_0} X_1 \\ X_3 & \xrightarrow{d_2 d_3, d_0 d_3, d_0 d_0} & X_1 \times_{X_0} X_1 \times_{X_0} X_1 \\ \vdots & & \vdots \\ X_n & \longrightarrow & X_1 \times_{X_0} X_1 \times_{X_0} \cdots \times_{X_0} X_1 \end{array}$$

are bijections for every $n \geq 2$. Such simplicial sets are the same thing as *categories*: X_0 is the set of objects, X_1 is the set of morphisms, and $d_1 \colon X_2 \to X_1$ provides a composition operation $X_1 \times_{X_0} X_1 \to X_1$ (see [**Gro61**, Proposition 4.1]). This situation can be fruitfully generalized to simplicial spaces by replacing bijections with weak equivalences, leading toward a model for $(\infty, 1)$-categories due to Rezk [**Rez01**]. This approach was extended by Cisinski and Moerdijk to provide a model for $(\infty, 1)$-operads in [**CM13a**].

In this note, we discuss further extensions the Segal condition (in both the weak and strict sense) from categories and (colored) operads to other sorts of operadic structures. Though generalizations of operads have appeared in many guises [**YJ15, KW17, BM15, BB17, Get09**], we take here an example-driven approach. Our focus will be on colored operads, dioperads/symmetric polycategories

2020 *Mathematics Subject Classification.* Primary 18M85, 18F20, 18M60, 55P48, 55U10, 05C20.

This work was supported by a grant from the Simons Foundation (#850849).

©2024 by Philip Hackney

[**Gan03, Gar08**], cyclic operads [**GK95**], properads [**Val07, Dun06**], wheeled pr-operads [**MMS09**], modular operads [**GK98**], and props [**ML65**]. Note that the generalizations of operads in the above sources are usually focused on the monochrome case (or at least the case where morphisms fix the color set), while we consider the 'category-like' version where operations are only composable if they have compatible colors/types. This setting is useful for applications, see for instance [**FBSD21, Spi13, VSL15**], and is formally clarifying.

Given a kind of generalized operad, the Segal condition is then formulated using some category of graphs; for categories, we can regard the simplicial category $\boldsymbol{\Delta}$ as a category of linear trees, and, for operads, the Moerdijk–Weiss dendroidal category $\boldsymbol{\Omega}$ is a category of rooted trees. We will present appropriate graph categories for each of the kinds of operadic structures we are focusing on (with the exception of props; see Question 5.9). Each such graph category has a natural notion of Segal presheaf. Additionally, these graph categories are generalized Reedy categories in the sense of [**BM11**], which has model-categorical implications for the homotopy-coherent situation.

This note is, in part, an expansion of the material from my talk "Interactions between different kinds of generalized operads" in the AMS *Special Session on Higher Structures in Topology, Geometry and Physics* in March 2022, and in part a distillation of the core ideas from [**Hac24**]. That paper introduced a new description of morphisms for several graph categories (most of which were introduced earlier by the author in joint work with Robertson and Yau), and was structured to allow efficient comparison with pre-existing definitions. In contrast, the present paper is designed as an overview for someone coming to the subject for the first time. Our goals are to give basic definitions of graph categories in elementary terms, explain the Segal condition for presheaves, and discuss the connections between the directed and undirected contexts. Pointers are provided to the relevant literature, and we highlight several interesting questions near the end. For particular graph categories, there are other introductory accounts, each having a different focus: [**Wei11**], [**Moe10**], and the first part of [**HM22**] are good starting points for the dendroidal theory (the book is a comprehensive account of the theory), [**HR18**] is focused on ∞-properads, while [**BR**] concerns modular ∞-operads.

1.1. What is a generalized operad? This is a question we don't wish to answer too directly or precisely. As a first approximation, 'generalized operad' should include all of the mathematical objects in MSC2020 code `18M85`,[1] as well as some closely related objects. At its core, a generalized operad should consist of a collection of *operations*, each of which comes equipped with some boundary information (for instance coloring by some set, an 'input' or 'output' tag, or so on), as well as specified ways of connecting (compatible) boundaries of operations to form new operations. See Figure 1.

In a moment we'll give a brief list of several of the structures that will be discussed, but first let's note some general features. Operad-like structures will always be *colored* or *typed* (by a set or involutive set), and, as with functors between categories, maps between them are not expected be the identity on color sets. Each color will have an associated *identity* operation. We will always be discussing the *symmetric* version of concepts, as for many structures (e.g., properads, modular

[1] **18M85** Polycategories/dioperads, properads, PROPs, cyclic operads, modular operads

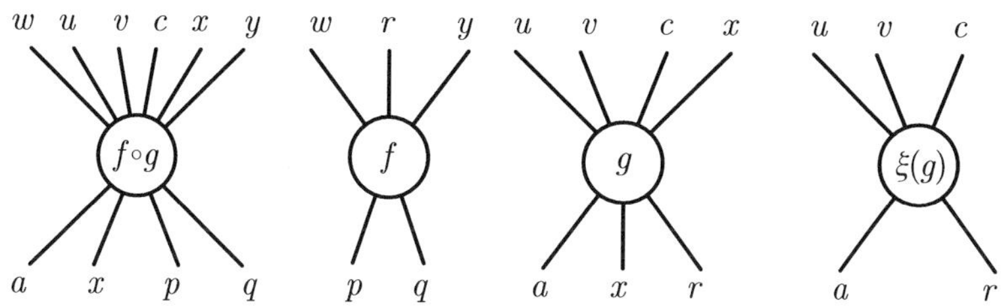

FIGURE 1. Operations $f\colon w,r,y \to p,q$ and $g\colon u,v,c,x \to a,x,r$ along with composition $f \circ_r g$ and contraction $\xi_x(g)$.

operads) the non-symmetric version is quite different. Some structures will be *directed* in the sense that there is a strong distinction between inputs and outputs of operations, while others will be *undirected* and may be considered to just have 'puts' (or 'ports').

In the following list of examples, we'll begin with directed operadic structures, roughly in order of increasing complexity, before turning to undirected operadic structures. Several references to precise definitions in the literature will be given just below the list.

Category: Morphisms in a category are of the form $f\colon c \to d$. Given two morphisms f and g, there are at most two ways to compose the morphisms: $g \circ f$ and $f \circ g$, provided they exist.

Operad: An operad is a generalization of a category, where operations now can have multiple (or no) inputs, they are of the form $f\colon c_1, \dots, c_n \to d$. There are compositions

$$P(c_1,\dots,c_n;d) \times P(d_1,\dots,d_m;c_i) \xrightarrow{\circ_i} P(c_1,\dots,c_{i-1},d_1,\dots,d_m,c_{i+1},\dots,c_n;d)$$

which may give many ways to compose two given operations.

Dioperad: A dioperad, also called a polycategory, expands operads to allow multiple outputs as well. To compose two operations, one chooses a single input of the first which agrees with a single input of the second, and connects them. The relevant structure map when $c_i = b_j$ is ${}_i\circ_j$ below.

$$\begin{array}{c} P(c_1,\dots,c_n;d_1,\dots,d_m) \times P(a_1,\dots,a_k;b_1,\dots,b_\ell) \\ \downarrow {}_i\circ_j \\ P(c_1,\dots,c_{i-1},a_1,\dots,a_k,c_{i+1},\dots,c_n;b_1,\dots,b_{j-1},d_1,\dots,d_m,b_{j+1},\dots,b_\ell) \end{array}$$

Properads: Arguably this is the most complicated structure on this list. The underlying operation sets are the same as those of dioperads, but now one can connect one *or more* inputs of the first operation to the same number of outputs of the second operation.

Props: These are like properads, but now one can connect *zero or more* inputs of the first operation to the same number of inputs of the second operation. More simply, props are just those symmetric monoidal categories whose set of objects is a free monoid under tensor. The set of colors is the free generating set for the monoid of objects.

Wheeled properads: A wheeled properad has additional *contractions* that act on a single operation. They take the form

$$P(c_1,\dots,c_n;d_1,\dots,d_m) \to P(c_1,\dots,\hat{c}_i,\dots,c_n;d_1,\dots,\hat{d}_j,\dots,d_m)$$

where $c_i = d_j$. The properadic compositions in a wheeled properad all decompose into simpler ones. Namely, a composition connecting $k > 0$ inputs of one operation to k outputs of another composes as a dioperadic composition, followed by $k-1$ contractions. Wheeled properads could instead be called 'wheeled dioperads.'

Involutive category: A *dagger category* is a category C equipped with an identity-on-objects functor $C \to C^{\mathrm{op}}$ so that $C \to C^{\mathrm{op}} \to (C^{\mathrm{op}})^{\mathrm{op}} = C$ is id_C. That is, every $f\colon c \to d$ has an associated $f^\dagger\colon d \to c$ (with $(f^\dagger)^\dagger = f$), and this allows you to exchange the role of input and output of morphisms. More generally, an *involutive category* relaxes the requirement that $C \to C^{\mathrm{op}}$ is the identity-on-objects, so every $f\colon c \to d$ has an associated $f^\dagger\colon d^\dagger \to c^\dagger$.

Cyclic operad: A cyclic operad P is an operad equipped with an involution $\dagger$ on its set of colors, along with additional structure maps

$$P(c_1,\dots,c_n;c_0) \xrightarrow{\cong} P(c_2,\dots,c_n,c_0^\dagger;c_1^\dagger)$$

that combine into a $\mathbb{Z}/(n+1)\mathbb{Z}$-action on operations of arity n. This allows one to exchange the roles of inputs and outputs. If the underlying operad P is just a category, this notion coincides with that of being an involutive category.

Augmented cyclic operad: If one can exchange the roles of inputs and outputs, is there any real distinction between the two? Operations in an augmented cyclic operad dispense with this, and regard inputs and outputs on entirely the same footing. Operation sets take the form $P(c_1,\dots,c_n)$ (for $n \geq 0$), and compositions take the form

$$\begin{array}{c} P(c_1,\dots,c_n) \times P(d_1,\dots,d_m) \\ \downarrow{\scriptstyle {}_i\circ_j} \\ P(c_1,\dots,c_{i-1},d_{j+1},\dots,d_m,d_0,\dots,d_{j-1},c_{i+1},\dots,c_n) \end{array}$$

whenever $c_i = d_j^\dagger$. The real difference is that in a cyclic operad you cannot compose two operations which only have outputs but no inputs, while in an augmented cyclic operad this is possible, and the result will land in the set $P(\,)$ which has no analogue for cyclic operads.

Modular operads: These are augmented cyclic operads, together with contraction operations

$$P(c_1,\dots,c_n) \to P(c_1,\dots,\hat{c}_i,\dots,\hat{c}_j,\dots,c_n)$$

for $i < j$ with $c_i = c_j^\dagger$.

Definitions in these terms for the directed structures may be found, for instance, in [**YJ15**, Chapter 11]; see also [**GH18**, §2.2], [**Dun06**, §6.1], and [**HR15**]. For cyclic operads see [**Shu20**, Definition 7.4] and [**CGR14**, Definition 3.3], for augmented cyclic operads see [**DCH21**, Definition 2.3], and for modular operads see [**Ray21**, Definition 1.24]. But please don't go and look all of this up: according to the contents of Section 5, many of these operadic structures could instead be defined as *Segal presheaves* over the graph categories we introduce here, just as in the case of categories. There are of course a number of other interesting and useful operadic structures that do not appear on the above list, some of which can even fit into the frameworks we lay out below.

2. Graphs

In operadic contexts, the correct notion of graphs to use is one with 'loose ends.' There are a number of different combinatorial models for this, and we now recall one that has a simple associated notion of morphism, due to Joyal and Kock [**JK11**, §3].

DEFINITION 2.1 (Undirected graphs). An **undirected graph** G is a diagram of finite sets

$$\dagger \circlearrowright A_G \hookleftarrow D_G \xrightarrow{t} V_G$$

with $\dagger$ a fixpoint-free involution and $D_G \to A_G$ a monomorphism. We call A_G the set of **arcs** and V_G the set of **vertices**.

A corresponding encoding of *directed* graph with loose ends was introduced by Kock as [**Koc16**, Definition 1.1.1]. (Graphs of this type have recently been used to give new perspectives on Petri nets [**BGMS21, Koc23**].)

DEFINITION 2.2 (Directed graphs). A **directed graph** G is a diagram of finite sets of the form

$$E_G \hookleftarrow I_G \longrightarrow V_G \longleftarrow O_G \hookrightarrow E_G,$$

where the two end maps are monomorphisms. We call elements of E_G the **edges** of the graph.

Every directed graph determines a canonical associated undirected graph, with $A_G := E_G \amalg E_G$ together with the swap involution, and $D_G := I_G \amalg O_G$. We will give important examples in a moment, but for now let us introduce some subsidiary notions.

DEFINITION 2.3. Suppose G is an undirected graph.

- If $v \in V_G$ is a vertex, then $\mathrm{nb}(v) := t^{-1}(v) \subseteq D_G$ is the **neighborhood** of v.
- The **boundary** of G is the set $\eth(G) = A_G \setminus D_G$.
- The set of **edges** is the set of $\dagger$-orbits.

$$E_G = \{[a, a^\dagger] \mid a \in A_G\} \cong A_G/\sim$$

- An edge $[a, a^\dagger]$ is an **internal edge** if neither of $a, a^\dagger$ are elements of $\eth(G)$.

Suppose G is a directed graph.

- If $v \in V_G$ is a vertex, then $\mathrm{in}(v) \subseteq I_G$ is the preimage of v under $I_G \to V_G$, called the set of **inputs of** v. Likewise, v has a set of **outputs** $\mathrm{out}(v) \subseteq O_G$.
- The set of **input edges of** G is defined to be $\mathrm{in}(G) := E_G \setminus O_G$ and the set of **output edges of** G is $\mathrm{out}(G) := E_G \setminus I_G$. All other edges are **internal edges**.

Each undirected graph G determines a topological pair $\mathscr{X}_G = (X_G, V_G)$, called the **geometric realization**. Each arc in A_G gets an open interval, each element of D_G gets a half-open interval, each vertex gets a point, and these are glued together in the manner described by the diagram for G (e.g., if $t \in (0,1)$ is in the interval associated to a, then it will be identified with $1-t \in (0,1)$ in the interval associated to $a^\dagger$). We mention this mostly in that we will draw pictures of our graphs, and we will frequently exaggerate the size of the vertices. We freely use the terms

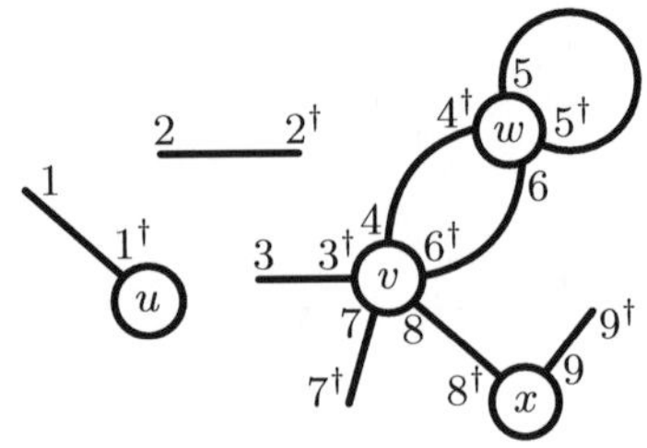

FIGURE 2. An undirected graph from Example 2.4

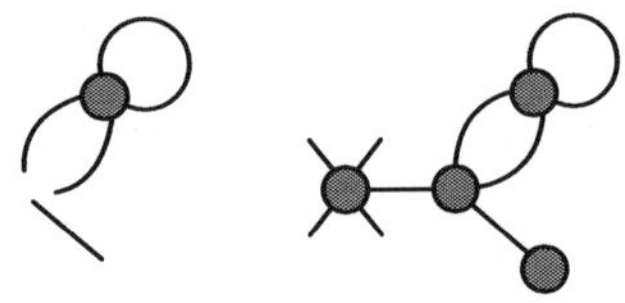

FIGURE 3. Subgraph determined by one vertex and four edges

'connected' and 'simply-connected' to refer to properties of the topological space X_G (though these can also be described combinatorially).

EXAMPLE 2.4. Figure 2 gives a picture of an undirected graph G with loose ends. The arc set $A_G = \{n, n^\dagger \mid 1 \leq n \leq 9\}$ has eighteen elements, together with the evident involution swapping n and $n^\dagger$, while the vertex set is $V_G = \{u, v, w, x\}$. The set D_G is given below

$$D_G = \{\underbrace{1^\dagger}_{\text{nb}(u)}, \underbrace{3^\dagger, 4, 6^\dagger, 8, 7}_{\text{nb}(v)}, \underbrace{4^\dagger, 5, 5^\dagger, 6}_{\text{nb}(w)}, \underbrace{8^\dagger, 9}_{\text{nb}(x)}\}$$

and the indicated partition specifies the function $t \colon D_G \to V_G$. The boundary set is $\eth(G) = \{1, 2, 2^\dagger, 3, 7^\dagger, 9^\dagger\}$, which has six elements.

REMARK 2.5. A different combinatorial definition of (undirected) graph with loose ends is common in the literature (see, e.g., [**KM94**, 6.6.1] or [**BM96**, Definition 1.1]). This is simply a diagram of finite sets

$$\circlearrowright F \to V$$

where the self-map of F is an involution. The fixed points of the involution represent the loose ends of the graph, and the free orbits represent edges between vertices. Given such a graph, we can form a graph as in Definition 2.1 by setting $D := F$, and to give A we double up the fixed points of the involution on F to turn it into a free involution. However, not all undirected graphs arise from this process: we never see graphs which have edges that float free. See [**BB17**, §15.3] for more details.

DEFINITION 2.6 (Subgraph). Suppose G is a graph, either directed or undirected. A **subgraph** of G is determined by a pair of subsets $E_H \subseteq E_G$ and $V_H \subseteq V_G$ subject to a single condition.

- Suppose G is undirected, and let $A_H \subseteq A_G$ denote the set of all arcs appearing in the edges in E_H; this is naturally an involutive subset of A_G. Further, let

$D_H := t^{-1}(V_H) \subseteq D_G$ be the set of those elements mapping into V_H. Then (E_H, V_H) determines a subgraph just when $D_H \to D_G \to A_G$ lands in the subset A_H.

$$\begin{array}{ccccc} \dagger \circlearrowright A_H & \leftarrow\dashrightarrow & D_H & \longrightarrow & V_H \\ \downarrow & & \downarrow & & \downarrow \\ \dagger \circlearrowright A_G & \hookleftarrow & D_G & \longrightarrow & V_G \end{array} \tag{1}$$

- Suppose G is directed, and let $I_H \subseteq I_G$ and $O_H \subseteq O_G$ be the preimages of the subset V_H along the functions $I_G \to V_G$ and $V_G \leftarrow O_G$. Then (E_H, V_H) determines a subgraph just when the dashed arrows in the following diagram exist.

$$\begin{array}{ccccccccc} E_H & \leftarrow\dashrightarrow & I_H & \longrightarrow & V_H & \longleftarrow & O_H & \dashrightarrow & E_H \\ \downarrow & & \downarrow & & \downarrow & & \downarrow & & \downarrow \\ E_G & \hookleftarrow & I_G & \longrightarrow & V_G & \longleftarrow & O_G & \hookrightarrow & E_G \end{array} \tag{2}$$

The top row in (1) or (2) exhibits H as an undirected or directed graph.

Of course we could write these requirements in other ways, such as declaring that a subgraph is a triple or quadruple of subsets satisfying some properties. But we prefer to emphasize that subgraphs are always determined by a set of edges and a set of vertices. Notice also that a subgraph of a directed graph is nothing but a subgraph of its underlying undirected graph.

We now describe two kinds of important graphs, which we term **elementary**. For a nonnegative integer n, we will write $\mathbf{n}$ for the set $\{1, 2, \dots, n\}$.

EXAMPLE 2.7 (Edge). The canonical **undirected edge**, here denoted $\updownarrow$, is

$$\dagger \circlearrowright \mathbf{2} \hookleftarrow \mathbf{0} \xrightarrow{t} \mathbf{0}.$$

That is, this graph has no vertices, and all arcs are in the boundary. The canonical **directed edge**, here denoted $\downarrow$, is

$$\mathbf{1} \hookleftarrow \mathbf{0} \longrightarrow \mathbf{0} \longleftarrow \mathbf{0} \hookrightarrow \mathbf{1}.$$

This graph has no vertices, and a single edge which is both an input and an output of the graph. We also refer to any connected graph without vertices as an edge.

EXAMPLE 2.8 (Stars or corollas). Let $n, m \geq 0$ be integers. The n-star, $\star_n$, is the following undirected graph

$$\dagger \circlearrowright \mathbf{n} + \mathbf{n} \hookleftarrow \mathbf{n} \xrightarrow{t} \mathbf{1},$$

which has a single vertex with an n-element neighborhood, and $\eth(\star_n) = \mathbf{n}$. The n, m-star, $\star_{n,m}$, is the following *directed* graph

$$\mathbf{n} + \mathbf{m} \hookleftarrow \mathbf{n} \longrightarrow \mathbf{1} \longleftarrow \mathbf{m} \hookrightarrow \mathbf{n} + \mathbf{m}.$$

The star $\star_{n,m}$ has a single vertex with n inputs and m outputs, and $\mathrm{in}(\star_{n,m}) = \mathbf{n}$, $\mathrm{out}(\star_{n,m}) = \mathbf{m}$. We will also refer to any connected graph with one vertex and no internal edges as a star.

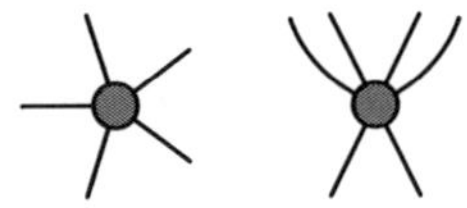

FIGURE 4. The stars $\star_5$ and $\star_{4,2}$

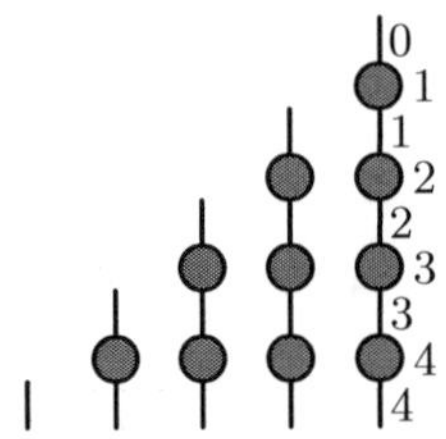

FIGURE 5. The first five linear graphs L_0, L_1, L_2, L_3, L_4

Every edge and star (including graphs isomorphic to these canonical forms) will be called **elementary**. The elementary graphs are precisely those graphs with no internal edges. Two examples of stars are given in Figure 4; we will always follow the convention that in a directed graph, the flow goes from top to bottom (in graphs with loops, this rule will apply locally at each vertex).

If $e \in E_G$ is any edge in a graph G, we may consider e as a subgraph as in Definition 2.6. But vertices in a graph need *not* span a star-shaped subgraph in the same way. For example, in Figure 2, the smallest subgraph containing the vertex w has one vertex and arc set $\{4, 4^\dagger, 6, 6^\dagger, 5, 5^\dagger\}$, and is not a star (the other three vertices all span star-shaped subgraphs). If v is a vertex in a graph G, we will write $\star_v$ for a star with $D_{\star_v} := \text{nb}(v)$ (when G is undirected) or $I_{\star_v} := \text{in}(v)$ and $O_{\star_v} := \text{out}(v)$ (when G is directed). Despite the fact that $\star_v$ might not be a subgraph of G, there is always a map (see Definition 2.12 below) $\iota_v \colon \star_v \to G$.

EXAMPLE 2.9 (Linear graphs). The linear directed graph L_n with n vertices (see Figure 5) is presented as follows.

$$\{0, 1, \ldots, n\} \xleftarrow{-1} \mathbf{n} \xrightarrow{\text{id}} \mathbf{n} \xleftarrow{\text{id}} \mathbf{n} \xrightarrow{+0} \{0, 1, \ldots, n\}$$

We have $\text{in}(L_n) = \{0\}$, $\text{out}(L_n) = \{n\}$, while the vertex k has $\text{in}(k) = \{k-1\}$ and $\text{out}(k) = \{k\}$. We also call the associated undirected graph a linear graph.

DEFINITION 2.10 (paths, cycles, trees). Let G be an undirected graph.

- A **weak path** in G is a finite alternating sequence of edges and vertices of G so that if e and v are adjacent then some arc of e appears in $\text{nb}(v)$, and so that the pattern vev can only appear in the path if *both* arcs of e are in $\text{nb}(v)$.
- A **path** in G is a finite alternating sequence of arcs and vertices of G so that if av appears in the sequence then $t(a) = v$, and if va appears in the sequence then $t(a^\dagger) = v$.
- A **cycle** is a path of length strictly greater than one that begins and ends at the same arc or same vertex.
- The graph G is a **tree** if and only if it is connected and does not have any paths which are cycles.

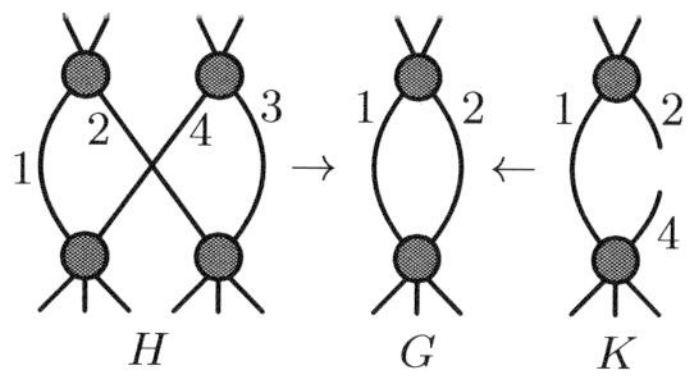

FIGURE 6. Example (directed) étale maps $H \to G \leftarrow K$

Each path gives a weak path by replacing the arcs with the edges they span, and each weak path has at least one associated path. The only ambiguity is how to handle the patterns of the form vev. In a tree, there is no real difference between paths and weak paths. The property of being connected is equivalent to having a weak path between every pair of distinct elements of $E_G \amalg V_G$.

DEFINITION 2.11. Let G be a directed graph.

- An **undirected path** in G is a weak path in the associated undirected graph.
- A **directed path** in G is a weak path in the associated undirected graph, so that the following hold:
  - If ev appears in the path, then $e \in \mathrm{in}(v)$.
  - If ve appears in the path, then $e \in \mathrm{out}(v)$.
- A **directed cycle** is a directed path of length strictly greater than one that begins and ends at the same edge or vertex.
- A directed graph G is **acyclic** if it is connected and does not contain any directed cycles.

2.1. Étale maps and embeddings. Étale maps are certain functions between graphs that preserve structure and preserve arity of vertices (that is, the cardinality of $\mathrm{nb}(v)$ and $\mathrm{nb}(f(v))$ coincide, and similarly for the directed case). The following definitions appear in [**JK11**] and [**Koc16**, 1.1.7].

DEFINITION 2.12. An **étale map** $G \to G'$ between undirected graphs is a triple of functions $A_G \to A_{G'}$, $D_G \to D_{G'}$, and $V_G \to V_{G'}$ so that the diagram below left commutes with right hand square a pullback.

$$\begin{array}{ccccccc} A_G & \xleftarrow{\dagger} & A_G & \hookleftarrow & D_G & \longrightarrow & V_G \\ \downarrow & & \downarrow & & \downarrow & \lrcorner & \downarrow \\ A_{G'} & \xleftarrow{\dagger'} & A_{G'} & \hookleftarrow & D_{G'} & \longrightarrow & V_{G'} \end{array} \qquad \begin{array}{ccccccccc} E_G & \hookleftarrow & I_G & \longrightarrow & V_G & \longleftarrow & O_G & \hookrightarrow & E_G \\ \downarrow & & \downarrow & \lrcorner & \downarrow & \llcorner & \downarrow & & \downarrow \\ E_{G'} & \hookleftarrow & I_{G'} & \longrightarrow & V_{G'} & \longleftarrow & O_{G'} & \hookrightarrow & E_{G'} \end{array}$$

Similarly, an étale map $G \to G'$ between *directed* graphs is a quadruple of functions $E_G \to E_{G'}$, $I_G \to I_{G'}$, $O_G \to O_{G'}$, and $V_G \to V_{G'}$ so that the diagram above right commutes and both middle squares are pullbacks.

Étale maps induce local homeomorphisms between geometric realizations. If G is an undirected graph, then the set of étale maps $\updownarrow \to G$ is in bijection with A_G; likewise if G is directed then $\{\downarrow \to G\} \cong E_G$. Similarly, étale maps from stars (modulo isomorphism in the domain) classify vertices of a graph. Every étale map between directed graphs induces an étale map between the associated undirected graphs. It is immediate that every subgraph in the sense of Definition 2.6 determines an étale map. But the latter are strictly more general.

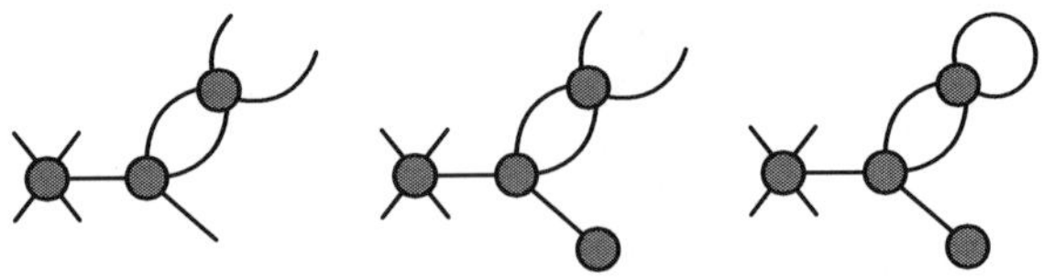

FIGURE 7. Two embeddings

EXAMPLE 2.13. In this example, we will refer to two étale maps $H \to G \leftarrow K$ between directed graphs appearing in Figure 6. Following our convention, the graphs have vertices with two inputs and two outputs, and also vertices with two inputs and three outputs. The edges 2 and 4 in H do not touch, and this graph is only drawn this way to make the étale map easier to see. Since G has only one $(2,2)$ vertex and one $(2,3)$ vertex, the actions of the maps on vertices are forced. The actions on the labelled middle edges are given by preservation of parity, and the action on the inputs and outputs of the graphs are the visually expected ones. We could likewise consider these as étale maps of undirected graphs. These examples show that étale maps do not need to be injective on vertices, and even if they are, they do not need to be injective on edges or arcs.

REMARK 2.14. Étale maps between undirected graphs take paths to paths (one does not even need the pullback condition for this). It follows that $G \to G'$ is an étale map between undirected graphs with G' a tree, then each connected component of G is a tree as well. Likewise, since étale maps between directed graphs preserve directed paths, if $G \to G'$ is such a map with G' acyclic, then each connected component of G is acyclic as well. Of course in this case if G' is simply-connected, then so is each component of G, by passing to the associated undirected graphs.

We are mainly interested in connected graphs in what follows, and we make the following definition.

DEFINITION 2.15. An **embedding** between connected graphs is an étale map which is injective on vertices. We denote an embedding by $G \rightarrowtail G'$.

Since monomorphisms are stable under pullback, if $G \rightarrowtail G'$ is an embedding between undirected graphs, then $D_G \to D'_G$ is a monomorphism (and likewise for $I_G \to I_{G'}$ and $O_G \to O_{G'}$ in the directed case). Embeddings between connected graphs yield injective local homeomorphisms between geometric realizations. Every subgraph inclusion in the sense of Definition 2.6 is an embedding, but embeddings are strictly more general. For instance, Figure 7 indicates two consecutive embeddings (from left to right), where the second indicates the kind of clutching behavior possible in embeddings, which can identify some arcs. The first is simply a subgraph inclusion, while the second is a bijection on D and V, but on arcs is not a monomorphism.

Any étale map between trees is a monomorphism (see, for instance, [**Ray21**, Corollary 4.23]), hence an embedding. For acyclic directed graphs the situation is more complicated: there can be étale maps between such which are not embeddings, and embeddings which are not injective, as we saw in Example 2.13/Figure 6.

CONSTRUCTION 2.16 (Lifting embeddings). Suppose G is an undirected graph and H is a directed graph, and temporarily let $U(H)$ be the undirected graph associated to H. If $f\colon G \rightarrowtail U(H)$ is an embedding, then there is a directed graph G' and an embedding $f'\colon G' \rightarrowtail H$, so that $G \cong U(G') \rightarrowtail U(H)$ is the map f. Moreover, the graph G' and embedding f' are essentially unique with respect to this property. We leave most details to the reader. Our starting data is the map f, as below.

$$\begin{array}{ccccccc} A_G & \xleftarrow{\dagger} & A_G & \hookleftarrow & D_G & \longrightarrow & V_G \\ \downarrow & & \downarrow & & \downarrow & \lrcorner & \downarrow \\ E_H \amalg E_H & \xleftarrow{\text{swap}} & E_H \amalg E_H & \hookleftarrow & I_H \amalg O_H & \longrightarrow & V_H \end{array}$$

We set $E_{G'} := E_G$ and $V_{G'} := V_G$, and define $I_{G'}$ and $O_{G'}$ as subsets of D_G fitting into the following pullbacks.

$$\begin{array}{ccccccccc} I_{G'} & \hookrightarrow & D_G & \longrightarrow & V_G & \longleftarrow & D_G & \hookleftarrow & O_{G'} \\ \downarrow & \lrcorner & \downarrow & \lrcorner & \downarrow & \llcorner & \downarrow & \llcorner & \downarrow \\ I_H & \hookrightarrow & I_H \amalg O_H & \longrightarrow & V_H & \longleftarrow & I_H \amalg O_H & \hookleftarrow & O_H \end{array}$$

Using that f is an embedding, one can show that G' is a directed graph (that is, $I_{G'} \to E_{G'}$ and $O_{G'} \to E_{G'}$ are injective). Further, $f'\colon G' \rightarrowtail H$ is an embedding.

We now introduce a set associated to graphs which will be used to describe all manner of graphical maps below.

DEFINITION 2.17. If G is an directed or undirected connected graph, we define $\mathrm{Emb}(G)$ to be the set of embeddings with codomain G, modulo isomorphisms in the domain.

There is a partial order on $\mathrm{Emb}(G)$ determined by factorization of embeddings. As a partially-ordered set, $\mathrm{Emb}(G)$ has a unique maximal element given by the identity of G, and each edge determines a minimal element (if G has at least one edge, that is, if it is not $\star_0$, then the set of minimal elements is precisely E_G).

In light of Construction 2.16, if G is a directed graph and $U(G)$ is its associated undirected graph, then $\mathrm{Emb}(G) \to \mathrm{Emb}(U(G))$ is an isomorphism. For this reason we do not distinguish notationally between the two: $\mathrm{Emb}(G)$ is fundamentally a set associated to an undirected graph.

Every connected subgraph (in the sense of Definition 2.6) of G determines an element of $\mathrm{Emb}(G)$, and if G is a tree, then $\mathrm{Emb}(G)$ coincides with the set of connected subgraphs. Thus if T is a subtree of a tree G, we will usually write $T \in \mathrm{Emb}(G)$ for the class of the associated inclusion $[T \rightarrowtail G]$.

For an arbitrary graph G, a vertex v determines a subgraph $\star_v$ just when there are no loops at the vertex v, but a vertex always determines an embedding $\iota_v\colon \star_v \rightarrowtail G$ picking out the vertex ($\star_v$ is a star with $D_{\star_v} = \mathrm{nb}(v) \subseteq D_G$). There are canonical inclusions

$$E_G \hookrightarrow \mathrm{Emb}(G) \qquad\qquad V_G \hookrightarrow \mathrm{Emb}(G)$$

of edges and vertices into the embedding set, which we regard as subset inclusions.

DEFINITION 2.18. Suppose that $h\colon H \rightarrowtail G$ and $k\colon K \rightarrowtail G$ are two embeddings. An embedding $\ell\colon L \rightarrowtail G$ is called **a union of h and k** if

(1) $[\ell]$ is an upper bound for both $[h]$ and $[k]$ in the poset $\mathrm{Emb}(G)$ (that is, there is a factorization $H \rightarrowtail L \rightarrowtail G$ of h and likewise for k), and

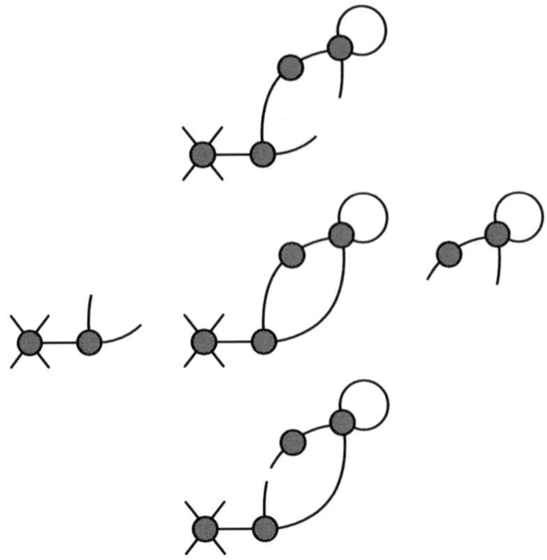

FIGURE 8. Example of non-uniqueness of unions

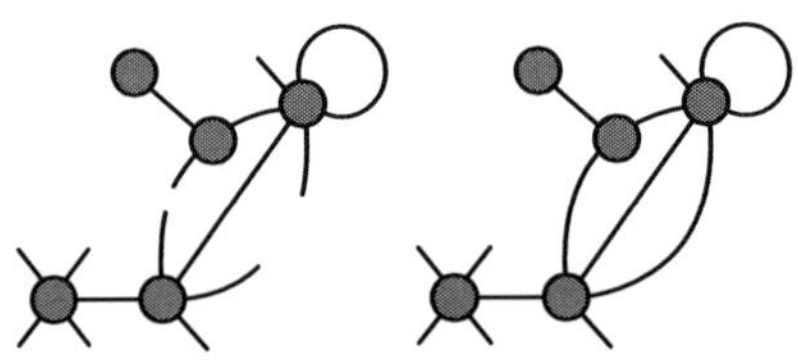

FIGURE 9. Even self-unions are not unique

(2) $\ell(V_L) = h(V_H) \cup k(V_K)$.

The two embeddings are **vertex disjoint** if

(3) $h(V_H) \cap k(V_K) = \varnothing$.

Unions need not be unique. In Figure 8, the two subgraphs to the left and right of the graph G in the center have three possible unions in $\mathrm{Emb}(G)$. These are pictured in the center column, and include the maximum $[\mathrm{id}_G] \in \mathrm{Emb}(G)$. Note that these subgraphs do not have a *least* upper bound in $\mathrm{Emb}(G)$; also note that least upper bounds that do exist may fail to be unions. As another example, the embedding pictured in Figure 9 has *four* possible self-unions, determined by which of the two snipped edges one decides to glue together.

If H and K are subgraphs of a graph G in the sense of Definition 2.6, then there is a natural notion of union subgraph obtained by setting $E_{H \cup K} = E_H \cup E_K \subseteq E_G$ and $V_{H \cup K} = V_H \cup V_K \subseteq V_G$. If H and K are both connected subgraphs of a connected graph G, then $H \cup K$ will be a union in the above sense just when H and K *overlap*, that is, just when at least one of $E_H \cap E_K$ or $V_H \cap V_K$ is inhabited. In fact, $H \cup K$ is the only union which is also a subgraph, by essentially the same proof as [**Hac24**, Lemma B.8]. In particular, if G is a tree then we know all embeddings come from subtrees, hence any pair $S, T \in \mathrm{Emb}(G)$ has at most one union.

2.2. Boundary-compatibility. We end this section with one further construction that we will use in our definition of graph maps, that of *boundary-compatibility* in Definition 2.19. If $f\colon H \rightarrowtail G$ is an embedding between undirected graphs, then the composite

$$\eth(H) \hookrightarrow A_H \to A_G$$

is a monomorphism (see [**HRY20a**, Lemma 1.20]), and we write $\eth(f) \subseteq A_G$ for its image, which we call the **boundary of** f. For example, if G is the graph from Example 2.4 (see Figure 2) and f is the star-embedding picking out the vertex w, then $\eth(f) = \{4, 6^\dagger, 5, 5^\dagger\}$. Likewise, if $f\colon H \rightarrowtail G$ is an embedding between directed graphs, then $\mathrm{in}(H) \to E_G$ and $\mathrm{out}(H) \to E_G$ are also monomorphisms, and we write $\mathrm{in}(f), \mathrm{out}(f) \subseteq E_G$ for the images of these sets.

The assignment $\eth$ on embedddings descends to a function $\mathrm{Emb}(G) \to \wp(A_G)$ landing in the power set of A_G. It is more convenient to think about a multi-set structure, so we instead consider the function

$$\eth\colon \mathrm{Emb}(G) \to \wp(A_G) \to \mathbb{N}A_G$$

landing in the free commutative monoid on A_G (whose elements are finite unordered lists of elements of A_G). Likewise, we have functions

$$\mathrm{in}, \mathrm{out}\colon \mathrm{Emb}(G) \to \wp(E_G) \to \mathbb{N}E_G$$

when G is directed.

DEFINITION 2.19. Suppose G and G' are two undirected connected graphs, $\varphi_0\colon A_G \to A'_G$ is an involutive function, and $\hat{\varphi}\colon \mathrm{Emb}(G) \nrightarrow \mathrm{Emb}(G')$ is a partially-defined function. We say that the pair $\varphi := (\varphi_0, \hat{\varphi})$ is **boundary-compatible** if the diagram

$$\begin{array}{ccc} \mathrm{Emb}(G) & \xrightarrow{\eth} & \mathbb{N}A_G \\ \big\downarrow{\scriptstyle\hat{\varphi}} & & \big\downarrow{\scriptstyle\mathbb{N}\varphi_0} \\ \mathrm{Emb}(G') & \xrightarrow{\eth} & \mathbb{N}A_{G'} \end{array}$$

commutes on the domain of definition of $\hat{\varphi}$. Likewise, if G and G' are directed connected graphs, $\varphi_0\colon E_G \to E'_G$ is a function, and $\hat{\varphi}\colon \mathrm{Emb}(G) \nrightarrow \mathrm{Emb}(G')$ is a partially-defined function, then we say that the pair $\varphi := (\varphi_0, \hat{\varphi})$ is **boundary-compatible** if the diagram

$$\begin{array}{ccccc} \mathbb{N}E_G & \xleftarrow{\mathrm{in}} & \mathrm{Emb}(G) & \xrightarrow{\mathrm{out}} & \mathbb{N}E_G \\ \big\downarrow{\scriptstyle\mathbb{N}\varphi_0} & & \big\downarrow{\scriptstyle\hat{\varphi}} & & \big\downarrow{\scriptstyle\mathbb{N}\varphi_0} \\ \mathbb{N}E_{G'} & \xleftarrow{\mathrm{in}} & \mathrm{Emb}(G') & \xrightarrow{\mathrm{out}} & \mathbb{N}E_{G'} \end{array}$$

commutes on the domain of definition of $\hat{\varphi}$.

We are writing the diagrams above using partial maps for conciseness, since we will often want to discuss functions between subsets of $\mathrm{Emb}(G)$ and $\mathrm{Emb}(G')$ (Definition 3.1, Definition 4.8, Proposition A.6), and work with the restrictions of $\eth$ (or in and out) to those subsets.

The standard example of a boundary compatible pair is one associated to an embedding $f\colon G \rightarrowtail G'$. The function φ_0 is defined to be $A_G \to A_{G'}$ (or $E_G \to E_{G'}$ if G, G' are directed), and the (total) function $\hat{\varphi}\colon \mathrm{Emb}(G) \to \mathrm{Emb}(G')$ is defined by post-composition with f.

3. Categories of simply-connected graphs

DEFINITION 3.1. Suppose G and G' are undirected trees.

(1) A **tree map** is a boundary-compatible pair (φ_0, φ_1) consisting of an involutive function $\varphi_0 \colon A_G \to A_{G'}$ and a function $\varphi_1 \colon V_G \to \mathrm{Emb}(G')$.
(2) Suppose $\varphi = (\varphi_0, \hat{\varphi})$ is a pair consisting of an involutive function $\varphi_0 \colon A_G \to A_{G'}$ and a function $\hat{\varphi} \colon \mathrm{Emb}(G) \to \mathrm{Emb}(G')$. We say that φ is a **full tree map** if is boundary-compatible, and, if two subtrees S and T of G overlap, then so do $\hat{\varphi}(S)$ and $\hat{\varphi}(T)$ and we have $\hat{\varphi}(S \cup T) = \hat{\varphi}(S) \cup \hat{\varphi}(T)$.

The full tree maps form a category via composition of pairs, which we denote by $\mathbf{U}_0$. Trees with inhabited boundary span the full subcategory $\mathbf{U}_{\mathrm{cyc}} \subset \mathbf{U}_0$.

Every full tree map determines a tree map by restriction of $\hat{\varphi}$ along $V_G \hookrightarrow \mathrm{Emb}(G)$, and we show in Proposition A.6 of Appendix A.2 that this gives a bijection between tree maps and full tree maps. In light of this, we will use the shorter term 'tree map' also for full tree maps. Since subtrees are uniquely determined by their boundaries, boundary-compatibility implies that tree maps send edges to edges.

It turns out that tree maps also preserve *intersections* between overlapping subtrees. This follows from [**Hac24**, Theorem A.7 and Proposition B.22], but we provide an elementary, direct proof in Appendix A.1.

We now give an alternative description of $\mathbf{U}_{\mathrm{cyc}}$ as a full subcategory of the category of *cyclic operads*, **Cyc**, and likewise of $\mathbf{U}_0$ as a full subcategory of *augmented cyclic operads*, $\mathbf{Cyc}^+$. The distinction between the two is that augmented cyclic operads P are allowed to have a meaningful set $P(\,)$ of operations, which cannot compose with any other operations. The term 'augmented cyclic operad' follows [**HV02**], and would be called 'entries-only $*$-polycategories' in [**Shu20**] or simply 'cyclic operads' in [**DCH21**, Definition 2.3]; see [**Shu20**, Example 7.7] for some discussion on terminology. If $\mathfrak{C}$ is an involutive set of colors, we write $\mathbf{Cyc}_{\mathfrak{C}} \subset \mathbf{Cyc}^+_{\mathfrak{C}}$ for categories of $\mathfrak{C}$-colored objects and identity-on-colors morphisms.

Each tree G determines a free object $\mathsf{C}(G)$ in $\mathbf{Cyc}^+_{A_G}$. The generating set of $\mathsf{C}(G)$ is precisely the set V_G, where $v \in V_G$ is considered as an element of $\mathsf{C}(G)(\eth(\star_v)) = \mathsf{C}(G)(\mathrm{nb}(v)^\dagger)$. A variation of [**HRY20b**, §2.2] allows one to show that this becomes a functor $\mathsf{C} \colon \mathbf{U}_0 \to \mathbf{Cyc}^+$.

PROPOSITION 3.2. *The category $\mathbf{U}_0$ is equivalent to the full subcategory of augmented cyclic operads $\mathbf{Cyc}^+$ on the objects $\mathsf{C}(G)$ (where G ranges over all trees). This equivalence restricts to an equivalence between $\mathbf{U}_{\mathrm{cyc}}$ and the full subcategory of cyclic operads $\mathbf{Cyc} \subset \mathbf{Cyc}^+$ on the objects $\mathsf{C}(G)$, where G ranges over all trees with inhabited boundary.*

The second part of this is simply the observation that $\mathsf{C}(G)$ is a cyclic operad if and only if G has inhabited boundary. The first part of the statement will be proved in Appendix A.3.

Turning now to directed trees, we introduce the 'dioperadic graphical category.'

DEFINITION 3.3. Suppose G and G' are directed trees. A **tree map** from G to G' is a boundary-compatible pair $\varphi = (\varphi_0, \hat{\varphi})$ consisting of a function $\varphi_0 \colon E_G \to E_{G'}$ and a function $\hat{\varphi} \colon \mathrm{Emb}(G) \to \mathrm{Emb}(G')$ which preserves unions: if S and T overlap, then $\hat{\varphi}(S \cup T) = \hat{\varphi}(S) \cup \hat{\varphi}(T)$. These form a category of all directed trees, which we denote by $\mathbf{O}_0$.[2]

[2]This was called $\mathbf{G}_{\mathrm{sc}}$ in [**CH22**], and is equivalent to the category $\mathbf{\Theta}$ in [**HRY15**, §6.3.5].

Of course one can equivalently define tree maps by replacing $\hat{\varphi}$ with a function $\varphi_1\colon V_G \to \mathrm{Emb}(G')$, as in Definition 3.1. A version of Proposition 3.2 then holds in this setting, with essentially the same proof: the category $\mathbf{O}_0$ is equivalent to a full subcategory (spanned by the directed trees) of the category of dioperads.

By restricting the objects further, we conclude that the dendroidal category from [**MW07**], originally defined as a category of the operads associated to rooted trees, is equivalent to a full subcategory of $\mathbf{O}_0$.

DEFINITION 3.4. The Moerdijk–Weiss **dendroidal category**, denoted by $\mathbf{\Omega}$, is the full subcategory of $\mathbf{O}_0$ consisting of those trees G so that $\mathrm{out}(v)$ is a singleton set for every $v \in V_G$. The **simplicial category** $\mathbf{\Delta}$ can be identified with the full subcategory of $\mathbf{O}_0$ on the linear graphs from Example 2.9.

Of course if one is only interested in rooted trees, simpler descriptions are possible; see, for example, [**Koc11**].

4. On several categories of graphs

We now extend Definition 3.1 and Definition 3.3 to maps between graphs that are not simply-connected. This formulation first appeared in [**Hac24**].

DEFINITION 4.1. Let G and G' be connected undirected (resp. directed) graphs. A **graph map** $\varphi\colon G \to G'$ is a pair $(\varphi_0, \hat{\varphi})$ consisting of a function $\hat{\varphi}\colon \mathrm{Emb}(G) \to \mathrm{Emb}(G')$ as well as

(1) an involutive function $\varphi_0\colon A_G \to A_{G'}$ if G and G' are undirected, or
(2) a function $\varphi_0\colon E_G \to E_{G'}$ if G and G' are directed.

The pair $(\varphi, \hat{\varphi})$ must satisfy the following four conditions:

(i) The function $\hat{\varphi}$ sends edges to edges.
(ii) The function $\hat{\varphi}$ preserves unions.
(iii) The function $\hat{\varphi}$ takes vertex disjoint pairs to vertex disjoint pairs.
(iv) The pair $(\varphi, \hat{\varphi})$ is boundary-compatible (in the sense of Definition 2.19).

Composition of graph maps is given by composition of the two constituent functions. This yields a category $\mathbf{U}$ of undirected connected graphs and their graph maps, as well as a category $\mathbf{O}$ of directed connected graphs and their graph maps.

PROPOSITION 4.2. *The category* $\mathbf{U}$ *of undirected connected graphs and their graph maps is equivalent to the graphical category from* [**HRY20a**, Definition 1.31]. *The category* $\mathbf{O}$ *of directed connected graphs and their graph maps is equivalent to the wheeled properadic graphical category from* [**HRY18**, §2].

PROOF. This result appears in [**Hac24**]: the first part is Theorem 4.15, and the second part combines Proposition 5.10 and Theorem 5.13. 'Wheeled properadic graphical category' refers to the category $\mathcal{B}$ from [**HRY18**]. □

The next proposition uses that tree maps send edges to edges and preserve vertex-disjoint pairs (Lemma A.2).

PROPOSITION 4.3. *The category* $\mathbf{U}_0$ *is the full subcategory of* $\mathbf{U}$ *on the undirected trees. The category* $\mathbf{O}_0$ *is the full subcategory of* $\mathbf{O}$ *on the directed trees.* □

REMARK 4.4. If $\mathbf{C}$ is a category, let $\widehat{\mathbf{C}} := \mathrm{Fun}(\mathbf{C}^{\mathrm{op}}, \mathbf{Set})$ be the category of set-valued presheaves. If $X \in \widehat{\mathbf{C}}$ is a presheaf and $G \in \mathbf{C}$ is an object, we will write X_G for the value of X at G. The Yoneda embedding $\mathbf{C} \to \widehat{\mathbf{C}}$ takes an object G to the representable presheaf $\mathbf{C}(-,G)$. An important example, when $\mathbf{C} = \mathbf{U}$ or some full subcategory, is the **orientation presheaf** $\mathfrak{o}$. This is defined by

$$\mathfrak{o}_G = \{\text{involutive functions } A_G \to \{+1,-1\}\}$$

where the set $\{+1,-1\}$ is equipped with the free involution. This additional data determines a directed structure on G, see [**Hac24**, Construction 5.4]. Meanwhile, every directed graph determines a canonical element of the orientation presheaf. In fact, we have equivalences of categories

$$\mathbf{O} \xrightarrow{\simeq} \mathbf{U} \times_{\widehat{\mathbf{U}}} \widehat{\mathbf{U}}_{/\mathfrak{o}}$$

$$\mathbf{O}_0 \xrightarrow{\simeq} \mathbf{U}_0 \times_{\widehat{\mathbf{U}_0}} \widehat{\mathbf{U}_0}_{/\mathfrak{o}}$$

whose first projections forget the directed structure. In other words, the category of elements of $\mathfrak{o} \in \widehat{\mathbf{U}}$ (resp. $\mathfrak{o} \in \widehat{\mathbf{U}_0}$) is $\mathbf{O}$ (resp. $\mathbf{O}_0$). See [**Hac24**, Proposition 5.10].

A similar construction appears in the context of Feynman categories and Borisov–Manin graph morphisms in [**KL17**, 6.4.1], and for étale maps in [**Ray21**, §4.5].

DEFINITION 4.5 (Active and inert maps). A graph map $\varphi = (\varphi_0, \hat{\varphi})\colon G \to G'$ is an **active map** if $\hat{\varphi}\colon \mathrm{Emb}(G) \to \mathrm{Emb}(G')$ preserves the maximum ($\hat{\varphi}([\mathrm{id}_G]) = [\mathrm{id}_{G'}]$). We write $\mathbf{U}_{\mathrm{act}} \subset \mathbf{U}$ and $\mathbf{O}_{\mathrm{act}} \subset \mathbf{O}$ for the wide subcategories of active maps, and decorate the arrows as $G \twoheadrightarrow G'$. A graph map φ is **inert** if it comes from an embedding; in other words, if the the dashed map in the diagram

$$\begin{array}{ccc} V_G & \dashrightarrow & V_{G'} \\ \downarrow & & \downarrow \\ \mathrm{Emb}(G) & \xrightarrow{\hat{\varphi}} & \mathrm{Emb}(G') \end{array}$$

exists. We write $\mathbf{U}_{\mathrm{int}} \subset \mathbf{U}$ and $\mathbf{O}_{\mathrm{int}} \subset \mathbf{O}$ for the subcategories of inert maps, and decorate the arrows by $G \rightarrowtail G'$.

Every graph map $\varphi\colon G \to G'$ factors as an active map followed by an inert map. Indeed, if $\hat{\varphi}([\mathrm{id}_G]) = [H \rightarrowtail G']$, then the factorization takes on the form $G \twoheadrightarrow H \rightarrowtail G'$. Constructing the active map $G \twoheadrightarrow H$ is delicate in general since $A_H \to A_{G'}$ (resp. $E_H \to E_{G'}$) need not be a monomorphism. Nevertheless, we have the following (see [**HRY20a**, Theorem 2.15]).

THEOREM 4.6. *The pair* $(\mathbf{U}_{\mathrm{act}}, \mathbf{U}_{\mathrm{int}})$ *is an orthogonal factorization system*[3] *on the category* $\mathbf{U}$, *as is* $(\mathbf{O}_{\mathrm{act}}, \mathbf{O}_{\mathrm{int}})$ *on* $\mathbf{O}$. *These factorization systems restrict to factorization systems on the subcategories*

$$\mathbf{U}_{\mathrm{cyc}} \subset \mathbf{U}_0 \subset \mathbf{U} \textit{ and } \mathbf{\Delta} \subset \mathbf{\Omega} \subset \mathbf{O}_0 \subset \mathbf{O}.$$

[3]An **orthogonal factorization system** on a category $\mathbf{C}$ is a pair (L, R) of classes of maps, each closed under composition and containing all of the isomorphisms. These have the property that any morphism of $\mathbf{C}$ factors as $r \circ \ell$, with $r \in R$ and $\ell \in L$, and this factorization is unique, up to unique isomorphism.

4.1. Properadic graphical category. We briefly describe the properadic graphical category $\mathbf{G}$, whose objects are connected *acyclic* directed graphs, in the sense of Definition 2.11. It requires a somewhat subtle notion of *structured subgraph*, which we give in a reformulation first appearing as [**Koc16**, 1.6.5]; see also [**CH22**, Remark 2.2.4].

DEFINITION 4.7 (Structured subgraph). A **naïve morphism of directed graphs** has the same data as an étale map Definition 2.12, but we do not require the middle squares to be pullbacks. Suppose G is an acyclic directed graph. A connected subgraph $H \subseteq G$ is a **structured subgraph** if the inclusion $H \rightarrowtail G$ is right orthogonal in the category of naïve morphisms of directed graphs to the maps $\downarrow \amalg \downarrow \to L_n$ (for all $n \geq 0$) picking out the input and output edge in the linear graph L_n. In other words, given a commutative square as below (whose horizontal arrows may just be naïve morphisms), there is a unique dashed lift.

$$\begin{array}{ccc} \downarrow \amalg \downarrow & \longrightarrow & H \\ {\scriptstyle 0,n}\downarrow & \overset{\exists!}{\nearrow} & \downarrow \\ L_n & \longrightarrow & G \end{array}$$

We write $\mathrm{sSb}(G) \subseteq \mathrm{Emb}(G)$ for the set of structured subgraphs of G.

This can alternately be phrased in terms of 'graph substitution' (see [**CH22**, Remark 2.2.4]). Every edge and star subgraph is a structured subgraph, so we have $E_G, V_G \subseteq \mathrm{sSb}(G)$. Note that if T is a *simply-connected* directed graph (i.e., an object in $\mathbf{O}_0$), then every connected subgraph is a structured subgraph, so we may identify $\mathrm{sSb}(T)$ with $\mathrm{Emb}(T)$.

The following formulation is due to Hongyi Chu and the author [**CH22**, Definition 2.2.11], who also showed it is equivalent to the graphical category from [**HRY15**, Chapter 6]. This description is akin to the simpler Definition 3.3, but predates it.

DEFINITION 4.8. The **properadic graphical category**, denoted $\mathbf{G}$, has ob jects the acyclic directed graphs. A morphism $\varphi \colon G \to G'$ consists of a pair of functions $\varphi_0 \colon E_G \to E_{G'}$ and $\hat{\varphi} \colon \mathrm{sSb}(G) \to \mathrm{sSb}(G')$ so that $(\varphi_0, \hat{\varphi})$ is boundary-compatible and $\hat{\varphi}$ preserves unions. By 'preserving unions,' we mean that if $H_1, H_2 \in \mathrm{sSb}(G)$ are such that $H_1 \cup H_2$ is a structured subgraph, then $\hat{\varphi}(H_1 \cup H_2) = \hat{\varphi}(H_1) \cup \hat{\varphi}(H_2)$. Composition in $\mathbf{G}$ is given by composition of pairs.

It turns out that given a map in $\mathbf{G}$, one can extend $\hat{\varphi} \colon \mathrm{sSb}(G) \to \mathrm{sSb}(G')$ to a function $\mathrm{Emb}(G) \to \mathrm{Emb}(G')$ in a canonical way. This is one ingredient in the following (see [**Hac24**] for a proof).

THEOREM 4.9. *The properadic graphical category* $\mathbf{G}$ *may be identified with the subcategory of* $\mathbf{O}$ *with*

(1) *objects the acyclic directed graphs, and*
(2) *morphisms those* $\varphi \colon G \to G'$ *so that* $\hat{\varphi}(G) \in \mathrm{sSb}(G')$.

Since $\mathrm{sSb}(G)$ and $\mathrm{Emb}(G)$ coincide when G is simply-connected, Proposition 4.3 implies the following (or directly compare Definition 3.3 and Definition 4.8).

COROLLARY 4.10. *The full subcategory of* $\mathbf{G}$ *on the directed trees is* $\mathbf{O}_0$. $\square$

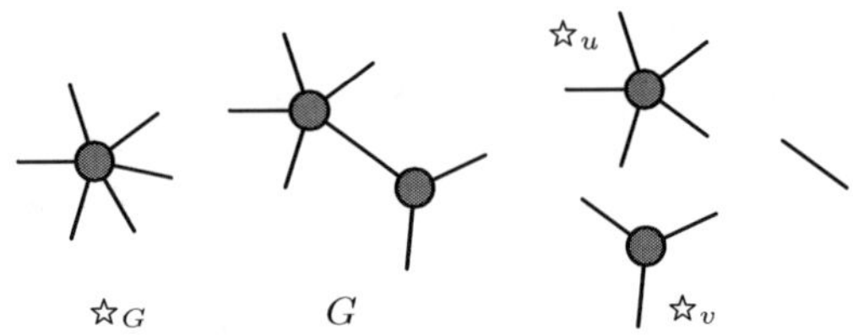

FIGURE 10. Graph G with two vertices and one internal edge

The notions of active and inert maps from Definition 4.5 can be modified in this context, where now maps in $\mathbf{G}_{\text{int}}$ are those maps which are isomorphic to a structured subgraph inclusion. This again is a factorization system $(\mathbf{G}_{\text{act}}, \mathbf{G}_{\text{int}})$, restricted from the one on $\mathbf{O}$ from Theorem 4.6 (see [**Koc16**, 2.4.14]).

5. The Segal condition and generalized operads

In this section, we turn to the main topic of the paper, which is the interpretation of generalized operads as Segal presheaves for appropriate graph categories.

First, let us give an idea about how to understand certain elements in a presheaf. Let $X \in \widehat{\mathbf{U}}$ be a $\mathbf{U}$-presheaf. We interpret the set $X_{\updownarrow}$ as a set of colors. Since $\updownarrow$ has a non-trivial automorphism in $\mathbf{U}$, the set $X_{\updownarrow}$ attains a potentially non-trivial involution. (If instead X was an $\mathbf{O}$-presheaf, then we would not expect to have an involution on $X_{\downarrow}$, as $\downarrow \in \mathbf{O}$ only possesses the identity automorphism.) We interpret $X_{☆_n}$ as operations with arity n. Using the maps $\updownarrow \to ☆_n$ classifying each boundary arc in the n-element set $\eth(☆_n) = \mathbf{n}$, we obtain a function

$$X_{☆_n} \to \prod_{\mathbf{n}} X_{\updownarrow}$$

giving the boundary profile of an operation. (Likewise, if $X \in \widehat{\mathbf{O}}$ then we have functions $X_{☆_{n,m}} \to \prod_{\mathbf{n}} X_{\downarrow}$ and $X_{☆_{n,m}} \to \prod_{\mathbf{m}} X_{\downarrow}$ giving the inputs and outputs.)

Suppose now that $G \in \mathbf{U}$ has two vertices u, v and exactly one internal edge, as in Figure 10. We then have a span of sets as follows.

$$\begin{array}{ccc} & X_G & \\ \swarrow & & \searrow \\ X_{☆_G} & & X_{☆_u} \underset{X_{\updownarrow}}{\times} X_{☆_v} \end{array} \tag{3}$$

The leftward leg is associated to an active map $☆_G \twoheadrightarrow G$ from a star graph (we may take $☆_G$ to be a star with $\eth(☆_G) = \eth(G)$), and the rightward leg is associated to the two inert maps $\iota_u \colon ☆_u \rightarrowtail G$ and $\iota_v \colon ☆_v \rightarrowtail G$. This span represents a kind of generalized way of 'composing' operations X. If the leg on the right happens to be a bijection, then this is a legitimate function

$$X_{☆_G} \leftarrow X_{☆_u} \times_{X_{\updownarrow}} X_{☆_v}$$

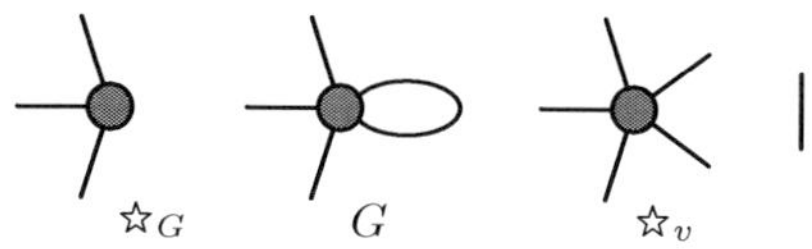

FIGURE 11. Graph with one vertex and one internal edge

giving *composition* of two operations in X. This is the kind of fundamental operation one has in a(n augmented) cyclic operad. In the directed setting, the evident variation gives a dioperadic composition.[4]

Instead, suppose that $G \in \mathbf{U}$ has one vertex v and exactly one internal edge, which is necessarily a loop at v, as in Figure 11. We can again form a span

$$\begin{array}{ccc} & X_G & \\ \swarrow & & \searrow \\ X_{\star_G} & & \widetilde{X_{\star_v}} \xrightarrow{\subseteq} X_{\star_v} \end{array} \tag{4}$$

whose right leg lands in the subset $\widetilde{X_{\star_v}}$ of $X_{\star_v}$ consisting of those elements which agree along the two embeddings $j_a, j_{a'} \colon \updownarrow \rightarrowtail \star_v$ with $\iota_v j_a = \iota_v j_{a'} \colon \updownarrow \rightarrowtail \star_v \rightarrowtail G$. If this rightward map is a bijection, this gives a *contraction* of suitable operations

$$X_{\star_G} \leftarrow \widetilde{X_{\star_v}}.$$

These are precisely the sort of contractions that one would expect to see in a modular operad. In the directed case, the loop must go from an output of the vertex to an input, and we arrive at the contractions for a wheeled properad.

There are of course axioms that these compositions and contractions should satisfy (unitality, associativity, etc.), which can likewise be governed by graphs. Let us give the full abstract definition.

DEFINITION 5.1 (Segal presheaves). Suppose $\mathbf{C}$ is one of the graph categories we have discussed above $(\mathbf{U}, \mathbf{U}_0, \mathbf{U}_{\mathrm{cyc}}, \mathbf{O}, \mathbf{G}, \mathbf{O}_0, \mathbf{\Omega}, \mathbf{\Delta})$, and let $\mathbf{C}_{\mathrm{el}} \subseteq \mathbf{C}_{\mathrm{int}}$ denote the category whose objects are *elementary graphs*, that is, edges and stars, and whose morphisms are inert maps.

- If $G \in \mathbf{C}$ is a graph, write $\mathbf{C}^{\mathrm{el}}_{/G} := \mathbf{C}_{\mathrm{el}} \times_{\mathbf{C}} \mathbf{C}^{\mathrm{int}}_{/G}$ for the category whose objects are inert maps $K \rightarrowtail G$ from an elementary object K to G, and whose morphisms are inert maps.
- If $X \in \widehat{\mathbf{C}}$ is a presheaf, then the **Segal map** at G is the function
$$X_G \to \lim_{\substack{K \rightarrowtail G \\ \in (\mathbf{C}^{\mathrm{el}}_{/G})^{\mathrm{op}}}} X_K.$$
- A presheaf $X \in \widehat{\mathbf{C}}$ is said to satisfy the **Segal condition** if the Segal map is a bijection for all $G \in \mathbf{C}$.

We also say that X is a **Segal presheaf** in this case. We write $\mathrm{Seg}(\mathbf{C}) \subset \widehat{\mathbf{C}}$ for the full subcategory on the Segal presheaves.

[4]When considering a $\mathbf{G}$-presheaf X, it is more appropriate to work with acyclic graphs G having two vertices, but an arbitrary number of internal edges, resulting in a span of the form $X_{\star_G} \leftarrow X_G \to X_{\star_u} \times_{\prod X_\downarrow} X_{\star_v}$, where $\prod X_\downarrow$ is the product indexed by the inner edges of G.

One can unravel that if $\mathbf{C} = \mathbf{U}$, then the rightward maps in the spans (3) and (4) are equivalent to the Segal maps for these graphs. This uses a final subcategory argument (we can omit the objects of $\mathbf{C}^{\mathrm{el}}_{/G}$ corresponding to boundary edges without changing the limit). Likewise, if $\mathbf{C} = \mathbf{\Delta}$, the Segal condition reduces to the one at the beginning of the introduction (the Segal maps on X_1 and X_0 are always identities). As an exercise, the reader should determine for which graph categories $\mathbf{C}$ all of the representable presheaves $\mathbf{C}(-, G)$ are Segal.

5.1. The general situation. There are now abstract settings for Segal conditions, which cover many situations of interest, including all of those above. Chu and Haugseng gave an extensive study of the abstract Segal condition for ∞-categories in [**CH21**, Definition 2.1], which is shown in Definition 5.2.

DEFINITION 5.2. An **algebraic pattern** is an ∞-category $\mathbf{P}$ equipped with an inert-active factorization system $(\mathbf{P}_{\mathrm{int}}, \mathbf{P}_{\mathrm{act}})$, along with some specified full subcategory $\mathbf{P}_{\mathrm{el}} \subseteq \mathbf{P}_{\mathrm{int}}$ whose objects are called **elementary**.

Given a functor $X \colon \mathbf{P} \to \mathcal{S}$, one says that X is Segal if $X(p) \to \lim_{e \in (\mathbf{P}_{\mathrm{el}})_{/p}} X(e)$ (where here the limit is in the ∞-categorical sense) is an equivalence for all $p \in \mathbf{P}$. If $\mathbf{C}$ is any one of the graph categories under discussion, then $\mathbf{P} = \mathbf{C}^{\mathrm{op}}$ is an algebraic pattern in this sense, and this condition agrees with that from Definition 5.1.

Less general are the hypermoment categories of Berger [**Ber22**], a framework that still includes the graph categories. To state the definition, recall that the skeletal category of finite pointed sets $\mathbf{F}_*$ has an inert-active factorization system [**Lur**, Remark 2.1.2.2]. A map $\{*, 1, 2, \dots, n\} = \langle n \rangle \to \langle m \rangle$ is *active* when only the basepoint maps to the basepoint, and *inert* if exactly one element of $\langle n \rangle$ maps to each non-basepoint element of $\langle m \rangle$. We are interested in the corresponding active-inert factorization system on $\mathbf{F}_*^{\mathrm{op}}$.

DEFINITION 5.3 ([**Ber22**, Definition 3.1]). A (unital) **hypermoment category** consists of a category $\mathbf{C}$, an active-inert factorization system on $\mathbf{C}$, and a functor $\gamma \colon \mathbf{C} \to \mathbf{F}_*^{\mathrm{op}}$ respecting the factorization systems. These data must satisfy the following:

(1) For each $c \in \mathbf{C}$ and each inert map $\langle 1 \rangle \rightarrowtail \gamma(c)$ in $\mathbf{F}_*^{\mathrm{op}}$, there is an essentially unique inert lift $u \rightarrowtail c$ where u is a unit. Here, *unit* means that $\gamma(u) = \langle 1 \rangle$ and any active map with codomain u has precisely one inert section.
(2) For each $c \in \mathbf{C}$, there is an essentially unique active map $u \twoheadrightarrow c$ whose domain is a unit.

Given a hypermoment category $\mathbf{C}$, one can define $\mathbf{C}_{\mathrm{el}} \subset \mathbf{C}_{\mathrm{int}}$ to be the full subcategory spanned by the units and the *nilobjects*, those objects living over $\langle 0 \rangle \in \mathbf{F}_*^{\mathrm{op}}$. Imitating Definition 5.1 gives a notion of Segal presheaf on $\mathbf{C}$.

The graph categories we have discussed so far become hypermoment categories by means of the functor $\mathrm{V} \colon \mathbf{C} \to \mathbf{FinSet}_*^{\mathrm{op}}$ that sends a graph G to $(V_G)_+$. If $\varphi \colon G \to G'$ is a morphism and $w \in V_{G'}$, then $\mathrm{V}(\varphi)(w) = v$ just when w is a vertex in $\hat{\varphi}(\star_v)$. Otherwise w is sent to the basepoint of $(V_G)_+$. By vertex-disjointness, this is a well-defined function $(V_{G'})_+ \to (V_G)_+$. The units are precisely the stars, and the maps $\star_G \twoheadrightarrow G$ are those from (2). See [**Ber22, Hac24**] for details.

5.2. Nerves. In Section 3, we saw a construction taking a tree G to the (augmented) cyclic operad $\mathsf{C}(G)$ which is freely generated by the vertices and arcs of G.

This induced a (fully faithful) functor $\mathbf{U}_0 \to \mathbf{Cyc}^+$. This construction be imitated for each of the graph categories, yielding Table 5.2 of functors into categories of generalized operads.

TABLE 1. Functors from graphs to generalized operads

Directed		Undirected	
Functor	Structure	Functor	Structure
$\mathbf{O} \to \mathbf{WPpd}$	Wheeled properad	$\mathbf{U} \to \mathbf{ModOp}$	Modular operad
$\mathbf{G} \to \mathbf{Ppd}$	Properad		
$\mathbf{O}_0 \hookrightarrow \mathbf{DiOp}$	Dioperad	$\mathbf{U}_0 \hookrightarrow \mathbf{Cyc}^+$	Aug. cyclic operad
		$\mathbf{U}_{\mathrm{cyc}} \hookrightarrow \mathbf{Cyc}$	Cyclic operad
$\boldsymbol{\Omega} \hookrightarrow \mathbf{Opd}$	Operad		
$\boldsymbol{\Delta} \hookrightarrow \mathbf{Cat}$	Category		

REMARK 5.4. Most of the functors in the preceding table are fully faithful, with the exceptions of $\mathbf{O} \to \mathbf{WPpd}$, $\mathbf{U} \to \mathbf{ModOp}$, and $\mathbf{G} \to \mathbf{Ppd}$. We can uniquely factor each of them into an identity-on-objects functor, followed by a fully faithful functor. Two of these,

$$\mathbf{U} \xrightarrow{\text{iden-on-obj}} \mathbf{R} \xhookrightarrow{\text{f.f.}} \mathbf{ModOp}$$
$$\mathbf{G} \xrightarrow{\text{iden-on-obj}} \mathbf{K} \xhookrightarrow{\text{f.f.}} \mathbf{Ppd}$$

have appeared in the work of Raynor (see [**Ray21**, §8.4], where $\mathbf{R}$ is called Ξ) and Kock (see [**Koc16**, 2.4.14], where $\mathbf{K}$ is called $\widetilde{\boldsymbol{Gr}}$). These sources include detailed descriptions of working in these categories, which are somewhat more complicated in structure than those they extend. For example, these categories possess *weak* factorization systems, in contrast with the orthogonal factorization systems present on $\mathbf{U}$ and $\mathbf{G}$. Further, there are not any clear functors into $\mathbf{F}_*^{\mathrm{op}}$, since vertex disjointness fails (see [**CH22**, Remark 7.1.10]).

Each of the functors in the table induces a **nerve functor**. For example, suppose $\mathsf{M}\colon \mathbf{U} \to \mathbf{ModOp}$ is the functor that takes an undirected graph G to the A_G-modular operad $\mathsf{M}(G)$ freely generated by its vertices. Then we can define a functor

$$N\colon \mathbf{ModOp} \to \widehat{\mathbf{U}}$$

by $NP = \mathbf{ModOp}(\mathsf{M}(-), P)\colon \mathbf{U}^{\mathrm{op}} \to \mathbf{Set}$. Practically by definition, we have that

$$\operatorname*{colim}_{\substack{K \rightarrowtail G \\ \in \mathbf{U}^{\mathrm{el}}_{/G}}} \mathsf{M}(K) \to \mathsf{M}(G)$$

is an isomorphism in $\mathbf{ModOp}$, which formally implies that NP is a Segal presheaf. This construction and argument applies to all of the graph categories and functors in the table above, giving nerve functors as follows.

$$\mathbf{ModOp} \to \widehat{\mathbf{U}} \qquad \mathbf{WPpd} \to \widehat{\mathbf{O}} \qquad \mathbf{Opd} \to \widehat{\boldsymbol{\Omega}}$$
$$\mathbf{Cyc}^+ \to \widehat{\mathbf{U}_0} \qquad \mathbf{Ppd} \to \widehat{\mathbf{G}} \qquad \mathbf{Cat} \to \widehat{\boldsymbol{\Delta}}$$
$$\mathbf{Cyc} \to \widehat{\mathbf{U}_{\mathrm{cyc}}} \qquad \mathbf{DiOp} \to \widehat{\mathbf{O}_0}$$

Each of these factors through a full subcategory $\mathrm{Seg}(\mathbf{C}) \hookrightarrow \widehat{\mathbf{C}}$ of Segal presheaves. Generalizing the classical equivalence $\mathbf{Cat} \simeq \mathrm{Seg}(\mathbf{\Delta})$ from the beginning of the introduction, we have the following.

THEOREM 5.5 (Nerve theorem). *The nerve functors*

$$\mathbf{WPpd} \to \mathrm{Seg}(\mathbf{O}), \mathbf{Ppd} \to \mathrm{Seg}(\mathbf{G}), \mathbf{Opd} \to \mathrm{Seg}(\mathbf{\Omega}), \textit{and } \mathbf{ModOp} \to \mathrm{Seg}(\mathbf{U})$$

are equivalences of categories.

PROOF. These appear in [**HRY18**], [**HRY15**], [**Web07**], and [**HRY20a**]. □

SLOGAN 5.6. In the presence of the nerve theorem, one can work entirely at the level of Segal presheaves, rather than with the original operadic structure.

REMARK 5.7. It is expected that the three missing cases $\mathbf{DiOp} \to \mathrm{Seg}(\mathbf{O}_0)$, $\mathbf{Cyc}^+ \to \mathrm{Seg}(\mathbf{U}_0)$, and $\mathbf{Cyc} \to \mathrm{Seg}(\mathbf{U}_{\mathrm{cyc}})$ are also equivalences. I have been told that the latter two equivalences will appear in the PhD thesis of Patrick Elliott, along with a comparison with the 'strict inner-Kan condition.' Meanwhile, the nerve theorem for dioperads follows from the nerve theorem for augmented cyclic operads (see Proposition 6.4). Alternatively, in light of Proposition 3.2 (and its dioperadic analogue), it is expected that one can bring to bear the tools of *abstract nerve theory* to recover these three cases.

The aforementioned abstract nerve theory originated in work of Weber [**Web07**] and unpublished work of Leinster. See also [**BMW12, BG19**], or the overview of the key points appearing in [**Ray21**, 2.1]. In our restricted setting, one can consider $\widehat{\mathbf{U}_{\mathrm{el}}}$ as a category of *graphical species* ([**JK11**, §4] and [**Ray21**, §1.1]) and $\widehat{\mathbf{O}_{\mathrm{el}}}$ as *digraphical species* [**Koc16**, §2.1], which are a kind of collection with morphisms built in for the coloring maps. Then **ModOp** can be considered as the Eilenberg–Moore category of algebras $\mathrm{Alg}(T)$ for the free modular operad monad T on $\widehat{\mathbf{U}_{\mathrm{el}}}$. For particularly nice monads (strongly cartesian), one can automatically produce from this set-up a full subcategory $\mathbf{\Theta}_T \subset \mathrm{Alg}(T)$ which satisfies a nerve theorem $\mathrm{Alg}(T) \xrightarrow{\simeq} \mathrm{Seg}(\mathbf{\Theta}_T) \subset \widehat{\mathbf{\Theta}_\mathbf{T}}$. This applies to produce, for instance, $\mathbf{\Delta}$, $\mathbf{\Omega}$ (see [**Web07**, Example 2.14 and Example 4.19]), and also other important categories such as Joyal's cell category $\mathbf{\Theta}_n$.

REMARK 5.8. There are also nerve theorems for the categories **K** and **R** from Remark 5.4. The nerve functors (with a similar definition as above)

$$N\colon \mathbf{ModOp} \to \widehat{\mathbf{R}} \qquad N\colon \mathbf{Ppd} \to \widehat{\mathbf{K}}$$

are fully faithful, and the essential image of each is the subcategory of Segal presheaves. The first of these is [**Ray21**, Theorem 8.2] (see also [**HRY20b**, §4]), while the second is [**Koc16**, Theorem 2.3.9]. Both of these results utilize the framework for abstract nerve theorems, though there are intricate subtleties in both instances, and neither is a straightforward application of existing theory.

QUESTION 5.9. So far in the main text of this paper, we have not mentioned props at all [**ML65, HR15**], yet these are one of the most widespread and expressive kinds of operadic structures. Nor have we mentioned wheeled props [**MMS09**] or nc (nonconnected) modular operads [**KW17**, 2.3.2]. All of these structures are based on disconnected graphs. A major motivation for the development of Definition 4.1 was to facilitate extensions of the categories **G**, **O**, and **U** to categories of disconnected graphs $\mathbf{G}_{\mathrm{dis}}$, $\mathbf{O}_{\mathrm{dis}}$, and $\mathbf{U}_{\mathrm{dis}}$. Specifically, we hope for an appropriate notion of *embedding* between disconnected graphs, so that we can simply imitate Definition 4.1 directly to obtain $\mathbf{O}_{\mathrm{dis}}$ and $\mathbf{U}_{\mathrm{dis}}$ (and similarly a disconnected version of structured subgraph, to obtain $\mathbf{G}_{\mathrm{dis}}$). We give a wish-list for such a construction:

(1) Each of the three categories would have a naturally occurring active-inert orthogonal factorization system, with inert maps the disconnected embeddings (or the appropriate maps for $\mathbf{G}_{\mathrm{dis}}$).
(2) For each of these categories, there is a hypermoment category structure as in Definition 5.3, given by a vertex functor V into $\mathbf{FinSet}_*^{\mathrm{op}}$. In particular, there is a natural notion of Segal presheaf.
(3) Each of these categories would be a dualizable generalized Reedy category in the sense of [**BM11**], as is the case for their connected counterparts ([**HRY15**, §6.4], [**HRY18**, Theorem 1.2], [**HRY20a**, §2.2])
(4) The nerve theorem holds, that is, we may regard the categories of props, wheeled props, and nc modular operads as the categories of Segal presheaves $\mathrm{Seg}(\mathbf{G}_{\mathrm{dis}})$, $\mathrm{Seg}(\mathbf{O}_{\mathrm{dis}})$, and $\mathrm{Seg}(\mathbf{U}_{\mathrm{dis}})$.

Such categories would provide a sound foundation for an approach to homotopy-coherent versions of each operadic structure mentioned, either by appropriately localizing the Reedy model structure on simplicial presheaves, or by localizing ∞-category of Segal presheaves $\mathrm{Seg}^{\infty}(\mathbf{C}) \subset \mathrm{Fun}(\mathbf{C}^{\mathrm{op}}, \mathcal{S})$ of [**CH21**] to impose a Rezk-completeness condition (as in [**CH20, CH22**]). We note that an entirely different approach to ∞-props has recently been proposed in [**HK24**].

These proposed categories of disconnected graphs should also be related to the double category of graphs of [**BK**, 2.5], which has as a key component the total dissection. There is also the category of forests from [**HHM16**] (where each component is a rooted tree), but we are unsure of the precise relationship to expect.

6. Interactions

In this final section, we turn to some interactions between the operadic structures that have appeared so far.

THEOREM 6.1. *Suppose $f\colon \mathbf{C} \to \mathbf{D}$ is one of the following three functors:*

$$\mathbf{\Omega} \to \mathbf{U}_{\mathrm{cyc}} \qquad \mathbf{O}_0 \to \mathbf{U}_0 \qquad \mathbf{O} \to \mathbf{U},$$

and consider the associated adjoint string

$$\widehat{\mathbf{D}} \; \begin{array}{c} \xleftarrow{\;f_!\;} \\ \bot \\ \xrightarrow{\;f^*\;} \\ \bot \\ \xleftarrow{\;f_*\;} \end{array} \; \widehat{\mathbf{C}}$$

where f^* is restriction along f, and $f_!$ and f_* are left and right Kan extension. The following hold:

(1) *If $X \in \widehat{\mathbf{D}}$ is a presheaf, then X is Segal if and only if $f^*X \in \widehat{\mathbf{C}}$ is Segal.*
(2) *If $Y \in \widehat{\mathbf{C}}$ is Segal, then so is $f_!Y \in \widehat{\mathbf{D}}$.*

PROOF. See [**Hac24**, Proposition 6.17, Theorem 6.23, and Proposition 6.24]. □

In light of the equivalence in Remark 4.4, a directed graph is essentially just the data of an undirected graph along with an element of the orientation presheaf $\mathfrak{o}$. If $x \in \mathfrak{o}_G$ where G is an undirected graph, for the moment we will write G_x for the associated directed graph. If G is an undirected tree and $r \in \eth(G)$, then there is a unique directed structure on G which is a rooted tree with output r; we write G_r for this rooted tree. The main component of Theorem 6.1(2) is the following description of the left Kan extension (see [**Hac24**, §6.1]).

PROPOSITION 6.2. *Let $f\colon \mathbf{C} \to \mathbf{D}$ be one of the forgetful functors $\mathbf{O}_0 \to \mathbf{U}_0$ or $\mathbf{O} \to \mathbf{U}$. If $Z \in \widehat{\mathbf{C}}$ is $\mathbf{C}$-presheaf, then the left Kan extension $f_!Z \in \widehat{\mathbf{D}}$ is given by the formula*

$$(f_!Z)_G = \coprod_{x \in \mathfrak{o}_G} Z_{G_x}.$$

For the functor $f\colon \mathbf{\Omega} \to \mathbf{U}_{\mathrm{cyc}}$, we have

$$(f_!Z)_G = \coprod_{r \in \eth(G)} Z_{G_r}.$$

QUESTION 6.3. Is the analogue of Theorem 6.1 also true in the ∞-categorical setting? (See Definition 5.2.) That is, for these f, does $f_!\colon \mathrm{Fun}(\mathbf{C}^{\mathrm{op}}, \mathcal{S}) \to \mathrm{Fun}(\mathbf{D}^{\mathrm{op}}, \mathcal{S})$ restrict to a functor $\mathrm{Seg}^\infty(\mathbf{C}) \to \mathrm{Seg}^\infty(\mathbf{D})$ between the ∞-categories of Segal presheaves from [**CH21**]?

As is explained in [**DCH21**, §4.3], the category of dioperads may be viewed as a certain (non-full) subcategory of augmented cyclic operads $\mathbf{DiOp} \to \mathbf{Cyc}^+$. It is better to describe this as a certain slice category, that is, $\mathbf{DiOp} \simeq \mathbf{Cyc}^+_{/\mathsf{IO}}$. Here, IO is the terminal object of $\mathbf{Cyc}^+_{\{i,o\}}$ where the color set $\{i,o\}$ possesses the nontrivial involution. Another way to say this is that IO is isomorphic to the image of the terminal dioperad under $\mathbf{DiOp} \to \mathbf{Cyc}^+$, and induces an equivalence

$$\mathbf{DiOp} \simeq \mathbf{DiOp}_{/*} \xrightarrow{\simeq} \mathbf{Cyc}^+_{/\mathsf{IO}}.$$

The map to IO of course forces all cyclic operads to have free involution on color sets, and also yields a splitting of the profile of an operation into 'inputs' and 'outputs.' This follows a similar approach for wheeled properads $\mathbf{WPpd} \simeq \mathbf{ModOp}_{/\mathsf{IO}}$ from [**Ray21**, Example 1.29].

We use this to establish the following provisional statement (see Remark 5.7).

PROPOSITION 6.4. *The nerve theorem for augmented cyclic operads implies that for dioperads.*

PROOF. Let $f\colon \mathbf{O}_0 \to \mathbf{U}_0$ be the functor forgetting the directed structure. Notice that $N(\mathsf{IO}) \cong \mathfrak{o}$ is nothing but the orientation $\mathbf{U}_0$-presheaf from Remark 4.4. We then have the following commutative diagram of categories, where on presheaves we use that $f_!(*) = \mathfrak{o}$ (by inspection, or application of Proposition 6.2).

$$\begin{array}{ccccc}
\mathbf{DiOp} & \xrightarrow{\simeq} & \mathbf{Cyc}^+_{/\mathsf{IO}} & \longrightarrow & \mathbf{Cyc}^+ \\
\downarrow N & & \downarrow & & \downarrow N \\
\mathrm{Seg}(\mathbf{O}_0) & \longrightarrow & \mathrm{Seg}(\mathbf{U}_0)_{/N(\mathsf{IO})} & \longrightarrow & \mathrm{Seg}(\mathbf{U}_0) \\
\downarrow \text{f.f.} & & \downarrow \text{f.f.} & & \downarrow \text{f.f.} \\
\widehat{\mathbf{O}_0} & \xrightarrow{\simeq} & \widehat{\mathbf{U}_0}_{/\mathfrak{o}} & \longrightarrow & \widehat{\mathbf{U}_0}
\end{array}$$

The equivalence on the top left was mentioned just above, while that on the bottom left is a standard fact on slices of presheaf categories (see, for instance, Exercise 8 in [**MLM92**, Chapter III]) using the second equivalence of categories appearing in Remark 4.4. If $N\colon \mathbf{Cyc}^+ \to \mathrm{Seg}(\mathbf{U}_0)$ assumed to be an equivalence (that is, if the nerve theorem holds for augmented cyclic operads), then the induced map on slices $\mathbf{Cyc}^+_{/\mathsf{IO}} \to \mathrm{Seg}(\mathbf{U}_0)_{/N(\mathsf{IO})}$ is also an equivalence. By the left cancellation property for fully faithful functors, $N\colon \mathbf{DiOp} \to \mathrm{Seg}(\mathbf{O}_0)$ is fully faithful. But this functor is also essentially surjective: if $X \in \mathrm{Seg}(\mathbf{O}_0)$, then we know $f_!X \to N(\mathsf{IO})$ in $\mathrm{Seg}(\mathbf{U}_0)_{/N(\mathsf{IO})}$ is isomorphic to the image of some dioperad $D \in \mathbf{DiOp}$. We can conclude that $X \cong N(D)$ since $\mathrm{Seg}(\mathbf{O}_0) \to \mathrm{Seg}(\mathbf{U}_0)_{/N(\mathsf{IO})}$ is injective on isomorphism classes of objects. Hence $N\colon \mathbf{DiOp} \to \mathrm{Seg}(\mathbf{U}_0)$ is an equivalence, as desired. □

Essentially the same argument shows that the nerve theorem for modular operads from [**HRY20b**] implies that for wheeled properads from [**HRY18**].

If $\mathbf{C}$ is a category, write $\mathrm{s}\widehat{\mathbf{C}} := \mathrm{Fun}(\mathbf{C}^{\mathrm{op}}, \widehat{\mathbf{\Delta}})$ for the category of *simplicial* presheaves. The following question was posed by Drummond–Cole and the author [**DCH21**, Remark 6.8], and has implications in the theory of cyclic 2-Segal objects [**Wal21**, Remark 5.0.20].

QUESTION 6.5. Let $f\colon \mathbf{\Omega} \to \mathbf{U}_{\mathrm{cyc}}$ be the functor which forgets the root. Consider the operadic model structure on dendroidal sets $\widehat{\mathbf{\Omega}}$ from [**CM11**] and the dendroidal Rezk model structure on $\mathrm{s}\widehat{\mathbf{\Omega}}$ from [**CM13a**, §6]. We have two adjoint strings associated to the functor f:

$$\widehat{\mathbf{U}_{\mathrm{cyc}}} \underset{f_*}{\overset{f_!}{\leftleftarrows}} \!\!\xrightarrow{\;f^*\;}\! \widehat{\mathbf{\Omega}} \qquad\qquad \mathrm{s}\widehat{\mathbf{U}_{\mathrm{cyc}}} \underset{f_*}{\overset{f_!}{\leftleftarrows}} \!\!\xrightarrow{\;f^*\;}\! \mathrm{s}\widehat{\mathbf{\Omega}}$$

Is the endo-adjunction $f^*f_! \dashv f^*f_*$ on $\widehat{\mathbf{\Omega}}$ (resp. on $\mathrm{s}\widehat{\mathbf{\Omega}}$) a *Quillen* adjunction? The formula from Proposition 6.2 may well help answer this question. If $f^*f_!$ is a left Quillen functor, then the adjoint string lifting results from [**DCH19**] will imply the existence of a lifted model structure on $\widehat{\mathbf{U}_{\mathrm{cyc}}}$ (resp. on $\mathrm{s}\widehat{\mathbf{U}_{\mathrm{cyc}}}$). Further, this should yield a Quillen equivalence with the category of simplicially-enriched cyclic operads, lifting those for simplicially-enriched operads from [**CM13b**].

For the potential relationship with cyclic 2-Segal objects mentioned in [**Wal21**], one should instead use planar and nonsymmetric versions of trees and (cyclic) operads everywhere (including the model structure on planar dendroidal sets from [**Moe10**, §8.2]). It is likely that the planar/nonsymmetric versions follow from the symmetric versions via an appropriate slicing, as in [**Gag15**].

QUESTION 6.6. In [**Ray21**, Corollary 8.14], Raynor gives a model structure on the category of simplicial presheaves $\mathrm{s}\widehat{\mathbf{R}}$ where the fibrant objects are the projectively fibrant presheaves which satisfy a weak version of the Segal condition. Similarly, the category $\mathrm{s}\widehat{\mathbf{U}}$ admits Segal-type model structure, obtained by taking a left Bousfield localization of the projective model structure (a variation of this appears in [**HRY20a**, Proposition 3.10]). It would be interesting to know if the restriction map

$$\mathrm{s}\widehat{\mathbf{R}} \to \mathrm{s}\widehat{\mathbf{U}}$$

is a right Quillen equivalence with respect to these model structures. This would reflect the equivalence of categories $\mathrm{Seg}(\mathbf{R}) \to \mathrm{Seg}(\mathbf{U})$ between ordinary simplicial presheaves (as follows from Remark 5.8 and Theorem 5.5).

Appendix A. Proofs

A.1. Full tree maps preserve intersections. In this section, we show that it is automatic that full tree maps preserve intersections. This was shown in a roundabout way in [**Hac24**], which went through the properadic graphical category. Here we give a elementary direct proof, that deals only with trees. It is important that our notion of boundary-compatibility refers to involutive maps $A_G \to A_{G'}$ rather than maps $E_G \to E_{G'}$ (compare [**HRY19**, Definition 1.12]).

LEMMA A.1. *Suppose G and G' are undirected trees, and (φ_0, φ_1) is a tree map from G to G'. If $u \neq v$ are distinct vertices of G, then the subtrees $\varphi_1(u)$ and $\varphi_1(v)$ of G' do not share any vertices in common.*

PROOF. By way of contradiction, suppose that w is a vertex in both subtrees $S = \varphi_1(u)$ and $T = \varphi_1(v)$. For simplicity, we first treat the case when u and v are adjacent vertices. That is, there is a (unique) arc $\tilde{a} \in D_G$ with $t(\tilde{a}) = v$ and $t(\tilde{a}^\dagger) = u$. Let $a = \varphi_0(\tilde{a})$. By boundary-compatibility we have $a \in \eth(S) = \eth(\varphi_1(u))$ and $a^\dagger \in \eth(T) = \eth(\varphi_1(v))$. Since S is a tree, there is a unique path in S

$$P_S := v_0 a_0 v_1 a_1 \cdots v_n a_n$$

with $v_0 = w$ and $a_n = a$. By *path in S* we mean that $v_i \in V_S$, $a_i \in A_S$, and these elements satisfy the equations $t(a_i) = v_{i+1}$ and $t(a_i^\dagger) = v_i$ whenever both sides are defined in S (see Definition 2.10). Likewise, there is a unique path in T

$$P_T := v_0' a_0' v_1' a_1' \cdots v_m' a_m'$$

with $v_0' = w$ and $a_m' = a^\dagger$.

Then we have a composite path in G

$$\begin{aligned} P_S P_T^\dagger &= v_0 a_0 v_1 a_1 \cdots v_n a_n v_m' (a_{m-1}')^\dagger v_{m-1}' \cdots (a_1')^\dagger v_1' (a_0')^\dagger v_0' \\ &= v_0 a_0 v_1 a_1 \cdots v_n a_n v_{n+1} a_{n+1} v_{n+2} \cdots a_{n+m-1} v_{n+m} a_{n+m} v_{n+m+1} \end{aligned}$$

where $v_{n+k} = v_{m-k+1}'$ for $1 \leq k \leq m+1$ and $a_{n+j} = (a_{m-j}')^\dagger$ for $0 \leq j \leq m$. Since $v_0 = w = v_{n+m+1}$, this composite path $P_S P_T^\dagger$ is a cycle in G'. This is impossible, since G' is a tree.

The general case, when u and v are not adjacent in G, is proved using similar ideas. Namely, there is a unique path in G

$$u_0 \tilde{a}_0 u_1 \tilde{a}_1 \ldots \tilde{a}_\ell u_{\ell+1}$$

with $u_0 = u$ and $u_{\ell+1} = v$. We define P_S to be the path in S starting at w and ending at $\varphi_0(\tilde{a}_0)$, and P_T to be the path in T starting at w and ending at $\varphi_0(\tilde{a}_\ell^\dagger)$. For $1 \leq j \leq \ell$ there is a unique path P_j in $\varphi_1(u_j)$ which starts with $\varphi_0(\tilde{a}_{j-1})$ and ends with $\varphi_0(\tilde{a}_j)$. The composite path $P_S P_1 \cdots P_\ell P_T^\dagger$ will be a cycle at w, which again is impossible since G' is a tree. □

LEMMA A.2. *Suppose G and G' are undirected trees, and $(\varphi_0, \hat{\varphi})$ is a full tree map from G to G'. If $S, T \in \mathrm{Emb}(G)$ are subtrees of G so that $\hat{\varphi}(S)$ and $\hat{\varphi}(T)$ have a vertex $w \in V_{G'}$ in common, then there is a unique vertex $v \in V_S \cap V_T \subset V_G$ so that $w \in \hat{\varphi}(\star_v)$.*

PROOF. Suppose w is a vertex of $\hat{\varphi}(S) \cap \hat{\varphi}(T)$. Writing S as an iterated union of distinct stars

$$S = ((\cdots((\star^1 \cup \star^2) \cup \star^3) \cup \cdots) \cup \star^{n-1}) \cup \star^n$$

we must have $w \in \hat{\varphi}(\star^k)$ for some k since $\hat{\varphi}(S)$ is the union of the $\hat{\varphi}(\star^k)$. Likewise, $w \in \hat{\varphi}(\star)$ for some $\star \rightarrowtail T$. By Lemma A.1 applied to $(\varphi_0, \hat{\varphi}|_{V_G})$, we conclude that $\star = \star^k$. Then v is the vertex of the star $\star = \star^k$ which is in both S and T. Uniqueness is guaranteed by Lemma A.1. □

For subtrees of a graph G, the graph $S \cap T$ is either empty or again a subtree. In particular, if $S \cap T$ does not contain a vertex, then it is either empty or a single edge.

PROPOSITION A.3. *Full tree maps preserve intersections between overlapping subtrees. That is, if $(\varphi_0, \hat{\varphi})\colon G \to G'$ is a full tree map and $S, T \in \mathrm{Emb}(G)$ are subtrees with $S \cap T$ non-empty, then $\hat{\varphi}(S \cap T) = \hat{\varphi}(S) \cap \hat{\varphi}(T)$.*

PROOF. Since $\hat{\varphi}\colon \mathrm{Emb}(G) \to \mathrm{Emb}(G')$ preserves unions, it also preserves the partial order. As $S \cap T$ is inhabited, it is a subtree of both S and T, so

$$\hat{\varphi}(S \cap T) \subseteq \hat{\varphi}(S) \cap \hat{\varphi}(T). \tag{5}$$

Notice that both sides are subtrees of G'. If w is a vertex in $\hat{\varphi}(S) \cap \hat{\varphi}(T)$, then w is also in $\hat{\varphi}(S \cap T)$ by Lemma A.2. Since subtrees containing a vertex are completely determined by their vertices, we conclude that (5) is an equality. If $\hat{\varphi}(S) \cap \hat{\varphi}(T)$ does not contain any vertices, then it is a single edge, so again (5) is an equality. □

A.2. Tree maps and full tree maps coincide. Our goal is to prove that there is no real difference between tree maps and full tree maps from Definition 3.1.

LEMMA A.4. *Suppose S and T are subtrees of a tree G, and that S and T intersect at a single edge $e = [a, a^\dagger]$, where $a \in \eth(S)$ and $a^\dagger \in \eth(T)$. If neither S nor T is an edge, then*

$$\eth(S \cup T) = (\eth(S) \backslash \{a\}) \amalg (\eth(T) \backslash \{a^\dagger\}).$$

If S is an edge then $\eth(S \cup T) = \eth(T)$ and if T is an edge then $\eth(S \cup T) = \eth(S)$.

PROOF. If S is an edge then $S \cup T = T$, and similarly $S \cup T = S$ when T is an edge, so the final statement is immediate. We suppose that neither S nor T is an edge and compute the boundary of the union.

Since S is not an edge and $a \in \eth(S)$, there is a vertex $v \in V_S$ with $t(a^\dagger) = v$. Since v is also in $V_{S \cup T}$, it follows that $a^\dagger \notin \eth(S \cup T)$. A symmetric argument shows that $a \notin \eth(S \cup T)$.

Suppose $a' \in \eth(S) \backslash \{a\}$, so that $t((a')^\dagger) \in V_S$. We either have $a' \in \eth(G)$, or $t(a') = v$ is a vertex of G which is not in S. If $a' \in \eth(G)$, then $a' \in \eth(S \cup T)$ as well. If $t(a') = v \in V_G \backslash V_S$, then either $v \in V_T$ or $v \notin V_T$. If $v \in V_T$, then $e' = [a', (a')^\dagger]$ is an edge in $S \cap T$ distinct from $e = [a, a^\dagger]$, contrary to our assumption. If $v \notin V_T$, then $v \notin V_{S \cup T}$, hence $a' \in \eth(S \cup T)$. Since a' was arbitrary, we have shown $\eth(S) \backslash \{a\} \subset \eth(S \cup T)$; reversing roles gives $\eth(T) \backslash \{a^\dagger\} \subset \eth(S \cup T)$.

It remains to observe that $\eth(S \cup T) \subset \eth(S) \cup \eth(T)$. Suppose $a' \in \eth(S \cup T)$; since $S \cup T$ is not an edge, we have $t((a')^\dagger) \in V_{S \cup T} = V_S \cup V_T$. Without loss of generality suppose $t((a')^\dagger) \in V_S$. Since a' is in $\eth(S \cup T)$, either $a' \in \eth(G)$ or $a' \in D_G$ and $t(a') = w \in V_G \backslash V_{S \cup T} \subseteq V_G \backslash V_S$. In both cases, we conclude that a' is in $\eth(S)$, as desired. □

LEMMA A.5. *Suppose G is a tree and T is a proper subtree. Then there exists a star subtree $\star \subset G$ and a proper subtree $R \subset G$ so that R overlaps with $\star$, the union $R \cup \star$ is equal to G, and $T \subset R$.*

PROOF. Note that an edge does not have proper subtrees, so G has at least one vertex. If G is a star, then T must be an edge, and one takes $\star = G$ and $R = T$. If G has more than one vertex and T is an edge, then one can take $\star$ to be any extremal vertex of G which does not contain T as a boundary, and R to be the complement of $\star$.

We now suppose T contains a vertex. Let $v \in V_G \backslash V_T$ be a vertex of minimum positive distance from T, that is, there is an edge e between v and some vertex w of T. Consider the collection of all paths of G which contain e but do not contain w, and let $S \subset G$ be the subtree consisting of all vertices appearing on such paths, and all arcs appearing on edges on such paths. Notice that $v \in V_S$. If S is a star then v is extremal in G, and we set $\star = S$ and let R be its complement. Otherwise, let $u \in V_S$ be an extremal vertex of S other than v (which may or may not be extremal in S). The vertex u is also extremal in G, and we let $\star$ be its star and R the complement. In either case, the three conditions hold. □

PROPOSITION A.6. *Suppose G and G' are undirected trees. The assignment $(\varphi_0, \hat{\varphi}) \mapsto (\varphi_0, \hat{\varphi}|_{V_G})$ is a bijection between tree maps and full tree maps from Definition 3.1.*

PROOF. Let $\mathrm{Emb}^n(G) \subseteq \mathrm{Emb}(G)$ denote the set of subtrees having n or fewer vertices. This gives a (finite) filtration of the set $\mathrm{Emb}(G)$, starting with

$$\mathrm{Emb}^0(G) = E_G \subseteq E_G \amalg V_G = \mathrm{Emb}^1(G).$$

Given a tree map (φ_0, φ_1), we inductively define $\hat{\varphi}_n \colon \mathrm{Emb}^n(G) \to \mathrm{Emb}(G')$ and show that it satisfies boundary-compatibility and preserves unions that exist in $\mathrm{Emb}^n(G)$. The first map $\hat{\varphi}_0 \colon E_G \to E_{G'} \subseteq \mathrm{Emb}(G')$ is induced from $\varphi_0 \colon A_G \to A_{G'}$, while $\hat{\varphi}_1$ is defined to be $\hat{\varphi}_0 \amalg \varphi_1$. These both satisfy boundary-compatibility. Any unions $S \cup T \in \mathrm{Emb}^1(G)$ for overlapping S, T which exist must either have $S = T$ or one of the two terms is an edge. If S is an edge or if $S = T$, then $S \cup T = T$, and also $\hat{\varphi}_1(S) \cup \hat{\varphi}_1(T) = \hat{\varphi}_1(T)$ since either $\hat{\varphi}_1(S)$ is an edge or $\hat{\varphi}_1(S) = \hat{\varphi}_1(T)$. Thus $\hat{\varphi}_1$ preserves unions.

We continue inductively. Suppose $\hat{\varphi}_n \colon \mathrm{Emb}^n(G) \to \mathrm{Emb}(G)$ is defined, preserves unions that exist in the nth filtration level, and $(\varphi_0, \hat{\varphi}_n)$ is boundary-compatible. Let $n \geq 1$. Any subtree R having $n+1$ vertices can be written as $S \cup \star$ where $\star$ is a star and S has n vertices. It is automatic that S and $\star$ intersect only at an edge $e = [a, a^\dagger]$. We make a provisional definition $\hat{\varphi}_{n+1}(R) \coloneqq \hat{\varphi}_n(S) \cup \hat{\varphi}_1(\star)$, which of course depends on a choice of $S \subset R$ having n vertices. If $n + 1 = 2$, then S is itself a star and by symmetry there is no actual choice involved in this definition. Given a different subtree $T \subset R$ having $n \geq 2$ vertices, the tree $S \cap T$ will have $n - 1$ vertices. We have $R = T \cup \star'$ for a star $\star' \subset S \subset R$, with T and $\star'$ intersecting at an edge e'. Note that e, e' are both in $S \cap T$. We then have

$$\begin{aligned}
\hat{\varphi}_n(T) \cup \hat{\varphi}_1(\star') &= \hat{\varphi}_n((S \cap T) \cup \star) \cup \hat{\varphi}_1(\star') \\
&= \hat{\varphi}_{n-1}(S \cap T) \cup \hat{\varphi}_1(\star) \cup \hat{\varphi}_1(\star') \\
&= \hat{\varphi}_n((S \cap T) \cup \star') \cup \hat{\varphi}_1(\star) \\
&= \hat{\varphi}_n(S) \cup \hat{\varphi}_1(\star)
\end{aligned}$$

which shows that $\hat{\varphi}_{n+1}$ is well-defined.

Further, observe that $\hat{\varphi}_n(S)$ and $\hat{\varphi}_1(\star)$ (as above) intersect only at the single edge $\bar{e} = [\varphi_0(a), \varphi_0(a)^\dagger]$. Indeed, the proof of Lemma A.2 works equally well for $\hat{\varphi}_n$ as it does for $\hat{\varphi}$, so $\hat{\varphi}_n(S)$ and $\hat{\varphi}_n(\star)$ cannot share any vertices.

We can use this to show that $\hat{\varphi}_{n+1}$ is boundary-compatible. As above, suppose that $R = S \cup \star$ has $n+1$ vertices, $\star$ is a star, S has n vertices, and S and $\star$ intersect only at the edge $e = [a, a^\dagger]$. Here we take $a \in \eth(S)$ and $a^\dagger \in \eth(\star)$; write $\bar{a} = \varphi_0(a)$. Boundary-compatibility implies that

$$\eth(S) \xrightarrow{\varphi_0} \eth(\hat{\varphi}_n(S)) \qquad \eth(\star) \xrightarrow{\varphi_0} \eth(\hat{\varphi}_1(\star))$$

are bijections. If neither $\hat{\varphi}_n(S)$ nor $\hat{\varphi}_1(\star)$ is an edge, then we have the desired bijection below by Lemma A.4.

$$\begin{array}{ccc} \eth(S \cup \star) & \xleftarrow{\quad = \quad} & (\eth(S)\backslash\{a\}) \amalg (\eth(\star)\backslash\{a^\dagger\}) \\ & & \downarrow \cong \\ \eth(\hat{\varphi}_{n+1}(S \cup \star)) \xleftrightarrow{=} \eth(\hat{\varphi}_n(S) \cup \hat{\varphi}_1(\star)) & \xleftrightarrow{=} & (\eth(\hat{\varphi}_n(S))\backslash\{\bar{a}\}) \amalg (\eth(\hat{\varphi}_1(\star))\backslash\{\bar{a}^\dagger\}) \end{array}$$

We must still address the case when $\hat{\varphi}_n(S)$ is an edge or when $\hat{\varphi}_1(\star)$ is an edge. These cases are entirely analogous, so we suppose that $\hat{\varphi}_1(\star)$ is an edge. By boundary-compatibility, $\eth(\star) = \{a^\dagger, a'\}$ with $\varphi_0((a')^\dagger) = \varphi_0(a^\dagger) = \bar{a}^\dagger$, so $\varphi_0(a') = \bar{a}$. Thus

$$\eth(S \cup \star) = (\eth(S)\backslash\{a\}) \amalg (\eth(\star)\backslash\{a^\dagger\}) = (\eth(S)\backslash\{a\}) \amalg \{a'\} \to \eth(\hat{\varphi}_n(S))$$

is a bijection, as required. We have thus shown that $(\varphi_0, \hat{\varphi}_{n+1})$ is boundary-compatible.

Finally, we must show that $\hat{\varphi}_{n+1}$ preserves unions that exist in $\mathrm{Emb}^{n+1}(G)$. Suppose that $S \cup T \in \mathrm{Emb}^{n+1}(G)$ is a union of overlapping subtrees S and T. We may assume that $S \cup T$ has $n+1$ vertices, for if it has n or fewer vertices then $\hat{\varphi}_{n+1}(S \cup T) = \hat{\varphi}_n(S \cup T) = \hat{\varphi}_n(S) \cup \hat{\varphi}_n(T)$ by the induction hypothesis. We further assume that $S \neq T$, since if $S = T$ then $\hat{\varphi}_{n+1}(S) \cup \hat{\varphi}_{n+1}(T) = \hat{\varphi}_{n+1}(S) = \hat{\varphi}_{n+1}(S \cup T)$. Without loss of generality, suppose there is a vertex of S which is not in T. By Lemma A.5, there is a decomposition $S \cup T = R \cup \star$ where T is a subtree of R, and R is a proper subtree of $S \cup T$. Notice also that $\star \subset S$. Of course $S \cap R$ is a subtree since it is non-empty (it contains $S \cap T$). We then have $S = S \cap (S \cup T) = S \cap (\star \cup R) = \star \cup (S \cap R)$. Now $S \cap R$ overlaps with T, so

$$S \cup T = (\star \cup (S \cap R)) \cup T = \star \cup ((S \cap R) \cup T).$$

The union $(S \cap R) \cup T$ contains exactly n vertices, and we find

$$\begin{aligned} \hat{\varphi}_{n+1}(S \cup T) &= \hat{\varphi}_1(\star) \cup \hat{\varphi}_n((S \cap R) \cup T) \\ &= \hat{\varphi}_1(\star) \cup (\hat{\varphi}_n(S \cap R) \cup \hat{\varphi}_n(T)) \\ &= (\hat{\varphi}_1(\star) \cup \hat{\varphi}_n(S \cap R)) \cup \hat{\varphi}_n(T) \\ &= \hat{\varphi}_{n+1}(\star \cup (S \cap R)) \cup \hat{\varphi}_n(T) \\ &= \hat{\varphi}_{n+1}(S) \cup \hat{\varphi}_{n+1}(T). \end{aligned}$$

This uses that $\hat{\varphi}_n$ preserves unions and the definition of $\hat{\varphi}_{n+1}$.

If G has k vertices, then $\mathrm{Emb}^k(G) = \mathrm{Emb}(G)$ and we have thus defined a full tree map from G to G'. This construction is a section to the assignment from the proposition statement. But notice that it is the *unique* such section: our $\hat{\varphi}$ must preserve unions, and we defined the section so that it would preserve particular unions. Thus the map in question is a bijection. □

A.3. The (augmented) cyclic operad associated to a tree. Our goal in this section is to prove Proposition 3.2. The reader should recall the category of augmented cyclic operads from [**DCH21**, Definition 2.3], which we have denoted by $\mathbf{Cyc}^+$. If G is an undirected tree, then the (augmented) cyclic operad $\mathsf{C}(G) \in \mathbf{Cyc}^+_{A_G}$ is obtained by choosing for each $v \in V_G$ an ordering $a_0, a_1, \dots, a_n$ of $\eth(\star_v) = \mathrm{nb}(v)^\dagger$, and considering v as a free generator in $\mathsf{C}(G)(a_0, a_1, \dots, a_n)$. This choice of order does not matter for the isomorphism class of $\mathsf{C}(G)$.

LEMMA A.7. *The operations in $\mathsf{C}(G)$ are in bijection with subtrees T of G equipped with a total ordering on the boundary $\eth(T)$.*

PROOF. Let $P = \mathsf{C}(G)$. We have identity elements $\mathrm{id}_a \in P(a^\dagger, a)$ and $\mathrm{id}_{a^\dagger} \in P(a, a^\dagger)$, and these correspond to the two possible orderings for the boundary of the edge subtree $[a, a^\dagger]$.

Working inductively by composing with one generator at a time, each nonedge subtree T determines an element of $P(\eth(T))$ for some ordering of $\eth(T)$ by Lemma A.4. The axioms of a cyclic operad guarantee that this element does not depend on the choice of order of attaching vertices to construct the tree T. This is an element $T \in P(\eth(T)) = P(a_0, \dots, a_n)$, where the ordering of $\eth(T)$ is determined by the chosen orderings on $\eth(\star_v)$. The cyclic operad axioms imply that applying a permutation at any intermediate step simply has the effect of applying some permutation at the last step

$$T \in P(a_0, a_1, \dots, a_n) \xrightarrow{\sigma^*} P(a_{\sigma(0)}, a_{\sigma(1)}, \dots, a_{\sigma(n)}).$$

Thus all subtrees of equipped with an ordering of the boundary may regarded as (necessarily distinct, by [**Hac24**, Lemma A.3]) operations of P; we refer to these as simply subtrees in the next paragraph.

Now suppose that S, T are two nonedge subtrees of G, considered as elements

$$S \in P(a'_0, \dots, a'_m) = P(\eth(S)) \qquad T \in P(a_0, \dots, a_n) = P(\eth(T))$$

with $a'_i = a_j^\dagger$, so that we can form the composition $S {}_i\circ_j T$. The paths argument in the first two paragraphs of the proof of Lemma A.1 shows that S and T do not share any vertices, hence the composite of S and T uses each generator at of P at most once. It follows that the set of generators used is $V_S \amalg V_T = V_{S \cup T}$, and the composite is just the union $S \cup T$, along with the relevant ordering of $\eth(S \cup T)$. Since composites of subtrees are again subtrees, all iterated composites of the generators of P are subtrees. □

PROOF OF PROPOSITION 3.2. If H and G are undirected trees, then a map of cyclic operads $F \colon \mathsf{C}(H) \to \mathsf{C}(G)$ consists of an involutive function $f_0 \colon A_H \to A_G$, and a function f from V_H to the operations of $\mathsf{C}(G)$. As we are regarding $v \in V_H$ as living in $\mathsf{C}(H)(\eth(\star_v)) = \mathsf{C}(H)(a_0, \dots, a_n)$, the pair of functions needs only satisfy color-compatibility, that is, $f(v)$ should live in $\mathsf{C}(G)(f_0(a_0), \dots, f_0(a_n))$. By Lemma A.7, $f(v)$ is a subtree T of G together with an ordering of its boundary $\eth(T) = \{f_0(a_0), \dots, f_0(a_n)\}$. But this ordering is not actual data of the morphism, rather is just serves to establish boundary-compatibility. Thus, this assignment is exactly the same thing as a tree map of Definition 3.1. Since C is a functor, this establishes the desired equivalence. Lemma A.7 also implies that $\mathsf{C}(G)$ is a cyclic operad (rather than an augmented cyclic operad) if and only if G has inhabited boundary, giving the statement about $\mathbf{U}_{\text{cyc}}$. □

Acknowledgments

I am grateful to the organizers of the special session for inviting me to present this work, and to the other speakers and participants for helping to make an invigorating meeting.

References

[BB17] M. A. Batanin and C. Berger, *Homotopy theory for algebras over polynomial monads*, Theory Appl. Categ. **32** (2017), Paper No. 6, 148–253. MR3607212

[Ber22] Clemens Berger, *Moment categories and operads*, Theory Appl. Categ. **38** (2022), Paper No. 39, 1485–1537. MR4541944

[BG19] John Bourke and Richard Garner, *Monads and theories*, Adv. Math. **351** (2019), 1024–1071, DOI 10.1016/j.aim.2019.05.016. MR3956765

[BGMS21] John C. Baez, Fabrizio Genovese, Jade Master, and Michael Shulman, *Categories of nets*, 2021 36th Annual ACM/IEEE Symposium on Logic in Computer Science (LICS), IEEE, [Piscataway], NJ, [2021] ©2021, pp. [13 pp.], DOI 10.1109/LICS52264.2021.9470566. MR4401577

[BK] Clemens Berger and Ralph M. Kaufmann, *Trees, graphs and aggregates: a categorical perspective on combinatorial surface topology, geometry, and algebra*, Preprint, `arXiv:2201.10537v1 [math.AT]`, (2022).

[BM96] K. Behrend and Yu. Manin, *Stacks of stable maps and Gromov-Witten invariants*, Duke Math. J. **85** (1996), no. 1, 1–60, DOI 10.1215/S0012-7094-96-08501-4. MR1412436

[BM11] Clemens Berger and Ieke Moerdijk, *On an extension of the notion of Reedy category*, Math. Z. **269** (2011), no. 3-4, 977–1004, DOI 10.1007/s00209-010-0770-x. MR2860274

[BM15] Michael Batanin and Martin Markl, *Operadic categories and duoidal Deligne's conjecture*, Adv. Math. **285** (2015), 1630–1687, DOI 10.1016/j.aim.2015.07.008. MR3406537

[BMW12] Clemens Berger, Paul-André Melliès, and Mark Weber, *Monads with arities and their associated theories*, J. Pure Appl. Algebra **216** (2012), no. 8-9, 2029–2048, DOI 10.1016/j.jpaa.2012.02.039. MR2925893

[BR] Olivia Borghi and Marcy Robertson, *Lecture notes on modular infinity operads and Grothendieck–Teichmüller theory*, Preprint, `arXiv:2210.13640 [math.AT]`, (2023).

[CGR14] Eugenia Cheng, Nick Gurski, and Emily Riehl, *Cyclic multicategories, multivariable adjunctions and mates*, J. K-Theory **13** (2014), no. 2, 337–396, DOI 10.1017/is013012007jkt250. MR3189430

[CH20] Hongyi Chu and Rune Haugseng, *Enriched ∞-operads*, Adv. Math. **361** (2020), 106913, 85, DOI 10.1016/j.aim.2019.106913. MR4038556

[CH21] Hongyi Chu and Rune Haugseng, *Homotopy-coherent algebra via Segal conditions*, Adv. Math. **385** (2021), Paper No. 107733, 95, DOI 10.1016/j.aim.2021.107733. MR4256131

[CH22] Hongyi Chu and Philip Hackney, *On rectification and enrichment of infinity properads*, J. Lond. Math. Soc. (2) **105** (2022), no. 3, 1418–1517, DOI 10.1112/jlms.12528. MR4389306

[CM11] Denis-Charles Cisinski and Ieke Moerdijk, *Dendroidal sets as models for homotopy operads*, J. Topol. **4** (2011), no. 2, 257–299, DOI 10.1112/jtopol/jtq039. MR2805991

[CM13a] Denis-Charles Cisinski and Ieke Moerdijk, *Dendroidal Segal spaces and ∞-operads*, J. Topol. **6** (2013), no. 3, 675–704, DOI 10.1112/jtopol/jtt004. MR3100887

[CM13b] Denis-Charles Cisinski and Ieke Moerdijk, *Dendroidal sets and simplicial operads*, J. Topol. **6** (2013), no. 3, 705–756, DOI 10.1112/jtopol/jtt006. MR3100888

[DCH19] Gabriel C. Drummond-Cole and Philip Hackney, *A criterion for existence of right-induced model structures*, Bull. Lond. Math. Soc. **51** (2019), no. 2, 309–326, DOI 10.1112/blms.12232. MR3937590

[DCH21] Gabriel C. Drummond-Cole and Philip Hackney, *Dwyer-Kan homotopy theory for cyclic operads*, Proc. Edinb. Math. Soc. (2) **64** (2021), no. 1, 29–58, DOI 10.1017/S0013091520000267. MR4249838

[Dun06] Ross Duncan, *Types for quantum computing*, Ph.D. thesis, Oxford University, 2006.

[FBSD21] John D. Foley, Spencer Breiner, Eswaran Subrahmanian, and John M. Dusel, *Operads for complex system design specification, analysis and synthesis*, Proc. A. **477** (2021), no. 2250, Paper No. 20210099, 35, DOI 10.1098/rspa.2021.0099. MR4291809

[Gag15] Andrea Gagna, *The Cisinski–Moerdijk model structure on planar dendroidal sets*, Master's thesis, Universiteit Leiden, 2015.

[Gan03] Wee Liang Gan, *Koszul duality for dioperads*, Math. Res. Lett. **10** (2003), no. 1, 109–124, DOI 10.4310/MRL.2003.v10.n1.a11. MR1960128

[Gar08] Richard Garner, *Polycategories via pseudo-distributive laws*, Adv. Math. **218** (2008), no. 3, 781–827, DOI 10.1016/j.aim.2008.02.001. MR2414322

[Get09] Ezra Getzler, *Operads revisited*, Algebra, arithmetic, and geometry: in honor of Yu. I. Manin. Vol. I, Progr. Math., vol. 269, Birkhäuser Boston, Boston, MA, 2009, pp. 675–698, DOI 10.1007/978-0-8176-4745-2_16. MR2641184

[GH18] Richard Garner and Tom Hirschowitz, *Shapely monads and analytic functors*, J. Logic Comput. **28** (2018), no. 1, 33–83, DOI 10.1093/logcom/exx029. MR3798968

[GK95] E. Getzler and M. M. Kapranov, *Cyclic operads and cyclic homology*, Geometry, topology, & physics, Conf. Proc. Lecture Notes Geom. Topology, IV, Int. Press, Cambridge, MA, 1995, pp. 167–201. MR1358617

[GK98] E. Getzler and M. M. Kapranov, *Modular operads*, Compositio Math. **110** (1998), no. 1, 65–126, DOI 10.1023/A:1000245600345. MR1601666

[Gro61] Alexander Grothendieck, *Techniques de construction et théorèmes d'existence en géométrie algébrique. III. Préschemas quotients* (French), Séminaire Bourbaki, Vol. 6, Soc. Math. France, Paris, 1995, pp. Exp. No. 212, 99–118. MR1611786

[Hac24] Philip Hackney, *Categories of graphs for operadic structures*, Math. Proc. Cambridge Philos. Soc. **176** (2024), no. 1, 155–212, DOI 10.1017/s0305004123000452. MR4680484

[HHM16] Gijs Heuts, Vladimir Hinich, and Ieke Moerdijk, *On the equivalence between Lurie's model and the dendroidal model for infinity-operads*, Adv. Math. **302** (2016), 869–1043, DOI 10.1016/j.aim.2016.07.021. MR3545944

[HK24] Rune Haugseng and Joachim Kock, *∞-operads as symmetric monoidal ∞-categories*, Publ. Mat. **68** (2024), no. 1, 111–137, DOI 10.5565/publmat6812406. MR4682726

[HM22] Gijs Heuts and Ieke Moerdijk, *Simplicial and dendroidal homotopy theory*, Ergebnisse der Mathematik und ihrer Grenzgebiete. 3. Folge. A Series of Modern Surveys in Mathematics [Results in Mathematics and Related Areas. 3rd Series. A Series of Modern Surveys in Mathematics], vol. 75, Springer, Cham, [2022] ©2022, DOI 10.1007/978-3-031-10447-3. MR4485749

[HR15] Philip Hackney and Marcy Robertson, *On the category of props*, Appl. Categ. Structures **23** (2015), no. 4, 543–573, DOI 10.1007/s10485-014-9369-4. MR3367131

[HR18] Philip Hackney and Marcy Robertson, *Lecture notes on infinity-properads*, 2016 MATRIX annals, MATRIX Book Ser., vol. 1, Springer, Cham, 2018, pp. 351–374, DOI 10.1007/978-3-319-72299-3_18. MR3792530

[HRY15] Philip Hackney, Marcy Robertson, and Donald Yau, *Infinity properads and infinity wheeled properads*, Lecture Notes in Mathematics, vol. 2147, Springer, Cham, 2015, DOI 10.1007/978-3-319-20547-2. MR3408444

[HRY18] Philip Hackney, Marcy Robertson, and Donald Yau, *On factorizations of graphical maps*, Homology Homotopy Appl. **20** (2018), no. 2, 217–238, DOI 10.4310/HHA.2018.v20.n2.a11. MR3812464

[HRY19] Philip Hackney, Marcy Robertson, and Donald Yau, *Higher cyclic operads*, Algebr. Geom. Topol. **19** (2019), no. 2, 863–940, DOI 10.2140/agt.2019.19.863. MR3924179

[HRY20a] Philip Hackney, Marcy Robertson, and Donald Yau, *A graphical category for higher modular operads*, Adv. Math. **365** (2020), 107044, 61, DOI 10.1016/j.aim.2020.107044. MR4064770

[HRY20b] Philip Hackney, Marcy Robertson, and Donald Yau, *Modular operads and the nerve theorem*, Adv. Math. **370** (2020), 107206, 39, DOI 10.1016/j.aim.2020.107206. MR4099828

[HV02] Vladimir Hinich and Arkady Vaintrob, *Cyclic operads and algebra of chord diagrams*, Selecta Math. (N.S.) **8** (2002), no. 2, 237–282, DOI 10.1007/s00029-002-8106-2. MR1913297

[JK11] André Joyal and Joachim Kock, *Feynman graphs, and nerve theorem for compact symmetric multicategories (extended abstract)*, Electronic Notes in Theoretical Computer Science **270** (2011), no. 2, 105–113. DOI:10.1016/j.entcs.2011.01.025.

[KL17] Ralph Kaufmann and Jason Lucas, *Decorated Feynman categories*, J. Noncommut. Geom. **11** (2017), no. 4, 1437–1464, DOI 10.4171/JNCG/11-4-8. MR3743229

[KM94] M. Kontsevich and Yu. Manin, *Gromov-Witten classes, quantum cohomology, and enumerative geometry*, Comm. Math. Phys. **164** (1994), no. 3, 525–562. MR1291244

[Koc11] Joachim Kock, *Polynomial functors and trees*, Int. Math. Res. Not. IMRN **3** (2011), 609–673, DOI 10.1093/imrn/rnq068. MR2764874

[Koc16] Joachim Kock, *Graphs, hypergraphs, and properads*, Collect. Math. **67** (2016), no. 2, 155–190, DOI 10.1007/s13348-015-0160-0. MR3484016

[Koc23] Joachim Kock, *Whole-grain Petri nets and processes*, J. ACM **70** (2023), no. 1, Art. 1, 58, DOI 10.1145/3559103. MR4529841

[KW17] Ralph M. Kaufmann and Benjamin C. Ward, *Feynman categories* (English, with English and French summaries), Astérisque **387** (2017), vii+161. MR3636409

[Lur] Jacob Lurie, *Higher algebra*, manuscript available at `https://www.math.ias.edu/~lurie/papers/HA.pdf`.

[ML65] Saunders Mac Lane, *Categorical algebra*, Bull. Amer. Math. Soc. **71** (1965), 40–106, DOI 10.1090/S0002-9904-1965-11234-4. MR171826

[MLM92] Saunders Mac Lane and Ieke Moerdijk, *Sheaves in geometry and logic*, Universitext, Springer-Verlag, New York, 1994. A first introduction to topos theory; Corrected reprint of the 1992 edition. MR1300636

[MMS09] M. Markl, S. Merkulov, and S. Shadrin, *Wheeled PROPs, graph complexes and the master equation*, J. Pure Appl. Algebra **213** (2009), no. 4, 496–535, DOI 10.1016/j.jpaa.2008.08.007. MR2483835

[Moe10] Ieke Moerdijk, *Lectures on dendroidal sets*, Simplicial methods for operads and algebraic geometry, Adv. Courses Math. CRM Barcelona, Birkhäuser/Springer Basel AG, Basel, 2010, pp. 1–118, DOI 10.1007/978-3-0348-0052-5. Notes written by Javier J. Gutiérrez. MR2778589

[MW07] Ieke Moerdijk and Ittay Weiss, *Dendroidal sets*, Algebr. Geom. Topol. **7** (2007), 1441–1470, DOI 10.2140/agt.2007.7.1441. MR2366165

[Ray21] Sophie Raynor, *Graphical combinatorics and a distributive law for modular operads*, Adv. Math. **392** (2021), Paper No. 108011, 87, DOI 10.1016/j.aim.2021.108011. MR4316667

[Rez01] Charles Rezk, *A model for the homotopy theory of homotopy theory*, Trans. Amer. Math. Soc. **353** (2001), no. 3, 973–1007, DOI 10.1090/S0002-9947-00-02653-2. MR1804411

[Shu20] Michael Shulman, *The 2-Chu-Dialectica construction and the polycategory of multivariable adjunctions*, Theory Appl. Categ. **35** (2020), Paper No. 4, 89–136. MR4063570

[Spi13] David I. Spivak, *The operad of wiring diagrams: formalizing a graphical language for databases, recursion, and plug-and-play circuits*, preprint, `arXiv:1305.0297 [cs.DB]`, (2013).

[Val07] Bruno Vallette, *A Koszul duality for PROPs*, Trans. Amer. Math. Soc. **359** (2007), no. 10, 4865–4943, DOI 10.1090/S0002-9947-07-04182-7. MR2320654

[VSL15] Dmitry Vagner, David I. Spivak, and Eugene Lerman, *Algebras of open dynamical systems on the operad of wiring diagrams*, Theory Appl. Categ. **30** (2015), Paper No. 51, 1793–1822. MR3431301

[Wal21] Tashi Walde, *2-Segal spaces as invertible infinity-operads*, Algebr. Geom. Topol. **21** (2021), no. 1, 211–246, DOI 10.2140/agt.2021.21.211. MR4224740

[Web07] Mark Weber, *Familial 2-functors and parametric right adjoints*, Theory Appl. Categ. **18** (2007), No. 22, 665–732. MR2369114

[Wei11] Ittay Weiss, *From operads to dendroidal sets*, Mathematical foundations of quantum field theory and perturbative string theory, Proc. Sympos. Pure Math., vol. 83, Amer. Math. Soc., Providence, RI, 2011, pp. 31–70, DOI 10.1090/pspum/083/2742425. MR2742425

[YJ15] Donald Yau and Mark W. Johnson, *A foundation for PROPs, algebras, and modules*, Mathematical Surveys and Monographs, vol. 203, American Mathematical Society, Providence, RI, 2015, DOI 10.1090/surv/203. MR3329226

Department of Mathematics, University of Louisiana at Lafayette, 104 E University Ave, Lafayette, Louisiana 70504
Email address: philip@phck.net
URL: http://phck.net

Contemporary Mathematics
Volume **802**, 2024
https://doi.org/10.1090/conm/802/16078

The topological Atiyah–Segal map

Daniel A. Ramras

Abstract. Associated to each finite dimensional linear representation of a group G, there is a vector bundle over the classifying space BG. This construction was studied extensively for compact groups by Atiyah and Segal. We introduce a homotopy theoretical framework for studying the Atiyah–Segal construction in the context of infinite discrete groups, taking into account the topology of representation spaces.

We explain how this framework relates to the Novikov conjecture, and we consider applications to spaces of flat connections on the over the 3-dimensional Heisenberg manifold and families of flat bundles over classifying spaces of groups satisfying Kazhdan's property (T).

1. Introduction

In the 1950s and 1960s, Atiyah and Segal studied a construction that associates a vector bundle to each (complex) representation of a group G. This construction yields a map

$$R[G] \longrightarrow K^*(BG),$$

from the complex representation ring of G to the complex K-theory of its classifying space BG. The classical Atiyah–Segal completion theorem [**3**] states that when G is a compact Lie group, this map becomes an isomorphism after completing $R[G]$ at its augmentation ideal.

We introduce an analogous map for infinite discrete groups, where the topology of the representation spaces $\mathrm{Hom}(G, U(n))$ plays a key role. Loosely speaking, this map assigns to a spherical family of representations

$$\rho\colon S^m \to \mathrm{Hom}(G, U(n))$$

the K-theory class of the associated vector bundle

$$E_\rho \to S^m \times BG$$

with holonomy ρ. See Section 5 for the definition of E_ρ, and a more precise statement along these lines (Theorem 5.3). The construction $\rho \mapsto E_\rho$ was considered previously in [**20**], where it was used to study the (strong) Novikov conjecture, and in [**4**], where it was used to obtain cohomological lower bounds on the homotopy groups of spaces of flat, unitary connections.

2020 *Mathematics Subject Classification.* Primary 55R37, 53C05; Secondary 55P43, 55R50.
The author was partially supported by the Simons Foundation (#279007 and #579789).

©2024 by the author

The appropriate context for this construction is that of *deformation K-theory.* The most down-to-earth description of the deformation K-theory of a group G is that it arises from the homotopy group completion $\Omega B\mathrm{Rep}(G)$, where $\mathrm{Rep}(G)$ is the topological monoid

$$\mathrm{Rep}(G) = \coprod_n \mathrm{Hom}(G, U(n)),$$

with block sum of matrices as the monoid operation. In particular, in positive degrees the reduced deformation K-theory groups $\widetilde{K}^{\mathrm{def}}_*(G)$ agree with the homotopy groups of $\Omega B\mathrm{Rep}(G)$. This is discussed in detail in Section 3, where $\Omega B\mathrm{Rep}(G)$ is compared to more highly structured models for deformation K-theory. General linear deformation K-theory is obtained by replacing $U(n)$ by $GL(n) = GL_n(\mathbb{C})$. The (reduced, unitary) topological Atiyah–Segal map is a homomorphism

$$\widetilde{\alpha}_*\colon \widetilde{K}^{\mathrm{def}}_*(G) := \pi_*\Omega B\mathrm{Rep}(G) \longrightarrow \widetilde{K}^{-*}(BG), \tag{1}$$

induced by a map of topological monoids arising from the natural maps

$$B\colon \mathrm{Hom}(G, U(n)) \longrightarrow \mathrm{Map}_*(BG, BU(n)). \tag{2}$$

The details of this construction appear in Section 5.

Deformation K-theory has proven difficult to compute, and we do not expect any uniform description of its relationship to topological K-theory. This stands in contrast to similar functors such as algebraic K-theory of the group ring, or the K-theory of group C^*-algebras, where there are general conjectures describing this functors in terms of homological data. We view this as a positive feature of the theory: it is subtle enough to capture delicate information about the group in question, so that when computation can be achieved, concrete consequences follow.

We consider several applications in this article.

- In Section 6, we reinterpret a result from **[20]** to show that rational surjectivity of $\widetilde{\alpha}_*$ in high dimensions implies the strong Novikov conjecture (Theorem 6.2). Thus surjectivity of $\widetilde{\alpha_*}$ should be viewed as a very strong Novikov-type property. Earlier work of the author implies that surjectivity holds for surface groups (Theorem 6.4).
- In Section 7, we use Tyler Lawson's calculations of deformation K-theory for the 3-dimensional Heisenberg group $\pi_1 N^3$ to produce huge families of homotopy classes in the space of flat, unitary connections on bundles over the corresponding 3-dimensional Heisenberg manifold (Theorem 7.6 and Corollary 7.10). This shows a marked difference between gauge theory in 2- and 3-dimensions: over surfaces, homotopy in the space of flat connections is tightly controlled by Yang–Mills theory and complex geometric considerations, but in 3-dimensions the flood gates open.

  We note that for these results, the spectrum-level description of the topological Atiyah–Segal map is not really needed; one could instead take Theorem 5.3 as a definition.
- In Section 8, we show that the topological Atiyah–Segal map is actually a map of E_∞ rings, and consider some applications. First, a result of S. P. Wang, combined with computational methods due to Lawson, allows us to calculate the deformation K-theory of groups satisfying Kazhdan's property (T), and to describe the topological Atiyah–Segal map for such groups (Propositions 8.4 and 8.6). We then deduce that for property

(T) groups admitting finite classifying spaces, the vector bundles E_ρ associated to spherical families of unitary representations always represent torsion classes in $\widetilde{K}^0(S^m \times BG)$ (Theorem 8.1). Finally, we show that the topological Atiyah–Segal map for the Heisenberg group fails to be surjective (Proposition 8.8).

2. Permutative categories arising from group actions

In this section we introduce a framework for producing permutative categories, internal to the category **CGTop** of compactly generated topological spaces, from certain sequences of group actions. All topological notions, such as topological groups and group actions, should be interpreted in the category **CGTop**.

In what follows, we will assume familiarity with the notion of permutative categories, as defined in [**14**]. We briefly recall the definition. A *permutative category* is a (small) internal category $\mathcal{C}$ in **CGTop** equipped with a distinguished unit object $*$ and a continuous functor $\oplus\colon \mathcal{C}\times\mathcal{C}\to\mathcal{C}$ which is *strictly* associative and unital, and admits a natural commutativity isomorphism satisfying certain coherence requirements. Permutative functors between permutative categories are continuous functors that are appropriately compatible with the permutative structures. The category of permutative categories and permutative functors will be denoted **PCat**.

The framework introduced here will be used in subsequent sections to give compatible descriptions of deformation K-theory and topological K-theory, facilitating the construction of the topological Atiyah–Segal map as a morphism of spectra. The proofs of the claims made in this section are all routine, and in fact relatively short, and will mostly be left to the reader. In Section 2.3, we briefly explain how this theory can be enhanced to produce *bipermutative categories* and hence ring spectra.

We will use the following terminology regarding (topological) group actions: if X is a G-space, Y is an H-space, and $\phi\colon G\to H$ is a (continuous) homomorphism, then a map $f\colon X\to Y$ is called ϕ-equivariant, or equivariant with respect to ϕ, if

$$f(g\cdot x) = \phi(g)\cdot f(x)$$

for all $g\in G$, $x\in X$.

2.1. Action sequences.

The canonical example to keep in mind when reading the following definition is the unitary (or general linear) groups acting on themselves by conjugation, with the usual matrix block sum operations (see Example 2.3).

DEFINITION 2.1. A permutative (left) action sequence is a sextuple

$$\mathcal{A} = (I, \{G_i\}_{i\in I}, \{X_i\}_{i\in I}, *, \oplus, \{C_{i,j}\}_{i,j\in I}),$$

where:

- I is a commutative monoid, with identity element 0 and operation $+$;
- Each G_i is a topological group with identity element $e_i\in G_i$, and each X_i is a left G_i-space;
- $* = *_0 \in X_0$ is a nondegenerate basepoint;
- $\oplus = (\oplus^{\mathrm{alg}}, \oplus^{\mathrm{top}})$;
- $\oplus^{\mathrm{alg}}$ is an associative collection of homomorphisms

$$\oplus^{\mathrm{alg}}_{i,j}\colon G_i\times G_j\to G_{i+j},\ \ i,j\in I;$$

- $\oplus^{\text{top}}$ is an associative collection of $\oplus^{\text{alg}}_{i,j}$-equivariant maps
$$\oplus^{\text{top}}_{i,j}\colon X_i \times X_j \to X_{i+j},\ \ i,j \in I,$$
where equivariance refers to the product action of $G_i \times G_j$ on $X_i \times X_j$ (and the action of G_{i+j} on X_{i+j});
- For each $i, j \in I$, we have $C_{i,j} \in G_{i+j}$.

From here on, we will usually simplify notation by writing $\oplus$ in place of $\oplus^{\text{alg}}_{i,j}$ or $\oplus^{\text{top}}_{i,j}$.

The elements $C_{i,j}$ are subject to the following further axioms for all $i, j, k \in I$.

- $C_{i,j} \cdot (x_i \oplus x_j) = x_j \oplus x_i$ for each $x_i \in X_i$, $x_j \in X_j$
- $C_{i,0} = C_{0,i} = e_i$;
- $C_{i,j}C_{j,i} = e_{i+j}$;
- If $g_i \in G_i$ and $g_j \in G_j$, then $C_{i,j}(g_i \oplus g_j) = (g_j \oplus g_i)C_{i,j}$;
- $(C_{i,k} \oplus e_j)(e_i \oplus C_{j,k}) = C_{i+j,k}$.

Note that it is not necessary to assume $*_0$ is fixed by the action of G_0.

We refer to the operations $\oplus$ as the *monoidal*, or *additive*, structure of $\mathcal{A}$, and we refer to the elements $C_{i,j}$ as the *commutativity operators.*

It will be convenient to use the notation $X = \coprod_i X_i$, $G = \coprod_i G_i$, as well as to set $C = \{C_{i,j}\}$. Then we can write an action sequence in the simplified notation $\mathcal{A} = (I, G, X, *, \oplus, C)$.

A morphism of permutative action sequences
$$\mathcal{A} = (I, G, X, *, \oplus, C) \longrightarrow \mathcal{B} = (J, H, Y, *, \oplus, D)$$
consists of a homomorphism of monoids $f\colon I \to J$ together with group homomorphisms
$$\phi_i\colon G_i \to H_{f(i)}$$
satisfying $\phi_{i+j}(C_{i,j}) = D_{f(i),f(j)}$, and ϕ_i-equivariant maps
$$\zeta_i\colon X_i \to Y_{f(i)}$$
for each $i \in I$, such that $\zeta_0(*) = *$. This defines the category **PAct** of permutative (left) action sequences.

We will always work with left actions, and we drop the adjective *left.*

CONSTRUCTION 2.2. There is a functor
$$\mathcal{T}\colon \mathbf{PAct} \longrightarrow \mathbf{PCat},$$
defined as follows.

Given an action sequence $\mathcal{A} = (I, G, X, *, \oplus, C)$, the object space of $\mathcal{T}(\mathcal{A})$ is simply $X = \coprod_{i\in I} X_i$, while the morphism space is $\coprod_{i\in I} G_i \times X_i$. The domain of (g, x) is x, the codomain is $g \cdot x$, and composition is given by
$$(h, g\cdot x) \circ (g, x) = (hg, x).$$
The operations $\oplus$ give rise to a continuous functor
$$\oplus\colon \mathcal{T}(\mathcal{A}) \times \mathcal{T}(\mathcal{A}) \longrightarrow \mathcal{T}(A),$$
which is (strictly) associative and has the object $*_0 \in X_0$ as (strict) unit. The commutativity isomorphisms are given by $(C_{i,j}, x_i \oplus x_j)$, and our axioms on the $C_{i,j}$ are exactly what is needed to make the coherence diagrams in $\mathcal{T}(\mathcal{A})$ commute.

We call $\mathcal{T}(\mathcal{A})$ as the *translation category* of $\mathcal{A}$. Note that $\mathcal{T}(\mathcal{A})$ is in fact a groupoid.

EXAMPLE 2.3. The tautological (additive) unitary permutative action sequence is given by setting $I = \mathbb{N}$, with ordinary addition as the monoid operation, and setting $X_n = G_n = U(n)$ for $n \in \mathbb{N}$. We define $U(0) = \{0\}$, the trivial group. Here we view $U(n)$ as a left $U(n)$-space via *conjugation*, and we use the usual matrix block sum operation to define both $\oplus^{\mathrm{alg}}$ and $\oplus^{\mathrm{top}}$, with $0 \in U(0)$ acting as the unit element. The commutativity operators are the (unitary) permutation matrices

$$I_{m,n} = \begin{bmatrix} 0_{nm} & I_n \\ I_m & 0_{mn} \end{bmatrix}, \tag{3}$$

where 0_{pq} denotes the $p \times q$ zero matrix. The tautological additive general linear action sequence is defined similarly, by replacing $U(n)$ by $GL(n)$.

DEFINITION 2.4. An (additive) *unitary permutative action sequence* is one in which the underlying monoid is $\mathbb{N}$, with its usual addition, and we have $G_n = U(n)$ and $C_{m,n} = I_{m,n}$ for all $m, n \in \mathbb{N}$. Note that such sequences are completely determined by their topological data, that is, the $U(n)$-spaces X_n (and the basepoint $x_0 \in X_0$) together with the maps $\oplus^{\mathrm{top}}$.

REMARK 2.5. The notion of a permutative action sequence can be generalized by allowing the elements $C_{i,j}$ to depend on $x_i \in X_i$ and $x_j \in X_j$ rather than just on $i, j \in I$, and a small modification again gives a functor from this larger category of sequences to **PCat**. Furthermore, there is no need to assume the G_i are groups; monoids would suffice.

2.2. The nerve of a permutative action sequence. Consider a permutative action sequence $\mathcal{A} = (I, G, X, *, \oplus, C)$. The continuous functor

$$\oplus\colon \mathcal{T}(\mathcal{A}) \times \mathcal{T}(\mathcal{A}) \longrightarrow \mathcal{T}(\mathcal{A})$$

makes $|N.\mathcal{T}(\mathcal{A})|$ into a topological monoid, since geometric realization commutes with products of simplicial spaces. May's infinite loop space machine [**14**] gives a functor $\mathbf{K}$ from **PCat** to the category of connective Ω-spectra. One key feature of this functor is that for each permutative category $\mathcal{C}$, the infinite loop space underlying the spectrum $\mathbf{K}(\mathcal{C})$ is naturally weakly equivalent to $\Omega B|N.\mathcal{C}|$. Our next goal is to give an explicit description of the monoid $|N.\mathcal{T}(\mathcal{A})|$.

For a space X with a left action of a topological group G, the homotopy orbit space, or Borel construction, is the quotient space

$$X_{hG} = (EG \times X)/G,$$

where EG is the geometric realization of the category $\overline{G}$, internal to **CGTop**, with object space G and morphism space $G \times G$ (here (g, h) is the unique morphism from h to g, and $(g, h) \circ (h, k) = (g, k)$). Note that $\overline{G}$ admits a right action of G (by functors), defined via right-multiplication in G. This induces a right action of G on EG, and now G acts on $EG \times X$ via $g \cdot (e, x) = (e \cdot g^{-1}, g \cdot x)$. Let BG denote the geometric realization of the category $\underline{G}$ (again internal to **CGTop**) with a single object $*$ and morphism space G, and with composition $g \circ h = gh$. When G is a Lie group, the natural map $EG \to BG$ (induced by the functor sending a morphism (g, h) in $\overline{G}$ to the morphism gh^{-1} in $\underline{G}$) is a universal principal G-bundle [**27**], and the natural map $X_{hG} \to BG$ is a fiber bundle with fiber X.

Given a permutative action sequence $\mathcal{A} = (I, G, X, *, \oplus, C)$, we can form

$$\mathcal{M}(\mathcal{A}) := \coprod_{i \in I} (X_i)_{hG_i}.$$

The maps $\oplus^{\text{alg}}_{i,j}$ induce continuous functors $\overline{G}_i \times \overline{G}_j \longrightarrow \overline{G}_{i+j}$ and hence continuous maps

$$EG_i \times EG_j \longrightarrow EG_{i+j}. \tag{4}$$

Since the maps $\oplus^{\text{alg}}_{i,j}$ are homomorphisms, the maps (4) are equivariant with respect to $\oplus\colon\ G_i \times G_j \to G_{i+j}$. Together with the equivariant maps $\oplus\colon\ X_i \times X_j \to X_{i+j}$, these maps induce a map

$$\mathcal{M}(\mathcal{A}) \times \mathcal{M}(\mathcal{A}) \longrightarrow \mathcal{M}(\mathcal{A}).$$

It is an exercise to check that this map makes $\mathcal{M}(\mathcal{A})$ into a topological monoid with $[*, *_0] \in (X_0)_{hG_0} \subset \mathcal{M}(\mathcal{A})$ as unit element, where $* \in EG$ corresponds to the object in $\overline{G}$ represented by the identity of G.

PROPOSITION 2.6. *There is a natural homeomorphism of topological monoids*

$$\mathcal{M}(\mathcal{A}) \longrightarrow |N.\mathcal{T}(\mathcal{A})|.$$

PROOF. (Sketch) A special case of this statement is proven in [**21**, Proposition 2.4] using an argument due to Tyler Lawson. That argument immediately generalizes to produce the desired map and to show that it is a continuous bijection. The argument proceeds by viewing each side as the geometric realization of a simplicial space, and providing a map of simplicial spaces that is a homeomorphism on each level. In [**21**], an appeal to compactness was made to deduce continuity of the inverse maps on each level, but it is in fact a simple matter to write down explicit formulas for these inverse maps, from which it is clear that the inverses are continuous. □

In [**25**], the notion of strongly anchored elements in a (homotopy commutative) monoid M was introduced, in order to give a concrete description of the homotopy groups $\pi_*\Omega BM$ of the homotopy group completion. We say that $x \in M$ is strongly anchored if there exists a homotopy between the maps

$$(m, n) \mapsto mn \text{ and } (m, n) \mapsto nm$$

that is constant on (x, x). We write $[X, Y]$ for the set of unbased homotopy classes of maps from X to Y; if $Y = M$ is a topological monoid, this set acquires a monoid structure. (We will also write $[f] \in [X, Y]$ for the class of $f\colon\ X \to Y$.)

THEOREM 2.7 ([**25**, Theorem 1.1, Proposition 2.8]). *Let M be a topological monoid such that each $x \in M$ is strongly anchored. Then for each $k \geq 0$, there is a natural isomorphism*

$$\frac{\mathrm{Gr}[S^k, M]}{\mathrm{Gr}(\nu_k M)} \xrightarrow{\cong} \pi_k(\Omega BM), \tag{5}$$

where Gr *denotes the Grothendieck group and $\nu_k M \leqslant [S^k, M]$ is the submonoid of nullhomotopic maps $S^k \to M$.*

In fact, for $k > 0$, the natural map

$$\pi\colon\ [S^k, M] \to \frac{\mathrm{Gr}[S^k, M]}{\mathrm{Gr}(\nu_k M)}$$

is surjective, and $\pi([\phi]) = 0$ *if and only if there exists a constant map* $c\colon S^k \to M$ *such that* $\phi \bullet c$ *is nullhomotopic (where* $\bullet$ *denotes pointwise multiplication in* M*).*

Note that for each $k \geqslant 0$, there is a natural isomorphism between $\nu_k M$ and the monoid of path components $\pi_0 M$.

One aspect of Theorem 2.7 is that it describes the product in the group $\pi_k(\Omega BM)$ via the operation in the underlying monoid M. As one might guess, the proof of Theorem 2.7 relies on a version of the Eckmann–Hilton Argument. Results of this type have a long history. Possibly the earliest such fact is that group operation in the fundamental group of a topological group, or more generally an H-space, can be defined either by loop concatenation or by point-wise multiplication. Another example is the fact that the group operation in the algebraic K-theory of a ring, defined for instance in terms of the homotopy groups of Quillen's Q-construction, is induced by the direct sum operation in the underlying category of finitely generated projective modules [**19**, p. 20].

It will be convenient to define

$$\widetilde{\Pi}_k M := \frac{\mathrm{Gr}[S^k, M]}{\mathrm{Gr}(\nu_k M)}.$$

The above result tells us, in particular, that $\widetilde{\Pi}_k M$ is the quotient of $[S^k, M]$ by $\nu_k M$ in the category of abelian monoids. In particular, each element of $[S^k, M]$ is invertible modulo constant maps.

If $\mathcal{A}$ is a unitary or general linear permutative action sequence, then the argument in [**21**, Proof of Corollary 4.4] shows that every element in the monoid $\mathcal{M}(\mathcal{A})$ is (strongly) anchored.

Corollary 2.8. *If* $\mathcal{A}$ *is a unitary or general linear permutative action sequence, then the natural map*

$$\widetilde{\Pi}_k \mathcal{M}(\mathcal{A}) \longrightarrow \Omega B\mathcal{M}(\mathcal{A})$$

is an isomorphism for each $k \geqslant 0$.

Remark 2.9. The commutativity operators $C_{i,j}$ induce a natural transformation between the functors $\oplus\colon \mathcal{T}(\mathcal{A}) \times \mathcal{T}(\mathcal{A}) \to \mathcal{T}(A)$ and $\oplus \circ \tau$, where τ is the twist functor on the product category. It then follows from basic categorical homotopy theory [**27**, Proposition 2.1] that $|N.(\mathcal{T}(\mathcal{A}))| \cong \mathcal{M}(\mathcal{A})$ is homotopy commutative. However, this homotopy does not anchor elements. The argument in [**21**, Proof of Corollary 4.4] involves constructing different homotopies, specific to each element we wish to anchor. The main point in the proof is that for each $x \in X_n$, the matrix $I_{n,n}$ lies in the identity component of the stabilizer of x^2. The fact that this requires no extra assumptions on the stabilizer appears to be a rather special feature of the unitary and general linear groups.

2.3. Bipermutative action sequences. We now extend the notion of permutative action sequence to a notion of *bipermutative action sequence*, in such a way that the translation category inherits the structure of a bipermutative category. As discussed in Section 8, this gives a relatively efficient way of verifying that the topological Atiyah–Segal map is a map of ring spectra.

In short, a bipermutative action sequence is a pair of permutative action sequences, sharing the same indexing set I, the same spaces X_i, and the same groups G_i. Additional coherence axioms relating the two permutative structures must hold, and these axioms imply that the two monoid structures on I give it the structure of a rig, or a "ring without negatives." We give the details in Definition 2.10 below; the axioms are just direct translations of the axioms for bipermutative categories. Maps of bipermutative action sequences are just maps that respect both permutative structures.

As an example, the Kronecker product of matrices endows the tautological unitary and general linear action sequences with a bipermutative structure. The details are just an elaboration of the discussion in [**18**, VI §5]. We note that some care must be taken when specifying the exact definition of Kronecker product, so that the coherence axioms hold.

DEFINITION 2.10. A bipermutative action sequence is rig R together with a pair of action sequences

$$((R,+), G, X, *_0, \oplus, C) \text{ and } ((R,\cdot), G, X, *_1, \otimes, D)$$

sharing the same groups G_r and the same G_r-spaces X_r for all $r \in R$. Let $0 \in R$ and $1 \in R$ denote the additive and multiplicative identity elements of R, respectively.

These data must satisfy the following additional axioms for all $r, s, t, u \in R$ and all $x, y, z, w \in X$ and all $g, h, k, l \in G$:

- Zero Axioms: $*_0 \otimes x = *_0 = x \otimes *_0$ and $e_0 \otimes g = e_0 = g \otimes e_0$ (recall that $e_0 \in G_0$ is the identity element);
- Right Distributivity Axioms: $(x \oplus y) \otimes z = (x \otimes z) \oplus (y \otimes z)$ and $(g \oplus h) \otimes k = (g \otimes k) \oplus (h \otimes k)$;
- Coherence Axioms: $C_{r,s} \otimes e_t = C_{r\cdot t, s\cdot t}$, and

$$\begin{aligned}(D_{t,r+s} \oplus D_{u,r+s})D_{r+s,t+u} \\ = (e_{r\cdot t} \oplus C_{r\cdot u, s\cdot t} \oplus e_{s\cdot u})\left[[(D_{t,r} \oplus D_{u,r})D_{r,t+u}] \oplus [(D_{t,s} \oplus D_{u,s})D_{s,t+u}]\right].\end{aligned}$$

We will sometimes denote these sequences in the simplified form

$$(R, G, X, *_0, *_1, \oplus, C, \otimes, D).$$

A morphism of bipermutative action sequences

$$(R, G, X, *_0, *_1, \oplus, C, \otimes, D) \longrightarrow (S, H, Y, *_0, *_1, \oplus, C', \otimes, D')$$

consists of a function $f\colon R \to S$ that is a monoid homomorphism for both $+$ and $\cdot$, together with homomorphisms $\phi_r\colon G_r \to H_{f(r)}$ and ϕ_r-equivariant maps $X_r \to Y_{f(r)}$ (preserving both basepoints) for all $r \in R$. The homomorphisms ϕ_r must satisfy $\phi_{s+t}(C_{s,t}) = C'_{f(s+t)}$ and $\phi_{s\cdot t}(D_{s,t}) = D'_{f(s\cdot t)}$.

Bipermutative action sequence now form a category **BPAct**, and the translation category construction provides a functor from **BPAct** to the category of bipermutative categories internal to **CGTop**, as defined in [**18**, Chapter VI].

Multiplicative infinite loop space theory, as developed in [**16–18**], provides a functor taking a bipermutative category $\mathcal{C}$ to a (connective) E_∞ ring spectrum $\mathbb{K}_\infty(\mathcal{C})$. Just as in the permutative case, the underlying infinite loop space $\Omega^\infty\mathbb{K}_\infty(\mathcal{C})$ is naturally weakly equivalent to the group completion $\Omega B\mathcal{C}$, where $B\mathcal{C}$ denotes the geometric realization of the bar construction applied to the nerve of $\mathcal{C}$, using its *additive* monoidal structure; the key step in the proof of this statement is [**17**, Theorem 9.3].

3. Deformation K-theory

Here we explain how to view the unitary and general linear deformation K-theory spectra associated to finitely generated discrete groups in terms of action sequences.

DEFINITION 3.1. Given a discrete group G, let $\mathcal{A}(G)$ denote the unitary permutative action sequence associated to the spaces $X_n = \mathrm{Hom}(G, U(n))$, which we topologize as subsets of the mapping spaces $\mathrm{Map}(G, U(n))$. Note that $\mathrm{Hom}(G, U(0))$ consists of a single point, which will serve as $*$. We let the unitary groups act on these spaces by conjugation, and the block sum operations are induced by block sum of matrices. Note that $\mathcal{A}(G)$ is contravariantly functorial in G.

The deformation K-theory spectrum $K^{\mathrm{def}}(G)$ is the connective Ω-spectrum associated to the permutative translation category $\mathcal{T}(\mathcal{A}(G))$, and we define

$$K_*^{\mathrm{def}}(G) = \pi_* K^{\mathrm{def}}(G) \cong \pi_* \Omega^\infty K^{\mathrm{def}}(G).$$

By Proposition 2.6, we have a natural homeomorphism

$$\Omega^\infty K^{\mathrm{def}}(G) \cong \Omega B \left(\coprod_{n=0}^{\infty} \mathrm{Hom}(G, U(n))_{hU(n)} \right),$$

where the coproduct on the right has the topological monoid structure induced from the additive structure of $\mathcal{A}(G)$.

REMARK 3.2. The spectra $K^{\mathrm{def}}(G)$ have been computed in a number of cases, and we briefly describe these results. When G is the trivial group, $K^{\mathrm{def}}(G)$ is simply the connective K-theory spectrum **ku** (this follows, for instance, from [**18**, VIII §2]). When G is finite, $K^{\mathrm{def}}(G)$ decomposes as a wedge of copies of **ku**, one for each irreducible complex representation of G [**11**, Chapter 6.3]. If G is the fundamental group of a closed surface (or the circle), then $K^{\mathrm{def}}(G)$ is a wedge of suspensions of **ku**, with $\beta_k(G)$ summands of the form $\Sigma^k\mathbf{ku}$, where $\beta_k(G)$ is the kth Betti number [**24**, Theorem 6.1]. That theorem also gives a slightly more complicated description for nonorientable surface groups. The behavior of deformation K-theory with respect to free products and direct products is also understood (see [**21**] and [**12**] respectively), allowing one to compute $K^{\mathrm{def}}(G)$ for finitely generated free and free abelian groups.

Underlying deformation K-theory, we have the following topological monoids.

DEFINITION 3.3. Given a discrete group G, define

$$\mathrm{Rep}(G) = \mathrm{Rep}(G, U) := \coprod_{n=0}^{\infty} \mathrm{Hom}(G, U(n))$$

and

$$\mathrm{Rep}(G)_{hU} = \coprod_{n=0}^{\infty} \mathrm{Hom}(G, U(n))_{hU(n)}.$$

Block sum of matrices makes $\mathrm{Rep}(G)$ into a topological monoid, with the unique element in $\mathrm{Hom}(G, U(0))$ as the identity. Replacing $U(n)$ by $GL(n)$, we obtain the monoids $\mathrm{Rep}(G, GL)$ and $\mathrm{Rep}(G, GL)_{hU}$.

The main results of [**25**] give a description of $K_*^{\mathrm{def}}(G)$ in terms of spherical families of representations.

THEOREM 3.4 ([**25**, Theorem 7.3]). *Let G be a discrete group. For each $m > 0$, there is are natural isomorphisms*

$$K_m^{\mathrm{def}}(G) \cong \widetilde{\Pi}_m(\mathrm{Rep}(G)) \oplus K^{-m}(*),$$
$$K_0^{\mathrm{def}}(G) \cong \widetilde{\Pi}_0(\mathrm{Rep}(G)) \cong \mathrm{Gr}(\pi_0\mathrm{Rep}(G)).$$

Moreover, for $k > 0$ the natural map of monoids

$$[S^k, \mathrm{Rep}(G)] \longrightarrow \widetilde{\Pi}_m(\mathrm{Rep}(G))$$

is surjective, *with kernel*

$$\{[\rho] : \exists \text{ a constant map } c\colon S^k \to \mathrm{Rep}(G) \text{ such that } \rho \oplus c \text{ is nullhomotopic}\}.$$

If G is finitely generated, the analogous result holds in the general linear case.

It will also be helpful to consider a reduced form of deformation K-theory. The unitary and general linear cases of this discussion are identical.

For each group G, the map $\{1\} \to G$ induces a map of spectra

$$q\colon K^{\mathrm{def}}(G) \to K^{\mathrm{def}}(\{1\}), \tag{6}$$

which admits a splitting induced by the map $G \to \{1\}$. As noted in Remark 3.2, we have $K^{\mathrm{def}}(\{1\}) \simeq \mathbf{ku}$.

DEFINITION 3.5. We define $\widetilde{K}^{\mathrm{def}}(G)$ to be the homotopy fiber of the natural map $K^{\mathrm{def}}(G) \to K^{\mathrm{def}}(\{1\})$, and we set $\widetilde{K}_*^{\mathrm{def}}(G) = \pi_* \widetilde{K}^{\mathrm{def}}(G)$.

COROLLARY 3.6. *There is a natural splitting*

$$K_m^{\mathrm{def}}(G) \cong \widetilde{K}_m^{\mathrm{def}}(G) \oplus K^{-m}(*)$$

for each $m \geqslant 0$, and for $m > 0$ there are natural isomorphisms

$$\widetilde{K}_m^{\mathrm{def}}(G) \cong \widetilde{\Pi}_m\mathrm{Rep}(G) \cong \pi_m \Omega B\mathrm{Rep}(G). \tag{7}$$

Additionally, $\widetilde{K}_0^{\mathrm{def}}(G)$ is naturally isomorphic to the quotient of

$$\mathrm{Gr}(\pi_0\mathrm{Rep}(G)) \cong \pi_0 \Omega B\mathrm{Rep}(G)$$

by the subgroup generated by the trivial 1-dimensional representation.

PROOF. The first statement is immediate from the fact that (6) splits. For $m > 0$, [**25**, Proof of Theorem 6.5] shows that the sequence

$$0 \longrightarrow \widetilde{\Pi}_m\mathrm{Rep}(G) \xrightarrow{i_*} \widetilde{\Pi}_m\left(\mathrm{Rep}(G)_{hU}\right) \xrightarrow{q_*} \widetilde{\Pi}_m\left(\mathrm{Rep}(\{1\})_{hU}\right) \longrightarrow 0. \tag{8}$$

is split exact, which implies (7).

When $m = 0$, the splitting of $q\colon K^{\mathrm{def}}(G) \to K^{\mathrm{def}}(\{1\})$ gives a natural isomorphism between $\widetilde{K}_0^{\mathrm{def}}(G)$ and the cokernel of the map $K_0^{\mathrm{def}}(\{1\}) \to K_0^{\mathrm{def}}(G)$, whose image is generated by the trivial 1-dimensional representation. □

4. Topological K-theory

In this section, $X = (X, x_0)$ will denote a (based) paracompact space having the homotopy type of compact Hausdorff space. The key case will be $X = B\pi_1(K)$ with K an aspherical finite CW complex. We will define a permutative category whose homotopy groups agree with the complex topological K-theory of X. Our construction, and the subsequent discussion, is designed to mirror the construction of deformation K-theory in the previous section. This will facilitate our construction and analysis of the topological Atiyah–Segal map in the next section. As in the previous section, there is both a unitary and a general linear version of the constructions given here. We focus on the unitary case; the general linear case is completely analogous (more so than for deformation K-theory).

DEFINITION 4.1. As before, let $BU(n)$ denote the geometric realization of the one-object category $\underline{U(n)}$. Then the (left) conjugation action of $U(n)$ on itself induces an action, by continuous functors, of $U(n)$ on the category $\underline{U(n)}$, and hence an action of $U(n)$ on $BU(n)$. This in turn induces an action of $U(n)$ on the based mapping space $\mathrm{Map}_*(X, BU(n))$. Throughout this section, we will view $\mathrm{Map}_*(X, BU(n))$ as a $U(n)$-space under this action.

The block sum operations $\oplus\colon U(m) \times U(n) \to U(m+n)$ are homomorphisms, and hence induce continuous functors $\underline{U(m)} \times \underline{U(n)} \to \underline{U(m+n)}$, maps

$$BU(m) \times BU(n) \longrightarrow BU(m+n),$$

and equivariant maps

$$\oplus\colon \mathrm{Map}_*(X, BU(m)) \times \mathrm{Map}_*(X, BU(n)) \longrightarrow \mathrm{Map}_*(X, BU(m+n)).$$

Since block sum is equivariant with respect to conjugation (that is, $(CAC^{-1}) \oplus (DBD^{-1}) = (C \oplus D)(A \oplus B)(C \oplus D)^{-1}$), functoriality implies that this data gives a unitary permutative action sequence $\mathcal{A}_K(X)$ with nth space $\mathrm{Map}_*(X, BU(n))$.

Note that when $X = *$, we recover the unitary permutative action sequence whose associated spectrum is **ku** (since we are taking homotopy orbits of $BU(n)$ with respect to the conjugation action of $U(n)$).

Let $\mathcal{C}_K(X) = \mathcal{T}(\mathcal{A}_K(X))$ be the translation category of $\mathcal{A}_K(X)$, and let $\mathcal{K}(X)$ denote the associated spectrum. Let $\widetilde{\mathcal{K}}(X)$ denote the homotopy fiber of the natural map $\mathcal{K}(X) \to \mathcal{K}(*)$. Finally, set $\mathcal{K}_*(X) = \pi_*\mathcal{K}(X)$ and $\widetilde{\mathcal{K}}_*(X) = \pi_*\widetilde{\mathcal{K}}(X)$.

We have the following consequence of Propositions 2.6 and 2.8.

COROLLARY 4.2. *The geometric realization of the nerve of $\mathcal{C}_K(X)$ is isomorphic, as a topological monoid, to the topological monoid*

$$\mathcal{V}(X)_{hU} := \coprod_n \mathrm{Map}_*(X, BU(n))_{hU(n)},$$

and for each $m \geqslant 0$, there are natural isomorphisms

$$\widetilde{\Pi}_m\left(\mathcal{V}(X)_{hU}\right) \xrightarrow{\cong} \pi_m \Omega B\left(\mathcal{V}(X)_{hU}\right) \cong \mathcal{K}_m(X).$$

Our goal in this section is to compare the homotopy groups $\mathcal{K}_*(X)$ with the (complex) topological K-theory of X (for $* \geqslant 0$).

We need to specify a definition of topological K-theory. Note that the Group Completion Theorem gives a natural homotopy equivalence

$$\mathbb{Z} \times BU \longrightarrow \Omega B\left(\coprod_n BU(n)\right),$$

where $BU = \mathrm{colim}_n BU(n)$, the colimit being formed with respect to the maps induced by block sum with the identity $I_1 \in U(1)$.

For based spaces Y and Z, let $\langle Y, Z\rangle$ be the set of based homotopy classes of based maps, and let $\langle f\rangle$ be the based homotopy class of f.

DEFINITION 4.3. For $m \geqslant 0$, the *reduced* topological K-theory of X is

$$\widetilde{K}^{-m}(X) = \widetilde{K}^0(S^m \wedge X) = \langle S^m \wedge X, \mathbb{Z} \times BU \rangle$$

and the unreduced K-theory of X is

$$K^{-m}(X) = \widetilde{K}^{-m}(X) \oplus K^{-m}(*) = \widetilde{K}^{-m}(X) \oplus \pi_m(\mathbf{ku}).$$

Maps $X \to Y$ induce maps on both reduced an unreduced K-theory (in the latter case, all maps act as the identity on $K^{-m}(*)$).

Note that by the group completion theorem, we have a natural zig-zag of homotopy equivalences connecting $\mathbb{Z} \times BU$ and $\Omega B(\coprod_n BU(n))$, so one may replace $\mathbb{Z} \times BU$ by $\Omega B(\coprod_n BU(n))$ in the above definition.

We will need to consider another topological monoid related to $\mathcal{K}(X)$.

DEFINITION 4.4. Define $\mathcal{V}(X) := \coprod_n \mathrm{Map}_*(X, BU(n))$, and equip $\mathcal{V}(X)$ with the monoid structure induced by the block sum operations on $\{BU(n)\}_n$.

We now have the following analogue of the results from Section 3.

COROLLARY 4.5. *Let (X, x_0) be a based, path-connected CW complex homotopy equivalent to a finite CW complex. For each $m \geqslant 0$, there is a natural splitting*

$$\mathcal{K}_m(X) \cong \widetilde{\mathcal{K}}_m(X) \oplus \mathcal{K}_m(*) = \widetilde{\mathcal{K}}_m(X) \oplus \pi_m \mathbf{ku} \tag{9}$$

and a natural isomorphism

$$\widetilde{\mathcal{K}}_m(X) \cong \widetilde{K}^{-m}(X).$$

Moreover, for $m > 0$ there are natural isomorphisms

$$\widetilde{\mathcal{K}}_m(X) \cong \widetilde{\Pi}_m \mathcal{V}(X) \cong \pi_m \Omega B \mathcal{V}(X),$$

while for $m = 0$, $\widetilde{\mathcal{K}}_0(X) \cong \widetilde{K}^0(X)$ is naturally isomorphic to the quotient of

$$\mathrm{Gr}(\pi_0 \mathcal{V}(X)) \cong \pi_0 \Omega B \mathcal{V}(X)$$

by the subgroup generated by the class of nullhomotopic maps $X \to BU(1)$.

PROOF. The splitting (9) is immediate since the inclusion $\{x_0\} \hookrightarrow X$ splits. Next, consider the sequence of groups

$$\widetilde{\Pi}_m(\mathcal{V}(X)) \longrightarrow \widetilde{\Pi}_m(\mathcal{V}(X)_{hU}) = \mathcal{K}_m(X) \longrightarrow \widetilde{\Pi}_m\left(\coprod_n (BU(n))\right) = \mathcal{K}_m(*) \tag{10}$$

induced by the maps

$$\mathrm{Map}_*(X, BU(n)) \longrightarrow \mathrm{Map}_*(X, BU(n))_{hU(n)} \longrightarrow BU(n). \tag{11}$$

Using the fact that the sequences (11) are (split) fibrations, one can check directly that (10) is a (split) short exact sequence for each $m > 0$. By naturality, it follows that $\widetilde{\Pi}_m(\mathcal{V}(X)) \cong \widetilde{\mathcal{K}}_m(X)$ for $m > 0$. We will return to the analogous description of $\widetilde{\mathcal{K}}_0(X)$ at the end of the proof.

Next we show that for $m > 0$, there is an isomorphism

$$\widetilde{K}^{-m}(X) = \langle S^m \wedge X, \mathbb{Z} \times BU \rangle \cong \langle S^m, \mathrm{Map}_*(X, BU) \rangle \xrightarrow[\cong]{\Phi} \widetilde{\Pi}_m \mathcal{V}(X).$$

To define the map Φ, note that our hypotheses on X guarantee that every map $f\colon S^m \wedge X \to BU$ factors, up to homotopy, through $BU(n)$ for some n, yielding a well-defined element of $\widetilde{\Pi}_m \mathcal{V}(X)$. To see that Φ is surjective, note that our hypotheses guarantee that every vector bundle over X is a direct summand of a trivial bundle, and hence every class in $\widetilde{\Pi}_m\mathcal{V}(X)$ is represented by a map

$$S^m \to \mathrm{Map}_*(X, BU(n))$$

whose value the basepoint in S^m is nullhomotopic. Since the basepoint of S^m is nondegenerate, we can in fact obtain a representative lying in

$$\mathrm{Map}_*(S^m, \mathrm{Map}_*(X, BU(n))),$$

as desired. To prove injectivity of Φ, say $\Phi(\langle f\rangle) = 0$ for some

$$f \in \mathrm{Map}_*(S^m, \mathrm{Map}_*(X, BU(n))).$$

Note that it since the basepoint of X is nondegenerate, will suffice to show that f is nullhomotopic in the *unbased* sense (see [**8**, Section 4A], for instance). Since $\Phi(\langle f\rangle) = 0$, there exist $h, g \in \mathrm{Map}_*(X, BU(k))$ (for some k) such that

$$[f \oplus c_g] = [c_h],$$

where c_g and c_h denote the constant maps out of S^m with values h and g. Now, there exists $g' \in \mathrm{Map}_*(X, BU(l))$ such that $g \oplus g'$ is nullhomotopic, so we have

$$[f] = [f \oplus c_g \oplus c_{g'}] = [c_{h+g'}]$$

in $[S^m, \mathrm{Map}_*(X, BU)]$, and hence f is nullhomotopic as a map into $\mathrm{Map}_*(X, BU)$.

For $m = 0$, connectedness of $U(n)$ yields bijections

$$\pi_0\mathrm{Map}_*(X, BU(n))_{hU(n)} \cong \pi_0\mathrm{Map}_*(X, BU(n)) \cong \langle X, BU(n)\rangle \cong [X, BU(n)]$$

so

$$\mathcal{K}_0(X) = \pi_0\mathcal{V}(X)_{hU} \cong \mathrm{Gr}\,(\pi_0\mathcal{V}(X)_{hU}) \cong \mathrm{Gr}\,(\pi_0\mathcal{V}(X)) = \mathrm{Gr}[X, \coprod_n BU(n)].$$

Now,

$$\mathrm{Gr}[X, \coprod_n BU(n)] \cong \mathrm{colim}([X, \coprod_n BU(n)] \to [X, \coprod_n BU(n)] \to \cdots),$$

where the colimit is taken with respect to addition by the basepoint in $BU(1)$. This colimit is precisely $\mathbb{Z} \times \mathrm{colim}_n[X, BU(n)] \cong \mathbb{Z} \times \langle X, BU\rangle \cong K^0(X)$.

The splitting (9) identifies $\widetilde{\mathcal{K}}_0(X)$ with the cokernel of $\mathcal{K}_0(*) \to \mathcal{K}_0(X)$. By the discussion above, this agrees with cokernel of $\mathrm{Gr}\,(\pi_0\mathcal{V}(*)) \to \mathrm{Gr}\,(\pi_0\mathcal{V}(X))$. The claimed description of $\widetilde{\mathcal{K}}_0(X)$ now follows from the fact that $\mathcal{V}(*)$ is generated by the constant map from $*$ to the basepoint of $BU(1)$. □

REMARK 4.6. Corollary 4.5 requires the assumption that X is path-connected. For instance, Corollary 4.2 gives

$$\mathcal{K}_0(S^0) \cong \mathrm{Gr}\left(\coprod_n \mathrm{Map}_*(S^0, BU(n))_{hU(n)}\right),$$

and since $BU(n)$ is simply connected, each set

$$\pi_0\mathrm{Map}_*(S^0, BU(n))_{hU(n)} \cong \pi_0\mathrm{Map}_*(S^0, BU(n))$$

is a singleton. So $\mathcal{K}_0(S^0) = \mathbb{Z}$, whereas $K^0(S^0) = \mathbb{Z} \oplus \mathbb{Z}$.

5. The topological Atiyah–Segal map

Let G be a discrete group whose classifying space BG has the homotopy type of a finite CW complex. We now define reduced and unreduced versions of the topological Atiyah–Segal map, which relates the deformation K-theory of G to the topological K-theory of BG. The unitary and general linear discussions are completely parallel, so we focus on the unitary case.

The classical Atiyah–Segal map associates to each representation $\rho\colon G \to U(n)$ the K-theory class represented by the vector bundle $E_\rho = (EG \times \mathbb{C}^n)/\pi_1 G \to BG$, where $\pi_1 G$ acts via $\gamma \cdot (e, v) = (e \cdot \gamma^{-1}, \rho(\gamma)v)$. We will see that in dimension zero, the classical Atiyah–Segal map factors as

$$R[G] = \mathrm{Gr}(\mathrm{Rep}(G)^\delta) \longrightarrow \mathrm{Gr}(\pi_0 \mathrm{Rep}(G)) \xrightarrow{\alpha_0} K^0(BG),$$

where $\mathrm{Rep}(G)^\delta$ is the discrete monoid underlying the topological monoid $\mathrm{Rep}(G)$, and α_0 is the topological Atiyah–Segal map (in dimension 0).

The simplicial classifying space functor B induces continuous, $U(n)$-equivariant maps

$$B = B_n\colon \mathrm{Hom}(G, U(n)) \longrightarrow \mathrm{Map}_*(BG, BU(n)) \qquad (12)$$
$$\rho \longmapsto B\rho$$

which combine to give a map between the associated unitary permutative action sequences. Recall that the spectra associated to these action sequences are $K^{\mathrm{def}}(G)$ and $\mathcal{K}(BG)$, respectively, and the homotopy groups of the latter are the complex K-theory groups of BG.

DEFINITION 5.1. The *topological Atiyah–Segal map*

$$\alpha = \alpha^G\colon K^{\mathrm{def}}(G) \longrightarrow \mathcal{K}(BG)$$

is the map of spectra induced by the above map of permutative action sequences. The reduced topological Atiyah–Segal map

$$\widetilde{\alpha}\colon \widetilde{K}^{\mathrm{def}}(G) \longrightarrow \widetilde{\mathcal{K}}(BG)$$

is the induced map

$$\mathrm{hofib}\left(K^{\mathrm{def}}(G) \to K^{\mathrm{def}}(\{1\})\right) \longrightarrow \mathrm{hofib}\left(\mathcal{K}(BG) \to \mathcal{K}(*)\right).$$

Note that we have $K^{\mathrm{def}}(\{1\}) = \mathcal{K}(*) \simeq \mathbf{ku}$, and $\alpha^{\{1\}}$ is the identity map.

The results in the previous sections combine to give the following descriptions of the induced maps α_* and $\widetilde{\alpha}_*$ on homotopy groups.

COROLLARY 5.2. *For $m \geqslant 0$, the topological Atiyah–Segal map is naturally isomorphic to the maps*

$$\pi_m \Omega B\mathrm{Rep}(G)_{hU} \longrightarrow \pi_m \Omega B\mathcal{V}(BG)_{hU}$$

and

$$\widetilde{\Pi}_m \mathrm{Rep}(G)_{hU} \longrightarrow \widetilde{\Pi}_m \Omega B\mathcal{V}(BG)_{hU}$$

induced by the simplicial classifying space functor B, and there is a natural splitting

$$\alpha_* = \widetilde{\alpha}_* \oplus \mathrm{Id}_{\pi_* \mathbf{ku}}. \qquad (13)$$

Moreover, for $m > 0$, $\widetilde{\alpha}_m$ is naturally isomorphic to the maps

$$\pi_m \Omega B\mathrm{Rep}(G) \longrightarrow \pi_m \Omega B\mathcal{V}(BG)$$

and

$$\widetilde{\Pi}_m \mathrm{Rep}(G) \longrightarrow \widetilde{\Pi}_m \mathcal{V}(BG)$$

induced by B.

Our goal in this section is to give an explicit description of the topological Atiyah–Segal map in terms of vector bundles. Consider the diagram

$$\begin{array}{ccc} \widetilde{\Pi}_m \mathrm{Rep}(G) & \xrightarrow{\widetilde{\alpha}_m} & \widetilde{\Pi}_m \mathcal{V}(BG) \\ & \searrow & \uparrow \Psi \,\cong \\ & & \pi_m \mathrm{Map}_*(BG, BU) \cong \widetilde{K}^{-m}(BG). \end{array} \tag{14}$$

where Ψ is defined by $\Psi(\langle f \rangle) = [f]$. We will describe the diagonal map $\Psi^{-1} \circ \widetilde{\alpha}_m$.

By Theorem 3.4, each class in $\widetilde{\Pi}_m \mathrm{Rep}(G) \cong K_m^{\mathrm{def}}(G)$ has a representative of the form $[\rho]$, with ρ a family of representations $\rho\colon S^m \to \mathrm{Rep}(G)$. Let E_ρ be the right principal $U(n)$-bundle over $S^m \times BG$ defined by

$$\begin{aligned} E_\rho = (S^m \times EG \times U(n))\,/G &\longrightarrow S^m \times BG \\ [z, e, A] &\longmapsto (z, q(e)), \end{aligned}$$

where $q\colon EG \to BG$ is the bundle projection and $g \in G$ acts via

$$g \cdot (z, x, A) := (z, x \cdot g^{-1}, (\rho(z)(g))A).$$

Basic properties of this construction are reviewed in [**4**, Section 3].

We will use $1 \in S^0 \subset S^m$ as the basepoint of S^m, and for any family

$$\rho\colon S^m \to \mathrm{Hom}(G, U(n)),$$

we let $\widetilde{\rho(1)}\colon S^m \to \mathrm{Hom}(G, U(n))$ denote the constant family at $\rho(1)\colon G \to U(n)$.

For based CW complexes X_1 and X_2, the long exact sequence in K-theory for the pair $(X_1 \times X_2, X_1 \vee X_2)$ yields a (naturally) split short exact sequence

$$0 \longrightarrow \widetilde{K}^0(X_1 \wedge X_2) \xrightarrow{\pi^*} \widetilde{K}^0(X_1 \times X_2) \xrightarrow{i^*} \widetilde{K}^0(X_1 \vee X_2) \longrightarrow 0.$$

If ρ is an S^m-family ($m > 0$), the bundle $E_\rho \to S^m \times BG$ is trivial when restricted to $S^m \times \{x\}$ (for each point $\{x\} \in BG$): indeed, each point $\widetilde{x} \in q^{-1}(x) \subset EG$ gives rise to a continuous section $z \mapsto [z, \widetilde{x}, I]$ of the restricted bundle. Thus we have

$$E_\rho|_{S^m \vee BG} \cong E_{\widetilde{\rho(1)}}|_{S^m \vee BG},$$

and hence

$$[E_\rho] - [E_{\widetilde{\rho(1)}}] \in \ker(i^*) = \mathrm{Im}(\pi^*).$$

Since π^* is injective, it follows that the class $[E_\rho] - [E_{\widetilde{\rho(1)}}]$ has a well-defined preimage under π^*, which we will denote by

$$(\pi^*)^{-1}([E_\rho] - [E_{\widetilde{\rho(1)}}]) \in \widetilde{K}^0(S^m \wedge BG) = \widetilde{K}^{-m}(BG). \tag{15}$$

By [**4**, Lemma 3.1(2)], the bundles E_ρ and $E_{\widetilde{\rho(1)}}$ only depend (up to isomorphism) on the unbased homotopy class of ρ. Hence the class (15) depends only the unbased homotopy class of ρ. Note that if ρ is constant, then $E_\rho = E_{\widetilde{\rho(1)}}$, so in this case the class (15) is trivial.

With this understood, we have the following explicit description of $\widetilde{\alpha}_*$ (or more precisely, of the map $\Psi^{-1} \circ \widetilde{\alpha}_*$ in Diagram (14)).

THEOREM 5.3. *Let G be a group whose classifying space BG is homotopy equivalent to a finite CW complex. Then for $m \geqslant 1$, the reduced topological Atiyah–Segal map, viewed as a map*

$$\widetilde{\Pi}_m \mathrm{Rep}(G) \longrightarrow \widetilde{K}^{-m}(BG)$$

via the diagram (14), has the form

$$[\rho] \mapsto (\pi^*)^{-1}\left([E_\rho] - [E_{\widetilde{\rho(1)}}]\right). \tag{16}$$

When $m = 0$, the map

$$\alpha_0\colon K_0^{\mathrm{def}}(G) \cong \mathrm{Gr}(\pi_0 \mathrm{Rep}(G)) \longrightarrow K^0(BG)$$

is given by $\alpha_0([\rho]) = [E_\rho]$.

Note that the statement for $m = 0$ shows that the classical Atiyah–Segal map factors through α_0, as claimed earlier.

PROOF. We assume $m > 0$; the proof for $m = 0$ is similar but simpler. By definition, $\widetilde{\alpha}_*([\rho]) = [B \circ \rho]$, where B is the map (12). Let $f\colon S^m \wedge BG \to BU(n)$ be a map classifying $(\pi^*)^{-1}\left([E_\rho] - [E_{\widetilde{\rho(1)}}]\right)$. To prove the proposition, we need to show that the adjoint map $f^\vee\colon S^m \to \mathrm{Map}_*(BG, BU(n))$ satisfies

$$\Psi(\langle f^\vee \rangle) = [B \circ \rho]$$

in $\widetilde{\Pi}_m \mathcal{V}(BG)$. (Recall that $\Psi(\langle f^\vee \rangle)$ is simply $[f^\vee]$.)

By choice of f, the composite $f \circ \pi$ classifies $[E_\rho] - [E_{\widetilde{\rho(1)}}]$, and by [**4**, Lemma 4.1], if c is the constant map $S^m \to \mathrm{Map}_*(BG, BU(p))$ with image $B(\rho(1))$, then $c^\vee \circ \pi$ classifies $[E_{\widetilde{\rho(1)}}]$ (here $c^\vee$ is the adjoint of c), while $(B \circ \rho)^\vee \circ \pi$ classifies $[E_\rho]$. Hence the maps

$$(f \circ \pi) \oplus (c^\vee \circ \pi) = (f \oplus c^\vee) \circ \pi \text{ and } (B \circ \rho)^\vee \circ \pi$$

represent the same class in $\widetilde{K}^0(S^m \wedge BG)$. Since π^* is injective, it follows that $f \oplus c^\vee$ and $(B \circ \rho)^\vee$ are homotopic as maps $S^m \wedge BG \to BU(N)$ (for sufficiently large N), and consequently $f^\vee \oplus c$ and $B \circ \rho$ are homotopic as maps $S^m \to \mathrm{Map}_*(BG, BU(N))$. Since c is constant, it follows that $[B \circ \rho] = [f^\vee] = \Psi(\langle f^\vee \rangle)$ in $\widetilde{\Pi}_m \mathcal{V}(BG)$. □

One may replace $U(n)$ by $GL(n)$ throughout the preceding discussion, yielding a general linear version α^{GL} of the topological Atiyah–Segal map. We note that the unitary topological Atiyah–Segal map factors through this general linear version, which leads to the following natural question.

QUESTION 5.4. Does there exist a group G for which the image of α_*^{GL} is strictly larger than the image of α_*?

6. Relations with previous work

In this section, we reinterpret some of the main results from [**4**], [**20**], and [**24**] in terms of the topological Atiyah–Segal map.

6.1. Bounds on the image of α_*. We now show that $\widetilde{\alpha}_*$ fails to be surjective in dimensions $\mathbb{Q}cd(G) - 2k$ $(k > 0)$, where $\mathbb{Q}cd(G)$ is the largest number n for which $H^n(G;\mathbb{Q})$ is nonzero (Theorem 6.1). This low-dimensional failure is closely analogous to the low-dimensional failure of the Quillen–Lichtenbaum conjectures (in the form discussed in [**26**], for instance), which the relate algebraic K-theory of a field k to its étale K-theory in dimensions greater than the virtual cohomological dimension of k minus 2.

THEOREM 6.1. *The image of $\alpha_*\colon K_m^{\mathrm{def}}(G) \to K^{-m}(BG)$ (and, in fact, of α_*^{GL}) has rank at most $\beta_m(G) + \beta_{m-2}(G) + \cdots$, where $\beta_i(G)$ is the rank of $H_i(BG;\mathbb{Z})$. Hence if $\beta_d(G) \neq 0$, then the maps α_{d-2}, α_{d-4}, ... are not surjective.*

PROOF. By Theorem 5.3 and [**4**, Theorem 3.5], the image of α_m lies in the subgroup of $K^{-m}(BG)$ on which the Chern classes c_{m+i}, $i = 1, 2, \ldots$, vanish rationally. It follows from [**4**, Lemma 4.2] that the rank of this subgroup is given by the above sum. The last statement follows from the fact that the Chern character is a rational isomorphism, which implies that the rank of $K^{-m}(BG)$ is equal to the sum of all the Betti numbers of BG in dimensions equivalent to m modulo 2. □

6.2. Relation with the Novikov conjecture. A group G satisfies the *strong Novikov conjecture* if the analytical assembly map, from the K-homology of BG to the K-theory of the maximal C^*-algebra of G, is injective after tensoring with $\mathbb{Q}$. For background on this conjecture, see [**20**] and the references therein.

THEOREM 6.2. *If G is a group such that BG is homotopy equivalent to a finite CW complex, and the unitary topological Atiyah–Segal map α_m is rationally surjective for all sufficiently large m, then G satisfies the strong Novikov conjecture.*

PROOF. Surjectivity of α_m is equivalent to surjectivity of $\widetilde{\alpha}_m$, and surjectivity of $\widetilde{\alpha}_m$ implies that for sufficiently large n, every element in $[S^m, \mathrm{Map}_*(BG, BU(n))]$ has the form $[B \circ \rho]$ for some ρ. It now follows from [**20**, Theorem 3.16] (or rather from the proof of that result) that G lies in the class of *flatly detectable groups*; all such groups satisfy the strong Novikov conjecture by [**20**, Corollary 4.3]. □

6.3. Surface groups. We now translate the results in [**24**] into information about the topological Atiyah–Segal map when $G = \pi_1(\Sigma)$ is the fundamental group of a compact, aspherical surface Σ. Note that we allow Σ to have boundary, in which case $\pi_1(\Sigma)$ is a finitely generated free group. We begin by recalling a result from [**24**], which was proven using Morse theory for the Yang–Mills functional.

THEOREM 6.3 ([**24**, Theorem 3.4]). *If Σ is a compact aspherical surface, possibly with boundary, then for each $M \geqslant 0$, there exists N such that for every $n > N$ the natural map*

$$B_*\colon \pi_m\mathrm{Hom}(\pi_1\Sigma, U(n)) \longrightarrow \pi_m\mathrm{Map}_*(B\pi_1\Sigma, BU(n))$$

induces an isomorphism on homotopy groups in dimensions $1 \leqslant m \leqslant M$ for all choices of compatible basepoints. If Σ is nonorientable, or has nonempty boundary, then this statement holds for $0 \leqslant m \leqslant M$.

THEOREM 6.4. *If Σ is a compact aspherical surface, possibly with boundary, then the topological Atiyah–Segal map*

$$\alpha_m\colon K_m^{\mathrm{def}}(\pi_1\Sigma) \longrightarrow K^{-m}(\Sigma)$$

is an isomorphism for $m \geqslant 1$. If Σ is nonorientable, or has nonempty boundary, then α_0 is an isomorphism as well.

PROOF. Note that it suffices to prove that $\widetilde{\alpha}_m$ is an isomorphism. We work in the case $m > 0$; the same reasoning will apply when $m = 0$ and Σ is nonorientable or has nonempty boundary, using the last part of Theorem 6.3.

We prove injectivity; the proof of surjectivity is similar but simpler. By Theorem 3.4, each element in the kernel of $\widetilde{\alpha}_m$ has the form $[\rho]$ for some

$$\rho\colon S^m \to \operatorname{Hom}(\pi_1\Sigma, U(n))$$

satisfying

$$(B \circ \rho) \oplus d \simeq c \tag{17}$$

for some constant maps $c, d\colon S^m \to \mathcal{V}(B\pi_1\Sigma)$. In fact, by [**22**, Corollaries 4.11 and 4.12] we may assume without loss of generality that $d = BI_p$ for some p, where I_p denotes the constant map $S^m \to \operatorname{Hom}(\pi_1\Sigma, U(p))$ with image the trivial representation. Indeed, for orientable surfaces the homomorphism spaces are all path connected, while in the nonorientable case each homomorphism space has two path components, and adding representations from the nontrivial components always yields a representation in the trivial component, so (for instance) we may simply replace d by $d \oplus d$ if necessary.

Equation (17) now implies that $B \circ (\rho \oplus I_p)$ is nullhomotopic, and the injectivity portion of Theorem 6.3 implies that $\rho \oplus I_p$ must be nullhomotopic. Hence $[\rho] = 0$ in $\widetilde{\Pi}_m(\operatorname{Rep}(\pi_1\Sigma))$, as desired. □

7. Families of flat connections over the Heisenberg manifold

In this section we study flat, unitary connections on complex vector bundles over the 3-dimensional Heisenberg manifold by combining the main results of this paper with computations due to Tyler Lawson. We begin with a review of the definition and the basic properties of this manifold.

7.1. Background. The discrete 3-dimensional Heisenberg group H is the group of 3×3 upper triangular integer matrices under ordinary matrix multiplication. This group sits as a (discrete) subgroup of the real Heisenberg group $H_{\mathbb{R}}$, which consists of all real upper triangular matrices. We will identify the real Heisenberg group with $\mathbb{R}^3$ via the function

$$\begin{bmatrix} 1 & x & z \\ 0 & 1 & y \\ 0 & 0 & 1 \end{bmatrix} \mapsto (x, y, z)$$

(note, though, that we are using matrix multiplication to define the operation in $H_{\mathbb{R}}$, not addition in the vector space $\mathbb{R}^3$). The Heisenberg manifold is defined by

$$N^3 = \mathbb{R}^3/H,$$

where H acts on $\mathbb{R}^3 \cong H_{\mathbb{R}}$ by (left) multiplication.

It is an elementary exercise to check that N^3 is Hausdorff, and that the map $\mathbb{R}^3 \to N^3$ is a covering. In particular, this means N^3 is an aspherical 3-dimensional manifold with fundamental group H (and hence $N^3 \simeq BH$), and N^3 is orientable since the action of H on $\mathbb{R}^3$ is orientation-preserving. Moreover, N^3 is compact;

this follows, for instance, from the fact that each closed unit cube in $\mathbb{R}^3$ surjects onto N^3. In fact, N^3 is a circle bundle over $\mathbb{R}^2/\mathbb{Z}^2$. Indeed, consider the mapping

$$N^3 = \mathbb{R}^3/H \xrightarrow{q} \mathbb{R}^2/\mathbb{Z}^2$$

given by sending $[(x,y,z)]$ to $[(x,y)]$. It is elementary to check that this map is a fiber bundle with circle fibers; indeed, for each $[(x,y)] \in \mathbb{R}^2/\mathbb{Z}^2$, there exists $\epsilon > 0$ such that the mapping

$$\begin{aligned} [x-\epsilon, x+\epsilon] \times [y-\epsilon, y+\epsilon] \times \mathbb{R}/\mathbb{Z} &\longrightarrow N^3 \\ (x', y', [z]) &\longmapsto [(x', y', z)] \end{aligned}$$

is a homeomorphism onto

$$q^{-1}\left(\pi([x-\epsilon, x+\epsilon] \times [y-\epsilon, y+\epsilon])\right),$$

where $\pi\colon \mathbb{R}^2 \to \mathbb{R}^2/\mathbb{Z}^2$ is the quotient map.

The fibration sequence $S^1 \to N^3 \to \mathbb{R}^2/\mathbb{Z}^2$ gives rise to a short-exact sequence on fundamental groups:

$$1 \longrightarrow \mathbb{Z} \longrightarrow H \xrightarrow{q_*} \mathbb{Z}^2 \longrightarrow 1.$$

Covering space theory gives canonical identifications of $\pi_1(\mathbb{R}^3/H, [(0,0,0)])$ and $\pi_1(\mathbb{R}^2/\mathbb{Z}^2, [(0,0)])$ with H and $\mathbb{Z}^2$ (respectively), and under these identifications the map q_* is simply

$$\begin{bmatrix} 1 & a & c \\ 0 & 1 & b \\ 0 & 0 & 1 \end{bmatrix} \mapsto (a,b).$$

The kernel of q_* is generated by

$$Z = \begin{bmatrix} 1 & 0 & 1 \\ 0 & 1 & 0 \\ 0 & 0 & 1 \end{bmatrix},$$

which is the commutator of the elements

$$X = \begin{bmatrix} 1 & 1 & 0 \\ 0 & 1 & 0 \\ 0 & 0 & 1 \end{bmatrix} \text{ and } Y = \begin{bmatrix} 1 & 0 & 0 \\ 0 & 1 & 1 \\ 0 & 0 & 1 \end{bmatrix}.$$

It is elementary to check that X and Y generate H, and it follows that $\ker(q_*)$ is precisely the commutator subgroup of H, giving

$$H_1(N^3;\mathbb{Z}) \cong \mathbb{Z}^2. \tag{18}$$

Since Z commutes with both X and Y, we see that $\ker(q_*)$ is central, and it follows that H is a nilpotent group. Poincaré Duality, together with (18), yields:

$$H_i(N^3;\mathbb{Z}) \cong H^i(N^3;\mathbb{Z}) \cong \begin{cases} \mathbb{Z}^2 & \text{if } i = 1,2, \\ \mathbb{Z}, & \text{if } i = 0,3. \end{cases}$$

In particular, the cohomology of N^3 is torsion-free.

7.2. Flat bundles over the Heisenberg manifold. To understand flat bundles over N^3, we will use the following fact.

PROPOSITION 7.1. *If X is a finite CW complex with torsion-free integral cohomology, and $E \to X$ is a complex vector bundle whose Chern classes $c_i(E) \in H^{2i}(X;\mathbb{Z})$ vanish for $i \geqslant 1$, then E is stably trivial.*

In particular, if M is a smooth manifold with torsion-free integral cohomology and $E \to M$ is a vector bundle admitting a flat connection, then E is stably trivial.

Note that for finite CW complexes, $H^*(X;\mathbb{Z})$ is torsion-free if and only if $H_*(X;\mathbb{Z})$ is torsion-free.

PROOF. The second statement follows from the first, since by Chern–Weil theory the Chern classes of a flat vector bundle over M map to zero in $H^*(M;\mathbb{Q})$, and when $H^*(M;\mathbb{Z})$ is torsion-free the natural map $H^*(M;\mathbb{Z}) \to H^*(M;\mathbb{Q})$ is injective.

To prove the first statement, consider a vector bundle $E \to X$ with $c_i(E) = 0$ for $i \geqslant 1$. By [**2**, Section 2.5] (see also [**9**, Proposition 6.10]) the complex X-theory of X is torsion-free (and finitely generated), so the natural map $\widetilde{K}^*(X) \to \widetilde{K}^*(X) \otimes \mathbb{Q}$ is injective. Composing with the Chern character gives an injection

$$\widetilde{K}^*(X) \hookrightarrow \widetilde{K}^*(X) \otimes \mathbb{Q} \xrightarrow{\cong} \widetilde{H}^*(X;\mathbb{Q}).$$

Since $[E]$ maps to zero under this injection, we have $[E] = 0$ in $\widetilde{K}^*(X)$. $\square$

COROLLARY 7.2. *Let G be a discrete group whose classifying space BG has torsion-free integral cohomology and has the homotopy type of a CW complex X of dimension at most d. Then if $n \geqslant d/2$, the bundle E_ρ associated to a representation $\rho\colon G \to GL(n)$ is always trivial.*

PROOF. By Proposition 7.1, we know that the vector bundle associated to E_ρ is stably trivial, meaning that its classifying map

$$B\rho\colon BG \simeq X \to BGL(n)$$

becomes nullhomotopic after composing with the natural map

$$j\colon BGL(n) \to BGL(n+n', \mathbb{C})$$

for sufficiently large n'. Since j induces an isomorphism on homotopy groups up to dimension $2n$ and a surjection in dimension $2n+1$, the Whitehead Theorem [**15**, Section 10.3] shows that for all CW complexes X of dimension at most $2n$, the map

$$[X, BGL(n)] \longrightarrow [X, BGL(\infty)]$$

is bijective. In particular, since $j \circ B\rho$ is nullhomotopic, we conclude that $B\rho$ is itself nullhomotopic so long as $d \leqslant 2n$. $\square$

COROLLARY 7.3. *Let $G = \pi_1 K$, where K is an aspherical, 3-dimensional CW complex with torsion-free integral cohomology. Then for each representation $\rho\colon G \to GL_n(\mathbb{C})$, the associated vector bundle E_ρ is trivial.*

PROOF. Since $BG \simeq K$, Corollary 7.2 shows that E_ρ is trivial whenever the degree of ρ is at least 2. Our assumptions imply that the abelianization of K is free abelian, so $\mathrm{Hom}(G, GL_1(\mathbb{C})) \cong (GL_1\mathbb{C})^k$ for some k, and in particular is path-connected. Hence for each $\rho\colon G \to GL_1(\mathbb{C})$, we have $E_\rho \cong E_1 \cong BG \times \mathbb{C}$. $\square$

7.3. Homotopy in the space of flat connections over N^3. Let $\mathcal{A}^\flat_n(N^3)$ denote the space of flat connections on the trivial bundle $N^3 \times U(n)$ (or, equivalently, the space of flat unitary connections on $N^3 \times \mathbb{C}^n$). More precisely, $\mathcal{A}^\flat_n(N^3)$ will denote the subspace of flat connections inside the Sobolev completion, with respect to a sufficiently strong Sobolev norm, of the space of all smooth connections on $N^3 \times U(n)$, as in [**4**, Section 5]. In this section we study the homotopy groups of $\mathcal{A}^\flat_n(N^3)$ as $n \to \infty$, using Lawson's calculation of $K^{\mathrm{def}}_*(H)$ [**11–13**].

It was proven in [**4**, Corollary 1.3] that if M^d is a closed, smooth, aspherical d-manifold with $H^3(M;\mathbb{Q}) \neq 0$, then $\mathcal{A}^\flat_n(M)$ has infinitely many path components so long as $n \geqslant (d+1)/2$. In particular, $\mathcal{A}^\flat_n(N^3)$ has infinitely many path components so long as $n \geqslant 2$. For manifolds M of dimension $d > 3$, [**4**, Corollary 1.3] also gives cohomological lower bounds on the rank of $\pi_m\mathcal{A}^\flat_n(M)$ for $0 < m \leqslant d-3$, but for 3-manifolds no information about the homotopy groups $\pi_m\mathcal{A}^\flat_n(M)$ ($m \geqslant 1$) is obtained through the methods of that paper.

In this section, we will show that the homotopy groups of $\mathcal{A}^\flat_n(N^3)$ are in fact very large. Moreover, while the classes in $\pi_m\mathcal{A}^\flat_n(M)$ produced by the methods in [**4**] all admit representatives lying inside a single gauge orbit, we produce classes in $\mathcal{A}^\flat_n(N^3)$ that do not admit such representatives (although see Question 7.7).

This result shows a sharp contrast between flat connections over 3-manifolds and over surfaces, where Morse theory for the Yang–Mills functional leads to the conclusion that spaces of flat connections are highly connected, with vanishing homotopy in a range tending to infinity with the rank of the underlying bundle. For details and precise results, see [**22, 23**] and the references therein.

DEFINITION 7.4. Given a space X together with a choice of representatives $\{x_C\}_{C\in\pi_0(X)}$ for the path components of X, we define

$$\widetilde{\pi}_n(X) = \bigoplus_{C\in\pi_0(X)} \pi_n(X, x_C).$$

Up to isomorphism, this group is independent of the chosen representatives x_C.

Recall that the gauge group $\mathcal{G} = \mathrm{Map}(N^3, U(n))$ acts on the space of all connections on $N^3 \times U(n)$, and this action preserves the subspace $\mathcal{A}^\flat_n(N^3)$. More precisely, $\mathcal{G}$ is the Sobolev completion of the space of smooth maps with respect to the appropriate Sobolev norm. The based gauge group $\mathcal{G}_0 \leqslant \mathcal{G}$ is the kernel of the restriction map $\mathcal{G} \to U(n)$ induced by evaluation at a fixed basepoint $x \in N^3$. The holonomy map induces a fibration sequence (in fact, a principal $\mathcal{G}_0$-bundle)

$$\mathcal{G}_0 \xrightarrow{i_A} \mathcal{A}^\flat_n(N^3) \xrightarrow{\mathrm{Hol}} \mathrm{Hom}(H, U(n)) \tag{19}$$

for each n. The first map in this sequence is simply the inclusion of the gauge orbit of some flat connection $A \in \mathcal{A}^\flat_n(N^3)$, and sends $g \in \mathcal{G}_0$ to $g \cdot A$. We call maps of the form i_A, and their induced maps on $\widetilde{\pi}_*$, *(based) gauge orbit inclusions.*

We need the following result of Lawson.

PROPOSITION 7.5 ([**13**, Section 4.2]). *For each $m \geqslant 0$, the group $K^{\mathrm{def}}_m(H)$ is free abelian of countably infinite rank.*

THEOREM 7.6. *Given $m, R \geqslant 1$, there exists $n_0 \geqslant 1$ such that for all $n \geqslant n_0$, $\widetilde{\pi}_m(\mathcal{A}^\flat_n(N^3))$ contains a subgroup F satisfying:*

(1) *The abelianization of F is free of infinite rank;*

(2) $\mathrm{Hol}_*(F) \leqslant \widetilde{\pi}_m \mathrm{Hom}(H, U(n))$ *has rank* R*; and*

(3) *Only the trivial element of* F *is in the image of a based gauge-orbit inclusion.*

When $m = 1$*, and when* m *is even,* (3) *can be strengthened by replacing the based gauge group by the full gauge group.*

PROOF. Since $\widetilde{K}^{-m}(BH) \cong K^{-m}(N^3)$ is finitely generated, Proposition 7.5 implies that the kernel of $\widetilde{\alpha}_m$ is free abelian of countably infinite rank (note here that subgroups of free abelian groups—of any rank—are free [**10**, Appendix 2]).

Now assume $m \geqslant 1$. By Theorem 3.4, the natural map

$$[S^m, \mathrm{Rep}(H)] \longrightarrow \widetilde{\Pi}_m(\mathrm{Rep}(H)) \cong \widetilde{K}_m^{\mathrm{def}}(H)$$

is surjective. This means that we can choose families

$$\rho_i \colon S^m \to \mathrm{Hom}(H, U(n_i)),$$

$i = 1, 2, \ldots$, such that the associated classes in $\widetilde{\Pi}_m(\mathrm{Rep}(H))$ are linearly independent. We have $\widetilde{\alpha}_*[\rho_i] = [B \circ \rho_i]$, and $B \circ \rho_i$ is nullhomotopic by Corollary 7.3.

Now fix an integer $R > 0$. Let $n_0 = \max(n_1, \ldots, n_R)$. Then for $i = 1, \ldots, R$, and for each $n \geqslant n_0$, the map

$$S^m \xrightarrow{B \circ \rho_i} \mathrm{Map}_*(BH, BU(n_i)) \longrightarrow \mathrm{Map}_*(BH, BU(n_0))$$

is nullhomotopic. Say $n \geqslant n_0$, and let G be the subgroup of $\widetilde{\pi}_m \mathrm{Hom}(H, U(n))$ generated by the elements $\{\langle \rho_i \rangle\}_{i=1}^R$. Then G surjects onto the subgroup of $\widetilde{\Pi}_m(\mathrm{Rep}(H))$ generated by $\{[\rho_i]\}_{i=1}^R$, which is free abelian of rank R, so the abelianization of G must have rank at least R (of course when $m > 1$, the group G is already abelian).

By [**4**, Remark 5.6], the boundary map on homotopy groups associated to the principal bundle (19) can be identified (up to isomorphism) with the map

$$B_* \colon \pi_* \mathrm{Hom}(H, U(n)) \longrightarrow \pi_* \mathrm{Map}_*(BH, BU(n)),$$

so each class $\langle \rho_i \rangle$ maps to zero under the boundary map for the fibration sequence (19) (with appropriately chosen basepoints). Hence there exist maps

$$\alpha_i \colon S^m \to \mathcal{A}_n^\flat(N^3)$$

such that $\mathrm{Hol}_*(\langle \alpha_i \rangle) = \langle \rho_i \rangle$.

The based gauge group $\mathcal{G}_0$ is weakly equivalent to the continuous mapping space

$$\mathrm{Map}_*(N^3, U(n)) \simeq \mathrm{Map}_*(BH, U(n)),$$

and $\mathrm{Map}_*(BH, BU(n))$ is the classifying space of $\mathrm{Map}_*(N^3, U(n))$ (this result is originally due to Gottlieb [**7**]). Hence for $* \geqslant 1$ we have

$$\pi_* \mathrm{Map}_*(BH, BU(n)) \cong \pi_{*-1} \mathcal{G}_0.$$

Let A_i denote a flat connection in the image of α_i, and let $\psi_i = \mathrm{Hol}(A_i)$. The long exact sequence associated the to the fibration sequence (19) ends with

$$\cdots \longrightarrow \pi_1(\mathrm{Hom}(H, U(n)), \psi_i) \xrightarrow{\partial} \pi_0 \mathcal{G}_0 \xrightarrow{(i_{A_i})_*} \pi_0 \mathcal{A}_n^\flat(N^3) \longrightarrow \pi_0(\mathrm{Hom}(H, U(n)).$$

Again, ∂ can be identified (up to isomorphism) with

$$B_* \colon \pi_1(\mathrm{Hom}(H, U(n)), \psi_i) \longrightarrow \pi_1(\mathrm{Map}_*(BH, BU(n)), B\psi_i),$$

and the cokernel of this map has rank 1 by [**4**, Corollary 1.3]. This means there exists an element $g_i \in \mathcal{G}_0$ such that the path component $[g_i]$ has infinite order in

$$\pi_0\mathcal{G}_0 \cong \pi_1(\mathrm{Map}_*(BH, BU(n)), B\psi_i)$$

and the subgroup of $\pi_0(\mathcal{G}_0)$ generated by $[g_i]$ intersects the image of ∂ trivially. In general, given a (left) principal bundle K-bundle

$$K \longrightarrow P \xrightarrow{q} B,$$

if $k_1, k_2 \in K$ and $p \in P$ are elements such that $[k_1 \cdot p] = [k_1 \cdot p]$ in $\pi_0 P$, then $[k_1^{-1}k_2]$ is in the image of the boundary map $\pi_1(B, q(p)) \xrightarrow{\partial} \pi_0(K)$ (where we identify K with $q^{-1}(p)$ via $p \cdot k \mapsto k$). This implies that for each fixed $i \in \{1, \ldots, R\}$, the path components $[A_i], [g_i \cdot A_i], [g_i^2 \cdot A_i], \cdots$ are distinct in $\pi_0\mathcal{A}_n^\flat(N^3)$.

Let $F \leqslant \widetilde{\pi}_m\mathcal{A}_n^\flat(N^3)$ be the subgroup generated by the elements $\langle g_i^k \cdot \alpha_i \rangle$, $i = 1, \ldots, R$, $k = 0, 1, \ldots$. We claim that F satisfies the conditions in the theorem. First, note that $\mathrm{Hol}_*(F) = G$, so F satisfies (2).

Next, we show that the abelianization of F is freely generated by the elements $\langle g_i^k \cdot \alpha_i \rangle$. For simplicity, we give the argument when $m > 1$, in which case F is already abelian; the argument for $m = 1$ just requires notational changes. If $\sum_{i,k} \lambda_{i,k}\langle g_i^k \cdot \alpha_i \rangle = 0$, then summing the terms whose images lie in a particular path component C of $\mathcal{A}_n^\flat(N^3)$ will also give zero. By choice of g_i, such a sum contains at most one term of the form $\lambda_{i,k}\langle g_i^k \alpha_i \rangle$ for each i. Thus we have $\sum_i \lambda_{i,k_i}\langle g_i^{k_i}\alpha_i \rangle = 0$ for some collection of natural numbers k_i ($i = 1, \ldots, R$), and every term from the original sum whose image lies in C appears in this new sum. But now

$$0 = \mathrm{Hol}_*(0) = \mathrm{Hol}_*\left(\sum_i \lambda_{i,k_i}\langle g_i^{k_i}\alpha_i \rangle\right) = \sum_i \lambda_{i,k_i}\langle \rho_i \rangle,$$

and linear independence of the elements $\langle \rho_i \rangle$ implies that $\lambda_{i,k_i} = 0$ for each i. Applying this argument to each path component, we see that all the coefficients $\lambda_{i,k}$ must be zero, as desired.

Finally, we consider gauge-orbit inclusions. Say $f \in F$ is in the image of a based gauge-orbit inclusion. We need to prove that $f = 0$. Again, for notational convenience we work in the case $m > 1$. We can write

$$f = \sum_{i,k} \lambda_{i,k}\langle g_i^k \cdot \alpha_i \rangle,$$

and it suffices to consider the case in which all the maps $g_i^k \cdot \alpha_i$ land in the same path component of $\mathcal{A}_n^\flat(N^3)$. Then, as before, there can be at most one term in this sum for each i, so we can write

$$f = \sum_i \lambda_{i,k_i}\langle g_i^k \cdot \alpha_i \rangle.$$

Since f is in the image of a based gauge-orbit inclusion $\mathcal{G}_0 \to \mathcal{A}_n^\flat(N^3)$, and the composite

$$\mathcal{G}_0 \longrightarrow \mathcal{A}_n^\flat(N^3) \longrightarrow \mathrm{Hom}(H, U(n))$$

is constant, we have $\sum_i \lambda_{i,k_i}(\langle \rho_i \rangle) = 0$. But this element maps to $\sum_i \lambda_{i,k_i}[\rho_i]$ in $\widetilde{\Pi}_m(\mathrm{Rep}(H))$, which implies that all of the coefficients λ_{i,k_i} are in fact zero.

Finally, we consider the orbits of the full gauge group. Evaluation at a point gives a split fibration

$$\mathcal{G}_0 \to \mathcal{G} \to U(n), \tag{20}$$

and since $U(n)$ is path-connected, $\widetilde{\pi}_*\mathcal{G} \cong \widetilde{\pi}_*\mathcal{G}_0 \times \pi_* U(n)$ for each $* \in \mathbb{N}$. Now consider the case where $m = 2l$ is even. In constructing the subgroup F, we are free to choose n_0 large with respect to $2l$, and hence we may assume that that $\pi_{2l}U(n) = 0$ whenever $n \geqslant n_0$. Thus when m is even, the image of a full gauge-orbit inclusion is the same (in homotopy) as the image of the corresponding based gauge-orbit inclusion.

To obtain the desired conclusion when $m = 1$, it suffices to show that for every gauge-orbit inclusion, the composite

$$U(n) \xrightarrow{s} \mathcal{G} \longrightarrow \mathcal{A}_n^\flat(N^3) \longrightarrow \mathrm{Hom}(H, U(n))$$

(where s is a splitting of (20)) induces the zero map on π_1. The image of this composite map lies inside a single conjugation orbit inside $\mathrm{Hom}(H, U(n))$, so this composite factors through the projection $U(n) \to U(n)/Z$, where $Z \cong U(1)$ denotes the center of $U(n)$. Since the inclusion $Z \hookrightarrow U(n)$ induces isomorphisms on π_0 and π_1, we see that $\pi_1(U(n)/Z) = 0$. ∎

Here are some natural questions regarding the above results.

QUESTION 7.7. Can the groups F in Theorem 7.6 be constructed more explicitly? When m is odd and greater than 1, can they be constructed so as to avoid nontrivial elements in the image of the full gauge-orbit inclusions $\mathcal{G} \to \mathcal{A}_n^\flat(N^3)$?

QUESTION 7.8. Do the results of this section extended to higher-dimensional Heisenberg manifolds?

7.4. Flat connections and the topological Atiyah–Segal map. The arguments above suggest that when G is a group such that BG has the homotopy type of a closed, smooth (aspherical) manifold M, the homotopy fiber of the topological Atiyah–Segal map should be closely related to spaces of flat connections on bundles over M. Here we give one result along these lines.

First we formulate a general result about topological monoids. Let $p\colon M \to N$ be a homomorphism of homotopy commutative topological monoids. We will write the operations in M and N, and the induced pointwise multiplication operations for maps into M and N, as $\bullet$.

Let $I \subset \pi_0 N$ denote the image of $p_*\colon \pi_0 M \to \pi_0 N$. Assume that I is cofinal in $\pi_0 N$ (that is, for each $c \in \pi_0 N$ there exists $d \in \pi_0 N$ such that $cd \in I$) and that there is a map of monoids $s\colon I \to N$ satisfying $[sc] = c$ for each component $c \in I$, where $[sc] \in \pi_0 N$ denotes the path component of $sc \in N$. In other words, s is a partially defined section of the tautological map $N \to \pi_0 N$. Then the collection of homotopy fibers $F := \coprod_{c\in I} \mathrm{hofib}_{sc}(p)$ inherits a topological monoid structure from M and N—here it is critical that the basepoints $\{sc : c \in I\} \subset N$ is a submonoid.

PROPOSITION 7.9. *In the setting above, for each $m > 0$, there is a surjection*

$$\widetilde{\Pi}_m F \longrightarrow \ker(\Omega Bp\colon \pi_*(\Omega BM, *) \to \pi_*(\Omega BN, *)).$$

Note that if all elements in M and N are strongly anchored, then the above kernel agrees with $\ker(\Omega Bp_*\colon \pi_*(\Omega BM, *) \to \pi_*(\Omega BN, *))$.

PROOF. The natural maps from the homotopy fibers to M induce maps of monoids $[S^m, F] \to [S^m, M]$ and $\widetilde{\Pi}_m F \to \widetilde{\Pi}_m M$, and it is immediate that the composite map $\widetilde{\Pi}_m F \to \widetilde{\Pi}_m M \to \widetilde{\Pi}_m M$ is zero. It remains only to show that if $\mu\colon S^m \to M$ maps to zero in $\widetilde{\Pi}_m N$, then it lifts, up to homotopy, to F. Without loss of generality, we may assume that $p(\mu(1)) = sc$ for some $c \in I$. Now $p_*[\mu] = 0$ and cofinality of I imply that $[\mu \bullet sc'] = [sd]$ for some $c', d \in I$ (here sc' and sd denote constant maps with these values). Letting $H\colon S^m \times I \to M$ be a nullhomotopy from $\mu \bullet sc'$ to sd, the pair $(\mu \bullet sc, H)$ defines a map $S^m \to \mathrm{hofib}_{sc}(p)$, whose image in $\widetilde{\Pi}_m M$ is $[\mu \bullet sc] \equiv [\mu]$. □

We note that while $\pi_k \mathrm{hofib}_*(\Omega Bp)$ also surjects onto the above kernel, we do not expect an isomorphism $\widetilde{\Pi}_m F \cong \pi_m \mathrm{hofib}_*(\Omega Bp)$ in general: one might try to construct a map $\widetilde{\Pi}_{m+1} N \to \widetilde{\Pi}_m F$ using the boundary maps for the homotopy fiber sequences $\mathrm{hofib}_{sc}(p) \to M \to N$, but for each component of N, there may be many such boundary maps, coming from different path components, and hence choices of basepoint, in $\mathrm{hofib}_{sc}(p)$.

We now consider what Proposition 7.9 tells us about flat connections. Let G be the fundamental group of a closed, aspherical 3-manifold X with torsion-free cohomology, e.g. G could be the Heisenberg group H. Then setting $M = \mathrm{Rep}(G)$, $N = \mathcal{V}(BG)$, and $p = B$, Corollary 7.3 tells us that the image of B_* on components is just the components of nullhomotopic maps in $\mathcal{V}(BG)$, which is cofinal in $\pi_0 \mathcal{V}(BG)$ since every bundle over BG is a direct summand of a trivial bundle. Now we may define s by choosing the constant maps $BG \to BU(n)$ to represent their components. By [**4**, Theorem 5.5], the homotopy fibers of B are weakly equivalent to spaces of flat connections over the manifold X, meaning that there are natural zig-zags of maps between these spaces that induce isomorphisms on π_0 and on homotopy groups for all compatible choices of basepoints. Since the unbased homotopy sets are naturally in bijection with the π_1-orbits of the unbased homotopy sets, we obtain a bijection $[S^m, F] \cong [S^m, \mathcal{A}^\flat(X)]$, where

$$\mathcal{A}^\flat(X) := \coprod_n \mathcal{A}^\flat_n(X).$$

Hence $\widetilde{\Pi}_m \mathcal{A}^\flat(X)$ surjects onto the kernel of the reduced topological Atiyah–Segal map. In the case of the Heisenberg manifold N^3, we saw above that this kernel is a nonfinitely generated free abelian group. In summary, we have Corollary 7.10.

COROLLARY 7.10. *The monoid* $\widetilde{\Pi}_m \mathcal{A}^\flat(N^3)$ *has infinite rank for each* $m > 0$.

8. Multiplicativity of the topological Atiyah–Segal map

In this section, we upgrade the topological Atiyah–Segal map to a map of E_∞ ring spectra, so that the induced map α_* on homotopy becomes a ring homomorphism. This additional structure allows us to deduce further constraints on the image of α_* for groups satisfying Kazhdan's property (T) and for the discrete Heisenberg group. In particular, in Section 8.3 we prove the following result regarding families of flat vector bundles.

THEOREM 8.1. *Let G be a discrete group satisfying Kazhdan's property* (T), *and assume that BG has the homotopy type of a finite CW complex. Then for* *every family* $\rho\colon S^m \to \mathrm{Hom}(G, U(n))$, *the bundle* E_ρ *represents a torsion class in* $\widetilde{K}^0(S^m \times BG)$.

Note that by Lemma 7.1, when $H^*(BG; \mathbb{Z})$ is torsion-free the conclusion of Theorem 8.1 can be strengthened: E_ρ is in fact stably trivial.

There are many interesting groups to which this result applies. All torsion-free word hyperbolic groups admit finite CW models for BG, built using Rips complexes [**1**, Corollary 4.12]. There are many such groups with property (T), including cocompact, torsion-free lattices in $\mathrm{Sp}(n,1)$ [**5**].

8.1. Bipermutative structures. Kronecker product of matrices makes the unitary permutative action sequences giving rise to $K^{\mathrm{def}}(G)$ and $\mathcal{K}(BG)$ into *bipermutative* action sequences, in the sense described in Section 2.3; the details are just a routine extension of the computations in May [**18**, VI §5]. We thus obtain functors $K^{\mathrm{def}}_{\otimes}$ and $\mathcal{K}_{\otimes}$ from the category of discrete groups to the category of E_∞ ring spectra, which become naturally equivalent to K^{def} and $\mathcal{K}$ after applying the forgetful functor to spectra.

THEOREM 8.2. *There is a natural transformation* $\alpha^{\otimes}$ *between the functors* $K^{\mathrm{def}}_{\otimes}$ *and* $\mathcal{K}_{\otimes}$, *which becomes equivalent to* α *after applying the forgetful functor from* E_∞ *ring spectra to spectra. In particular,* α_* *is a homomorphism of unital rings, and* $\widetilde{\alpha}_*$ *is a homomorphism of nonunital rings.*

PROOF. The desired natural transformation is again induced by the simplicial classifying space functor B, which respects the multiplicative structure as well as the additive structures (by functoriality, essentially). The statement regarding $\widetilde{\alpha}_*$ follows from the fact that $\widetilde{K}_*(G)$ and $\widetilde{\mathcal{K}}_*(BG)$ are simply the kernels of the compatible (ring) homomorphisms induced by the inclusion $\{1\} \to G$. □

The defect in this construction is that while the homotopy groups $\mathcal{K}_*(X)$ agree *additively* with the complex topological K-theory of X, the ring structure is not immediately accessible in general, and we postpone discussion of this point to future work. Nevertheless, applications of Theorem 8.2 are provided in Section 8 below, based on the following (rather limited) information regarding the rings $\mathcal{K}_*(X)$. When $X = \{*\}$, May [**18**, VIII §2] showed that the ring $\mathcal{K}_*(*)$ is isomorphic to $\pi_*\mathbf{ku} = \mathbb{Z}[\beta]$, where $\beta \in \pi_2(\mathbf{ku}) \cong \mathbb{Z}$ is a generator. In other words, this is the standard Bott-periodic connective K-theory ring of a point. For each finite CW complex X, the projection $X \to *$ induces an injective ring map $\mathcal{K}_*(*) \to \mathcal{K}_*(X)$, which embeds the ring $\pi_*\mathbf{ku}$ in $\mathcal{K}_*(X)$.

8.2. Deformation K-theory and spaces of irreducible representations. We need to review some of Lawson's results from [**12, 13**], which allow one to compute (unitary) deformation K-theory from homological information about spaces of irreducible representations.

First, consider the space

$$\overline{\mathrm{Rep}}(G) = \coprod_n \mathrm{Hom}(G, U(n))/U(n),$$

where the quotient on the right is taken with respect to the conjugation action. Block sum makes this into a strictly commutative topological monoid. In fact, the sequence of spaces

$$\mathrm{Rep}_n(G) = \mathrm{Hom}(G, U(n))/U(n)$$

form a *bipermutative* action sequence for the trivial groups $G_n = \{1\}$, using Kronecker product to define the multiplicative structure. The associated E_∞ ring spectrum $R^{\mathrm{def}}(G)$ satisfies $\Omega^\infty R^{\mathrm{def}}(G) \simeq \overline{\mathrm{Rep}}(G)$ by Proposition 2.6. We define

$$R^{\mathrm{def}}_*(G) = \pi_* R^{\mathrm{def}}(G).$$

The quotient maps $\mathrm{Hom}(G, U(n)) \to \mathrm{Hom}(G, U(n))/U(n)$ respect block sum and Kronecker product, so we obtain an induced map of E_∞ ring spectra

$$K^{\mathrm{def}}(G) \longrightarrow R^{\mathrm{def}}(G).$$

At this point, we need to pass from the category of E_∞ ring spectra to the category of $\mathbf{S}$-algebras, as constructed in [**6**]. The desired functor is discussed in [**6**, II.3], and for us the important point is that it induces an isomorphism on the underlying homotopy rings. We will continue to use the same notation for our E_∞ ring spectra and their associated $\mathbf{S}$-algebras, but it should be noted that smash products will be formed in the derived category of $\mathbf{S}$ modules or $\mathbf{ku}$-modules, as appropriate.

In [**12**], it is shown that when G is finitely generated, there is an equivalence $H\mathbb{Z} \wedge K^{\mathrm{def}}(G) \simeq R^{\mathrm{def}}(G)$. Fix a generator of $\pi_2\mathbf{ku}$ and a map $\beta\colon \mathbf{S}^2 \to \mathbf{ku}$ representing it. We call β the Bott element. The natural map $G \to \{1\}$ induces a map $\mathbf{ku} \to K^{\mathrm{def}}(G)$, and the image of β (which we still denote simply by β) is the Bott element in $K^{\mathrm{def}}_2(G)$. Smashing $\mathbf{S}^2 \xrightarrow{\beta} \mathbf{ku}$ with $\mathbf{ku}$ induces a map

$$\Sigma^2\mathbf{ku} \longrightarrow \mathbf{ku}$$

which we call the Bott map (and also denote by β). Bott periodicity implies that the homotopy cofiber of β is the Eilenberg–MacLane spectrum $H\mathbb{Z}$.

Smashing the homotopy cofiber sequence

$$\Sigma^2\mathbf{ku} \longrightarrow \mathbf{ku} \longrightarrow H\mathbb{Z}$$

with $K^{\mathrm{def}}(G)$ (as $\mathbf{ku}$-modules; that is, applying $\wedge_{\mathbf{ku}}$) and taking homotopy groups now gives a long exact sequence of the form

$$(21) \qquad \cdots \xrightarrow{\partial} K^{\mathrm{def}}_*(G) \xrightarrow{\beta} K^{\mathrm{def}}_{*+2}(G) \longrightarrow R^{\mathrm{def}}_{*+2}(G) \xrightarrow{\partial} K^{\mathrm{def}}_{*-3}(G) \xrightarrow{\beta} \cdots,$$

wherein the map β is multiplication by the Bott element

Lawson also developed a spectral sequence for computing $R^{\mathrm{def}}_*(G)$ from the integral homology of the spaces

$$\mathrm{Irr}^+_n(G) := \mathrm{Rep}_n(G)/\mathrm{Sum}_n(G),$$

where $\mathrm{Sum}_n(G)$ denotes the subspace of reducible representations. Note that $\mathrm{Irr}^+_n(G)$ is the one-point compactification of complement of $\mathrm{Sum}_n(G)$ in $\mathrm{Rep}_n(G)$, and this complement is precisely the subspace of irreducible representations. The spectral sequence is constructed by considering the tower of spectra

$$* = R^{\mathrm{def}}_{\leqslant 0}(G) \longrightarrow R^{\mathrm{def}}_{\leqslant 1}(G) \longrightarrow R^{\mathrm{def}}_{\leqslant 2}(G) \longrightarrow \cdots,$$

where $R^{\mathrm{def}}_{\leqslant k}$ is the spectrum associated to the subspaces of $\mathrm{Rep}_n(G)$ consisting of representations whose irreducible summands all have dimension at most k; note that these subspaces provide a submonoid of $\overline{\mathrm{Rep}(G)}$, and in fact a permutative subsequence of $(\mathrm{Rep}_n(G))_{n=0}^\infty$. The homotopy colimit of this sequence is $R^{\mathrm{def}}(G)$, and Lawson proves that there are homotopy cofiber sequences of spectra

$$R^{\mathrm{def}}_{k-1}(G) \longrightarrow R^{\mathrm{def}}_k(G) \longrightarrow H\mathbb{Z} \wedge \mathrm{Irr}^+_k(G)$$

for each $k \geqslant 1$.

In general, a sequence of spectra

$$X_0 \xrightarrow{f_0} X_1 \xrightarrow{f_1} X_2 \xrightarrow{f_2} \cdots$$

gives rise to an exact couple

$$\begin{array}{ccc} \bigoplus_{q,p} \pi_q X_p & \xrightarrow{\quad \oplus (f_p)_* \quad} & \bigoplus_{q,p} \pi_q X_p \\ & \nwarrow^{\partial} \qquad \swarrow & \\ & \bigoplus_{q,p} \pi_q (\mathrm{hocofib}\, f_p) & \end{array}$$

and hence to a spectral sequence of the form

$$E^1_{p,q} = \pi_{p+q}(\mathrm{hocofib}\, f_p) \implies \pi_{p+q} \operatorname*{hocolim}_i X_i,$$

with differentials

$$d^r_{p,q}\colon E^r_{p,q} \longrightarrow E^r_{p-r,q+r-1}.$$

Since $H\mathbb{Z}$ is the spectrum representing integral homology, in the case at hand we obtain a spectral sequence

$$E^1_{p,q} = \widetilde{H}_{p+q}(\mathrm{Irr}^+_p(G);\mathbb{Z}) \implies \pi_{p+q} R^{\mathrm{def}}(G). \tag{22}$$

Note that with this indexing, the spectral sequence can be nonzero in the quadrant where $p, q \geqslant 0$ and in the region where $-p \leqslant q < 0$.

8.3. Groups satisfying Kazhdan's property (T). Property (T) has been widely studied since its introduction by Kazhdan in the late 1960s. Loosely speaking, property (T) is a weak rigidity property for (possibly infinite-dimensional) unitary representations of locally compact groups. See [**5**] for background.

We need a lemma regarding unitary representations of property (T) groups.[1]

LEMMA 8.3. *Let G be a discrete group with property* (T). *Then the space* $\mathrm{Hom}(G, U(n))/U(n)$ *is a finite, discrete space.*

PROOF. By Wang [**28**, Theorem 2.5], if $\rho\colon G \to U(n)$ is irreducible, then the path component of ρ in $\mathrm{Hom}(G, U(n))$ coincides with its conjugation orbit O_ρ. Since $\mathrm{Hom}(G, U(n))$ is compact and triangulable, it has finitely many path components, so we conclude that there are only finitely many irreducible unitary representations in each dimension. Since every unitary representation is a direct sum of irreducibles, finiteness of $\mathrm{Hom}(G, U(n))/U(n)$ follows immediately, and discreteness follows as well since this space is Hausdorff. □

PROPOSITION 8.4. *Let G be a discrete group satisfying property* (T). *Then* $K^{\mathrm{def}}_{2m+1}(G)$ *is trivial for all $m \geqslant 0$, and the iterated Bott map*

$$\beta^m_*\colon K^{\mathrm{def}}_0(G) \cong \mathrm{Gr}(\pi_0 \mathrm{Rep}(G)) \longrightarrow K^{\mathrm{def}}_{2m}(G)$$

is an isomorphism for all $m \geqslant 1$.

---

[1]The author learned this result from Rufus Willett.

Proof. Lemma 8.3 implies that $\mathrm{Hom}(G, U(n))/U(n)$ is *discrete* for all n, so its cohomology vanishes in positive dimensions. Discrete groups with property (T) are finitely generated [**5**, Theorem 1.3.1], so the spectral sequence (22) yields $\pi_*(R^{\mathrm{def}}(G)) = 0$ for $* > 0$. The long exact sequence (21) shows that $K_1^{\mathrm{def}}(G) \cong R_1^{\mathrm{def}}(G) = 0$ and that $\beta_*\colon K_m^{\mathrm{def}}(G) \to K_{m+2}(G)$ is an isomorphism for $m \geqslant 0$. □

We need a standard fact about flat bundles, coming from Chern–Weil theory.

Lemma 8.5. *Let G be a discrete group such that BG has the homotopy type of a finite CW complex, and let ϵ^n denote the trivial bundle $BG \times U(n)$. Then for every representation $\rho\colon G \to U(n)$, the class $[E_\rho] - [\epsilon^n]$ is torsion in $\widetilde{K}^0(BG)$.*

For a complete proof (of a much more general statement) see [**4**, Theorem 3.5].

Proposition 8.6. *Let G be a discrete group satisfying property* (T), *and assume that BG has the homotopy type of a finite CW complex. Then the reduced unitary topological Atiyah–Segal map*

$$\widetilde{\alpha}_m\colon \widetilde{K}_*^{\mathrm{def}}(G) \longrightarrow \widetilde{K}^{-m}(G)$$

is zero when m is odd, and its image is torsion when m is even.

Proof. For m odd, this is immediate from Proposition 8.4, so we consider the even case. By Theorem 5.3, the image of α_0 consists of K-theory classes of the form $[E_\rho]$, where $\rho\colon G \to U(n)$ is a single representation. Lemma 8.5 implies that the image of α_0 becomes torsion after modding out the summand $\pi_0\mathbf{ku} \cong \mathbb{Z}$ corresponding to the trivial bundles. But since $\alpha_0 = \widetilde{\alpha}_0 \oplus \mathrm{Id}_{\pi_0\mathbf{ku}}$ (see (13)), this quotient is isomorphic to the image of $\widetilde{\alpha}_0$.

Now consider the commutative diagram

$$\begin{array}{ccc} K_{2m}^{\mathrm{def}}(G) & \xrightarrow{\ \alpha_{2m}\ } & K_{2m}(BG) \\ {\cdot\beta^m}\Big\uparrow{\cong} & & \Big\uparrow{\cdot\alpha_{2m}(\beta^m)} \\ K_0^{\mathrm{def}}(G) & \xrightarrow{\ \alpha_0\ } & K_0(BG), \end{array} \tag{23}$$

where the vertical maps are given by multiplication. Each group in the diagram contains a $\mathbb{Z}$ summand arising from the homotopy of $\mathbf{ku}$, via the maps induced by $G \to \{1\}$ and $BG \to \{*\}$, and these summands are complementary to the reduced subgroups. All four maps in the diagram are isomorphisms when restricted to these $\mathbb{Z}$ summands, so Proposition 8.6 implies that the image of the lower composite, and hence also the upper composite, has rank 1. Since $\cdot\beta^m\colon K_0^{\mathrm{def}}(G) \to K_{2m}^{\mathrm{def}}(G)$ is an isomorphism, we conclude that the image of α_{2m} has rank 1, and finally that the image of $\widetilde{\alpha}_{2m}$ is torsion, as desired. □

Proof of Theorem 8.1. By Proposition 8.6 and Theorem 5.3, we know that

$$\widetilde{\alpha}_m([\rho]) = \pi_*^{-1}\left([E_\rho] - [E_{\widetilde{\rho(1)}}]\right)$$

is torsion in $\widetilde{K}^0(S^m \wedge BG)$. But the projection $\pi\colon S^m \times BG \to S^m \wedge BG$ induces a (split) injection on reduced K-theory, so for some $k \geqslant 1$ we have $k[E_\rho] = k[E_{\widetilde{\rho(1)}}]$ in $K^0(S^m \times BG)$. Hence it will suffice to show that $l[E_{\widetilde{\rho(1)}}]$ represents a torsion class in reduced K-theory for some $l \geqslant 1$. This holds for the bundle $E_{\rho(1)} \to BG$ by Lemma 8.5, and since $E_{\widetilde{\rho(1)}}$ is a pullback of $E_{\rho(1)}$, the proof is complete. □

REMARK 8.7. Theorem 8.1 provides a partial answer to [**20**, Question 3.20]. If G satisfies the hypotheses of Theorem 8.1, then no nontrivial class in the rational K-homology of BG can be detected, in the sense of [**20**, Definition 3.4], by a spherical family of representations. However, it remains possible that nontrivial classes can be detected by nonspherical families of representations.

8.4. The topological Atiyah–Segal map for the Heisenberg group. In the case of the 3-dimensional integral Heisenberg group H, Lawson showed in [**11**, **12**] that $R^{\mathrm{def}}_m(H) = 0$ for $m \geqslant 3$. In dimension 1, Theorem 6.1 tells us that the image of $\alpha_1 = \alpha_1^H$ has rank at most $\beta_1(N^3) = 2$. Reasoning similar to the proof of Proposition 8.6 yields the following result.

PROPOSITION 8.8. *The image of the unitary topological Atiyah–Segal map*

$$\alpha_{2m+1}\colon K^{\mathrm{def}}_{2m+1}(H) \longrightarrow K^{-(2m+1)}(BH) \cong K^{-(2m+1)}(N^3)$$

has rank at most 2 for each $m \geqslant 0$. Hence α_* is not surjective in odd dimensions.

Acknowledgements

The results in Section 5 arose from conversations among the author, Willett, and Yu surrounding the work in [**20**]. Proposition 7.1 arose from discussions with Robert Lipshitz regarding work of his student Kristen Hendricks [**9**]. The author thanks Baird, Lipshitz, Willett, and Yu, as well as Markus Szymik and Peter May, for helpful discussions.

References

[1] J. M. Alonso, T. Brady, D. Cooper, V. Ferlini, M. Lustig, M. Mihalik, M. Shapiro, and H. Short, *Notes on word hyperbolic groups*, Group theory from a geometrical viewpoint (Trieste, 1990), World Sci. Publ., River Edge, NJ, 1991, pp. 3–63. Edited by Short. MR1170363

[2] M. F. Atiyah and F. Hirzebruch, *Vector bundles and homogeneous spaces*, Proc. Sympos. Pure Math., Vol. III, Amer. Math. Soc., Providence, RI, 1961, pp. 7–38. MR139181

[3] M. F. Atiyah and G. B. Segal, *Equivariant K-theory and completion*, J. Differential Geometry **3** (1969), 1–18. MR259946

[4] Thomas Baird and Daniel A. Ramras, *Smoothing maps into algebraic sets and spaces of flat connections*, Geom. Dedicata **174** (2015), 359–374, DOI 10.1007/s10711-014-0022-z. MR3303057

[5] Bachir Bekka, Pierre de la Harpe, and Alain Valette, *Kazhdan's property (T)*, New Mathematical Monographs, vol. 11, Cambridge University Press, Cambridge, 2008, DOI 10.1017/CBO9780511542749. MR2415834

[6] A. D. Elmendorf, I. Kriz, M. A. Mandell, and J. P. May, *Rings, modules, and algebras in stable homotopy theory*, Mathematical Surveys and Monographs, vol. 47, American Mathematical Society, Providence, RI, 1997. With an appendix by M. Cole, DOI 10.1090/surv/047. MR1417719

[7] Daniel Henry Gottlieb, *Applications of bundle map theory*, Trans. Amer. Math. Soc. **171** (1972), 23–50, DOI 10.2307/1996373. MR309111

[8] Allen Hatcher, *Algebraic topology*, Cambridge University Press, Cambridge, 2002. MR1867354

[9] Kristen Hendricks, *A rank inequality for the knot Floer homology of double branched covers*, Algebr. Geom. Topol. **12** (2012), no. 4, 2127–2178, DOI 10.2140/agt.2012.12.2127. MR3020203

[10] Serge Lang, *Algebra*, 3rd ed., Graduate Texts in Mathematics, vol. 211, Springer-Verlag, New York, 2002, DOI 10.1007/978-1-4613-0041-0. MR1878556

[11] Tyler Lawson, *Derived representation theory of nilpotent groups*, ProQuest LLC, Ann Arbor, MI, 2004. Thesis (Ph.D.)–Stanford University. MR2705866

[12] Tyler Lawson, *The product formula in unitary deformation K-theory*, K-Theory **37** (2006), no. 4, 395–422, DOI 10.1007/s10977-006-9003-9. MR2269819

[13] Tyler Lawson, *The Bott cofiber sequence in deformation K-theory and simultaneous similarity in* $\mathrm{U}(n)$, Math. Proc. Cambridge Philos. Soc. **146** (2009), no. 2, 379–393, DOI 10.1017/S0305004108001928. MR2475972

[14] J. P. May, *The spectra associated to permutative categories*, Topology **17** (1978), no. 3, 225–228, DOI 10.1016/0040-9383(78)90027-7. MR508886

[15] J. P. May, *A concise course in algebraic topology*, Chicago Lectures in Mathematics, University of Chicago Press, Chicago, IL, 1999. MR1702278

[16] J. P. May, *The construction of E_∞ ring spaces from bipermutative categories*, New topological contexts for Galois theory and algebraic geometry (BIRS 2008), Geom. Topol. Monogr., vol. 16, Geom. Topol. Publ., Coventry, 2009, pp. 283–330, DOI 10.2140/gtm.2009.16.283. MR2544392

[17] J. P. May, *What precisely are E_∞ ring spaces and E_∞ ring spectra?*, New topological contexts for Galois theory and algebraic geometry (BIRS 2008), Geom. Topol. Monogr., vol. 16, Geom. Topol. Publ., Coventry, 2009, pp. 215–282, DOI 10.2140/gtm.2009.16.215. MR2544391

[18] J. Peter May, *E_∞ ring spaces and E_∞ ring spectra*, Lecture Notes in Mathematics, Vol. 577, Springer-Verlag, Berlin-New York, 1977. With contributions by Frank Quinn, Nigel Ray, and Jørgen Tornehave. MR494077

[19] Daniel Quillen, *Higher algebraic K-theory. I*, Algebraic K-theory, I: Higher K-theories (Proc. Conf., Battelle Memorial Inst., Seattle, Wash., 1972), Lecture Notes in Math., Vol. 341, Springer, Berlin-New York, 1973, pp. 85–147. MR338129

[20] Daniel Ramras, Rufus Willett, and Guoliang Yu, *A finite-dimensional approach to the strong Novikov conjecture*, Algebr. Geom. Topol. **13** (2013), no. 4, 2283–2316, DOI 10.2140/agt.2013.13.2283. MR3073917

[21] Daniel A. Ramras, *Excision for deformation K-theory of free products*, Algebr. Geom. Topol. **7** (2007), 2239–2270, DOI 10.2140/agt.2007.7.2239. MR2366192

[22] Daniel A. Ramras, *Yang-Mills theory over surfaces and the Atiyah-Segal theorem*, Algebr. Geom. Topol. **8** (2008), no. 4, 2209–2251, DOI 10.2140/agt.2008.8.2209. MR2465739

[23] Daniel A. Ramras, *On the Yang-Mills stratification for surfaces*, Proc. Amer. Math. Soc. **139** (2011), no. 5, 1851–1863, DOI 10.1090/S0002-9939-2010-10614-7. MR2763772

[24] Daniel A. Ramras, *The stable moduli space of flat connections over a surface*, Trans. Amer. Math. Soc. **363** (2011), no. 2, 1061–1100, DOI 10.1090/S0002-9947-2010-05218-3. MR2728597

[25] Daniel A. Ramras, *The homotopy groups of a homotopy group completion*, Israel J. Math. **234** (2019), no. 1, 81–124, DOI 10.1007/s11856-019-1914-2. MR4040822

[26] Andreas Rosenschon and Paul Arne Østvær, *The homotopy limit problem for two-primary algebraic K-theory*, Topology **44** (2005), no. 6, 1159–1179, DOI 10.1016/j.top.2005.04.004. MR2168573

[27] Graeme Segal, *Classifying spaces and spectral sequences*, Inst. Hautes Études Sci. Publ. Math. **34** (1968), 105–112. MR232393

[28] P. S. Wang, *On isolated points in the dual spaces of locally compact groups*, Math. Ann. **218** (1975), no. 1, 19–34, DOI 10.1007/BF01350065. MR384993

Department of Mathematics, Indiana University Indianapolis, 402 N Blackford St, Indianapolis, Indiana 46032

Email address: dramras@iu.edu

Contemporary Mathematics
Volume **802**, 2024
https://doi.org/10.1090/conm/802/16079

Bordism for the 2-group symmetries of the heterotic and CHL strings

Arun Debray

ABSTRACT. In the presence of a nonzero B-field, the symmetries of the $E_8 \times E_8$ heterotic string form a 2-group, or a categorified group, as do the symmetries of the CHL string. We express the bordism groups of the corresponding tangential structures as twisted string bordism groups, then compute them through dimension 11 modulo a few unresolved ambiguities. Then, we use these bordism groups to study anomalies and defects for these two string theories.

Introduction

String theory has long been a place where higher-categorical structures in mathematics meet their applications. This is true for a few different reasons, but one crucial reason is that many fields in superstring and supergravity theories have mathematical incarnations that are higher-categorical objects, and so even precisely setting up mathematical questions coming out of string theory, let alone solving them, often requires engaging with or developing the foundations of various kinds of geometric objects with higher structure. This paper is concerned with the appearance of a higher structure called a *2-group* in two specific string theories, and how including this structure affects computations of bordism groups for the tangential structures of these theories. These bordism groups control anomalies and extended objects for these theories. The main results of this paper are computations of bordism groups and their generating manifolds through dimension 11, except for a few ambiguities we did not address, for the tangential structures underlying these two string theories.

For the higher structures we investigate in this paper, the story begins with the *Kalb–Ramond field*, or the *B-field*. This is an analogue of the field strength of an electromagnetic field, represented as a closed differential 2-form with a quantization condition. Locality of quantum field theory means expressing the field strength of the electromagnetic field as a section of a sheaf, specifically as a connection on a principal $\mathbb{T}$-bundle, where $\mathbb{T}$ is the circle group. For the B-field, everything is one degree higher: it comes to us as a closed differential 3-form with a quantization condition, which we would like to express as a geometric object that sheafifies. This cannot be a connection on a principal G-bundle for a finite-dimensional Lie group G; instead, one models the B-field as a connection on a $\mathbb{T}$-*gerbe*, which is a categorification of a principal $\mathbb{T}$-bundle. A $\mathbb{T}$-gerbe on a manifold M is, roughly

2020 *Mathematics Subject Classification.* Primary 57R90, 81T30; Secondary 55T15.

©2024 American Mathematical Society

speaking, a bundle of groupoids on M which is locally equivalent to $\mathrm{pt}/\mathbb{T}$. There are several ways to make this precise; we discuss one, Murray's *bundle gerbes* [**Mur96**], in Definition 1.1.

In this article, we consider higher structures in two string theories: the $E_8 \times E_8$ heterotic string, and the Chaudhuri–Hockney–Lykken (CHL) string. The former is a ten-dimensional superstring theory whose low-energy limit is ten-dimensional $\mathcal{N} = 1$ supergravity, and the latter is a nine-dimensional theory obtained from the $E_8 \times E_8$ heterotic string theory by compactifying on a circle. Both of these theories have B-fields, but Green and Schwarz [**GS84**] showed that in order to cancel an anomaly, the B-field and the gauge field must satisfy a relation known as a *Bianchi identity*. Fiorenza, Schreiber, and Stasheff [**FSS12**] and Sati, Schreiber, and Stasheff [**SSS12**] describe how the Bianchi identity mixes the data of the B-field and the gauge field into data that can be interpreted as a connection on a principal bundle for a 2-group $\mathbb{G}$, specifically a *string 2-group* $\mathcal{S}tr(G, \mu)$ associated to the data of a compact Lie group G and a class $\mu \in H^4(BG; \mathbb{Z})$; typically, G is the gauge group and μ is determined by the anomaly polynomial.

2-groups have been used in the theoretical and mathematical physics literature for some time now. This program began in earnest with work of Baez, Crans, Lauda, Stevenson, and Schreiber [**Bae02, BC04, BL04, BSCS07, BS07**]; more recently, 2-groups, their symmetries, and their anomalies have made a resurgence in quantum field theory following work of Córdova, Dumitriescu, and Intrilligator [**CDI19**] and Benini, Córdova, and Hsin [**BCH19**] identifying many examples of 2-group symmetries in commonly studied QFTs. See also Sharpe [**Sha15**] and the references therein.

In the first part of this article, we introduce the Bianchi identity and 2-groups, then review work of [**FSS12**] and [**SSS12**] mentioned above. These authors work in the setting of stacks on the site $\mathcal{M}an$ of smooth manifolds; the data of the B-field (Q, Θ_Q) and the principal G-bundle with connection (P, Θ_P) on a manifold M refine to maps from M to classifying stacks of these data. The data of an identification of two differential characteristic classes associated to Θ_P and Θ_Q gives rise to

(1) a principal $\mathcal{S}tr(G, \mu)$-bundle lifting P for a specified choice of μ (Proposition 1.35), and
(2) local data of solutions to the Bianchi identity (Proposition 1.37, [**FSS12**, §6.3]).

Inspired by this, we introduce the tangential structures ξ^{het} and ξ^{CHL}, which are special cases of a general construction of [**SSS12**, Definition 2.8]: a ξ_n^{het}-structure on a spin manifold M is data of a principal $\mathbb{G}_n^{\mathrm{het}}$-bundle, where $\mathbb{G}_n^{\mathrm{het}} := \mathcal{S}tr(\mathrm{Spin}_n \times (E_8 \times E_8 \rtimes \mathbb{Z}/2), c_1 + c_2 - \lambda)$ (1.42), whose associated Spin_n-bundle via the quotient $\mathbb{G}_n^{\mathrm{het}} \to \mathrm{Spin}_n$ is the principal Spin_n-bundle of spin frames (Definition 1.41). This is compatible as n varies, allowing us to stabilize and define a ξ^{het}-structure as usual. The definition of ξ^{CHL} in Definition 1.52, which coincides with $B\mathrm{String}^{2a}$ in [**SSS12**, (2.18), §2.2.3], is analogous. Related tangential structures appear in [**Sat11b, FSS15a, FSS15b, FSS21**].

Given a tangential structure, we can compute bordism groups, and indeed the point of this paper is to compute ξ^{het} and ξ^{CHL} bordism groups in low dimensions. These bordism groups can then be used to learn more about the $E_8 \times E_8$ heterotic and CHL strings. We have two primary applications in mind.

(1) The *cobordism conjecture* of McNamara and Vafa [**MV19**] is an application to the question of what kinds of spacetime backgrounds are summed over in quantum gravity. Such backgrounds are often taken to be manifolds or something closely related equipped with data of a tangential structure ξ. The cobordism conjecture says that if ξ is the most general tangential structure which can appear in this way in any particular d-dimensional theory of quantum gravity, then $\Omega_k^\xi = 0$ for $3 \le k \le d-1$. We will see that $\Omega_k^{\xi^{\mathrm{het}}}$ and $\Omega_k^{\xi^{\mathrm{CHL}}}$ are often nonzero in that range. This is consistent with the cobordism conjecture: it suggests that ξ^{het} and ξ^{CHL} are not the most general tangential structures that can be summed over. Typically these bordism groups are killed by allowing singular manifolds corresponding to considering the theory with branes or other defects, so one can use bordism computations to predict new defects in string theories.

(2) A broad class of n-dimensional quantum field theories come with data of an *anomaly*, which in many cases can roughly be described an $(n+1)$-dimensional invertible field theory α. In some cases one wants to trivialize α, meaning exhibiting an isomorphism from α to the trivial field theory. By work of Freed, Hopkins, and Teleman [**FHT10**] and Freed and Hopkins [**FH21b**], invertible field theories can be classified using bordism group computations. For both the $\mathrm{E}_8 \times \mathrm{E}_8$ heterotic string and the CHL string, the bordism groups indicating a potential anomaly are nonzero, and it would be interesting to check whether the corresponding anomalies are nontrivial.

See §3, as well as Questions 0.1 to 0.3, for more on these applications and what we can learn from our bordism computations.

Our main theorems are the following two computations of the ξ^{het} and ξ^{CHL} bordism groups in low dimensions.

THEOREM A. *For $k \le 10$, the ξ^{het}-bordism groups are:*

$$\begin{aligned}
\Omega_0^{\xi^{\mathrm{het}}} &\cong \mathbb{Z},\\
\Omega_1^{\xi^{\mathrm{het}}} &\cong \mathbb{Z}/2 \oplus \mathbb{Z}/2,\\
\Omega_2^{\xi^{\mathrm{het}}} &\cong \mathbb{Z}/2 \oplus \mathbb{Z}/2,\\
\Omega_3^{\xi^{\mathrm{het}}} &\cong \mathbb{Z}/8,\\
\Omega_4^{\xi^{\mathrm{het}}} &\cong \mathbb{Z} \oplus \mathbb{Z}/2,\\
\Omega_5^{\xi^{\mathrm{het}}} &\cong 0,\\
\Omega_6^{\xi^{\mathrm{het}}} &\cong \mathbb{Z}/2,\\
\Omega_7^{\xi^{\mathrm{het}}} &\cong \mathbb{Z}/16,\\
\Omega_8^{\xi^{\mathrm{het}}} &\cong \mathbb{Z}^3 \oplus (\mathbb{Z}/2)^{\oplus i},\\
\Omega_9^{\xi^{\mathrm{het}}} &\cong (\mathbb{Z}/2)^{\oplus j},\\
\Omega_{10}^{\xi^{\mathrm{het}}} &\cong (\mathbb{Z}/2)^{\oplus k}.
\end{aligned}$$

Here, either $i = 1$, $j = 4$, *and* $k = 4$, *or* $i = 2$, $j = 6$, *and* $k = 5$.

$\Omega_{11}^{\xi^{\mathrm{het}}}$ *is an abelian group of order* 64 *isomorphic to one of* $\mathbb{Z}/8 \oplus \mathbb{Z}/8$, $\mathbb{Z}/16 \oplus \mathbb{Z}/4$, $\mathbb{Z}/32 \oplus \mathbb{Z}/2$, *or* $\mathbb{Z}/64$.

This is a combination of Theorems 2.62 and 2.74. In §2.2.1, we find manifold representatives for all classes in $\Omega_k^{\xi^{\mathrm{het}}}$ for $k \leq 10$ except potentially for two missing classes X_8 and X_9 of dimensions 8, resp. 9 and their products with S^1_{nb}. These classes may or may not be zero depending on the fate of an Adams differential. In §2.2.2, we find a manifold representing X_8: if the unaddressed Adams differential vanishes, X_8 should be added to the list of generators in §2.2.1, and if the differential does not vanish, then X_8 bounds as a ξ^{het}-manifold.

Our calculation of ξ^{CHL}-bordism builds on work of Hill [**Hil09**, Theorem 1.1], who computes $\Omega_*^{\mathrm{String}}(B\mathrm{E}_8)$ in dimensions 14 and below.

THEOREM B. *For* $k \leq 11$, *there is an abstract isomorphism from* $\Omega_*^{\xi^{\mathrm{CHL}}}$ *to the free and* 2*-torsion summands of* $\Omega_*^{\mathrm{String}}(B\mathrm{E}_8)$. *Therefore, by Hill's computation* [**Hil09**], *there are isomorphisms*

$$\begin{aligned} \Omega_0^{\xi^{\mathrm{CHL}}} &\cong \mathbb{Z}, & \Omega_6^{\xi^{\mathrm{CHL}}} &\cong \mathbb{Z}/2,\\ \Omega_1^{\xi^{\mathrm{CHL}}} &\cong \mathbb{Z}/2, & \Omega_7^{\xi^{\mathrm{CHL}}} &\cong 0,\\ \Omega_2^{\xi^{\mathrm{CHL}}} &\cong \mathbb{Z}/2, & \Omega_8^{\xi^{\mathrm{CHL}}} &\cong \mathbb{Z} \oplus \mathbb{Z} \oplus \mathbb{Z}/2,\\ \Omega_3^{\xi^{\mathrm{CHL}}} &\cong \mathbb{Z}/8, & \Omega_9^{\xi^{\mathrm{CHL}}} &\cong \mathbb{Z}/2 \oplus \mathbb{Z}/2 \oplus \mathbb{Z}/2,\\ \Omega_4^{\xi^{\mathrm{CHL}}} &\cong \mathbb{Z}, & \Omega_{10}^{\xi^{\mathrm{CHL}}} &\cong \mathbb{Z}/2 \oplus \mathbb{Z}/2,\\ \Omega_5^{\xi^{\mathrm{CHL}}} &\cong 0, & \Omega_{11}^{\xi^{\mathrm{CHL}}} &\cong \mathbb{Z}/8. \end{aligned}$$

This is a combination of Theorems 2.90 and 2.92. We also obtain some information about manifold representatives of generators of these groups.

The computational tool we use to prove Theorems A and B is standard: the Adams spectral sequence. This spectral sequence has seen plenty of applications in the mathematical physics literature, and there is a standard procedure reviewed by Beaudry and Campbell [**BC18**] for simplifying the E_2-page for a wide class of tangential structures, namely those which can be described as oriented, spin^c, spin, or string bordism twisted by a virtual vector bundle. For example, the twisted string bordism computations of [**FK96**, **Fan99**, **FW10**] make use of this simplifying technique. Unfortunately, this procedure is unavailable to us: in Lemma 2.2, we prove that ξ^{het} and ξ^{CHL} cannot be described as twists of this sort. However, we are still able to describe them as twists in a more general sense due to Ando, Blumberg, Gepner, Hopkins, and Rezk [**ABG$^+$14a**, **ABG$^+$14b**]: adapting an argument of Hebestreit and Joachim [**HJ20**], one learns that the Thom spectra for ξ^{het} and ξ^{CHL} can be produced as the *MTString*-module Thom spectra associated to certain maps to $B\mathrm{GL}_1(\mathit{MTString})$. Using this structure, in joint work with Matthew Yu, we are able to prove a theorem simplifying the calculation of the E_2-page.

THEOREM C (Debray and Yu [**DY23**]). *In topological degrees* 15 *and below, the* E_2*-pages of the Adams spectral sequences computing* 2*-completed twisted string bordism for a class of twists including those for* ξ^{het} *and* ξ^{CHL} *can be computed as* Ext *over the subalgebra* $\mathcal{A}(2)$ *of the Steenrod algebra.*

What we prove is more precise and holds in more generality; see Theorem 2.20 and Corollary 2.22 for that version of the result.[1]

The $\mathcal{A}(2)$-module Ext groups we have to compute are simpler than what one a priori has to work with over the entire Steenrod algebra $\mathcal{A}$. We do not need this simplification at odd primes; there the full Adams spectral sequence is easier to work with, and the absence of a simplification does not hinder us (though see also [**DY23**, §3.2]).

The reason we computed these bordism groups in this paper is with applications to physics, specifically to anomalies and the cobordism conjecture, in mind. We discuss some implications of our calculations in §3; for example, one of the $\mathbb{Z}/2$ summands of $\Omega_1^{\xi^{\mathrm{het}}}$ corresponds to the non-supersymmetric 7-brane recently discovered by Kaidi, Ohmori, Tachikawa, and Yonekura [**KOTY23**]. We end this section of the introduction with some questions related to these physics predictions.

QUESTION 0.1. What does the Kaidi–Ohmori–Tachikawa–Yonekura 7-brane correspond to in Hořava–Witten theory, and what does this look like in bordism? Hořava and Witten [**HW96a, HW96b, Wit96**] proposed that the $\mathrm{E}_8 \times \mathrm{E}_8$ heterotic string can be identified with a certain limit of M-theory compactified on an interval; thus this ought to correspond to a notion of bordism of manifolds with boundary. Conner and Floyd [**CF66**, §16] define a notion bordism of compact manifolds with boundary—is this the correct kind of bordism for applications to McNamara and Vafa's conjecture?

We discuss some additional extended objects predicted by our bordism computations to exist in the $\mathrm{E}_8 \times \mathrm{E}_8$ heterotic and CHL strings in §3.1.

QUESTION 0.2. Is the $\mathbb{Z}/2$ symmetry exchanging the two E_8-bundles in $\mathrm{E}_8 \times \mathrm{E}_8$ heterotic string theory anomalous? Because $\Omega_{11}^{\xi^{\mathrm{het}}}$ is nonzero, we were unable to rule out this anomaly.

Witten [**Wit86**, §4] and Tachikawa and Yonekura [**TY21**] show that the $\mathrm{E}_8 \times \mathrm{E}_8$ heterotic string is anomaly-free in certain cases, but they do not address the $\mathbb{Z}/2$ symmetry.

QUESTION 0.3. Does the CHL string have an anomaly? This anomaly could be nontrivial, because $\Omega_{10}^{\xi^{\mathrm{CHL}}} \cong \mathbb{Z}/2 \oplus \mathbb{Z}/2$.

There is another application of twisted string bordism to physics that we did not address in this paper: studying elliptic genera, the Witten genus and related invariants, along the lines of, e.g., Bunke and Naumann [**BN14**], McTague [**McT14**], Han, Huang, and Duan [**HHD21**], and Berwick-Evans [**BE23**]. It would be interesting to study whether the calculations in this paper could be applied in similar contexts.

Outline. We begin in §1.1 by introducing the fields present in 10d $\mathcal{N} = 1$ supergravity, the low-energy limit of heterotic string theory. We discuss how the Green-Schwarz anomaly cancellation condition imposes an equation called the *Bianchi identity* (1.10) on the fields in this theory. We then generalize this to a *twisted Bianchi identity* (1.12) associated to data of a Lie group G and a class $\mu \in H^4(BG; \mathbb{Z})$. In §1.2, we relate these Bianchi identities to the presence of a

[1] We also provide a proof sketch of the case we need in Remark 2.26.

2-group symmetry in this field theory. We begin by reviewing 2-groups, their principal bundles, and their connections, and in Example 1.22 define the string cover $\mathcal{S}(G, \mu)$ corresponding to a group G and a class $\mu \in H^4(BG; \mathbb{Z})$. Then we review [**FSS12**] and [**SSS12**] relating the Bianchi identity to twisted string structures. Using this, we define the heterotic tangential structure in Definition 1.41, which is the topological part of the structure necessary for defining $\mathcal{N} = 1$ supergravity. Then, in §1.3, we introduce the CHL string and define the CHL tangential structure using what we learned in §1.2.

In §2, we compute the bordism groups $\Omega_*^{\xi^{\mathrm{het}}}$ and $\Omega_*^{\xi^{\mathrm{CHL}}}$ in low degrees. For the latter we are able to completely compute them in dimensions 11 and below, but for the former, we have only partial information above dimension 7, occluded by Adams differentials and an extension problem we could not solve. We begin in §2.1 by discussing how to simplify the Thom spectra $MT\xi^{\mathrm{het}}$ and $MT\xi^{\mathrm{CHL}}$; we prove in Lemma 2.2 that a standard approach does not work, and so we use a different idea: construct $MT\xi^{\mathrm{het}}$ and $MT\xi^{\mathrm{CHL}}$ as *MTString*-module Thom spectra using machinery developed by Ando, Blumberg, Gepner, Hopkins, and Rezk. We review this machinery and discuss how it leads to Corollary 2.22, a special case of the main theorem of our work [**DY23**] joint with Matthew Yu, simplifying the calculation of the E_2-page of the Adams spectral sequence at 2 for a wide class of twisted string bordism groups. Next, in §2.2, we undertake this computation for ξ^{het}. We do not have such a simplification at odd primes, so in §2.3 we press ahead directly with the Adams spectral sequence for ξ^{het}, proving in Theorem 2.74 that $\Omega_*^{\xi^{\mathrm{het}}}$ lacks odd-primary torsion in degrees 11 and below. Finally, in §2.4 we run the analogous calculations for the CHL string, again using Corollary 2.22 at $p = 2$ and arguing more directly at odd primes.

The final section, §3, is about applications to string theory. We first discuss the cobordism conjecture of McNamara and Vafa [**MV19**] in §3.1, and go over a few predictions that follow from the bordism group computations in §2. In §3.2, we briefly introduce anomalies of quantum field theories and their bordism-theoretic classification, and touch on questions raised by our bordism computations.

1. Tangential structures for heterotic and CHL string theories

The goal of this section is to define the tangential structures ξ^{het} and ξ^{CHL} that are necessary to formulate the (low-energy limits of) the $\mathrm{E}_8 \times \mathrm{E}_8$ heterotic string and the CHL string. By "tangential structure" we mean the topological part of the structure needed on a manifold to define a given field theory; see Definition 1.40 for the precise definition. The presence of a B-field in both theories means that these tangential structures arise as classifying spaces of higher groups. First, we introduce the heterotic string in §1.1, and see what data and conditions are told to us by Green–Schwarz anomaly cancellation; then in §1.2, we reinterpret that data as combining the gauge field and the B-field into a connection for a principal bundle for a higher group. Finally, in §1.3, we use the general theory from §1.2 to determine the tangential structure for the CHL string.

The material in this section is not new, though it was not always stated in this form before. The fact that a Bianchi identity/Green–Schwarz mechanism is expressing a lift to a connection for a higher-group principal bundle is well-known; see [**FSS12**] and [**SSS12**].

1.1. The $E_8 \times E_8$ heterotic string. Heterotic string theories are ten-dimensional superstring theories whose low-energy limits are 10d $\mathcal{N} = 1$ supergravity theories. These supergravity theories can have Yang-Mills terms, and so are parametrized by the data of the gauge group G, a compact Lie group. However, not all choices of G yield valid supergravity theories; there is the potential for an anomaly that must be trivialized, and this is quite a strong constraint, implying that the connected component of the identity in G must be either $E_8 \times E_8$ or $G = \mathrm{SemiSpin}_{32}$[2] [**GS84, ATD10**]. The anomaly cancellation mechanism itself, due to Green-Schwarz [**GS84**], combines the different fields in the theory into a connection for a principal $\mathbb{G}$-bundle, where $\mathbb{G}$ is a higher group;[3] we use this subsection to discuss the fields and the Green-Schwarz condition, and the next subsection to discuss the role of higher group. In this paper, we will focus solely on the $E_8 \times E_8$ case; it would be interesting to study the analogues of the computations and applications in this paper in the $\mathrm{SemiSpin}_{32}$ case.

The group $\mathbb{Z}/2$ acts on $E_8 \times E_8$ by exchanging the two factors, and the setup of heterotic string theory, including the low-energy supergravity limit and Green-Schwarz' anomaly cancellation, is invariant under this symmetry, so we can expand the gauge group to $G := (E_8 \times E_8) \rtimes \mathbb{Z}/2$.[4] This appears to have first been noticed by McInnes [**McI99**, §I]; see also [**dBDH$^+$00**, §2.2.1].

Enlarging the gauge group from $E_8 \times E_8$ to $(E_8 \times E_8) \rtimes \mathbb{Z}/2$ is a choice, and requires justification—why this semidirect product, and not other or larger expansions? The answer is that $\mathbb{Z}/2$ is acting through the isomorphism $\mathbb{Z}/2 \cong \mathrm{Out}(E_8 \times E_8)$, meaning both that it acts nontrivially on the gauge group (and in turn, on the fields of the theory) and that no larger group acts faithfully in this way. Similar outer automorphism extensions have been considered in, for example, [**HLMM22, BG23**].

The fields of 10d $\mathcal{N} = 1$ supergravity on a manifold M include:

- a metric g,
- a spin structure on M,
- a principal G-bundle $P \to M$ with connection Θ_P,
- a *B-field* or *Kalb–Ramond field*, a gerbe $Q \to M$ with connection Θ_Q, and
- several additional fields (the dilaton, dilatino, gravitino, and gaugino) which will not be directly relevant to this paper.

Let us say more about the B-field, since its model as a gerbe with connection may be less familiar. A *gerbe* is a categorification of the idea of a principal $\mathbb{T}$-bundle; here $\mathbb{T}$ is the circle group. Thus, for example, a principal $\mathbb{T}$-bundle $P \to M$ is classified by its first Chern class $c_1(P) \in H^2(M; \mathbb{Z})$, and a gerbe $Q \to M$ is classified by its *Dixmier-Douady class* $\mathrm{DD}(Q) \in H^3(M; \mathbb{Z})$ [**DD63, Bry93**]. A connection on a principal $\mathbb{T}$-bundle has holonomy around loops; a connection on a gerbe has holonomy on closed surfaces. And so on.

Gerbes were first introduced by Giraud [**Gir71**]. There are several different and equivalent ways to precisely define gerbes and their connections; heuristically

[2]The center of Spin_{4k} is isomorphic to $\mathbb{Z}/2 \times \mathbb{Z}/2$. Quotienting by one copy of $\mathbb{Z}/2$ yields SO_{4k}; the quotients by the two other $\mathbb{Z}/2$ subgroups are isomorphic, and are called $\mathrm{SemiSpin}_{4k}$. See [**McI99**].

[3]Green and Schwarz's work only cancels the perturbative part of the anomaly; see §3.2 for more information.

[4]Though we often use the standard name "the $E_8 \times E_8$ heterotic string" to refer to this theory, we will always consider the larger gauge group $(E_8 \times E_8) \rtimes \mathbb{Z}/2$.

you can think of a gerbe on M as a sheaf of groupoids on M locally equivalent to the trivial sheaf with fiber $\mathrm{pt}/\mathbb{T}$. One way to make this precise is the following.

If $f\colon Y \to X$ is a map, we let $Y^{[n]} := Y \times_X Y \times_X \cdots \times_X Y$; $Y^{[n]}$ is the space of n-simplices in the Čech nerve for f.

Definition 1.1 (Murray [**Mur96**]). A *bundle gerbe* over a manifold M is a surjective submersion $\pi\colon Y \to M$, a $\mathbb{T}$-bundle $P \to Y^{[2]}$, and an isomorphism $\mu\colon \pi_{12}^*P \otimes \pi_{23}^*P \xrightarrow{\cong} \pi_{13}^*P$ of $\mathbb{T}$-bundles over $Y^{[3]}$ satisfying the natural associativity condition (see below) over $Y^{[4]}$.

Given two $\mathbb{T}$-bundles $P_1, P_2 \to X$, their *tensor product* $P_1 \otimes P_2$ is the unit circle bundle inside the tensor product of the Hermitian line bundles $L_1, L_2 \to X$ associated to P_1, resp. P_2. The maps $\pi_{12}, \pi_{23}, \pi_{13}\colon Y^{[3]} \rightrightarrows Y^{[2]}$ are the three face maps in the Čech nerve $Y^\bullet$ associated to f, given explicitly by contracting two of the three copies of Y via $Y \times_X Y \to Y$.

The associativity condition in Definition 1.1 is a little unwieldy to state explicitly, but can be found in in [**Mur10**, Definition 4.1(2)].

Definition 1.2 ([**Mur96**]). A *connection* Θ_Q on a bundle gerbe $Q = (Y, P, \mu)$ is data of a 2-form $B \in \Omega^2(Y)$ and a connection Θ_P on P such that if $\Omega_P \in \Omega^2(P)$ denotes the curvature of P and $\pi_1, \pi_2\colon Y^{[2]} \to Y$ are the two projections, then

$$\Omega_P = \pi_2^*B - \pi_1^*B. \tag{1.3}$$

The *curvature* of Θ_Q is $\Omega_Q := \mathrm{d}B$, which is a closed 3-form.

The key thing to know about this definition is that, just like a principal $\mathbb{T}$-bundle $P \to M$ with connection locally has a connection 1-form A and globally has a curvature 2-form Ω_P which locally satisfies $\Omega_P = \mathrm{d}A$, a gerbe with connection Q locally has a connection 2-form B and globally has a curvature 3-form Ω_Q which locally satisfies $\Omega_Q = \mathrm{d}B$. For more information, see, e.g., Brylinski [**Bry93**, §5.3].

Definition 1.4. Because E_8 is a simple, connected, simply connected, compact Lie group, there is a canonical isomorphism $H^4(B\mathrm{E}_8;\mathbb{Z}) \xrightarrow{\cong} \mathbb{Z}$. Let c denote the generator corresponding to $1 \in \mathbb{Z}$. In $B(\mathrm{E}_8 \times \mathrm{E}_8) \simeq B\mathrm{E}_8 \times B\mathrm{E}_8$, let c_1 and c_2 denote the copies of c coming from the first, resp. second copies of $B\mathrm{E}_8$ via the Künneth map.

The class $c_1 + c_2$ is invariant under the $\mathbb{Z}/2$ swapping action, so descends via the Serre spectral sequence to a class in $H^4(B((\mathrm{E}_8 \times \mathrm{E}_8) \rtimes \mathbb{Z}/2);\mathbb{Z})$, which we also call $c_1 + c_2$.

Definition 1.5. Spin_n is also a compact, connected, simply connected simple Lie group when $n \geq 3$, and the generator of $H^4(B\mathrm{Spin}_n;\mathbb{Z}) \xrightarrow{\cong} \mathbb{Z}$ corresponding to 1 is denoted λ.

The class λ is preserved under the standard embeddings $\mathrm{Spin}_n \hookrightarrow \mathrm{Spin}_{n+k}$, so we often work with its stabilized avatar $\lambda \in H^4(B\mathrm{Spin};\mathbb{Z})$. We use this to define λ for Spin_n when $n < 3$. Because $2\lambda = p_1$, the class λ is often denoted $\frac{1}{2}p_1$. The mod 2 reduction of λ is the Stiefel-Whitney class w_4.

Lemma 1.6 (Whitney sum formula). *Let X be a topological space and $E_1, E_2 \to X$ be two vector bundles with spin structure. Then $\lambda(E_1 \oplus E_2) = \lambda(E_1) + \lambda(E_2)$.*

PROOF. It suffices to prove the universal case, which amounts to the calculation of the pullback of λ by the map

$$\oplus\colon B\mathrm{Spin}_{k_1}\times B\mathrm{Spin}_{k_2}\longrightarrow B\mathrm{Spin}_{k_1+k_2}. \tag{1.7}$$

For $n\geq 3$, Spin_n is a connected, simply connected, compact simple Lie group, so $H^\ell(B\mathrm{Spin}_n;\mathbb{Z})$ vanishes for $\ell=1,2,3$ and is isomorphic to $\mathbb{Z}$ for $\ell=0,4$. For $n<3$, $H^*(B\mathrm{Spin}_n;\mathbb{Z})$ is still trivial or free abelian in degrees 4 and below. Therefore by the Künneth formula, for all k_1,k_2, $H^4(B\mathrm{Spin}_{k_1}\times B\mathrm{Spin}_{k_2};\mathbb{Z})$ is a free abelian group, meaning that if we can show $2\lambda(E_1\oplus E_2)=2\lambda(E_1)+2\lambda(E_2)$, then we can deduce $\lambda(E_1\oplus E_2)=\lambda(E_1)+\lambda(E_2)$.

As $2\lambda=p_1$, we have reduced to the Whitney sum formula for p_1. The Whitney sum formula $p_1(E_1\oplus E_2)=p_1(E_1)+p_1(E_2)$ does not actually hold for all vector bundles, but Brown [**Bro82**, Theorem 1.6] (see also Thomas [**Tho62**]) showed that the difference $p_1(E_1\oplus E_2)-p_1(E_1)-p_1(E_2)$ vanishes when E_1 and E_2 are orientable, so in our setting of spin vector bundles, we can conclude. □

REMARK 1.8. There are other ways to prove Lemma 1.6: for example, it follows immediately from a result of Jenquin [**Jen05**, Corollary 4.9] in a simple generalized cohomology theory. Johnson-Freyd and Treumann [**JFT20**, §1.4] sketch another proof of Lemma 1.6.

Next, we introduce the Chern-Weil homomorphism. Let G be a Lie group with Lie algebra $\mathfrak{g}$, and let $f\in\mathrm{Sym}^k(\mathfrak{g}^\vee)$, i.e., f is a degree-k polynomial function on $\mathfrak{g}$ which is invariant under the adjoint G-action on $\mathfrak{g}$. Given a manifold M, a principal G-bundle $P\to M$, and a connection Θ on P, let $\Omega\in\Omega_P^2(\mathfrak{g})$ denote the curvature 2-form. Then one can evaluate f on $\Omega^{\wedge k}\in\Omega_P^{2k}(\mathfrak{g}^{\otimes k})$, producing a form $f(\Omega^{\wedge k})\in\Omega_P^{2k}$; because f is Ad-invariant, $f(\Omega^{\wedge k})$ descends to a form $w(\Theta)\in\Omega_M^{2k}$, which is always closed. This defines a ring homomorphism, called the *Chern-Weil homomorphism* [**Car50, Che52**],

$$w\colon \mathrm{Sym}^\bullet(\mathfrak{g}^\vee)\longrightarrow H^*_{\mathrm{dR}}(M), \tag{1.9a}$$

which doubles the degree and is natural in M; moreover, the de Rham class of $w(\Theta)$ depends on P but not on the connection. Using de Rham's theorem and naturality, w upgrades to a ring homomorphism

$$w\colon \mathrm{Sym}^\bullet(\mathfrak{g}^\vee)\longrightarrow H^*(BG;\mathbb{R}), \tag{1.9b}$$

which Chern and Weil showed is an isomorphism when G is compact [**Che52, Wei49**]. Thus, when G is compact, a class $x\in H^{2*}(BG;\mathbb{Z})$ defines a polynomial $\mathrm{CW}_x\in\mathrm{Sym}^*(\mathfrak{g}^\vee)$, the w-pre-image of the de Rham class of x. We will also write $\mathrm{CW}_x(\Theta)$ to denote the form defined by evaluating the polynomial CW_x on the curvature form of Θ.

Returning to 10d $\mathcal{N}=1$ supergravity, Green-Schwarz [**GS84**] noticed that in order to trivialize an anomaly, one has to impose a relation between P and Q and their connections, so that Q is not quite a gerbe, but instead something twisted. Specifically, the curvature Ω_Q is no longer closed, but instead satisfies the equation

$$\mathrm{d}\Omega_Q=\mathrm{CW}_{c_1+c_2}(\Theta_P)-\mathrm{CW}_\lambda(\Theta^{\mathrm{LC}}), \tag{1.10}$$

where Θ^{LC} is the Levi-Civita connection on the principal Spin_n-bundle of frames of M.[5] This is called a *Bianchi identity* in the physics literature, motivating the following definition.

DEFINITION 1.11. Given data of a compact Lie group G and a class $\mu \in H^4(BG;\mathbb{Z})$, the *twisted Bianchi identity* is the equation

$$\mathrm{d}H = \mathrm{CW}_\mu(\Theta_P), \tag{1.12}$$

where H is a 3-form and Θ_P is a connection on a principal G-bundle.

As in the case of (1.10), we think of this as mixing the data of two connections, one on a principal G-bundle and one on a gerbe. In the next section, we interpret twisted Bianchi identities as coming from connections on higher groups.

1.2. From the Bianchi identity to higher groups. In this section, we show that the twisted Bianchi identity (1.12) is a natural consequence of combining a principal G-bundle and a gerbe, each with connections, into a principal $\mathbb{G}$-bundle, where $\mathbb{G}$ is a certain Lie 2-group built from G and μ, together with additional data that we think of as a connection on $\mathbb{G}$. First we introduce 2-groups and their principal bundles; then, following [**FSS12, SSS12**], we recover the twisted Bianchi identity. As a result, we can precisely define the tangential structure for the $\mathrm{E}_8 \times \mathrm{E}_8$ heterotic string, i.e., the topological part of the data which, when put on a manifold M, allows one to study $\mathrm{E}_8 \times \mathrm{E}_8$ heterotic string theory on that manifold.

DEFINITION 1.13. A *2-group* $\mathbb{G}$ is a group object in the bicategory of small categories.

DEFINITION 1.14. A *Lie 2-group* is a 2-group $\mathbb{G}$ whose underlying category has been given the structure of a category object in smooth manifolds.

This means that the sets of objects and morphisms are smooth manifolds, and assignments such as the source of a map or the composition of two maps are smooth. 2-groups were first introduced by Hoàng Xuân Sính in her thesis [**Hoà75**], and Lie 2-groups were introduced by Baez [**Bae02**, §2].

We call a 2-group *strict* if it is strict as a monoidal category, i.e., its associators and unitors are all identity maps. MacLane's coherence theorem [**Mac71**, Chapter 7] implies every 2-group is equivalent to a strict 2-group, but the analogous statement is false for Lie 2-groups; see Remark 1.25.

EXAMPLE 1.15. If G is a group, it defines a monoidal groupoid with G as its set of objects, tensor product $g \otimes h := gh$, and only the identity morphisms. This is a 2-group, and inherits the structure of a Lie 2-group if G is a Lie group.

This procedure embeds the bicategory of groups, group homomorphisms, and identity 2-morphisms into the bicategory of 2-groups, and we will therefore abuse notation and call this 2-group G again.

EXAMPLE 1.16. Let A be an abelian group, and let $A[1]$ denote the monoidal groupoid with a single object $*$ and $\mathrm{Hom}_{A[1]}(*,*) := A$. This is a 2-group, and if A is Lie, $A[1]$ is a Lie 2-group.

[5] Before Green and Schwarz, it was already known that $\mathrm{CW}_{c_1+c_2}(\Theta_P)$ and $\mathrm{d}\Omega_Q$ had to mix in order to preserve supersymmetry, thanks to work of Bergshoeff, de Roo, de Wit, and van Nieuwenhuizen [**BdRdWvN82**] and Chapline and Manton [**CM83**].

It turns out every 2-group $\mathbb{G}$ factors as an extension of these examples. Let e be the identity object of $\mathbb{G}$ and $\pi_0(\mathbb{G})$ be the group of isomorphism classes of objects in $\mathbb{G}$. Then there is a short exact sequence of 2-groups

$$0 \longrightarrow \mathrm{Aut}_{\mathbb{G}}(e)[1] \longrightarrow \mathbb{G} \longrightarrow \pi_0(\mathbb{G}) \longrightarrow 0. \tag{1.17}$$

The Eckmann–Hilton theorem guarantees $\mathrm{Aut}_{\mathbb{G}}(e)$ is abelian. Extensions (1.17) are classified by the data of:

(1) an action of $\pi_0(\mathbb{G})$ on $\mathrm{Aut}_{\mathbb{G}}(e)$, and
(2) a cohomology class $k \in H^3(B\pi_0(\mathbb{G}); \mathrm{Aut}_{\mathbb{G}}(e))$, called the *k-invariant* of $\mathbb{G}$.

When $\mathbb{G}$ has the discrete topology, this is unambiguous, but when $\mathbb{G}$ is a Lie 2-group, one must be careful what kind of cohomology is used here. The correct notion of cohomology is the *Segal-Mitchison cohomology* [**Seg70, Seg75**] of $\pi_0(\mathbb{G})$ valued in the abelian Lie group $\mathrm{Aut}_{\mathbb{G}}(e)$, as shown by Schommer-Pries [**SP11**, Theorem 1].

Now we want to discuss principal $\mathbb{G}$-bundles. The idea is that if G is a group, a principal G-bundle is a submersion which is locally trivial, and whose fibers are G-torsors. For a Lie 2-group $\mathbb{G}$, we need the fibers to locally look like $\mathbb{G}$, meaning they must be categorified somehow.

DEFINITION 1.18 ([**Bar06**], [**NW13**, Definition 6.1.5]). Let $\mathbb{G}$ be a Lie 2-group. A *principal* $\mathbb{G}$*-bundle* over a smooth manifold M is a Lie groupoid $\mathcal{P}$ with a surjective submersion $\mathrm{obj}(P) \to M$ and a smooth right action ρ of $\mathbb{G}$ on $\mathcal{P}$ such that the map

$$(\mathrm{pr}_1, \rho)\colon \mathcal{P} \times \mathbb{G} \longrightarrow \mathcal{P} \times_M \mathcal{P} \tag{1.19}$$

is a weak equivalence of Lie groupoids.

See Nikolaus and Waldorf [**NW13**, §6] for more details. The principal $\mathbb{G}$-bundles on a manifold M form a 2-groupoid $\mathcal{B}un_{\mathbb{G}}(X)$ [**NW13**, Theorem 6.2.1].

DEFINITION 1.20. Let $\mathbb{G}$ be a 2-group, and let $C_{\mathbb{G}}$ be the bicategory with a single object $*$ and morphism category $\mathrm{Hom}_{C_{\mathbb{G}}}(*, *) := \mathbb{G}$. The *classifying space* of $\mathbb{G}$, denoted $B\mathbb{G}$, is the geometric realization of the nerve of $C_{\mathbb{G}}$.[6]

When $\mathbb{G}$ is a Lie 2-group, we make the same definition. This time $C_{\mathbb{G}}$ is a topological bicategory, so its nerve is a simplicial space, and geometrically realizing, we obtain the space $B\mathbb{G}$.

THEOREM 1.21 ([**NW13**, Theorems 4.6, 5.3.2, 7.1]). *If $\mathbb{G}$ is a strict Lie 2-group, then there is a natural equivalence* $[X, B\mathbb{G}] \xrightarrow{\cong} \pi_0(\mathcal{B}un_{\mathbb{G}}(X))$.

Nikolaus and Waldorf's proof builds on Baez and Stevenson's related but distinct characterization of $[X, B\mathbb{G}]$ [**BS09**, Theorem 1] in terms of nonabelian Čech cohomology.

When G is an ordinary group, if G has the discrete topology, BG has only one nonzero homotopy group, which is $\pi_1(BG) = G$; likewise if $\mathbb{G}$ is a discrete 2-group, $\pi_i(B\mathbb{G})$ is nontrivial only for $i = 1, 2$; $\pi_1(B\mathbb{G}) = \pi_0(\mathbb{G})$ and $\pi_2(B\mathbb{G}) = \mathrm{Aut}_{\mathbb{G}}(e)$.

[6] There are many different definitions of the nerve of a bicategory; the fact that their geometric realizations are canonically homotopy equivalent is a theorem of Carrasco, Cegarra, and Garzón [**CCG10**], allowing us to speak about $B\mathbb{G}$ without specifying which kind of bicategorical nerve to use.

When $\mathbb{G}$ is a Lie 2-group, we have no control over its homotopy groups in general, just like BG when G is positive-dimensional.

If $\mathbb{G}$ has the discrete topology, the data classifying (1.17), namely the action of $\pi_0(\mathbb{G})$ on $\mathrm{Aut}_{\mathbb{G}}(e)$ and the k-invariant, is equivalent to the Postnikov data of $B\mathbb{G}$, worked out by MacLane and Whitehead [**MLW50**]: this data classifies fibrations over BG with fiber the Eilenberg–MacLane space $K(\mathrm{Aut}_{\mathbb{G}}(e), 2)$. The total space of the fibration with this Postnikov data is homotopy equivalent to $B\mathbb{G}$.

EXAMPLE 1.22. Let G be a compact Lie group; then, the Segal–Mitchison cohomology group $H^3_{\mathrm{SM}}(G;\mathbb{T})$ classifying Lie 2-group extensions of G by $\mathbb{T}[1]$ is naturally isomorphic to $H^4(BG;\mathbb{Z})$ [**SP11**, Corollary 97]. Therefore given a class $\mu \in H^4(BG;\mathbb{Z})$, we obtain a Lie 2-group $\mathcal{S}tr(G,\mu)$ fitting into a central extension

$$0 \longrightarrow \mathbb{T}[1] \longrightarrow \mathcal{S}tr(G,\mu) \longrightarrow G \longrightarrow 0, \tag{1.23}$$

which is sometimes called the *string 2-group* or *string cover* associated to G and λ. Of all the string covers, the most commonly studied one is $\mathrm{String}_n := \mathcal{S}tr(\mathrm{Spin}_n, \lambda)$, which is called *the* string 2-group.

This class of 2-groups was first studied by Baez and Lauda [**BL04**, §8.5].

The sequence (1.23) implies that upon taking classifying spaces,

$$B\mathbb{G} \longrightarrow BG \xrightarrow{\ \mu\ } K(\mathbb{Z},4) \tag{1.24}$$

is a fibration.

REMARK 1.25. Theorem 1.21 classified principal $\mathbb{G}$-bundles when $\mathbb{G}$ is a strict 2-group, but it is a theorem of Baez and Lauda [**BL04**, Corollary 60] that there is no strict Lie 2-group model for $\mathcal{S}tr(G,\mu)$ when G is simply connected and $\mu \neq 0$. However, there is a fix: in the setting of *Fréchet Lie 2-groups*, where we allow the spaces of objects and morphisms of $\mathbb{G}$ to be Fréchet manifolds, there is a strict model for $\mathcal{S}tr(G,\mu)$ [**BSCS07, LW23**], so $B\mathcal{S}tr(G,\mu)$ actually classifies principal $\mathcal{S}tr(G,\mu)$-bundles. This suffices for studying bordism groups.

Following [**SSS12**], we will now relate $\mathcal{S}tr(G,\mu)$ to the twisted Bianchi identity for G and μ. To do so, we use the language of stacks and differential cohomology, following [**HS05, FH13, Sch13, BNV16, ADH21**]. Make the category $\mathcal{M}an$ into a site by defining the covers to be surjective submersions, and define a *stack* to be a functor of ∞-categories $\mathcal{M}an^{op} \to \mathcal{T}op$ which satisfies descent for hypercovers. This defines a presentable ∞-category $\mathcal{S}t$ of stacks [**Lur09**, Proposition 6.5.2.14], and the Yoneda embedding $h\colon \mathcal{M}an \to \mathcal{S}t$ embeds $\mathcal{M}an$ as a full subcategory. We will often simply write M for the stack $h(M)$; we never compare these two notions directly, so this will not introduce confusion.

For any space X, the functor $\mathrm{Map}(-,X)\colon \mathcal{M}an \to \mathcal{T}op$ is a sheaf, and this procedure defines a functor of ∞-categories $\Gamma^*\colon \mathcal{T}op \to \mathcal{S}t$. The values of the stacks produced by Γ^* evaluated on manifolds M are homotopy-invariant in M. Γ^* has a left adjoint $\Gamma_\sharp\colon \mathcal{S}t \to \mathcal{T}op$ (see [**Dug01**, Proposition 8.3], [**MV99**, Proposition 3.3.3], and [**ADH21**, Proposition 4.3.1]); $\Gamma_\sharp(\mathbf{X})$ for a stack $\mathbf{X}$ can be thought of as the best approximation to $\mathbf{X}$ by a stack whose values on manifolds are homotopy-invariant.

Let $\Delta^n_{\mathrm{alg}} := \{(t_0,\dots,t_n) \mid t_0+\dots+t_n = 1\} \subset \mathbb{R}^{n+1}$. These "algebraic n-simplices" assemble into a cosimplicial manifold $\Delta^\bullet_{\mathrm{alg}}$, and [**ADH21**, Corollary

5.1.4] there is a natural homotopy equivalence $\Gamma_\sharp(\mathbf{X}) \simeq |\mathbf{X}(\Delta^\bullet_{\mathrm{alg}})|$, where as usual $|{-}|$ denotes geometric realization.

Thus, for a manifold M, there is a natural homotopy equivalence $\Gamma_\sharp(M) \xrightarrow{\simeq} M$, so a map $M \to \mathbf{X}$ naturally induces a map $M \to \Gamma_\sharp(\mathbf{X})$.

LEMMA 1.26. *Suppose $\mathbf{X} \to \mathbf{Y} \leftarrow \mathbf{Z}$ is a diagram in $\mathcal{S}t$, and that $\mathbf{Y}(\Delta^n_{\mathrm{alg}})$ and $\mathbf{Z}(\Delta^n_{\mathrm{alg}})$ are connected for all n. Then*

$$\Gamma_\sharp(\mathbf{X} \times_{\mathbf{Y}} \mathbf{Z}) \simeq \Gamma_\sharp(\mathbf{X}) \times_{\Gamma_\sharp(\mathbf{Y})} \Gamma_\sharp(\mathbf{Z}). \tag{1.27}$$

PROOF. Pullbacks of sheaves can be computed pointwise, then sheafifying, so given a pullback $\mathbf{X} \to \mathbf{Y} \leftarrow \mathbf{Z}$ in $\mathcal{S}t$, for each $n \geq 0$ the pullback of

$$\mathbf{X}(\Delta^n_{\mathrm{alg}}) \longrightarrow \mathbf{Y}(\Delta^n_{\mathrm{alg}}) \longleftarrow \mathbf{Z}(\Delta^n_{\mathrm{alg}}) \tag{1.28}$$

is $(\mathbf{X} \times_{\mathbf{Y}} \mathbf{Z})(\Delta^n_{\mathrm{alg}})$. The Bousfield–Friedlander theorem [**BF78**, **Bou01**] implies that, given the hypotheses on $\mathbf{Y}$ and $\mathbf{Z}$ in the theorem statement, the homotopy pullback of the geometric realizations of $\mathbf{X}$, $\mathbf{Y}$, and $\mathbf{Z}$ is the geometric realization of the levelwise homotopy pullback (1.28) (see [**War20**, p. 14-9] for this specific consequence of the Bousfield–Friedlander theorem). □

EXAMPLE 1.29. For G a Lie group, there is a stack $B_\nabla G$ whose value on a manifold M is the geometric realization of the nerve of the groupoid of principal G-bundles on M with connection [**FH13**]. This object is denoted $\mathbf{B}G_{\mathrm{conn}}$ in [**FSS12**, **SSS12**, **Sch13**], $\mathbb{B}G^\nabla$ in [**BNV16**, §5], and Bun_G^∇ in [**ADH21**].

There is a natural homotopy equivalence $\Gamma_\sharp(B_\nabla G) \xrightarrow{\simeq} BG$ [**ADH21**, Corollary 13.3.29], which can be interpreted as forgetting from a principal bundle with connection to a principal bundle.

EXAMPLE 1.30. For $k \geq 0$, there is a stack $B^k_\nabla \mathbb{T}$ whose value on a manifold M is the geometric realization of the nerve of the ∞-groupoid of cocycles for the differential cohomology group $\check{H}^{k+1}(M;\mathbb{Z})$. This object is studied in [**FSS12**, **SSS12**, **Sch13**], where it is denoted $\mathbf{B}^k U(1)_{\mathrm{conn}}$.

LEMMA 1.31. *There is a homotopy equivalence $\Gamma_\sharp(B^k_\nabla \mathbb{T}) \simeq K(\mathbb{Z}, k+1)$.*

PROOF. Schreiber [**Sch13**, Observation 1.2.134] produces the following pullback square in $\mathcal{S}t$:

$$\begin{array}{ccc} B^k_\nabla \mathbb{T} & \longrightarrow & \Omega^{k+1}_{c\ell} \\ \downarrow & \lrcorner & \downarrow \\ K(\mathbb{Z}, k+1) & \longrightarrow & K(\mathbb{R}, k+1), \end{array} \tag{1.32}$$

where $\Omega^{k+1}_{c\ell}$ is the stack of closed $(k+1)$-forms. For that stack and $K(\mathbb{R}, n+1)$, the values on each Δ^n_{alg} are connected spaces, so Lemma 1.26 identifies $\Gamma_\sharp(B^k_\nabla \mathbb{T}) \simeq K(\mathbb{Z}, k+1) \times_{K(\mathbb{R},k+1)} \Gamma_\sharp(\Omega^{k+1}_{c\ell})$. To finish, observe that, essentially by the de Rham theorem, the map $\Omega^{k+1}_{c\ell} \to K(\mathbb{R}, k+1)$ passes to a homotopy equivalence after applying $\Gamma_\sharp$. This follows from [**BNV16**, Lemma 7.15] together with the Dold–Kan theorem. □

These stacks are the universal setting for the Chern–Weil map.

THEOREM 1.33 ([**CS85**], [**BNV16**]). *Let G be a compact Lie group and $c \in H^k(BG;\mathbb{Z})$, where k is even. Then there is a map $\check{c}\colon B_\nabla G \to B_\nabla^{k-1}\mathbb{T}$ natural in (G,c) such that for any manifold M and map $f\colon M \to B_\nabla G$, interpreted as a principal G-bundle $P \to M$ with connection Θ,*

(1) *if $char\colon \check{H}^*(-;\mathbb{Z}) \to H^*(-;\mathbb{Z})$ denotes the characteristic class map, then $char(\check{c}\circ f) = c(P)$, and*
(2) *if $curv\colon \check{H}^*(-;\mathbb{Z}) \to \Omega^*_{c\ell}$ denotes the curvature map, then $curv(\check{c}\circ f) = \mathrm{CW}_c(\Theta)$.*

Cheeger and Simons lifted the Chern–Weil map to differential cohomology; Bunke, Nikolaus, and Völkl recast it in terms of $B_\nabla G$. The map *char* in Theorem 1.33 is the map down the left of the square (1.32); *curv* is the map across the top of (1.32).

DEFINITION 1.34 ([**FSS12**, §6.2]). Given a compact Lie group G and a class $\mu \in H^4(BG;\mathbb{Z})$, let $\mathbf{BStr}(G,\mu)$ denote the fiber of the map $\check{\mu}\colon B_\nabla G \to B_\nabla^3\mathbb{T}$.

We will see momentarily that maps to $\mathbf{BStr}(G,\mu)$ lead to solutions to the twisted Bianchi identity for G and μ.

PROPOSITION 1.35. *There is a natural homotopy equivalence $\Gamma_\sharp(\mathbf{BStr}(G,\mu)) \simeq B\mathcal{S}tr(G,\mu)$.*

For this reason we think of $\mathbf{BStr}(G,\mu)$ as the classifying stack of principal $\mathcal{S}tr(G,\mu)$-bundles with connection, though this is only a heuristic.[7]

PROOF. Apply Lemma 1.26 to the diagram

$$B_\nabla G \xrightarrow{\check{\mu}} B_\nabla^3\mathbb{T} \longleftarrow *, \tag{1.36}$$

as the values of both $*$ and $B_\nabla^3\mathbb{T}$ are connected on Δ^n_{alg} for each n. This implies that $\Gamma_\sharp(\mathbf{BStr}(G,\mu))$ is the fiber of $\mu\colon BG \to K(\mathbb{Z},4)$, which we identified with $B\mathbb{G}$ in (1.24). □

PROPOSITION 1.37 ([**FSS12**, §6.3]). *Let G be a compact Lie group, $U \subset \mathbb{R}^n$ be an open set, and $P \to U$ be a principal G-bundle with connection Θ. A lift of the corresponding map $f_{P,\Theta}\colon U \to B_\nabla G$ to a map $\widetilde{f}_{P,\Theta}\colon U \to \mathbf{BStr}(G,\mu)$ induces a form $H \in \Omega^3(U)$ such that H and Θ satisfy the twisted Bianchi identity* (1.12).

The idea here is that we have specified a trivialization of the differential characteristic class $\check{\mu}(P,\theta)$. Applying the curvature map $curv\colon B_\nabla^3\mathbb{T} \to \Omega^4_{c\ell}$, we have also specified a trivialization of $\mathrm{CW}_\mu(\Theta)$, which locally is the data H showing that $\mathrm{CW}_\mu(\Theta)$ is exact.

A map to $\mathbf{BStr}(G,\mu)$ is more data than what we get from Proposition 1.37, as we have trivialized not just the Chern–Weil form, but also the differential characteristic class. This can be interpreted as saying the data H specifying the trivialization is quantized to form a twisted version of a gerbe with connection.

To summarize, given a map $M \to \mathbf{BStr}(G,\mu)$, the stack which we think of as modeling $\mathcal{S}tr(G,\mu)$-bundles with connection, we obtain:

(1) a principal $\mathcal{S}tr(G,\mu)$-bundle $\mathcal{P} \to M$ by Proposition 1.35, and

[7] There are at least *five* notions of a connection on principal $\mathbb{G}$-bundles for $\mathbb{G}$ a 2-group: three are discussed by Waldorf [**Wal18**, §5], a fourth by Rist, Saemann, and Wolf [**RSW22**], and a fifth, defined only for $\mathbb{G} = \mathrm{String}_n$, by Waldorf [**Wal13**].

(2) a "twisted gerbe with connection," i.e., local data of a gerbe $Q \to M$ such that Ω_Q and the G-connection Θ induced by the map $\mathbf{BStr}(G,\mu) \to B_\nabla G$ satisfy the twisted Bianchi identity (1.12) by Proposition 1.37.

Motivated by this, we define of the tangential structure for the $\mathrm{E}_8 \times \mathrm{E}_8$ heterotic string. This first appears in [**SSS12**, §3.2], with [**Sat11b, FSS15a**] considering some related examples.

DEFINITION 1.38. Let $G := (\mathrm{E}_8 \times \mathrm{E}_8) \rtimes \mathbb{Z}/2$. A *differential* ξ_n^{het}*-structure* on a manifold M is the following data:

(1) a Riemannian metric and spin structure on M,
(2) a principal G-bundle $P \to M$ with connection Θ, and
(3) a lift of

$$((B_{\mathrm{Spin}}(M), \Theta^{\mathrm{LC}}), (P,\Theta))\colon M \longrightarrow B_\nabla(\mathrm{Spin}_n \times G) \tag{1.39}$$

to a map $M \to \mathbf{BStr}(\mathrm{Spin}_n \times G, c_1 + c_2 - \lambda)$.

Here $B_{\mathrm{Spin}}(M) \to M$ is the principal Spin_n-bundle of frames of M, and Θ^{LC} denotes its Levi–Civita connection.

For bordism groups we want the topological version of this.

DEFINITION 1.40. A *tangential structure* is a space B and a map $\xi\colon B \to BO$. Given a tangential structure ξ, a ξ*-structure* on a virtual vector bundle $E \to X$ is a lift of the classifying map $f_E\colon X \to BO$ to a map $\widetilde{f}_E\colon X \to B$ such that $\xi\circ\widetilde{f}_E = f_E$. A ξ*-structure* on a manifold M is a ξ-structure on its tangent bundle.

We make the analogous definition with maps $\xi_n\colon B_n \to BO_n$; in this case, we only refer to ξ_n-structures on n-manifolds.

Lashof [**Las63**] defined bordism groups Ω_*^ξ of manifolds with ξ-structure, and Boardman [**Boa65**, §V.1] defined a Thom spectrum $MT\xi$ whose homotopy groups are naturally isomorphic to Ω_*^ξ via the Pontrjagin–Thom construction.[8] We think of the category of tangential structures as the category of spaces over BO, and bordism groups and Thom spectra are functorial in this category. That is, taking bordism groups and Thom spectra is functorial as long as one commutes with the map down to BO.

The following definition is a special case of a definition due to [**SSS12**, Definition 2.8]. See [**Sat11b, FSS15a, FSS21**] for other related examples.

DEFINITION 1.41. Let $G_n := \mathrm{Spin}_n \times (\mathrm{E}_8 \times \mathrm{E}_8) \rtimes \mathbb{Z}/2$ and

$$\mathbb{G}_n^{\mathrm{het}} := \mathcal{S}tr(G_n, c_1 + c_2 - \lambda). \tag{1.42}$$

The $\mathrm{E}_8 \times \mathrm{E}_8$ *heterotic tangential structure* is the tangential structure

$$\xi_n^{\mathrm{het}}\colon B\mathbb{G}_n^{\mathrm{het}} \longrightarrow B\mathrm{Spin}_n \longrightarrow BO_n, \tag{1.43}$$

[8]In homotopy theory, it is common to study the Thom spectra $M\xi$ representing ξ-structures on the stable *normal* bundle ν_M of a manifold M, and indeed many of the results we cite about $MTSO$, $MTString$, etc., are stated for MSO, $MString$, etc., or about Thom spectra $M\xi$ in general. This is not a problem: for any tangential structure ξ, there is a tangential structure $\xi^\perp$ such that a ξ-structure on TM is equivalent data to a $\xi^\perp$-structure on ν_M and vice versa, so that $MT\xi \simeq M\xi^\perp$, so the general theory is the same. And for $\xi = \mathrm{O}$, SO, Spin, Spin^c, and String, $\xi \simeq \xi^\perp$ and in those cases we can ignore the difference between $M\xi$ and $MT\xi$.

where the first map comes from the quotient of $\mathbb{G}^{\mathrm{het}}$ by $\mathbb{T}[1]$, followed by projection onto the Spin_n factor in G_n. We also define $\mathbb{G}^{\mathrm{het}}$ and ξ^{het} analogously by stabilizing in n.

In other words: a differential ξ_n^{het}-structure is a lift of a map to $B_\nabla(\mathrm{Spin}\times G)$ to $\mathbf{BStr}(G,\mu)$; by Proposition 1.35, a topological ξ_n^{het}-structure is the image of this data under $\Gamma_\sharp$. In particular, a ξ_n^{het}-structure on an n-manifold M includes data of a principal $\mathbb{G}_n^{\mathrm{het}}$-bundle $\mathcal{P}\to M$.

Taking the quotient of $\mathbb{G}^{\mathrm{het}}$ by $\mathbb{T}[1]$ induces a map of tangential structures

$$\phi\colon B\mathbb{G}^{\mathrm{het}} \longrightarrow B\mathrm{Spin}\times B(\mathrm{E}_8^2\rtimes\mathbb{Z}/2). \tag{1.44}$$

Thus, much like a spin^c manifold M has an associated $\mathbb{T}$-bundle P with $c_1(P)$ mod $2 = w_2(M)$, a ξ^{het}-manifold has associated $(\mathrm{E}_8^2\rtimes\mathbb{Z}/2)$-bundle P. From this perspective, a ξ^{het}-structure on a manifold M is the following data:

- a spin structure on M,
- a double cover $\pi\colon \widetilde{M}\to M$,
- two principal E_8-bundles $P,Q\to\widetilde{M}$ which are exchanged by the nonidentity deck transformation of π, and
- a trivialization of the class $\lambda(M)-(c(P)+c(Q))\in H^4(M;\mathbb{Z})$.

By a *trivialization* of a cohomology class $\alpha\in H^n(M;A)$ we mean a null-homotopy of the classifying map $f_\alpha\colon M\to K(A,n)$. Thus orientations are identified with trivializations of w_1, etc. To make the trivialization of $\lambda(M)-(c(P)+c(Q))$ precise, we have to descend the class $c(P)+c(Q)$, a priori an element of $H^4(\widetilde{M};\mathbb{Z})$, to $H^4(M;\mathbb{Z})$. We can do this because, as noted in Definition 1.4, the class c_1+c_2 descends through the Serre spectral sequence to the base.

REMARK 1.45. We can combine some the data of a ξ^{het} structure on M into a twisted characteristic class. Let $\mathbb{Z}^\sigma$ be the $\mathbb{Z}[\mathbb{Z}/2]$-module isomorphic to $\mathbb{Z}^2$ as an abelian group, and in which the nontrivial element of $\mathbb{Z}/2$ swaps the two factors. Then, let $\mathbb{Z}_\pi^\sigma$ denote the local system on M which is the associated bundle $\widetilde{M}\times_{\mathbb{Z}/2}\mathbb{Z}^\sigma$. A pair of classes $x,y\in H^k(\widetilde{M};\mathbb{Z})$ exchanged by the deck transformation thus define a class in $H^k(M;\mathbb{Z}_\pi^\sigma)$, so, the classes $c(P)$ and $c(Q)$ in $H^4(\widetilde{M};\mathbb{Z})$ together define a class $\widetilde{c}(P,Q)\in H^4(M;\mathbb{Z}_\pi^\sigma)$, which is a characteristic class of an $((\mathrm{E}_8\times\mathrm{E}_8)\rtimes\mathbb{Z}/2)$-bundle.

If $\mathbb{Z}$ denotes the $\mathbb{Z}[\mathbb{Z}/2]$-module isomorphic to $\mathbb{Z}$ as an abelian group and with trivial $\mathbb{Z}/2$-action, then taking the quotient of $\mathbb{Z}^\sigma$ by the submodule generated by $(1,-1)$ defines a map of $\mathbb{Z}[\mathbb{Z}/2]$-modules $q\colon\mathbb{Z}^\sigma\to\mathbb{Z}$, hence also a map between the corresponding twisted cohomology groups, and this map sends $\widetilde{c}(P,Q)\mapsto c(P)+c(Q)$. Therefore one could recast a ξ^{het}-structure on a spin manifold M as the data of a principal $((\mathrm{E}_8\times\mathrm{E}_8)\rtimes\mathbb{Z}/2)$-bundle (P,Q,π) together with a trivialization of $\lambda(M)-q(\widetilde{c}(P,Q))$.

Bott and Samelson [**BS58**, Theorems IV, V(e)] showed that the map $B\mathrm{E}_8\to K(\mathbb{Z},4)$ defined by the characteristic class c is 15-connected. This implies that up to isomorphism, a principal $((\mathrm{E}_8\times\mathrm{E}_8)\rtimes\mathbb{Z}/2)$-bundle on a manifold of dimension 15 or lower is equivalent data to its characteristic class $\widetilde{c}$.

REMARK 1.46. One might want to simplify by restricting to the special case where $\pi\colon\widetilde{M}\to M$ is trivial (as done in, e.g., [**Wit86**]), in which case the data of a ξ^{het}-structure simplifies to the data of a spin structure on M, two principal E_8-bundles $P, Q \to M$, and a trivialization of $\lambda(M) - c(P) - c(Q)$. This corresponds to the tangential structure $\xi^{r,\mathrm{het}}\colon B\mathcal{S}tr(\mathrm{Spin}\times \mathrm{E}_8\times\mathrm{E}_8, c_1+c_2-\lambda) \to B\mathrm{Spin} \to BO$.

1.3. The CHL string. Eleven-dimensional $\mathcal{N} = 1$ supergravity admits a time-reversal symmetry, allowing it to be defined on pin$^+$ 11-manifolds.[9] Therefore we can compactify it on a Möbius strip with certain boundary data to obtain a nine-dimensional supergravity theory; the goal of this subsection is to determine the tangential structure of this theory. Eleven-dimensional $\mathcal{N} = 1$ supergravity is expected to be the low-energy limit of a theory called M-theory,[10] and compactifying M-theory on the Möbius strip is expected to produce a string theory called the *Chaudhuri–Hockney–Lykken (CHL) string* [**CHL95**] whose low-energy limit is the 9-dimensional supergravity theory described above; we study the tangential structure of this supergravity theory in this subsection with the aim of also learning about the CHL string.

However, we do not want our perspective on the CHL string to be overly one-sided. Once and for all, choose a section $s\colon \mathbb{Z}/2 \hookrightarrow (\mathrm{E}_8\times\mathrm{E}_8)\rtimes\mathbb{Z}/2$ of the quotient $(\mathrm{E}_8\times\mathrm{E}_8)\rtimes\mathbb{Z}/2 \to \mathbb{Z}/2$ by the normal $\mathrm{E}_8\times\mathrm{E}_8$ subgroup. Then there is another way to produce the CHL string by compactifying: consider the circle with its nontrivial principal $\mathbb{Z}/2$-bundle $P \to S^1$. Via the map $\mathbb{Z}/2 \hookrightarrow \mathrm{Spin}\times((\mathrm{E}_8\times\mathrm{E}_8)\rtimes\mathbb{Z}/2)$ sending $1 \mapsto (\mathrm{id}, s(1))$, this bundle defines a $\mathrm{Spin}\times((\mathrm{E}_8\times\mathrm{E}_8)\rtimes\mathbb{Z}/2)$-structure on S^1 for which λ and c_1+c_2 are both trivial, so this structure lifts to define a ξ^{het}-structure on S^1. We will call the circle with this ξ^{het}-structure $\mathbb{RP}^1$, as $S^1\cong\mathbb{RP}^1$ as manifolds and the ξ^{het}-structure comes from the double cover $S^1\to\mathbb{RP}^1$. The CHL string is precisely what one obtains by compactifying the E_8^2 heterotic string on $\mathbb{RP}^1$.

We want to determine the tangential structure ξ^{CHL} such that the product of $\mathbb{RP}^1$ with a manifold with ξ^{CHL}-structure has an induced ξ^{het}-structure. In general, keeping track of how the tangential structure changes under compactification can be subtle; for a careful analysis, see [**SP18**, §9]. But for the CHL string, we can get away with a more ad hoc approach: following [**CP95**] (see also [**dBDH$^+$00**, §2.2.1]) we restrict to the case where the principal $\mathbb{Z}/2$-bundle on $\mathbb{RP}^1\times M$ obtained by the quotient map (1.44) is the pullback of the Möbius bundle $S^1\to\mathbb{RP}^1$ along the projection $\mathrm{pr}_1\colon\mathbb{RP}^1\times M\to\mathbb{RP}^1$.

PROPOSITION 1.47. *Let M be a spin manifold and $P\to M$ be a principal E_8-bundle. The data of a trivialization $\mathfrak{s}$ of $\lambda(M)-2c(P)$ induces a ξ^{het}-structure on $\mathbb{RP}^1\times M$ whose associated principal $\mathbb{Z}/2$-bundle is the Möbius bundle $S^1\times M\to\mathbb{RP}^1\times M$. Moreover, if $\dim(M)\le 14$, this assignment is a natural bijection from the set of isomorphism classes of data $(P,\mathfrak{s})$ to the set of ξ^{het}-structures on $\mathbb{RP}^1\times M$ whose associated $\mathbb{Z}/2$-bundle is $S^1\times M\to\mathbb{RP}^1\times M$.*

PROOF. Let $\pi\colon S^1\times M\to M$ be the projection onto the second factor. Given $P\to M$ and $\mathfrak{s}$, the pair of E_8-bundles $(\pi^*P,\pi^*P)\to S^1\times M$ are exchanged by the deck transformation for $S^1\times M\to\mathbb{RP}^1\times M$, and (c_1+c_2) evaluated on the pair (π^*P,π^*P) is $2c(P)\in H^4(\mathbb{RP}^1\times M;\mathbb{Z}/2)$. Choosing the string structure on

[9]In addition to the pin$^+$ structure, one needs the additional data of a lift of $w_4(TM)$ to $w_1(TM)$-twisted integral cohomology. See [**Wit97, Wit16, FH21a**].

[10]M-theory is expected to require additional data on top of the tangential structure described above for 11-dimensional $\mathcal{N}=1$ supergravity. See [**FSS20**, Table 1] and the references listed there.

$\mathbb{RP}^1$ induced from the bounding framing, we obtain a canonical trivialization of $\lambda(\mathbb{RP}^1 \times M) - \lambda(M) \in H^4(\mathbb{RP}^1 \times M; \mathbb{Z})$ from the two-out-of-three property of string structures. Putting all of this together, we see that we have data of two E_8-bundles on $S^1 \times M$ exchanged by the deck transformation, and a trivialization of $\lambda - (c_1 + c_2)$ on $\mathbb{RP}^1 \times M$, thus defining a ξ^{het}-structure as claimed.

To see that this produces all ξ^{het}-structures associated with $S^1 \times M \to \mathbb{RP}^1 \times M$, recall from Remark 1.45 that the $((\mathrm{E}_8 \times \mathrm{E}_8) \rtimes \mathbb{Z}/2)$-bundle associated to a ξ^{het}-structure is classified by a characteristic class in twisted cohomology. The assumption that the associated $\mathbb{Z}/2$-bundle is $S^1 \times M \to \mathbb{RP}^1 \times M$ implies this class belongs to $H^4(\mathbb{RP}^1 \times M; \underline{\mathbb{Z} \oplus \mathbb{Z}})$, where a generator of $\pi_1(\mathbb{RP}^1)$ acts on $\underline{\mathbb{Z} \oplus \mathbb{Z}}$ by swapping the two factors, and $\pi_1(M)$ acts trivially. The twisted Künneth formula [**Gre06**, Theorem 1.7] gives us an isomorphism

$$H^4(\mathbb{RP}^1 \times M; \underline{\mathbb{Z} \oplus \mathbb{Z}}) \xrightarrow{\cong} H^4(M; \mathbb{Z}), \tag{1.48}$$

meaning that the pair of E_8-bundles on the orientation double cover $S^1 \times M$ pull back from bundles on M, which must be isomorphic in order to be exchanged by the $\mathbb{Z}/2$-action. ☐

The Bianchi identity corresponding to this data can therefore be simplified to use a single bundle $P \to M$ and the class $c(P) + c(P)$: we obtain

$$\mathrm{d}H = \mathrm{CW}_{2c}(\Theta_P) - \mathrm{CW}_\lambda(\Theta^{\mathrm{LC}}), \tag{1.49}$$

i.e., the twisted Bianchi identity for $G = \mathrm{Spin} \times \mathrm{E}_8$ and $\mu = 2c - \lambda$. Then, following Definitions 1.38 and 1.41, we make the following definitions.

DEFINITION 1.50. A *differential* ξ_n^{CHL}*-structure* on a manifold M is the following data:

(1) a Riemannian metric and spin structure on M,
(2) a principal E_8-bundle $P \to M$ with connection Θ, and
(3) a lift of

$$((B_{\mathrm{Spin}}(M), \Theta^{\mathrm{LC}}), (P, \Theta)) \colon M \longrightarrow B_\nabla(\mathrm{Spin}_n \times \mathrm{E}_8) \tag{1.51}$$

to a map $M \to \mathbf{BStr}(\mathrm{Spin}_n \times \mathrm{E}_8, 2c - \lambda)$.

What we call $B\mathbb{G}_n^{\mathrm{CHL}}$ coincides with what Sati, Schreiber, and Stasheff call $B\mathrm{String}^{2a}$ [**SSS12**, (2.18), §2.3.3] and which also appears in [**FSS15a**, Remark 4.1.1], though those papers do not discuss its relationship with the CHL string.

DEFINITION 1.52 ([**SSS12**, (2.18), §2.3.3]). Let

$$\mathbb{G}_n^{\mathrm{CHL}} := \mathcal{S}tr(\mathrm{Spin}_n \times \mathrm{E}_8, 2c - \lambda). \tag{1.53}$$

The *CHL tangential structure* is the tangential structure

$$\xi_n^{\mathrm{CHL}} \colon B\mathbb{G}_n^{\mathrm{CHL}} \longrightarrow B\mathrm{Spin}_n \longrightarrow B\mathrm{O}_n, \tag{1.54}$$

where the first map comes from the quotient of $\mathbb{G}^{\mathrm{CHL}}$ by $\mathbb{T}[1]$, followed by projection onto the Spin_n factor. Stabilizing in n, we also obtain $\mathbb{G}^{\mathrm{CHL}}$ and a tangential structure ξ^{CHL}.

A ξ^{CHL}-structure on an n-manifold M in particular comes with data of a principal $\mathbb{G}_n^{\mathrm{CHL}}$-bundle $\mathcal{P} \to M$, and can be formulated as the data of a principal E_8-bundle $P \to M$ and a trivialization of $\lambda(M) - 2c(P) \in H^4(M; \mathbb{Z})$.

REMARK 1.55. Since a ξ^{CHL} structure includes data identifying λ as twice another class, it induces a trivialization of the mod 2 reduction of λ, which is w_4. That is, a ξ^{CHL} structure induces a $\mathrm{Spin}\langle w_4\rangle$ structure, where $B\mathrm{Spin}\langle w_4\rangle$ is the homotopy fiber of $w_4\colon B\mathrm{Spin}\to K(\mathbb{Z},4)$. This structure has been studied in, e.g., [**Wit97, KS04, FH21a**] for applications to M-theory.

REMARK 1.56 (Variation of the tangential structure along the moduli space). There is a moduli space of CHL string theories, not just one, and the gauge group depends on where in the moduli space one is; this moduli space was first studied by Chaudhuri and Polchinski [**CP95**]. At a generic point, the gauge group is broken to $\mathbb{T}^8$, and at various special points the gauge group enhances to E_8 or other nonabelian groups: see [**FFG$^+$21**, Table 3]. We work only at the E_8 point of the moduli space in this paper; it would be interesting to apply the techniques in this paper to other points in the CHL moduli space.

There has been quite a bit of recent research studying the moduli spaces of compactifications of the $\mathrm{E}_8\times\mathrm{E}_8$ heterotic string and the CHL string, and investigating which gauge groups can occur [**FGN18, CDLZ20, FFG$^+$20, CDLZ21, FFG$^+$21, FPDF21, MV21, CDLZ22, CGH22, CMM22, FPDF22, PDF23, MPDF23**].

2. Bordism computations

Now it is time to compute. We will use the Adams spectral sequence to compute $\Omega_*^{\xi^{\mathrm{het}}}$ and $\Omega_*^{\xi^{\mathrm{CHL}}}$; this is a standard tool in computational homotopy theory and more recently appears frequently in the mathematical physics literature, and we point the interested reader to Beaudry and Campbell's introductory article [**BC18**].

Applications of the Adams spectral sequence to mathematical physics questions tend to follow the same formula. Suppose that we want to compute Ω_*^ξ for some tangential structure ξ.

(1) First, express ξ as a "twisted ξ'-structure," where ξ' is one of SO, Spin, Spin^c, or String: prove that a ξ-structure on a vector bundle $E\to M$ is equivalent data to an auxiliary vector bundle $V\to M$ and a ξ'-structure on $E\oplus V$.

This implies that $MT\xi\simeq MT\xi'\wedge X$ for some Thom spectrum X that is usually not too complicated.

(2) Next, invoke a change-of-rings theorem to greatly simplify the calculation of the E_2-page for ξ'-bordism of spaces or spectra. Then run the Adams spectral sequence, taking advantage of the extra structure afforded by the change-of-rings theorem.

This recipe goes back to work of Anderson, Brown, and Peterson [**ABP69**] and Giambalvo [**Gia73b, Gia73a, Gia76**] computing twisted spin bordism. It is most commonly used in the case $\xi'=\mathrm{Spin}$, where it has been frequently used to compute bordism groups for tangential structures representing field theories with fermions; $\xi'=\mathrm{String}$ is less common but still appears in physically motivated examples, including the tangential structure of the Sugimoto string [**Sug99**] and $\xi=\mathrm{String}^c$ [**CHZ11, Sat11b**].

Unfortunately, ξ^{het} and ξ^{CHL} do not belong to this class of examples: we will see in Lemma 2.2 that there is no way to write these tangential structures as twisted string structures in the sense above.[11,12] So we have to do something different.

At odd primes, we plow ahead with the unsimplified Adams spectral sequence, though since we only care about dimensions 11 and below the computations are very tractable. At $p = 2$, though, we can modify the above strategy to simplify the computation: in §2.1, we generalize the notion of "twisted string bordism" for which the change-of-rings trick works to include string covers (in the sense of Example 1.22) of groups of the form $\mathrm{Spin} \times G$. This applies to both ξ^{het} and ξ^{CHL}, and so we are off to the races.

REMARK 2.1. We are far from the first to compute bordism groups for a tangential structure $\xi\colon B \to BO$ where B is the classifying space of a 2-group. For example, $\Omega_*^{\mathrm{String}}$ has been calculated in a range of degrees by [**Gia71, HR95, MG95, Hov08**]; other examples include [**Hil09, KT17, WW19a, WW19b, WWZ19, Tho20, LT21, Yu21, DL23**].

2.1. Twists of string bordism.

"Started out with a twist, how did it end up like this?
It was only a twist, it was only a twist..."

Once the tangential structure for a bordism question is known, the next step is typically to prove a "shearing" theorem simplifying the tangential structure. For example, the usual route to computing pin$^-$ bordism [**Pet68**, §7] first establishes an isomorphism between pin$^-$ bordism and the spin bordism of the Thom spectrum $\Sigma^{-1}MO_1$, and then computes the latter groups using something like the Adams or Atiyah–Hirzebruch spectral sequence.

There are a few different approaches to shearing theorems, such as those in [**FH21b, DDHM23**], but generally they work with Thom spectra of vector bundles; for example, the above simplification of pin$^-$ bordism begins with the observation that a pin$^-$ structure on a bundle $E \to M$ is equivalent data to a real line bundle $L \to M$ and a spin structure on $E \oplus L$, which follows from a characteristic class computation, and then passes the data of "L and a spin structure on $E \oplus L$" through the Pontrjagin–Thom theorem.

This approach does not work for the heterotic and CHL tangential structures.

LEMMA 2.2. *There is no spin vector bundle V on $B((\mathrm{E}_8 \times \mathrm{E}_8) \rtimes \mathbb{Z}/2)$ such that $\lambda(V) = c_1 + c_2$, and there is no spin vector bundle W on $B\mathrm{E}_8$ such that $\lambda(V) = 2c$.*

This means there is no way to express a ξ^{het}-structure as "a G-bundle and a string structure on E plus some associated bundle," and likewise for ξ^{CHL}.

PROOF. Let G be a compact, simple, simply connected Lie group and $\rho\colon G \to \mathrm{SU}_n$ be a representation. $H^4(BG;\mathbb{Z})$ and $H^4(B\mathrm{SU}_n;\mathbb{Z})$ are both canonically isomorphic to $\mathbb{Z}$, so the pullback map ρ^* on H^4 is a map $\mathbb{Z} \to \mathbb{Z}$, necessarily multiplication by some integer $\delta(\rho)$. Because SU_n is compact, connected, and simply connected, the standard inclusion $\mathrm{SU}_n \to \mathrm{GL}_{2n}(\mathbb{R})$ lifts to a map $\mathrm{SU}_n \to \mathrm{Spin}_{2n}$. Choices of this lift are a torsor over $H^1(B\mathrm{SU}_n;\mathbb{Z}/2) = 0$, meaning that the characteristic class λ is uniquely defined for SU_n-representations. Moreover, λ of the

[11]The presence of the B-field, and how the Bianchi identity mixes it with the principal Spin_n-bundle of frames, rules out $\xi' = \mathrm{SO}$, Spin, or Spin^c.

[12]This problem also happens to the tangential structures studied in [**FH21a, DY22**].

defining representation is a generator of $H^4(B\mathrm{SU}_n;\mathbb{Z})$; because $H^4(B\mathrm{SU}_n;\mathbb{Z})$ is torsion-free, it suffices to show $2\lambda = p_1$ is twice a generator, which is standard. The *Dynkin index* of G is the minimum value of $|\delta(\rho)|$ over all such representations ρ. Laszlo-Sorger [**LS97**, Proposition 2.6] show that the Dynkin index of E_8 is 60, meaning that for any vector bundle $V \to B\mathrm{E}_8$ with SU-structure induced from a representation, $\lambda(V)$ is at least 60 times a generator.

We would like to generalize to real representations.

LEMMA 2.3. *The complexification map* $\mathrm{Spin}_n \to \mathrm{O}_n \to \mathrm{U}_n$ *has image contained in* SU_n.

PROOF. A lift of a representation $\rho\colon G \to \mathrm{U}_n$ has image contained in SU_n if and only if c_1 of the complex vector bundle associated to ρ vanishes. When one pulls back across the complexification map $B\mathrm{O}_n \to B\mathrm{U}_n$, c_1 is sent to the image of w_1 under the Bockstein map $\beta\colon H^1(B\mathrm{O}_n;\mathbb{Z}/2) \to H^2(B\mathrm{O}_n;\mathbb{Z})$; when we pull back further to $B\mathrm{Spin}_n$, $w_1 \mapsto 0$, so $c_1 = \beta w_1 \mapsto 0$ too. □

Thus the Dynkin index fact we mentioned above applies to complexifications of representations landing in Spin_n.

If V is a real representation of a group G, $V \otimes \mathbb{C} \cong V \oplus V$ as real representations, so using the Whitney sum formula for λ (Lemma 1.6), $\lambda(V \otimes \mathbb{C}) = 2\lambda(V)$. Therefore if V is any real spin representation of E_8, $\lambda(V \otimes \mathbb{C})$ is at least 60 times a generator, so $\lambda(V)$ is at least 30 times a generator. Thus the class defining $\mathbb{G}^{\mathrm{CHL}}$, which is twice a generator, is not λ of any spin representation of E_8; likewise for $\mathbb{G}^{\mathrm{het}}$, as one could restrict to either factor of E_8 inside $\mathrm{E}_8^2 \rtimes \mathbb{Z}/2$ and obtain a representation with λ equal to the generator.

Finally, the Atiyah–Segal completion theorem extends this from representations to all vector bundles. Because λ is additive (Lemma 1.6), it factors through the Grothendieck group $KSpin(BG)$ of spin vector bundles on BG, and similarly, evaluated on spin representations, λ factors through the corresponding Grothendieck group $RSpin(G)$. Atiyah and Segal [**AS69**, §7, §8] show that taking the associated bundle of an arbitrary representation exhibits the Grothendieck ring $KO^0(BG)$ of all vector bundles on BG as the completion of the representation ring $RO(G)$ at its augmentation ideal. Thus given a $\mathbb{Z}$-valued characteristic class c of arbitrary vector bundles of G which satisfies the Whitney sum formula, passing from representations of G to vector bundles on BG does not decrease the minimal value of $|c|$.

In order to use the Atiyah–Segal theorem, we need to get from spin representations and vector bundles to arbitrary ones. We will do so, at the cost of lowering the minimum value of λ a little bit. For any vector bundle V, $V^{\oplus 4}$ admits a canonical spin structure: the Whitney sum formula for Stiefel–Whitney classes shows a spin structure exists; then choose a spin structure universally over BO. Therefore we can define λ of an arbitrary representation of E_8 or vector bundle on $B\mathrm{E}_8$ by $\lambda(V) := \frac{1}{4}\lambda(V^{\oplus 4})$, valued in $\frac{1}{4}\mathbb{Z}$. Therefore passing from $RO(\mathrm{E}_8) \to KO^0(B\mathrm{E}_8)$ to $RSpin(\mathrm{E}_8) \to KSpin(B\mathrm{E}_8)$ divides the minimal value of λ by at most 4, and now we can invoke Atiyah–Segal, so it is still not possible to get $2c$ and ξ^{CHL}; and likewise for $\mathrm{E}_8^2 \rtimes \mathbb{Z}/2$ in place of E_8 to show that the characteristic class for ξ^{het} cannot be achieved. □

So we take a different approach: we cannot get Thom spectra corresponding to vector bundles, but we can still obtain $MTString$-module Thom spectra. We accomplish this using the theory of Ando, Blumberg, Gepner, Hopkins, and Rezk [**ABG$^+$14a, ABG$^+$14b**] (ABGHR), which we briefly summarize.

The idea behind the ABGHR perspective on Thom spectra is to generalize the notion of local coefficients to generalized cohomology theories. Given a based, connected space X and a homomorphism $\rho\colon \pi_1(X) \to \mathrm{GL}_1(\mathbb{Z}) \cong \{\pm 1\}$, one obtains a local coefficient system $\mathbb{Z}_\rho$ on X: this is a bundle on X with fiber $\mathbb{Z}$, and whose monodromy around a loop $\gamma \in \pi_1(X)$ is precisely $\rho(\gamma)$. Given $\mathbb{Z}_\rho$, we can take twisted cohomology groups: if $\widetilde{X} \to X$ denotes the universal cover, then the cochain complex $C^*(\widetilde{X};\mathbb{Z})$ has a $\pi_1(X)$-action induced from the $\pi_1(X)$-action on $\widetilde{X}$. If $C^*(X;\mathbb{Z}_\rho)$ denotes the subcomplex of $C^*(\widetilde{X};\mathbb{Z})$ of cochains which transform under this $\pi_1(X)$-action by ρ, then $H^*(X;\mathbb{Z}_\rho) := H^*(C^*(X;\mathbb{Z}_\rho))$.

Another way to say this is that if pt/G denotes the category with one object $*$ and $\mathrm{Hom}(*,*) = G$, ρ defines a $\mathrm{pt}/\pi_1(X)$-shaped diagram of chain complexes of abelian groups:

$$\mathrm{pt}/\pi_1(X) \xrightarrow{\rho} \mathrm{pt}/\{\pm 1\} \longrightarrow \mathcal{C}h_{\mathbb{Z}}, \tag{2.4}$$

sending pt to $C^*(\widetilde{X};\mathbb{Z})$, and sending $g \in \pi_1(X)$ to the action by $\rho(g)$. The subcomplex of cochains that transform by ρ is precisely the limit of this diagram. For functoriality reasons, we envision this complex as cochains on some object $\mathcal{X}$ which is a *co*limit of a diagram akin to (2.4).

To summarize, twisted cohomology, i.e., cohomology of the Thom spectrum, is expressed as a colimit of a diagram of chain complexes of $\mathbb{Z}$-modules induced from a map $X \to B\mathrm{Aut}(\mathbb{Z})$. Ando, Blumberg, Gepner, Hopkins, and Rezk lift this to spectra. Specifically, given a ring spectrum R, the authors naturally associate a topological group[13] $\mathrm{GL}_1(R)$, thought of as the group of units or group of automorphisms of R. The classifying space $B\mathrm{GL}_1(R)$ carries the universal local system of R-lines; a local system of R-lines over X is equivalent data to a map $X \to B\mathrm{GL}_1(R)$.

DEFINITION 2.5 ([**ABG$^+$14a**, Definition 2.20]). The *Thom spectrum Mf* associated to a map $f\colon X \to B\mathrm{GL}_1(R)$ is the colimit of the diagram $X \to B\mathrm{GL}_1(R) \to \mathcal{M}od_R$, where we think of X as its fundamental ∞-groupoid.

When $R = \mathbb{S}$, this is due to Lewis [**LMSM86**, Chapter IX]. In Definition 2.5, we have to consider the fundamental ∞-groupoid, rather than just π_1, because R can have higher automorphisms, because spectra are derived objects.

The Thom spectrum of a map to $B\mathrm{GL}_1(R)$ is an R-module.

EXAMPLE 2.6 (Twisted ordinary cohomology). There is a homotopy equivalence $B\mathrm{GL}_1(H\mathbb{Z}) \simeq K(\mathbb{Z}/2,1)$, so the ABGHR viewpoint recovers $\mathrm{Aut}(\mathbb{Z})$ and the usual notion of cohomology twisted by a local system. To prove this homotopy equivalence, use the homotopy pullback square of E_∞-spaces [**ABG$^+$14b**, Definition 2.1]

$$\begin{array}{ccc} \mathrm{GL}_1(H\mathbb{Z}) & \longrightarrow & \Omega^\infty H\mathbb{Z} \\ \big\downarrow{\scriptstyle\varphi} & & \big\downarrow{\scriptstyle\psi} \\ (\pi_0(H\mathbb{Z}))^\times & \longrightarrow & \pi_0(H\mathbb{Z}). \end{array} \tag{2.7}$$

[13] $\mathrm{GL}_1(R)$ is not exactly a topological group, but the homotopy-coherent version thereof: a grouplike A_∞-space.

$\Omega^\infty H\mathbb{Z} \simeq \mathbb{Z}$ as E_∞-spaces, and ψ is a homotopy equivalence of E_∞-spaces. Therefore φ is also a homotopy equivalence of E_∞-spaces, and we conclude.

EXAMPLE 2.8 (Thom spectra from vector bundles). Boardman's original definition of Thom spectra [**Boa65**, §V.1] associates them to virtual vector bundles $V \to X$. Let us connect this to the ABGHR definition. Virtual vector bundles are classified by maps $f_V : X \to BO$, and one avatar of the J-homomorphism [**Whi42**] is a map $J\colon \mathrm{O} \to \mathrm{GL}_1(\mathbb{S})$ [**ABG10**, Example 3.15], which deloops to a map of spaces $BJ\colon BO \to B\mathrm{GL}_1(\mathbb{S})$. A map with this signature is a natural assignment from virtual vector bundles $V \to X$ to local systems of invertible $\mathbb{S}$-modules, and BJ assigns to V the local system with fiber $\mathbb{S}^{V_x}$ at each $x \in X$. Putting these maps together, we have an X-shaped diagram

$$X \xrightarrow{f_V} BO \xrightarrow{BJ} B\mathrm{GL}_1(\mathbb{S}) \longrightarrow \mathcal{S}p, \tag{2.9}$$

and the colimit of this diagram, which is a Thom spectrum in the ABGHR sense, coincides with the Thom spectrum X^V in the usual sense. This is a combination of theorems of Lewis [**LMSM86**, Chapter IX] and Ando et al. [**ABG⁺14a**, Corollary 3.24].

This approach to Thom spectra plays well with multiplicative structures. If R is an E_∞-ring spectrum, then the grouplike A_∞-structure on $\mathrm{GL}_1(R)$ refines to a grouplike E_∞-structure, making $\mathrm{GL}_1(R)$ and therefore $B\mathrm{GL}_1(R)$ into infinite loop spaces. For $1 \le k \le \infty$, if X is a k-fold loop space and $f\colon X \to B\mathrm{GL}_1(R)$ is a k-fold loop map, then the Thom spectrum Mf inherits the structure of an E_k-ring spectrum. This is a theorem of Lewis [**LMSM86**, Theorem IX.7.1] for $R = \mathbb{S}$ and Ando, Blumberg, and Gepner [**ABG18**, Theorem 1.7] for more general R.

BO has an infinite loop space structure coming from the addition-like operation on BO of direct sum of vector bundles. The J-homomorphism $BJ\colon BO \to B\mathrm{GL}_1(\mathbb{S})$ is an infinite loop map, so we get an E_∞-ring structure on $MT\xi$ if ξ is a tangential structure satisfying a *2-out-of-3 property*, i.e., whenever any two of E, F, and $E \oplus F$ have a ξ-structure, the third has an induced ξ-structure. The idea is that the 2-out-of-3 property implies that $\xi\colon B \to BO$ is an infinite loop map, so passing to $B\mathrm{GL}_1(\mathbb{S})$ and taking the Thom spectrum, we obtain an E_∞-ring spectrum. This applies to MTO, $MTSO$, $MTSpin^c$, $MTSpin$, and $MTString$; however, some commonly considered tangential structures appearing in physics do not have this property, including $B\mathrm{Pin}^\pm$.

PROPOSITION 2.10. *Let B and X be infinite loop spaces and $\xi\colon B \to BO$ and $f\colon B \to X$ be infinite loop maps, so that the fiber $\eta\colon F \to B$ of f is also a map of infinite loop spaces. This data naturally defines twists of the Thom spectrum $M(\xi\circ\eta)$ over X, i.e., a map $X \to B\mathrm{GL}_1(M(\xi\circ\eta))$.*

PROOF. The fiber of $\eta\colon F \to B$ is another infinite loop map $\zeta\colon \Omega X \to F$, so the induced map of Thom spectra (where the maps down to BO are $\xi\circ\eta\circ\zeta$ and $\xi\circ\eta$ respectively) is a map of E_∞-ring spectra. Because $\xi\circ\eta\circ\zeta$ is nullhomotopic, its Thom spectrum is a suspension spectrum, so we have a map of E_∞-ring spectra $\Sigma^\infty_+\Omega X \to M(\xi\circ\eta)$.

Ando, Blumberg, Gepner, Hopkins, and Rezk [**ABG⁺14b**, (1.4), (1.7)] prove that Σ^∞_+ and GL_1 are an adjoint pair on the categories of infinite loop spaces and E_∞-ring spectra. Applying this adjunction, we have a map of infinite loop spaces $\Omega X \to \mathrm{GL}_1(M(\xi\circ\eta))$; deloop to obtain the map in the theorem statement. ☐

THEOREM 2.11 (Beardsley [**Bea17**, Theorem 1]). *With notation as in Proposition* 2.10, *the Thom spectrum of the "universal twist" $X \to \mathrm{BGL}_1(M(\xi \circ \eta))$ is canonically equivalent to $M\xi$.*

COROLLARY 2.12.

(1) *There is a map $\widehat{w}_1\colon K(\mathbb{Z}/2,1) \to \mathrm{BGL}_1(MTSO)$ which, after taking the quotient $MTSO \to H\mathbb{Z}$, passes to the homotopy equivalence $K(\mathbb{Z}/2,1) \to \mathrm{BGL}_1(H\mathbb{Z})$ from Example* 2.6.
(2) *There is a map $\widehat{w}_2\colon K(\mathbb{Z}/2,2) \to \mathrm{BGL}_1(MTSpin)$ which, after composing with the Atiyah–Bott–Shapiro map $MTSpin \to ko$* [**ABS64, Joa04**], *is the usual map $K(\mathbb{Z}/2,2) \hookrightarrow \mathrm{BGL}_1(ko)$* [**DK70, HJ20**].
(3) *There is a map $\widehat{\beta w_2}\colon K(\mathbb{Z},3) \to \mathrm{BGL}_1(MTSpin^c)$ which, after composing with the Atiyah–Bott–Shapiro map $MTSpin^c \to ku$* [**ABS64, Joa04, AHR10**], *is the usual twist of K-theory by degree-*3 *classes $K(\mathbb{Z},3) \to \mathrm{BGL}_1(ku)$* [**DK70, Ros89, AS04, ABG10**].
(4) *There is a map $\widehat{\lambda}\colon K(\mathbb{Z},4) \to \mathrm{BGL}_1(MTString)$ which, when composed with the Ando–Hopkins–Rezk orientation $MTString \to tmf$* [**AHR10**], *is the Ando–Blumberg–Gepner map $K(\mathbb{Z},4) \to \mathrm{BGL}_1(tmf)$* [**ABG10**, Proposition 8.2].

Part (3) is a theorem of Hebestreit and Joachim [**HJ20**, Appendix C]. The other parts are surely known, though we were unable to find them in the literature.

PROOF. Apply Proposition 2.10 to the four maps

(1) $w_1\colon BO \to K(\mathbb{Z}/2,1)$, whose fiber is BSO;
(2) $w_2\colon B\mathrm{SO} \to K(\mathbb{Z}/2,2)$, whose fiber is BSpin;
(3) $\beta \circ w_2\colon B\mathrm{SO} \to K(\mathbb{Z},3)$, whose fiber is $B\mathrm{Spin}^c$, where $\beta\colon H^k(-;\mathbb{Z}/2) \to H^{k+1}(-;\mathbb{Z})$ is the Bockstein; and
(4) $\lambda\colon B\mathrm{Spin} \to K(\mathbb{Z},4)$, whose fiber is BString.

All four of these are infinite loop maps, because these characteristic classes are additive in direct sums. For compatibility with preexisting twists, we use the fact that in the spin^c and string cases, Ando, Blumberg, and Gepner [**ABG10**, §7, §8] construct the desired twists $K(\mathbb{Z},3) \to \mathrm{BGL}_1(ku)$ and $K(\mathbb{Z},4) \to \mathrm{BGL}_1(tmf)$ in the same way as we construct the twists of $MTSpin^c$ and $MTString$, so compatibility follows from functoriality. The cases of ko and $H\mathbb{Z}$ are analogous. □

The homotopy groups of the Thom spectra of the twists Corollary 2.12 have bordism interpretations. Looking at $\widehat{w}_2$ for example, a spin structure on an oriented manifold is a trivialization of $w_2(TM)$, but given a space X and a degree-2 cohomology class B, thought of as a map $f_B\colon X \to K(\mathbb{Z}/2,2)$, the homotopy groups of $MT(\widehat{w}_2 \circ f_B)$ are the bordism groups of oriented manifolds M together with a map $g\colon M \to X$ and a trivialization of $w_2(TM) + g^*B$, as was shown by [**HJ20**, Corollary 3.3.8]. The other three cases are analogous; in particular, we have described the Thom spectra for ξ^{het} and ξ^{CHL} as $MTString$-module Thom spectra.

These kinds of twisted bordism have been studied before: spin^c structures twisted by a degree-3 cohomology class were first studied by Douglas [**Dou06**, §5], and they appear implicitly in work of Freed and Witten [**FW99**] on anomaly cancellation. Twisted spin and string structures of the sort appearing in Corollary 2.12 were first considered by B. L. Wang [**Wan08**, Definitions 8.2, 8.4]. See [**DFM11a**,

DFM11b, Sat11a, Sat11c, Sat12, Sat15, SW15, LSW20, SY21] for more examples of twisted generalized cohomology theories from a similar point of view and some applications in physics.

The first case, involving twists of $MTSO$ by degree-1 $\mathbb{Z}/2$-cohomology classes, is the notion of a twisted orientation from the beginning of this section: given a real line bundle $L \to X$, we ask for data of a map $g\colon M \to X$ and an orientation on $TM \oplus g^*(L)$. In the ABGHR perspective this says that the map $\widehat{w}_1$ factors through BO_1 as

$$K(\mathbb{Z}/2,1) \overset{\simeq}{\to} BO_1 \hookrightarrow BO \to BGL_1(\mathbb{S}) \to BGL_1(MTSO). \tag{2.13}$$

But the others do not factor this way.

REMARK 2.14. There is a complex version of (2.13). Let $\mathcal{W}$ denote *Wall's bordism spectrum* [**Wal60**], whose homotopy groups are the bordism groups of manifolds with an integral lift of w_1. Explicitly, if $\xi\colon F \to BO$ is the fiber of $\beta w_1\colon BO \to K(\mathbb{Z},2)$, then $\mathcal{W} := MT\xi$. Proposition 2.10 then produces a map $\widehat{\beta w_1}\colon K(\mathbb{Z},2) \to BGL_1(\mathcal{W})$, but degree-2 cohomology classes are equivalent to complex line bundles, and $\widehat{\beta w_1}$ factors as

$$K(\mathbb{Z},2) \overset{\simeq}{\to} B\mathbb{T} \to BO_2 \to BO \to BGL_1(\mathbb{S}) \to BGL_1(\mathcal{W}). \tag{2.15}$$

REMARK 2.16. One consequence of the fact that $\widehat{w}_1$ (resp. $\widehat{\beta w_1}$) factors as in (2.13) (resp. (2.15)), i.e., as a twist associated to a real (resp. complex) line bundle $L \to X$ is that the associated $MTSO$-module (resp. $\mathcal{W}$-module) Thom spectrum splits as $MTSO \wedge X^{L-1}$ (resp. $\mathcal{W} \wedge X^{L-2}$). Working universally over BO_1 and $B\mathbb{T}$, Theorem 2.11 gives us homotopy equivalences $MTSO \wedge (BO_1)^{L-1} \simeq MTO$ and $\mathcal{W} \wedge (B\mathbb{T})^{L-2} \simeq MTO$; the former is a theorem of Atiyah [**Ati61**, Proposition 4.1].

We will apply Corollary 2.12 to the degree-4 characteristic classes that the Bianchi identity told us for the heterotic and CHL tangential structures. Given a space X with a class $\mu \in H^4(X;\mathbb{Z})$, let $\mathcal{B}(X)$ denote the homotopy fiber of $\lambda + \mu\colon B\mathrm{Spin} \times X \to K(\mathbb{Z},4)$, and let ξ^μ denote the tangential structure

$$\xi^\mu\colon \mathcal{B}(X) \longrightarrow B\mathrm{Spin} \times X \longrightarrow BO. \tag{2.17}$$

$MT\xi^\mu$ is equivalent to the $MTString$-module Thom spectrum associated to the twist $\widehat{\lambda} \circ \mu\colon X \to BGL_1(MTString)$. If $X = BG$ for a Lie group G, $\mathcal{B}(X)$ is the classifying space of the string 2-group $\mathcal{S}(\mathrm{Spin} \times G, \lambda + \mu)$. Let $\mathcal{A}$ denote the 2-primary Steenrod algebra and for $n \geq 0$, let $\mathcal{A}(n)$ denote the subalgebra of $\mathcal{A}$ generated by $\mathrm{Sq}^1, \ldots, \mathrm{Sq}^{2^n}$. In joint work with Matthew Yu [**DY23**], we compute the $\mathcal{A}$-module structure on $H^*(MT\xi^\mu;\mathbb{Z}/2)$.

DEFINITION 2.18. Let R denote the $\mathbb{Z}/2$-algebra $\mathcal{A}(1)[S]$, i.e., the algebra with generators Sq^1, Sq^2, and S, and with Adem relations for Sq^1 and Sq^2. Given X and μ as above, define the $\mathcal{A}(1)$-module $T(X,\mu) := H^*(X;\mathbb{Z}/2)$, and give $T(X,\mu)$ an R-module structure by defining

$$S(x) := \mu x + \mathrm{Sq}^4(x). \tag{2.19}$$

We want to think of S as Sq^4 and $T(X,\mu)$ as an $\mathcal{A}(2)$-module, but a priori it is not clear that this S-action satisfies the Adem relations.

THEOREM 2.20 ([**DY23**]).

(1) *The R-module structure on $T(X,\mu)$ satisfies the Adem relations for* Sq^1, Sq^2, *and* $\mathrm{Sq}^4 = S$, *hence induces an $\mathcal{A}(2)$-module structure on $T(X,\mu)$.*

(2) *There is an map of $\mathcal{A}$-modules*

$$H^*(MT\xi^\mu;\mathbb{Z}/2) \longrightarrow \mathcal{A}\otimes_{\mathcal{A}(2)} T(X,\mu), \tag{2.21}$$

natural in the data (X,μ), which is an isomorphism in degrees 15 *and below.*

We describe a proof of this theorem in Remark 2.26 different from the one in [**DY23**].

Corollary 2.22. *For $t-s\le 15$, the E_2-page of the Adams spectral sequence computing* 2*-completed ξ^μ-bordism is*

$$E_2^{t,s} = \mathrm{Ext}^{s,t}_{\mathcal{A}(2)}(T(X,\mu),\mathbb{Z}/2). \tag{2.23}$$

As $\mathcal{A}(2)$ is much smaller than $\mathcal{A}$, this is much easier to work with.

Proof. This follows from the change-of-rings formula: if $\mathcal{B}$ is a graded Hopf algebra, $\mathcal{C}$ is a graded Hopf subalgebra of $\mathcal{B}$, and M and N are graded $\mathcal{B}$-modules, then there is a natural isomorphism

$$\mathrm{Ext}^{s,t}_{\mathcal{B}}(\mathcal{B}\otimes_{\mathcal{C}} M, N) \xrightarrow{\cong} \mathrm{Ext}^{s,t}_{\mathcal{C}}(M,N). \tag{2.24}$$

This you can think of as the derived version of a maybe more familiar isomorphism

$$\mathrm{Hom}_{\mathcal{B}}(\mathcal{B}\otimes_{\mathcal{C}} M, N) \xrightarrow{\cong} \mathrm{Hom}_{\mathcal{C}}(M,N). \tag{2.25}$$

In our example, $\mathcal{B}$ is the Steenrod algebra, which is a Hopf algebra, and $\mathcal{C}$ is $\mathcal{A}(2)$, which is indeed a Hopf subalgebra of $\mathcal{A}$, so we can invoke (2.24) and conclude. □

We will use this simplification in the cases $\xi^\mu = \xi^{\mathrm{het}}, \xi^{\mathrm{CHL}}$ to run the Adams spectral sequences computing $\Omega_*^{\xi^{\mathrm{het}}}$ and $\Omega_*^{\xi^{\mathrm{CHL}}}$ at $p=2$.

Remark 2.26 (Proof sketch of Theorem 2.20). To prove (1), check the Adem relations for $\mathcal{A}(2)$ directly. The first step in proving part (2) is to establish a Thom isomorphism for mod 2 cohomology. We make use of the *Thom diagonal*, a map of *MTString*-modules

$$MT\xi^\mu \xrightarrow{\Delta^t} MT\xi^\mu \wedge \mathit{MTString}\wedge \Sigma^\infty_+ X \tag{2.27}$$

defined as follows: the diagonal map $\Delta\colon X\to X\times X$ is a map of spaces over $B\mathrm{GL}_1(\mathit{MTString})$, if we give X the map $\widehat{\lambda}\circ\mu$ to $B\mathrm{GL}_1(\mathit{MTString})$ and we give $X\times X$ the map $(\widehat{\lambda}\circ\mu, *)$. Applying the *MTString*-module Thom spectrum functor to Δ produces (2.27). Smash (2.27) with $H\mathbb{Z}/2$. The result is the Thom diagonal for a twist of $H\mathbb{Z}/2$, but all such twists are trivializable (i.e., all $H\mathbb{Z}/2$-bundles admit an orientation). Therefore by [**ABG$^+$14b**, Proposition 3.26] the following composition is an equivalence:

$$MT\xi^\mu\wedge H\mathbb{Z}/2 \xrightarrow{\Delta^t} MT\xi^\mu\wedge\Sigma^\infty_+ X\wedge H\mathbb{Z}/2 \longrightarrow \mathit{MTString}\wedge\Sigma^\infty_+X\wedge H\mathbb{Z}/2, \tag{2.28}$$

which is the $\mathbb{Z}/2$-homology Thom isomorphism. The analogous fact is true for mod 2 cohomology.

The Thom diagonal makes $H^*(MT\xi^\mu;\mathbb{Z}/2)$ into a free, rank-1 module over $H^*(\mathcal{B}(X);\mathbb{Z}/2)$, generated by the Thom class U. As the Thom diagonal is a map of spectra, we may use the Cartan formula to compute the Steenrod squares of an arbitrary element of $H^*(MT\xi^\mu;\mathbb{Z}/2)$ in terms of Steenrod squares in $\mathcal{B}(X)$ and $\mathrm{Sq}(U)$. As both $\mathrm{Sq}(U)$ and our desired isomorphism in (2.21) are natural in X and μ, it suffices to understand the universal case, where $X = K(\mathbb{Z},4)$ and μ is the tautological class $\tau \in H^4(K(\mathbb{Z},4);\mathbb{Z})$. In this case, Theorem 2.11 implies $MT\xi^\mu \simeq MTSpin$. By work of Anderson, Brown, and Peterson [**ABP67**], if J is the $\mathcal{A}(1)$-module $\mathcal{A}(1)/\mathrm{Sq}^3$ and M is the $\mathcal{A}(1)$-module $\mathbb{Z}/2 \oplus \Sigma^8\mathbb{Z}/2 \oplus \Sigma^{10}J$, then there is a map of $\mathcal{A}$-modules

$$H^*(MTSpin;\mathbb{Z}/2) \longrightarrow \mathcal{A} \otimes_{\mathcal{A}(1)} M \tag{2.29}$$

which is an isomorphism in degrees 15 and below. And Giambalvo [**Gia71**, Corollary 2.3] shows that there is a map $H^*(MTString;\mathbb{Z}/2) \to \mathcal{A} \otimes_{\mathcal{A}(2)} \mathbb{Z}/2$ which is also an isomorphism in degrees 15 and below. Therefore by the change-of-rings theorem (2.24) it suffices to exhibit a map of $\mathcal{A}(2)$-modules

$$T(K(\mathbb{Z},4),\tau) \longrightarrow \mathcal{A}(2) \otimes_{\mathcal{A}(1)} M \tag{2.30}$$

which is an isomorphism in degrees 15 and below. This can be verified directly, using as input the $\mathcal{A}(2)$-module structure on $H^*(K(\mathbb{Z},4);\mathbb{Z}/2)$ calculated by Serre [**Ser53**, §10].

2.2. ξ^{het} bordism at $p = 2$. In this section we will first compute $H^*(BG;\mathbb{Z}/2)$ as an $\mathcal{A}(2)$-module in low degrees, where $G := \mathrm{E}_8^2 \rtimes \mathbb{Z}/2$; then, using Corollary 2.22, we run the Adams spectral sequence computing 2-completed ξ^{het} bordism in degrees 11 and below.

First, though, we reformulate the problem slightly. Consider the tangential structure $\xi^{\mathrm{het}\prime}\colon B^{\mathrm{het}\prime} \to BO$ defined in the same manner as ξ^{het}, but with $K(\mathbb{Z},4)$ replacing $B\mathrm{E}_8$. In a little more detail, $\mathbb{Z}/2$ acts on $K(\mathbb{Z},4) \times K(\mathbb{Z},4)$ by swapping the two factors; taking the Borel construction

$$B := (K(\mathbb{Z},4) \times K(\mathbb{Z},4)) \times_{\mathbb{Z}/2} E\mathbb{Z}/2 \tag{2.31}$$

produces a fiber bundle

$$K(\mathbb{Z},4) \times K(\mathbb{Z},4) \longrightarrow B \longrightarrow B\mathbb{Z}/2. \tag{2.32}$$

For $i = 1, 2$, let $c_i \in H^4(K(\mathbb{Z},4) \times K(\mathbb{Z},4);\mathbb{Z})$ be the tautological class for the ith $K(\mathbb{Z},4)$ factor. The class $c_1 + c_2$ is invariant under the $\mathbb{Z}/2$-action, so we can follow it through the Serre spectral sequence to learn that it defines a nonzero class $c_1 + c_2 \in H^4(B;\mathbb{Z}/2)$. Define $f\colon B^{\mathrm{het}\prime} \to B\mathrm{Spin} \times B$ to be the fiber of $\lambda - (c_1 + c_2)\colon B\mathrm{Spin} \times B \to K(\mathbb{Z},4)$; then the tangential structure $\xi^{\mathrm{het}\prime}$ is the composition

$$\underbrace{B^{\mathrm{het}\prime} \xrightarrow{\;f\;} B\mathrm{Spin} \times B \xrightarrow{\;\mathrm{pr}_1\;} B\mathrm{Spin} \longrightarrow BO}_{\xi^{\mathrm{het}\prime}}. \tag{2.33}$$

That is, a $\xi^{\mathrm{het}\prime}$ structure on a manifold M is a spin structure, a principal $\mathbb{Z}/2$-bundle $P \to M$, two classes $c_1, c_2 \in H^4(P;\mathbb{Z})$ which are exchanged under the deck transformation, and a trivialization of $\lambda(M) - (c_1 + c_2)$ (where the latter class is descended to M). This is the same data as a ξ^{het} structure, except that we do not ask for c_1 or c_2 to come from principal E_8-bundles; therefore there is a map of tangential structures $\widetilde{c}\colon \xi^{\mathrm{het}} \to \xi^{\mathrm{het}\prime}$, i.e., a map of spaces $B\mathbb{G}^{\mathrm{het}} \to B^{\mathrm{het}\prime}$ commuting with the maps down to BO. Like for ξ^{het}, a $\xi^{\mathrm{het}\prime}$-structure is a twisted string structure in the sense of Corollary 2.12, via the class $\lambda - (c_1 + c_2)\colon B \to K(\mathbb{Z}, 4)$.

Bott and Samelson [**BS58**, Theorems IV, V(e)] showed that the characteristic class $c \in H^4(B\mathrm{E}_8;\mathbb{Z})$ we defined in Definition 1.4, interpreted as a map $c\colon B\mathrm{E}_8 \to K(\mathbb{Z}, 4)$, is 15-connected. This means that the homomorphism $\widetilde{c}$ induces on bordism groups,

$$\widetilde{c}\colon \Omega_k^{\xi^{\mathrm{het}}} \to \Omega_k^{\xi^{\mathrm{het}\prime}},$$

is an isomorphism in degrees 14 and below. For our string-theoretic purposes, we only care about $k \le 12$, so we may as well compute $\xi^{\mathrm{het}\prime}$-bordism. In the rest of this subsection, we often blur the distinction between ξ^{het} and $\xi^{\mathrm{het}\prime}$; we will point out where it matters which one we are looking at.

REMARK 2.34. Turning off the $\mathbb{Z}/2$ symmetry switching the two E_8 factors, i.e., passing to a $\xi^{r,\mathrm{het}}$-structure as in Remark 1.46, simplifies this story considerably: the bordism groups were known decades ago. Specifically, replace $B\mathrm{E}_8$ with $K(\mathbb{Z}, 4)$ in the definition of $\xi^{r,\mathrm{het}}$ to define a tangential structure $\xi^{r,\mathrm{het}\prime}$, which on a manifold M consists of a spin structure on M, two classes $c_1, c_2 \in H^4(M;\mathbb{Z})$, and a trivialization of $\lambda(M) - c_1 - c_2$. As Witten [**Wit86**, §4] noticed, this data is equivalent to a spin structure and the single class c_1, which may be freely chosen; then c_2 must be $\lambda(M) - c_1$. Therefore the tangential structure $\xi^{r,\mathrm{het}\prime}$-structure is simply $B\mathrm{Spin} \times K(\mathbb{Z}, 4) \to B\mathrm{O}$, and just as for ξ^{het}, the map $MT\xi^{r,\mathrm{het}} \to MT\xi^{r,\mathrm{het}\prime} \simeq MT\mathit{Spin} \wedge K(\mathbb{Z}, 4)_+$ is an isomorphism on homotopy groups in degrees 14 and below. Stong [**Sto86**] computes $\Omega_*^{\mathrm{Spin}}(K(\mathbb{Z}, 4))$ in degrees 12 and below.

As we discussed in §1.2, the data of a trivial principal $\mathbb{Z}/2$-bundle on a manifold M and two principal E_8-bundles $P, Q \to M$ define a principal $\mathrm{E}_8^2 \rtimes \mathbb{Z}/2$-bundle on M with $c_1 + c_2$ equal to $c(P) + c(Q)$; data trivializing $c(P) + c(Q) - \lambda(M)$ therefore defines a ξ^{het} structure. Analogously, the trivial $\mathbb{Z}/2$-bundle and a pair $c_1, c_2 \in H^4(M;\mathbb{Z})$ with a trivialization of $c_1 + c_2 - \lambda$ define a $\xi^{\mathrm{het}\prime}$ structure.

LEMMA 2.35. *A spin manifold M has a canonical $\xi^{\mathrm{het}\prime}$ structure specified as above by the trivial principal $\mathbb{Z}/2$-bundle, the cohomology classes $c_1 = \lambda$ and $c_2 = 0$, and the canonical trivialization of $\lambda - \lambda = 0 \in H^4(M;\mathbb{Z})$.*

This defines a map of tangential structures and therefore a map of Thom spectra $s_1\colon MT\mathit{Spin} \to MT\xi^{\mathrm{het}\prime}$. A $\xi^{\mathrm{het}\prime}$-structure includes data of a spin structure; forgetting the rest of the $\xi^{\mathrm{het}\prime}$-structure defines a map $s_2\colon MT\xi^{\mathrm{het}\prime} \to MT\mathit{Spin}$. The composition of s_1 and s_2 is homotopy equivalent to the identity, because the underlying spin structure of the $\xi^{\mathrm{het}\prime}$ manifold built in Lemma 2.35 is the same spin structure we began with.

COROLLARY 2.36. *There is a spectrum Q and a splitting*

$$(s_2, q)\colon MT\xi^{\mathrm{het}\prime} \xrightarrow{\simeq} MT\mathit{Spin} \vee Q. \tag{2.37}$$

We will use this later to reduce the amount of spectral sequence computations we have to make.

Both Lemma 2.35 and Corollary 2.36 require us to use $\xi^{\mathrm{het}\prime}$ and not ξ^{het}, though of course the consequence on low-degree bordism groups is true for both.

When K is a finite group, Nakaoka [**Nak61**, Theorem 3.3] proved that there is a ring isomorphism from the mod 2 cohomology of $B(\mathbb{Z}/2 \ltimes (K \times K))$ to the

E_2-page of the Serre spectral sequence

$$E_2^{p,q} = H^p(B\mathbb{Z}/2; \underline{H^q(BK \times BK; \mathbb{Z}/2)}) \Longrightarrow H^{p+q}(B(\mathbb{Z}/2 \ltimes (K \times K)); \mathbb{Z}/2). \tag{2.38}$$

Here the underline denotes the local coefficient system arising from the $\mathbb{Z}/2$-action on $BK \times BK$ by switching the two factors. Since this local coefficient system can be nontrivial, one has to be careful defining the multiplicative structure on the E_2-page of (2.38), but here it can be made explicit. As a $\mathbb{Z}/2[\mathbb{Z}/2]$-module, $\underline{H^*(BK \times BK; \mathbb{Z}/2)}$ is a direct sum of:

- the subalgebra $\mathcal{H}_1$ of classes fixed by $\mathbb{Z}/2$, which are of the form $x \otimes x$ for $x \in H^*(BK; \mathbb{Z}/2)$; and
- the submodule $\mathcal{H}_2$ spanned by classes of the form $x \otimes y$ where x and y are linearly independent.

Since $\mathbb{Z}/2$ acts trivially on $\mathcal{H}_1$ and $\mathcal{H}_1$ is a ring, $H^*(B\mathbb{Z}/2; \mathcal{H}_1)$ has a ring structure. And as a $\mathbb{Z}/2[\mathbb{Z}/2]$-module, $\mathcal{H}_2$ is of the form $M \oplus M$ where $\mathbb{Z}/2$ acts by swapping the two factors, so $H^*(B\mathbb{Z}/2; \mathcal{H}_2)$ vanishes in positive degrees.[14] In degree zero, we obtain invariants, spanned by elements of the form $x \otimes y + y \otimes x$, with $x, y \in H^*(BK; \mathbb{Z}/2)$. $\mathcal{H}_1 \oplus (\mathcal{H}_2)^{\mathbb{Z}/2} = E_2^{0,\bullet}$ is a subalgebra of $H^*(BK \times BK; \mathbb{Z}/2)$.

So far we have specified ring structures on $H^*(B\mathbb{Z}/2; \mathcal{H}_1) \supsetneq E_2^{>0,\bullet}$ and $\mathcal{H}_1 \oplus (\mathcal{H}_2)^{\mathbb{Z}/2} = E_2^{0,\bullet}$, and these ring structures agree where they overlap. Therefore to specify a ring structure on the entirety of the E_2-page, it suffices to write down the product of an element in $(\mathcal{H}_2)^{\mathbb{Z}/2}$ and an element in positive p-degree. We say that all such products vanish; this is the ring structure that appears in Nakaoka's theorem.

Of course, E_8 is not a finite group. Nakaoka's theorem is true in quite great generality [**Eve65**, **Kah84**, **Lea97**]; the version we need is proven by Evens [**Eve65**], who proves the same ring isomorphism when K is a compact Lie group. Thus this applies to ξ^{het}, and not necessarily to $\xi^{\mathrm{het}'}$, but since their cohomology rings are isomorphic in degrees 14 and below, it does not matter which one we use in this calculation.

Now we make this ring structure and $\mathcal{A}(2)$-module structure explicit. Since $c\colon B\mathrm{E}_8 \to K(\mathbb{Z}, 4)$ is 15-connected, it induces an isomorphism in cohomology in degrees 14 and below, so we can use the cohomology of $K(\mathbb{Z}, 4)$ as a stand-in for the cohomology of $B\mathrm{E}_8$. Serre [**Ser53**, §10] computed the mod 2 cohomology of $K(\mathbb{Z}, 4)$. It is an infinitely generated polynomial algebra; in degrees 12 and below the generators are: the tautological class $D \in H^4(K(\mathbb{Z}, 4); \mathbb{Z}/2)$, $F := \mathrm{Sq}^2 D$, $G := \mathrm{Sq}^3 D$, $J := \mathrm{Sq}^4 F$, and $K := \mathrm{Sq}^5 F$.

If C is one of D, F, G, J, or K, we let C_1 denote the class coming from the first copy of $B\mathrm{E}_8$ and C_2 denote the class coming from the second copy. Thus we have the following additive basis for the low-degree cohomology of BG:

(1) In $\mathcal{H}_1$, $D_1 D_2 x^k$ and $F_1 F_2 x^k$ for $k \geq 0$.

[14] To see this, first observe that mod 2 group cohomology for G is additive in the $\mathbb{Z}/2[G]$-module of coefficients, so it suffices to prove that $H^*(B\mathbb{Z}/2; M \oplus M)$ vanishes in positive degrees when $M = \mathbb{Z}/2$. But $\mathbb{Z}/2 \oplus \mathbb{Z}/2$ is isomorphic to $\mathbb{Z}/2[\mathbb{Z}/2]$ as $\mathbb{Z}/2[\mathbb{Z}/2]$-modules (i.e., as vector spaces with $\mathbb{Z}/2$-representations, $\mathbb{Z}/2 \oplus \mathbb{Z}/2$ is isomorphic to the vector space of functions on the group $\mathbb{Z}/2$), and group cohomology valued in the group ring is trivial, e.g., because the group ring is its own free resolution.

(2) In $(\mathcal{H}_2)^{\mathbb{Z}/2}$, D_1+D_2, F_1+F_2, G_1+G_2, $D_1^2+D_2^2$, J_1+J_2, $D_1F_1+D_2F_2$, $D_1F_2+D_2F_1$, $D_1G_1+D_2G_2$, $D_1G_2+D_2G_1$, K_1+K_2, $F_1^2+F_2^2$, $D_1^3+D_2^3$, and $D_1^2D_2+D_1D_2^2$.

Next, we determine the $\mathcal{A}(2)$-module structure using a theorem of Quillen.

THEOREM 2.39 (Quillen's detection theorem [**Qui71**, Proposition 3.1]). *Let X be a space and let $\mathbb{Z}/k$ act on X^k by cyclic permutations. Let $Y := E\mathbb{Z}/k \times_{\mathbb{Z}/k} X^k$, which is a fiber bundle over $B\mathbb{Z}/k$ with fiber X^k. Let $i_1\colon X^k \to Y$ be inclusion of the fiber at the basepoint and $i_2\colon B\mathbb{Z}/k \times X \to Y$ be induced by the diagonal map; then*

$$(i_1^*, i_2^*)\colon H^*(Y;\mathbb{Z}/k) \longrightarrow H^*(X^k;\mathbb{Z}/k) \oplus H^*(B\mathbb{Z}/k \times X;\mathbb{Z}/k) \tag{2.40}$$

is injective.

For us, $k=2$, $X = B\mathrm{E}_8$, and $Y = BG$. Thus, to compute Steenrod squares for classes in $H^*(BG;\mathbb{Z}/2)$, we can assume we are in $B\mathrm{E}_8^2$ if the class is in $(\mathcal{H}_2)^{\mathbb{Z}/2}$; for $\mathcal{H}^1$, we also need to know $\mathrm{Sq}(x)$, and i_2^* tells us $\mathrm{Sq}(x) = x + x^2$. Thus we can compute the $\mathcal{A}(2)$-module structure on $H^*(BG;\mathbb{Z}/2)$, hence also on $T(-(c_1+c_2))$; we focus on the latter. Like most calculations of this form, it is a little tedious but straightforward, and can be done by hand in a reasonable length of time. After working through the calculation, we have learned the following.

PROPOSITION 2.41. *Let $\mathcal{M}$ be the quotient of $T(-(c_1+c_2))$ by all elements in degrees 14 and higher. Then $\mathcal{M}$ is the direct sum of the following submodules.*

(1) *M_1, the summand containing the Thom class U.*
(2) *$M_2 := \widetilde{H}^*(\mathbb{RP}^\infty;\mathbb{Z}/2)$ modulo those elements in degrees 13 and above.*
(3) *M_3, the summand containing $U(D_1^2+D_2^2)$.*
(4) *M_4, the summand containing UD_1D_2.*
(5) *M_5, the summand containing UD_1D_2x.*
(6) *M_6, the summand containing $U(D_1F_1+D_2F_2)$.*
(7) *M_7, the summand containing $U(D_1D_2^2+D_1^2D_2)$.*

We draw this decomposition in Figure 1.

Recall from Corollary 2.36 that $MT\xi^{\mathrm{het}'}$ splits as $MTSpin \vee \mathcal{Q}$. Since $\Omega_*^{\xi^{\mathrm{het}}} \cong \Omega_*^{\xi^{\mathrm{het}'}}$ in the range we need and Ω_*^{Spin} is known thanks to work of Anderson, Brown, and Peterson [**ABP67**], we focus on $\pi_*(\mathcal{Q})$. To do so, we will identify the submodule of the E_2-page of the Adams spectral sequence for $\xi^{\mathrm{het}'}$ coming from spin bordism via $s_1\colon MTSpin \to MT\xi^{\mathrm{het}'}$; the E_2-page for $\mathcal{Q}$ is then a complementary submodule.

The canonical $\xi^{\mathrm{het}'}$-structure on a spin manifold from Lemma 2.35 can be rephrased as follows: a spin structure on a manifold M is equivalent data to: a spin structure on M, a map $c\colon M \to K(\mathbb{Z},4)$, and a trivialization of $c - \lambda(M)$. Thus spin structures are twisted string structures in the sense of Corollary 2.12 (in fact the universal twist in the sense of Remark 2.16), so the map

$$(1,0)\colon K(\mathbb{Z},4) \longrightarrow (K(\mathbb{Z},4)\times K(\mathbb{Z},4)) \times_{\mathbb{Z}/2} E\mathbb{Z}/2 = B \tag{2.42}$$

lifts to a map of *MTString*-module Thom spectra $s_1\colon MTSpin \to MT\xi^{\mathrm{het}'}$. Naturality of Theorem 2.20 then tells us the image of s_1^* on mod 2 cohomology, allowing us to determine which of the summands in Proposition 2.41 correspond to *MTSpin* and which correspond to $\mathcal{Q}$. Specifically, the pullback map sends $x \mapsto 0$, is nonzero

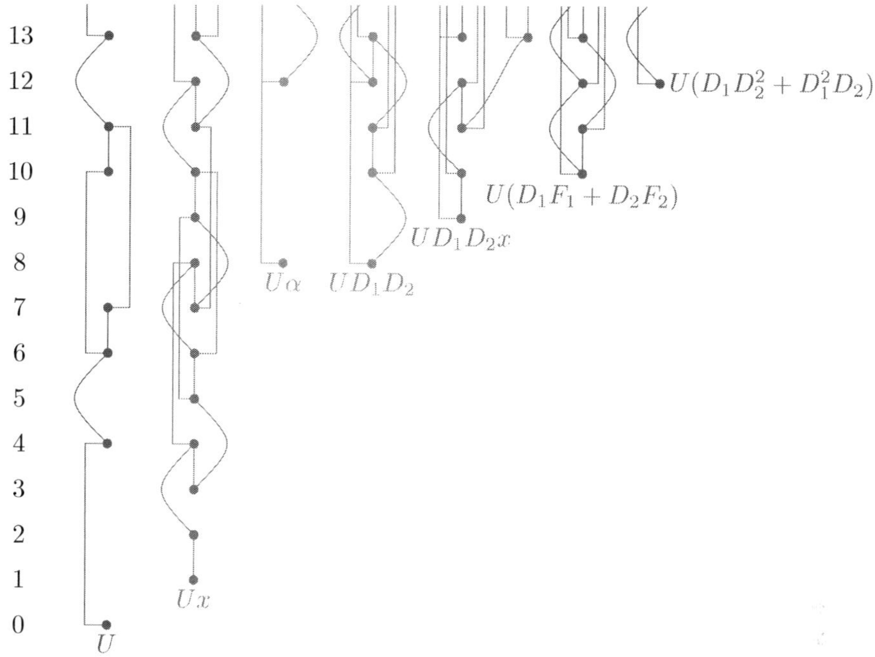

FIGURE 1. The $\mathcal{A}(2)$-module $T(-(c_1+c_2))$ in low degrees. Here $\alpha := D_1^2 + D_2^2$. The pictured submodule contains all classes in degrees 12 and below.

on D_1, F_1, G_1, etc., and sends D_2, F_2, G_2, etc., to zero. This implies that in the direct-sum decomposition $MT\xi^{\mathrm{het}'} \simeq MTSpin \vee \mathcal{Q}$, the summands M_1, M_3, and M_6 come from the cohomology of $MTSpin$, and the remaining summands come from the cohomology of $\mathcal{Q}$.

In order to run the Adams spectral sequence for $\mathcal{Q}$, we need to compute the Ext of M_2, M_4, M_5, and M_7 over $\mathcal{A}(2)$. After we compute this, we will display the E_2-page in Figure 3. For an $\mathcal{A}(2)$-module M, $\mathrm{Ext}_{\mathcal{A}(2)}^{*,*}(M, \mathbb{Z}/2)$, which we will usually denote $\mathrm{Ext}_{\mathcal{A}(2)}(M)$ or $\mathrm{Ext}(M)$, is a bigraded module over the bigraded $\mathbb{Z}/2$-algebra $\mathrm{Ext}_{\mathcal{A}(2)}(\mathbb{Z}/2)$; both the algebra and module structures arise from the *Yoneda product* [**Yon54**, §4] (see [**BC18**, §4.2] for a review). This module structure is helpful for determining differentials in the Adams spectral sequence: differentials are equivariant with respect to the action. The module structure also constrains extensions on its E_∞-page.

May (unpublished) and Shimada–Iwai [**SI67**, §8] determined $\mathrm{Ext}_{\mathcal{A}(2)}(\mathbb{Z}/2)$. We will only need to track the actions of three elements: $h_0 \in \mathrm{Ext}_{\mathcal{A}(2)}^{1,1}(\mathbb{Z}/2)$, $h_1 \in \mathrm{Ext}_{\mathcal{A}(2)}^{1,2}(\mathbb{Z}/2)$, and $h_2 \in \mathrm{Ext}_{\mathcal{A}(2)}(\mathbb{Z}/2)$. These elements are in the image of the map $\mathrm{Ext}_{\mathcal{A}}(\mathbb{Z}/2) \to \mathrm{Ext}_{\mathcal{A}(2)}(\mathbb{Z}/2)$ induced by the quotient $\mathcal{A} \to \mathcal{A}(2)$, so we do not have to worry about whether Corollary 2.22 is compatible with the $\mathrm{Ext}_{\mathcal{A}(2)}(\mathbb{Z}/2)$-action on the E_2-page of the Adams spectral sequence. (It is, though.) When we draw Ext charts as in Figure 3, we denote h_0-actions as vertical lines, h_1-actions as diagonal

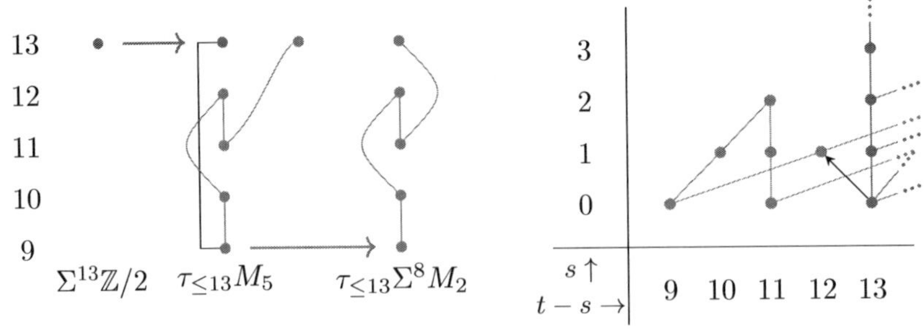

FIGURE 2. Left: the short exact sequence (2.44) of $\mathcal{A}(2)$-modules. Right: the associated long exact sequence in Ext. See the discussion after (2.45) for why the pictured boundary map (black arrow) is nonzero.

lines with slope 1, and h_2-actions as diagonal lines with slope 1/3. When one of these lines is not present, the corresponding h_i acts as 0.

Often one computes Ext groups of $\mathcal{A}(2)$-modules using computer programs developed by Bruner [**Bru18**] and Chatham and Chua [**CC21**], or tools such as the May spectral sequence [**May66**] or the Davis–Mahowald spectral sequence [**DM82**, **MS87**] (see also [**BR21**, Chapter 2]) to compute Ext groups of $\mathcal{A}(2)$-modules, but for the four modules we care about, we can get away using simpler calculations by hand and computations already in the literature.

(1) Davis and Mahowald [**DM78**, Table 3.2] compute $\mathrm{Ext}_{\mathcal{A}(2)}(M_2)$ in the degrees we need.
(2) In degrees 13 and below, M_4 is isomorphic to $\Sigma^8(\mathcal{A}(2)\otimes_{\mathcal{A}(0)}\mathbb{Z}/2)$; therefore the Ext groups of these two $\mathcal{A}(2)$-modules, as algebras over $\mathrm{Ext}_{\mathcal{A}(2)}(\mathbb{Z}/2)$, are isomorphic in topological degrees 12 and below. Thus we can compute with the change-of-rings theorem (2.24): as $\mathrm{Ext}_{\mathcal{A}(2)}(\mathbb{Z}/2)$-algebras,

$$\mathrm{Ext}_{\mathcal{A}(2)}(\mathcal{A}(2)\otimes_{\mathcal{A}(0)}\mathbb{Z}/2)\cong \mathrm{Ext}_{\mathcal{A}(0)}(\mathbb{Z}/2)\cong \mathbb{Z}/2[h_0], \tag{2.43}$$

with $h_0\in \mathrm{Ext}^{1,1}$. This identification of $\mathrm{Ext}_{\mathcal{A}(0)}(\mathbb{Z}/2)$ follows from Koszul duality [**BC18**, Example 4.5.5].
(3) M_5 looks a lot like M_2, which gives us a way to compute $\mathrm{Ext}_{\mathcal{A}(2)}(M_5)$. Specifically, if $\tau_{\leq k}M$ denotes the quotient of an $\mathcal{A}(2)$-module M by the submodule of elements in degrees greater than k, then there is a short exact sequence of $\mathcal{A}(2)$-modules

$$0\longrightarrow \Sigma^{13}\mathbb{Z}/2\longrightarrow \tau_{\leq 13}M_5\longrightarrow \tau_{\leq 13}\Sigma^8 M_2\longrightarrow 0. \tag{2.44}$$

We draw this sequence in Figure 2, left. (2.44) induces a long exact sequence in Ext groups; passage between M and $\tau_{\leq 13}M$ does not change Ext groups in degrees 12 and below, and since we only care about degrees 12 and below, we can and do pass between $\tau_{\leq 13}M$ and M without comment.

We already know $\mathrm{Ext}_{\mathcal{A}(2)}(\mathbb{Z}/2)$ and $\mathrm{Ext}_{\mathcal{A}(2)}(M_2)$, so we can run the long exact sequence associated to (2.44) to compute $\mathrm{Ext}_{\mathcal{A}(2)}(M_5)$ in degrees 12 and below; we draw this long exact sequence in Figure 2, right. In the range we care about, there is exactly one boundary map that is not

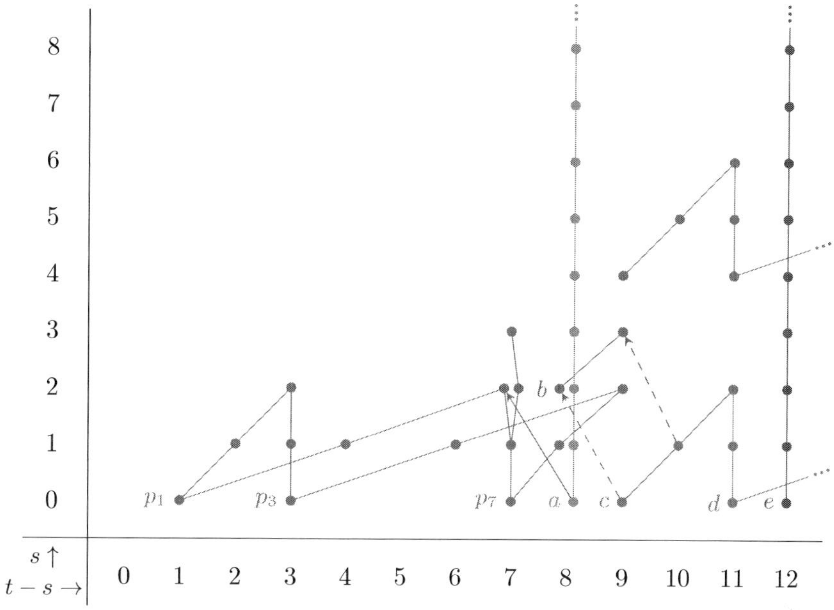

FIGURE 3. In Corollary 2.36, we showed $MT\xi^{\mathrm{het}'} \simeq MTSpin \vee \mathcal{Q}$; this figure denotes the E_2-page of the Adams spectral sequence computing $\pi_*(\mathcal{Q})$ in degrees 12 and below. This corresponds to a subset of the summands in Figure 1. In Lemma 2.50, we show that the solid gray differential beginning at a is nonzero; we leave open the other two differentials, which are dashed in this figure.

forced to be zero for degree reasons, namely

$$\partial\colon \operatorname{Ext}^{0,13}_{\mathcal{A}(2)}(\Sigma^{13}\mathbb{Z}/2) \longrightarrow \operatorname{Ext}^{1,13}_{\mathcal{A}(2)}(\Sigma^8 M_2); \tag{2.45}$$

it must be nonzero, because that is the only way to obtain $\operatorname{Ext}^{0,13}_{\mathcal{A}(2)}(M_5) = \operatorname{Hom}_{\mathcal{A}(2)}(M_5, \Sigma^{13}\mathbb{Z}/2) = 0$, and by inspection of Figure 1 this Hom group vanishes.

(4) If $C\eta := \Sigma^{-2}\widetilde{H}^*(\mathbb{CP}^2;\mathbb{Z}/2)$, there is a 14-connected quotient map $M_7 \to \Sigma^{12}C\eta$, so $\operatorname{Ext}_{\mathcal{A}(2)}(\Sigma^{12}C\eta)$ and $\operatorname{Ext}_{\mathcal{A}(2)}(M_7)$ do not differ in the range we care about. Bruner and Rognes [**BR21**, Figure 0.15] compute $\operatorname{Ext}_{\mathcal{A}(2)}(C\eta)$.

Using these computations, we obtain the following description of the E_2-page of the Adams spectral sequence for the summand $\mathcal{Q}$ of $MT\xi^{\mathrm{het}'}$.

PROPOSITION 2.46. *The E_2-page of the Adams spectral sequence for $\mathcal{Q}$ in topological degrees* 12 *and below is as given in Figure* 3. *In particular, in this range, the E_2-page is generated as an $\operatorname{Ext}_{\mathcal{A}(2)}(\mathbb{Z}/2)$-module by eight elements: $p_1 \in \operatorname{Ext}^{0,1}$, $p_3 \in \operatorname{Ext}^{0,3}$, $p_7 \in \operatorname{Ext}^{0,7}$, $a \in \operatorname{Ext}^{0,8}$, $b \in \operatorname{Ext}^{2,10}$, $c \in \operatorname{Ext}^{0,9}$, $d \in \operatorname{Ext}^{0,11}$, and $e \in \operatorname{Ext}^{0,12}$.*

There are plenty of differentials in this Adams spectral sequence which could be nonzero, even when we take into account the fact that Adams differentials commute with h_0, h_1, and h_2:

(D1) $d_2\colon E_2^{0,8} \to E_2^{2,9}$, whose value on a could be $h_2^2 p_1$, $h_0^2 p_7$, or a linear combination of those two elements.
(D2) $d_2\colon E_2^{1,9} \to E_2^{3,10}$, which could send $h_0 a$ or $h_1 p_7$ to $h_0^3 p_7$.
(D3) $d_2\colon E_2^{0,9} \to E_2^{2,8}$ and $d_2\colon E_2^{1,11} \to E_2^{3,12}$, intertwined by an h_1-action, which could send $c \mapsto b$ and $h_1 c \mapsto h_1 b$.
(D4) $d_2\colon E_2^{0,12} \to E_2^{2,13}$, which could send $e \mapsto h_1^2 c = h_0^2 d$.
(D5) If the differentials in (D1) and (D2) vanish, $d_3\colon E_3^{0,8} \to E_3^{3,10}$ could be nonzero on a.
(D6) If the differential in (D4) vanishes, $d_5\colon E_5^{0,12} \to E_5^{5,16}$ (and its image under h_0) or $d_6\colon E_6^{0,12} \to E_6^{6,17}$ could be nonzero.

LEMMA 2.47. *The differentials* (D2), (D5), *and* (D6) *vanish.*

PROOF. Our strategy is to use the fact that $\mathbb{G}^{\mathrm{het}} \to \mathbb{Z}/2$ splits to zero out differentials. This splitting does not extend to a splitting of $MT\xi^{\mathrm{het}}$, but it will be close enough.

The inclusion $\iota\colon \mathbb{Z}/2 \hookrightarrow \mathbb{G}^{\mathrm{het}}$ defines a map $\iota'\colon \mathit{MTString} \wedge B\mathbb{Z}/2 \to MT\xi^{\mathrm{het}}$ which on Adams E_2-pages is precisely the inclusion of the summand $\mathrm{Ext}(M_2)$. Quotienting $\mathbb{G}^{\mathrm{het}}$ by $\mathbb{T}[1]$, then by $\mathrm{E}_8 \times \mathrm{E}_8$, produces a map

$$p\colon MT\xi^{\mathrm{het}} \xrightarrow[(1.44)]{\phi} \mathit{MTSpin} \wedge (B((\mathrm{E}_8 \times \mathrm{E}_8) \rtimes \mathbb{Z}/2))_+ \longrightarrow \mathit{MTSpin} \wedge (B\mathbb{Z}/2)_+, \tag{2.48}$$

and $p \circ \iota\colon \mathit{MTString} \wedge (B\mathbb{Z}/2) \to \mathit{MTSpin} \wedge (B\mathbb{Z}/2)_+$ is the usual map $\mathit{MTString} \to \mathit{MTSpin}$ together with the addition of a basepoint. This means that any element of $\widetilde{\Omega}_*^{\mathrm{String}}(B\mathbb{Z}/2)$ whose image in $\widetilde{\Omega}_*^{\mathrm{Spin}}(B\mathbb{Z}/2)$ is nonzero must also be nonzero in $\Omega_*^{\xi^{\mathrm{het}}}$, which kills many differentials to or from $\mathrm{Ext}(M_2)$. To produce such elements, study the map of Adams spectral sequences induced by $p \circ \iota$, which on E_2-pages is the map

$$\mathrm{Ext}_{\mathcal{A}(2)}(\widetilde{H}^*(B\mathbb{Z}/2;\mathbb{Z}/2)) \longrightarrow \mathrm{Ext}_{\mathcal{A}(1)}(H^*(B\mathbb{Z}/2;\mathbb{Z}/2)). \tag{2.49}$$

Davis and Mahowald [**DM78**, Table 3.2] compute $\mathrm{Ext}_{\mathcal{A}(2)}(H^*(B\mathbb{Z}/2;\mathbb{Z}/2))$ in the degrees we need, and Gitler, Mahowald, and Milgram [**GMM68**, §2] provide a computation of $\mathrm{Ext}_{\mathcal{A}(1)}(H^*(B\mathbb{Z}/2;\mathbb{Z}/2))$. We draw the map (2.49) in Figure 4. All differentials in the spectral sequence over $\mathcal{A}(1)$ vanish using h_0- and h_1-equivariance, and by inspection there are no hidden extensions. Therefore we can identify some classes which survive $p \circ \iota$ and use this to trivialize some differentials in Figure 3.

- By computing the image of $p \circ \iota$ on Ext groups, we learn that the map $\widetilde{\Omega}_7^{\mathrm{String}}(B\mathbb{Z}/2) \to \widetilde{\Omega}_7^{\mathrm{Spin}}(B\mathbb{Z}/2)$ can be identified with the map $\mathbb{Z}/16 \oplus \mathbb{Z}/2 \to \mathbb{Z}/16$ sending $(1,0) \mapsto 1$ and $(0,1) \mapsto 0$.[15] Therefore, any differential to or from the four summands in topological degree 7 linked by h_0-actions must vanish, including (D2) and (D5).

[15] Alternatively, one could show that the $\mathbb{Z}/16 \subset \widetilde{\Omega}_7^{\mathrm{String}}(B\mathbb{Z}/2)$ is mapped injectively into $\widetilde{\Omega}_7^{\mathrm{Spin}}(B\mathbb{Z}/2)$ by checking on a generator. One can show that $\mathbb{RP}^7$ admits a string structure; then the generator of that $\mathbb{Z}/16$ subgroup of $\widetilde{\Omega}_7^{\mathrm{String}}(B\mathbb{Z}/2)$ is $\mathbb{RP}^7$ with its nontrivial principal $\mathbb{Z}/2$-bundle. Its image in $\widetilde{\Omega}_7^{\mathrm{Spin}}(B\mathbb{Z}/2)$ has order at least 16, because the η-invariant of a suitable twisted Dirac operator associated to the $\mathbb{Z}/2$-bundle defines a bordism invariant $\Omega_7^{\mathrm{Spin}}(B\mathbb{Z}/2) \to \mathbb{R}/\mathbb{Z}$, and on $(\mathbb{RP}^7, S^7 \to \mathbb{RP}^7)$, this η-invariant is $\ell/16 \bmod 1$ for some odd ℓ, as follows from a formula of Donnelly [**Don78**, Proposition 4.1].

- Similarly, the map $\widetilde{\Omega}_{11}^{\mathrm{String}}(B\mathbb{Z}/2) \to \widetilde{\Omega}_{11}^{\mathrm{Spin}}(B\mathbb{Z}/2)$ can be identified with the inclusion $\mathbb{Z}/8 \hookrightarrow \mathbb{Z}/128 \oplus \mathbb{Z}/8 \oplus \mathbb{Z}/2$ sending $1 \mapsto (16, 0, 0)$, which follows either by computing $p \circ \iota$ on Ext groups or computing η-invariants on the generator of $\widetilde{\Omega}_{11}^{\mathrm{String}}(B\mathbb{Z}/2)$, which can be taken to be the product of $\mathbb{RP}^3$ with a Bott manifold.[16] Thus (D6) vanishes. $\square$

LEMMA 2.50. *The differential* (D1) *is nonzero; specifically,* $d_2(a) = h_2^2 p_1$.

We will deduce this from the following fact.

PROPOSITION 2.51. *The map* $\Omega_4^{\xi^{r,\mathrm{het}}} \to \Omega_4^{\xi^{\mathrm{het}}}$ *is surjective after* 2*-completion.*

Recall that $\xi^{r,\mathrm{het}}$ is the analogue of ξ^{het} but with $(\mathrm{E}_8 \times \mathrm{E}_8) \rtimes \mathbb{Z}/2$ replaced with $\mathrm{E}_8 \times \mathrm{E}_8$.

PROOF OF LEMMA 2.50 ASSUMING PROPOSITION 2.51. In this proof, implicitly 2-complete all abelian groups. If $d_2(a) = 0$, then $h_2^2 p_1 \in E_2^{2,9}$ survives to the E_∞-page, so the h_2-action $E_\infty^{1,5} \to E_\infty^{2,9}$ is nonzero. This lifts to imply that taking the product with S^3 with string structure induced from its Lie group framing, which defines a map $\Omega_4^{\xi^{\mathrm{het}}} \to \Omega_7^{\xi^{\mathrm{het}}}$, is also nonzero. Direct products with framed manifolds correspond to action by elements of $\pi_*(\mathbb{S})$ on homotopy groups, so this product with S^3 is natural with respect to maps of spectra.

Since $\Omega_4^{\xi^{r,\mathrm{het}}} \to \Omega_4^{\xi^{\mathrm{het}}}$ is surjective, we may compute the product with S^3 as a map

$$- \times S^3 \colon \Omega_4^{\xi^{r,\mathrm{het}}} \longrightarrow \Omega_7^{\xi^{r,\mathrm{het}}} \tag{2.52}$$

and then map back to $\Omega_7^{\xi^{\mathrm{het}}}$. However, as we noted in Remark 2.34, $\Omega_7^{\xi^{r,\mathrm{het}}} \cong \Omega_7^{\mathrm{Spin}}(K(\mathbb{Z}, 4))$, and Stong [**Sto86**] showed $\Omega_7^{\mathrm{Spin}}(K(\mathbb{Z}, 4)) = 0$. Thus taking the product with S^3 is the zero map $\Omega_4^{\xi^{\mathrm{het}}} \to \Omega_7^{\xi^{\mathrm{het}}}$, which is incompatible with $d_2(a)$ vanishing. $\square$

PROOF OF PROPOSITION 2.51. Let F be the fiber of the map $\phi\colon MT\xi^{r,\mathrm{het}} \to MT\xi^{\mathrm{het}}$, so that there is a long exact sequence

$$\cdots \longrightarrow \Omega_4^{\xi^{r,\mathrm{het}}} \xrightarrow{\ \phi\ } \Omega_4^{\xi^{\mathrm{het}}} \xrightarrow{\ \partial\ } \pi_3(F) \longrightarrow \Omega_3^{\xi^{r,\mathrm{het}}} \longrightarrow \cdots \tag{2.53}$$

We will show $\pi_3(F)_2^\wedge = 0$, which implies the proposition statement by exactness. To do so, we must understand F.

(F is the fiber.)

We use a standard technique.

Let V be the rank-zero stable vector bundle on $B\mathbb{G}^{\mathrm{het}}$ classified by the map $\xi^{\mathrm{het}}\colon B\mathbb{G}^{\mathrm{het}} \to BO$ and let $\sigma \to B\mathbb{G}^{\mathrm{het}}$ be the line bundle classified by the map quotienting by $\mathbb{T}[1]$, then by Spin, then by E_8^2:

$$B\mathbb{G}^{\mathrm{het}} \longrightarrow B(\mathrm{E}_8^2 \rtimes \mathbb{Z}/2) \longrightarrow B\mathbb{Z}/2. \tag{2.54}$$

Then, inclusion of the zero section of σ defines a map of spaces over $\mathbb{Z} \times BO$: $\phi\colon (B\mathbb{G}^{\mathrm{het}}, V) \to (B\mathbb{G}^{\mathrm{het}}, V \oplus \sigma)$. Here we use the notation (B, ξ) to denote a space B and a map $\xi\colon B \to \mathbb{Z} \times BO$, and we use $\mathbb{Z} \times BO$ instead of BO because σ is not rank 0. Let M^- denote the Thom spectrum of $V \oplus \sigma\colon B\mathbb{G}^{\mathrm{het}} \to \mathbb{Z} \times BO$,

[16] All orientable 3-manifolds have trivializable tangent bundles, hence string structures; for a construction of a Bott manifold with string structure, see [**FH21a**, §5.3].

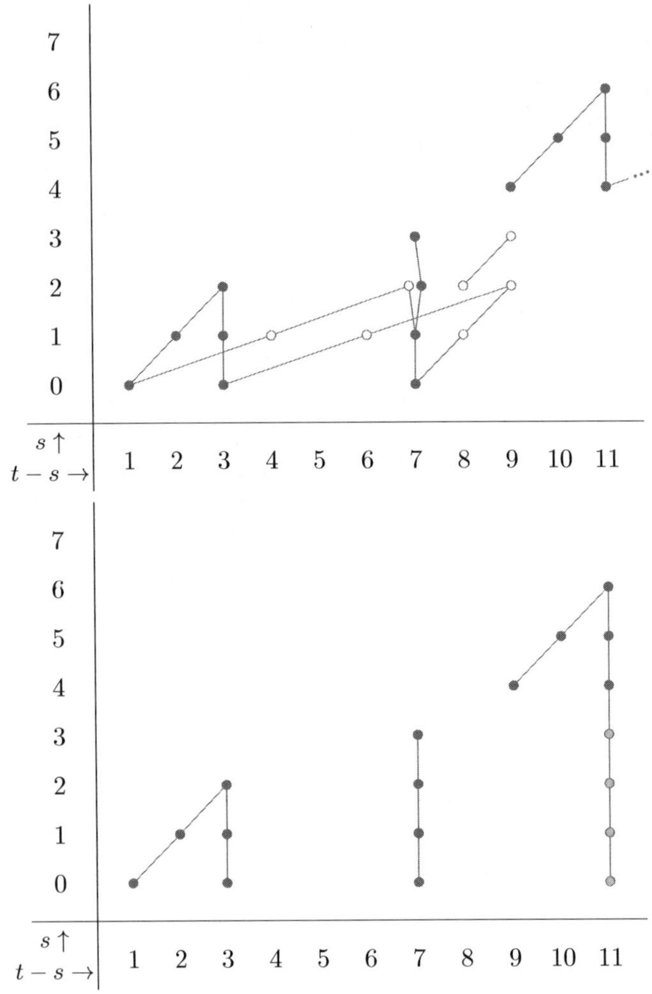

FIGURE 4. Top: $\mathrm{Ext}_{\mathcal{A}(2)}(\widetilde{H}^*(B\mathbb{Z}/2;\mathbb{Z}/2),\mathbb{Z}/2)$, the E_2-page of the Adams spectral sequence computing $\widetilde{\Omega}_*^{\mathrm{String}}(B\mathbb{Z}/2)_2^\wedge$. Filled dots have nonzero image in $\mathrm{Ext}_{\mathcal{A}(1)}$; unfilled dots are the kernel. Bottom: $\mathrm{Ext}_{\mathcal{A}(1)}(\widetilde{H}^*(B\mathbb{Z}/2;\mathbb{Z}/2),\mathbb{Z}/2)$, a summand of the E_2-page of the Adams spectral sequence computing $\widetilde{\Omega}_*^{\mathrm{Spin}}(B\mathbb{Z}/2)_2^\wedge$. Filled dots are in the image of the map from $\mathrm{Ext}_{\mathcal{A}(2)}$; gray dots are the cokernel. This map of spectral sequences is used in the proof of Lemma 2.47.

and let $\widetilde{\phi}\colon MT\xi \to M^-$ denote the map of Thom spectra induced by ϕ; we claim $F \simeq \Sigma^{-1}M^-$. To see this, we will use a theorem in [**DDK**$^+$] which identifies the fiber of $\widetilde{\phi}$ as the map $MT\xi^{r,\mathrm{het}} \to MT\xi^{\mathrm{het}}$. Specifically, [**DDK**$^+$] shows that the fiber of $\widetilde{\phi}$ is the Thom spectrum of the pullback of V to the sphere bundle $S(\sigma)$ of σ. This sphere bundle is the pullback of the universal sphere bundle over $B\mathbb{Z}/2$ by the classifying map of σ.

$$\begin{array}{ccc} S(\sigma) & \longrightarrow & S(L)\simeq E\mathbb{Z}/2 \\ \downarrow & \lrcorner & \downarrow \\ B\mathbb{G}^{\mathrm{het}} & \longrightarrow & B\mathbb{Z}/2 \end{array} \tag{2.55}$$

The sphere bundle of the tautological line bundle $L \to B\mathbb{Z}/2$ is $E\mathbb{Z}/2 \to B\mathbb{Z}/2$, which is contractible, so the pullback diagram (2.55) simplifies to a fiber diagram, and the sphere bundle is the fiber of (2.54). Since (2.54) was induced from a group homomorphism by taking classifying spaces, one can compute its fiber by taking the classifying space of the kernel of the homomorphism, which is $\mathcal{S}(\mathrm{Spin}\times \mathrm{E}_8^2, c_1 + c_2 - \lambda)$. In Remark 1.46 we saw that applying the Thom spectrum functor to $B\mathcal{S}(\mathrm{Spin}\times \mathrm{E}_8^2, c_1+c_2-\lambda) \to B\mathbb{G}^{\mathrm{het}}$, i.e., to the map $S(\sigma) \to B\mathbb{G}^{\mathrm{het}}$, produces the map $MT\xi^{r,\mathrm{het}} \to MT\xi^{\mathrm{het}}$, and therefore the fiber of this map is $\Sigma^{-1}M^-$.

To finish the proof, attack F with the Adams spectral sequence, using its description as the Thom spectrum $\Sigma^{-1}M$ to get a description in terms of Ext of an $\mathcal{A}(2)$-module by using [**DY23**] again. Recall from Figure 2, left, the $\mathcal{A}(2)$-module $\tau_{\le 13}M_5$; the result of the computation here is that the $\mathcal{A}(2)$-module relevant for computing $\pi_*(F)^\wedge_2$ agrees with $\Sigma^{-9}(\tau_{\le 13}M_5)$ in degrees 4 and below. Then, Figure 2, right, computes $\mathrm{Ext}_{\mathcal{A}(2)}(\Sigma^{-9}(\tau_{\le 13}M_5))$, which is the E_2-page of the Adams spectral sequence computing $\pi_*(F)^\wedge_2$, in degrees 3 and below (shift the topological degree of everything in Figure 2, right, down by 9). The E_2-page vanishes in topological degree 3, which implies $\pi_3(F)^\wedge_2 = 0$. □

LEMMA 2.56. *The differential* (D4) *vanishes.*

PROOF. The source of this differential is $E_2^{0,12} \cong \mathbb{Z}/2\cdot e$ in Adams filtration zero. Classes α in Adams filtration 0 are canonically identified with classes c_α forming a subgroup of mod 2 cohomology, and α survives to the E_∞-page if and only if the bordism invariant $\int c_\alpha$ is nonzero. Here, $\alpha = e$ and $c_\alpha = D_1D_2^2 + D_1^2D_2$, so our differential vanishes if and only if e survives to the E_∞-page if and only if the following invariant is nonzero:

$$\int (D_1D_2^2 + D_1^2D_2) : \Omega_{12}^{\xi^{\mathrm{het}}} \longrightarrow \mathbb{Z}/2. \tag{2.57}$$

We will produce a manifold on which this invariant is nonzero.

The quaternionic projective plane $\mathbb{HP}^2$ has $H^*(\mathbb{HP}^2;\mathbb{Z}) \cong \mathbb{Z}[x]/(x^3)$ with $|x| = 4$ and $\lambda(\mathbb{HP}^2) = x$ [**BH58**, §15.5, §15.6] (see also [**FH21a**, §5.2]). The Künneth formula tells us $H^*(\mathbb{HP}^2\times S^4;\mathbb{Z}) \cong \mathbb{Z}[x,y]/(x^3,y^2)$, with $|y| = 4$; since TS^4 is stably trivial, $\lambda(S^4)$ vanishes and the Whitney sum formula (Lemma 1.6) implies $\lambda(\mathbb{HP}^2\times S^4) = x$.

To define a ξ^{het}-structure on $\mathbb{HP}^2\times S^4$, it suffices to produce two E_8-bundles $P, Q \to \mathbb{HP}^2\times S^4$ and a trivialization of $\lambda(\mathbb{HP}^2\times S^4) - c(P) - c(Q)$. Since we can freely prescribe $c(P)$ and $c(Q)$, choose P and Q such that $c(P) = y$ and $c(Q) = x-y$; then $\lambda(\mathbb{HP}^2\times S^4) - c(P) - c(Q) = 0$, so we can choose a trivialization. Since $D_1 = c(P) \bmod 2$ and $D_2 = c(Q) \bmod 2$,

$$\int_{\mathbb{HP}^2\times S^4} (D_1D_2^2 + D_1^2D_2) = \left(\int_{\mathbb{HP}^2\times S^4} (yx^2+xy^2)\right) \bmod 2 = 1. \tag{2.58}$$

□

Now we have to tackle extension questions. In this part of the computation, it will be helpful to reference Figure 3, as we will use the description of the E_∞-page of this spectral sequence several times while addressing extension questions.

LEMMA 2.59. *In degrees* 10 *and below, all extension questions in the Adams spectral sequence for* $\pi_*(\mathcal{Q})_2^\wedge$ *either split or are detected by* h_0 *on the* E_∞*-page, except possibly for the extensions involving the classes* $c \in E_\infty^{0,9}$, $h_1^2 p_7 \in E_\infty^{2,11}$, *and* $h_1 b \in E_\infty^{3,12}$.

The classes $h_1 b$ and c may vanish on the E_∞-page, depending on the fate of the differentials in (D3).

PROOF. The h_0-action alone solves all extensions in this range except in degrees 8, 9, and 10.

If the d_2s in (D3) vanish, there is an extension question in degree 8. The h_0-actions in the tower generated by $h_0 a$ lift to produce a $\mathbb{Z}$ in $\Omega_8^{\xi^{\mathrm{het}}}$, so the only question is whether there is an extension involving $h_1 p_7$ and b. Suppose this extension does not split, so $\pi_8(\mathcal{Q})_2^\wedge \cong \mathbb{Z} \oplus \mathbb{Z}/4$. We can choose a generator x of this $\mathbb{Z}/4$ such that the image of x in the Adams E_∞-page is $h_1 p_7 \in E_\infty^{1,9}$; since this is h_1 times another class on the E_∞-page, x is η times a class $y \in \pi_7(\mathcal{Q})_2^\wedge$, where η is the generator of $\pi_1(\mathbb{S}) \cong \mathbb{Z}/2$. Since $2\eta = 0$, $2x = 2\eta y = 0$; since x was supposed to generate a $\mathbb{Z}/4$, this is a contradiction, and therefore this extension splits.

The same trick splits all extensions in degree 10, and all extensions involving the class in $E_\infty^{4,13}$. □

PROPOSITION 2.60. *All extension questions in* $\pi_9(\mathcal{Q})_2^\wedge$ *split, so* $\pi_9(\mathcal{Q})_2^\wedge \cong (\mathbb{Z}/2)^{\oplus 4}$ *if the differentials in* (D3) *vanish, and* $\pi_9(\mathcal{Q})_2^\wedge \cong (\mathbb{Z}/2)^{\oplus 2}$ *if they do not vanish.*

PROOF. If the differentials in (D3) do not vanish, this is a consequence of Lemma 2.59, so assume that those differentials vanish.

First suppose we can split all extensions involving c. Then the only extension remaining is between $h_1^2 p_7$ and $h_1 b$. In Lemma 2.59, we split the extension between $h_1 p_7$ and b, so the classes $h_1 p_7$ and b lift to classes $\underline{h_1 p_7}$, resp. $\underline{b}$, which generate a $\mathbb{Z}/2 \oplus \mathbb{Z}/2 \subset \pi_8(\mathcal{Q})_2^\wedge$. The action by h_1 lifts to imply that the images of $\eta \cdot \underline{h_1 p_7}$ and $\eta \cdot \underline{b}$ in the E_∞-page are $h_1^2 p_7$, resp. $h_1 b$, and η carries the $\mathbb{Z}/2 \oplus \mathbb{Z}/2$ generated by $\underline{h_1 p_7}$ and $\underline{b}$ to a $\mathbb{Z}/2 \oplus \mathbb{Z}/2 \subset \pi_9(\mathcal{Q})_2^\wedge$ generated by $\eta \underline{h_1 p_7}$ and $\eta \underline{b}$, thus splitting the extension between $h_1^2 p_7$ and $h_1 b$.

Now we need to prove that c lifts to a class $\underline{c}$ such that $2\underline{c} = 0$. Let X be the pullback

$$\begin{array}{ccc} X & \longrightarrow & B\mathbb{G}^{\mathrm{het}} \\ \big\downarrow & \lrcorner & \big\downarrow \\ \mathbb{RP}^2 & \longrightarrow & B\mathbb{Z}/2 \end{array} \tag{2.61}$$

and let $\xi \colon X \to BO$ be the pullback of ξ^{het} to X. Both vertical arrows in (2.61) are fibrations with fiber $B\mathrm{E}_8^2$; using the induced map of Serre spectral sequences, we learn $H^*(X;\mathbb{Z}/2) \cong H^*(B\mathbb{G}^{\mathrm{het}};\mathbb{Z}/2)/(x^3)$, where $x \in H^1(B\mathbb{G}^{\mathrm{het}};\mathbb{Z}/2)$ is the generator. One can replay the whole argument we ran with ξ in place of ξ^{het}, defining ξ' analogously to $\xi^{\mathrm{het}\prime}$, and deduce the following.

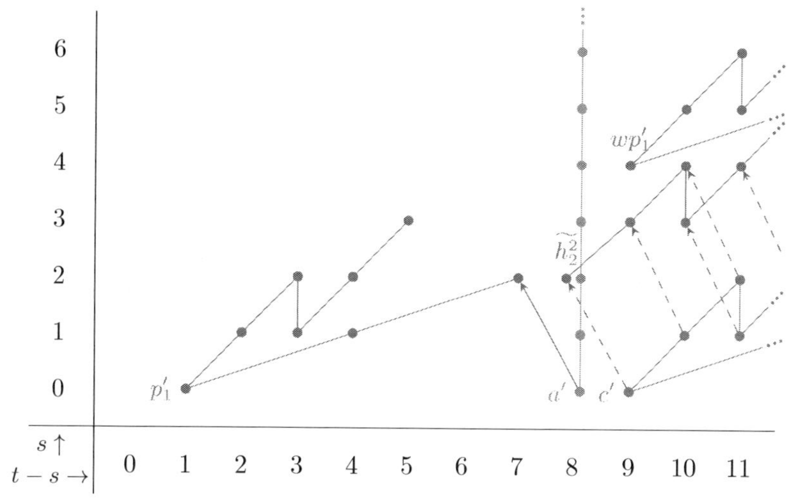

FIGURE 5. The E_2-page of the Adams spectral sequence computing $\pi_*(\mathcal{Q}')^\wedge_2$, where $\mathcal{Q}'$ is the spectrum defined in the proof of Proposition 2.60. By comparing with the Adams spectral sequence for $\mathcal{Q}$, we learn $d_2(a') = h_2^2 p_1'$ from Lemma 2.50, and that the dashed differentials (e.g., $d_2(c')$, $d_2(h_1 c')$) vanish if and only if the differentials in (D3) vanish.

(1) The map $c\colon B\mathrm{E}_8 \to K(\mathbb{Z}, 4)$ induces an isomorphism $\Omega_*^\xi \to \Omega_*^{\xi'}$ in degrees 14 and below,
(2) there is a spectrum $\mathcal{Q}'$ and a splitting $MT\xi \simeq MTSpin \vee \mathcal{Q}'$, and
(3) the map $X \to B\mathbb{G}^{\mathrm{het}}$ induces a map $MT\xi' \to MT\xi^{\mathrm{het}'}$ which is the identity on the $MTSpin$ factors and sends $\mathcal{Q} \to \mathcal{Q}'$.

The analogue of Proposition 2.41 for ξ' is exactly the same, except replacing M_2 with $\Sigma C2$ and M_5 with $\Sigma^9 C2$, where $C2$ is the $\mathcal{A}(2)$-module $\Sigma^{-1}\widetilde{H}^*(\mathbb{RP}^2; \mathbb{Z}/2)$. Bruner-Rognes [**BR21**, §6.1] compute $\mathrm{Ext}_{\mathcal{A}(2)}(C2)$, and using that we can draw the E_2-page of the Adams spectral sequence computing $\pi_*(\mathcal{Q}')^\wedge_2$ in Figure 5. For the classes p_1, a, and c we considered in the E_2-page of the Adams spectral sequence for $\mathcal{Q}$, let p_1', a', and c' be the corresponding classes in the E_2-page for $\mathcal{Q}'$: they live in the same bidegrees and the map $\mathcal{Q}' \to \mathcal{Q}$ carries $x' \to x$ for $x \in \{p_1, a, c\}$.

The point of all of this is that if the differentials in (D3) vanish, then both c and $h_1^2 p_7$ live to the E_∞-page for $\mathcal{Q}$, then both c and $h_1^2 p_7$ are in the image of the map Φ on E_∞-pages induced by $\mathcal{Q}' \to \mathcal{Q}$: $c = \Phi(c')$, and Bruner-Rognes [**BR21**, Corollary 4.3] define a class $h_2^2 \in \mathrm{Ext}^{2,9}_{\mathcal{A}(2)}(C2) = \mathrm{Ext}^{2,10}_{\mathcal{A}(2)}(\Sigma C2)$ such that $h_1^2 p_7 = \Phi(\widetilde{h_2^2})$. And looking at Figure 5, in the E_∞-page for $\mathcal{Q}'$, $h_1(\widetilde{h_2^2}) \neq 0$ and $h_1(wp_1') \neq 0$, so the $2\eta = 0$ trick from Lemma 2.59 splits the extensions in $\pi_9(\mathcal{Q}')^\wedge_2$. Thus there is a class $\underline{c}' \in \pi_9(\mathcal{Q}')^\wedge_2$ such that $2\underline{c}' = 0$ and the image of $\underline{c}'$ in the E_∞-page is c'. Applying $\Phi(c') = c$, we learn c lifts to $\Phi(\underline{c}')$ in $\pi_9(\mathcal{Q})^\wedge_2$, and twice this class is 0, as we wanted to prove. □

We have therefore proven the following theorem.

THEOREM 2.62. *Ignoring odd-primary torsion, there are isomorphisms*

$$\begin{aligned} \Omega_0^{\xi^{\mathrm{het}}} &\cong \mathbb{Z}, & \Omega_6^{\xi^{\mathrm{het}}} &\cong \mathbb{Z}/2,\\ \Omega_1^{\xi^{\mathrm{het}}} &\cong \mathbb{Z}/2\oplus\mathbb{Z}/2, & \Omega_7^{\xi^{\mathrm{het}}} &\cong \mathbb{Z}/16,\\ \Omega_2^{\xi^{\mathrm{het}}} &\cong \mathbb{Z}/2\oplus\mathbb{Z}/2, & \Omega_8^{\xi^{\mathrm{het}}} &\cong \mathbb{Z}^3\oplus(\mathbb{Z}/2)^{\oplus i},\\ \Omega_3^{\xi^{\mathrm{het}}} &\cong \mathbb{Z}/8, & \Omega_9^{\xi^{\mathrm{het}}} &\cong (\mathbb{Z}/2)^{\oplus j},\\ \Omega_4^{\xi^{\mathrm{het}}} &\cong \mathbb{Z}\oplus\mathbb{Z}/2, & \Omega_{10}^{\xi^{\mathrm{het}}} &\cong (\mathbb{Z}/2)^{\oplus k},\\ \Omega_5^{\xi^{\mathrm{het}}} &\cong 0, & \Omega_{11}^{\xi^{\mathrm{het}}} &\cong A, \end{aligned}$$

where:

- *A is an abelian group of order 64 isomorphic to one of $\mathbb{Z}/8\oplus\mathbb{Z}/8$, $\mathbb{Z}/16\oplus\mathbb{Z}/4$, $\mathbb{Z}/32\oplus\mathbb{Z}/2$, or $\mathbb{Z}/64$, and*
- *either $i=1$, $j=4$, and $j=4$, or $i=2$, $j=6$, and $k=5$.*

2.2.1. *Some manifold generators.* We finish this section by giving manifold representatives for all the generators for the groups we found in dimensions 10 and below, except possibly for two classes in degrees 9 and 10 if the differentials in (D3) vanish. We also give partial information in dimension 11. In this list, we implicitly localize at 2, though we will soon see in Theorem 2.74 that this does not lose any information.

The map $MTSpin \vee (MTString \wedge B\mathbb{Z}/2) \to MT\xi^{\mathrm{het}}$ is surjective on homotopy groups in degrees 7 and below, quickly giving us many of the generators we need. The low-dimensional generators of spin bordism are standard; for $\widetilde{\Omega}_*^{\mathrm{String}}(B\mathbb{Z}/2)$, we use the h_2-action on the E_∞-page together with the map $\widetilde{\Omega}_*^{\mathrm{String}}(B\mathbb{Z}/2) \to \widetilde{\Omega}_*^{\mathrm{Spin}}(B\mathbb{Z}/2)$, as in the proof of Lemma 2.47 (see Figure 4), to deduce generators.

(0) $\Omega_0^{\xi^{\mathrm{het}}} \cong \mathbb{Z}$, generated by the point.

(1) $\Omega_1^{\xi^{\mathrm{het}}} \cong \mathbb{Z}/2\oplus\mathbb{Z}/2$. The first summand comes from Ω_1^{Spin}, hence is generated by S^1_{nb}, the circle with ξ^{het}-structure induced from its nonbounding framing. The other summand, corresponding to $p_1 \in E_\infty^{0,1}$ of the Adams spectral sequence for $\mathcal{Q}$, is in Adams filtration zero, hence corresponds to a mod 2 cohomology class and is detected by that class. Looking at Figure 1, this class is the generator of $H^1(B\mathbb{Z}/2;\mathbb{Z}/2)$ evaluated on the principal $\mathbb{Z}/2$-bundle associated to a ξ^{het}-structure. Thus we can take as our generator S^1 with ξ^{het}-structure induced by the nontrivial $\mathbb{Z}/2$-bundle and the inclusion $\mathbb{Z}/2 \hookrightarrow \mathrm{E}_8^2 \rtimes \mathbb{Z}/2$. We will call this generator $\mathbb{RP}^1$, so that we can represent its $\mathbb{Z}/2$-bundle by $S^1 \to \mathbb{RP}^1$.

(2) An action by h_1 in the E_∞-page of an Adams spectral sequence calculating bordism lifts to taking the product with S^1_{nb} on manifold generators. Acting by h_1 defines an isomorphism from the 1-line of the E_∞-page to the 2-line, so we can take $S^1_{nb}\times S^1_{nb}$ and $\mathbb{RP}^1\times S^1_{nb}$ to be our two generators of $\Omega_2^{\xi^{\mathrm{het}}}$.

(3) $\Omega_3^{\xi^{\mathrm{het}}} \cong \mathbb{Z}/8$; there is a generator whose image in the Adams E_∞-page is p_3. The sequence of maps

$$\widetilde{\Omega}_3^{\mathrm{String}}(B\mathbb{Z}/2) \xrightarrow{\iota} \Omega_3^{\xi^{\mathrm{het}}} \xrightarrow{p} \Omega_3^{\mathrm{Spin}}(B\mathbb{Z}/2) \tag{2.63}$$

consists of two isomorphisms $\mathbb{Z}/8 \xrightarrow{\cong} \mathbb{Z}/8 \xrightarrow{\cong} \mathbb{Z}/8$, so it suffices to find a generator of $\Omega_3^{\mathrm{Spin}}(B\mathbb{Z}/2)$ that admits a string structure. The standard generator is $\mathbb{RP}^3$ with principal $\mathbb{Z}/2$-bundle $S^3 \to \mathbb{RP}^3$, and because $\mathbb{RP}^3$ is parallelizable, it admits a string structure.

(4) $\Omega_4^{\xi^{\mathrm{het}}} \cong \mathbb{Z}\oplus\mathbb{Z}/2$. The free summand comes from Ω_4^{Spin}, hence is generated by the K3 surface with trivial $\mathbb{Z}/2$-bundle, and E_8-bundles with characteristic classes $-\lambda(\mathrm{K3})$ and 0. $\mathbb{Z}/2$ corresponds to $E_\infty^{1,5} \cong \mathbb{Z}/2\cdot h_2p_1$. Action by h_2 lifts to the product with S^3 with its Lie group framing, so we can generate this summand with $S^3\times\mathbb{RP}^1$.

REMARK 2.64. In Proposition 2.51, we established that the homomorphism $\Omega_4^{\mathrm{Spin}}(K(\mathbb{Z},4)) \cong \Omega_4^{\xi^{r,\mathrm{het}}} \to \Omega_4^{\xi^{\mathrm{het}}}$ is surjective; using this, we can replace $S^3\times\mathbb{RP}^1$, which we will need later. Stong [**Sto86**] showed $\Omega_4^{\mathrm{Spin}}(K(\mathbb{Z},4)) \cong \mathbb{Z}\oplus\mathbb{Z}$; one $\mathbb{Z}$ factor comes from Ω_4^{Spin}, hence is represented by the K3 surface with trivial map to $K(\mathbb{Z},4)$. The other is detected by the bordism invariant which, given a 4-dimensional spin manifold X and a map $f\colon X\to K(\mathbb{Z},4)$, sends $X\mapsto \int_X f^*c$, where $c\in H^4(K(\mathbb{Z},4);\mathbb{Z})$ is the tautological class. For example, this invariant equals 1 on S^4 with its standard orientation and unique spin structure inducing that orientation, with the map to $K(\mathbb{Z},4)$ given by the class $1\in H^4(S^4;\mathbb{Z})\xrightarrow{\cong}\mathbb{Z}$.

The images of the two classes $(\mathrm{K3},0)$ and $(S^4,1)$ in $\Omega_4^{\xi^{\mathrm{het}}}$ must generate. Unsurprisingly, the K3 surface is sent to a generator of the $\mathbb{Z}$ summand we described above; this summand is detected by $\int p_1$. As this invariant vanishes on $(S^4,1)$, surjectivity of the map on Ω_4 implies that $(S^4,1)$ maps to the class of $\mathbb{RP}^1\times S^3$.[17] Thus the $\mathbb{Z}/2$ summand in $\Omega_4^{\xi^{\mathrm{het}}}$ can be generated by S^4 with trivial $\mathbb{Z}/2$-bundle and two E_8-bundles with characteristic classes $c=\pm1\in H^4(S^4;\mathbb{Z})$.

The map on Adams spectral sequences induced from the map of spectra $MT\xi^{r,\mathrm{het}}\to MT\xi^{\mathrm{het}}$ sends the class in the E_∞-page representing $(S^4,1)$ to 0 (see Francis [**Fra11**, §2] or Lee and Yonekura [**LY22**, §3.5] for the Adams spectral sequence for $\Omega_*^{\xi^{r,\mathrm{het}}} = \Omega_*^{\mathrm{Spin}}(K(\mathbb{Z},4))$), so the fact that the image of $(S^4,1)$ is nonzero in $\Omega_4^{\xi^{\mathrm{het}}}$ is analogous to a hidden extension.

(5) $\Omega_5^{\xi^{\mathrm{het}}} = 0$.

(6) $\Omega_6^{\xi^{\mathrm{het}}} \cong \mathbb{Z}/2$, and the image of a generator on the E_∞-page is h_2p_3, which lifts to imply that we can take $S^3\times\mathbb{RP}^3$ as a generator.

(7) $\Omega_7^{\xi^{\mathrm{het}}} \cong \mathbb{Z}/16$. This $\mathbb{Z}/16$ is detected by $\Omega_7^{\mathrm{Spin}}(B\mathbb{Z}/2)$ much like $\mathbb{RP}^3$ was, and we learn that this summand is generated by $\mathbb{RP}^7$ with $\mathbb{G}^{\mathrm{het}}$-bundle induced from the $\mathbb{Z}/2$-bundle $S^7\to\mathbb{RP}^7$, and is detected in the E_∞-page by p_7.

(8) $\Omega_8^{\xi^{\mathrm{het}}} \cong \mathbb{Z}^2\oplus\mathbb{Z}\oplus\mathbb{Z}/2$ together with an additional $\mathbb{Z}/2$ summand if the differentials in (D3) do not vanish.

- The first two free summands come from Ω_*^{Spin}; their generators may be taken to be the quaternionic projective plane $\mathbb{HP}^2$ and a Bott manifold B. One can choose B to have a string structure [**FH21a**,

[17]We thank Justin Kaidi for informing of us of this fact.

§5.3] and we do so. In both cases, the $\mathbb{Z}/2$-bundle associated to the ξ^{het}-structure is trivial; since B is string, we give it the ξ^{het}-structure in which both principal E_8-bundles are trivial. For $\mathbb{HP}^2$, $H^4(\mathbb{HP}^2;\mathbb{Z}) \cong \mathbb{Z}$ with generator x, as we discussed in the proof of Lemma 2.56; we choose a ξ^{het}-structure on $\mathbb{HP}^2$ with principal E_8-bundles $P, Q \to \mathbb{HP}^2$ with $c(P) = -x$ and Q trivial.

- The third free summand comes from the green h_0-tower in topological degree 8 in the Adams spectral sequence for $\pi_*(\mathcal{Q})$. This summand is detected by the bordism invariant

(2.65)
$$f := \int c(P)c(Q)\colon \Omega_8^{\xi^{\mathrm{het}}} \longrightarrow \mathbb{Z},$$

because this quantity can be nonzero (as we show below), it vanishes on the two generators we discovered for the other two free summands, and because it must vanish on the remaining, torsion summand. It is a consequence of Lemma 2.50 that the mod 2 reduction of (2.65), which is $\int D_1D_2$, vanishes. This is because every class $x \in E_2^{0,t}$ has an associated degree-t $\mathbb{Z}/2$ cohomology class c_x, and x lives to the E_∞-page if and only if the bordism invariant $\int c_x$ is nonvanishing. Thus the minimum nonzero value of $|f(M)|$, where M is a closed, 8-dimensional ξ^{het}-manifold, is at least 2.
Recall from the proof of Lemma 2.56 that $H^*(\mathbb{HP}^2;\mathbb{Z}) \cong \mathbb{Z}[x]/(x^3)$ with $|x| = 4$ and $\lambda(\mathbb{HP}^2) = x$. Consider the two E_8-bundles $P, Q \to \mathbb{HP}^2$ prescribed by $c(P) = 2x$ and $c(Q) = -x$; then $\lambda(\mathbb{HP}^2) - c(P) - c(Q) = 0$, so this data lifts to a ξ^{het}-structure, and

(2.66)
$$\int_{\mathbb{HP}^2} c(P)c(Q) = 2,$$

achieving the minimum. Therefore $\mathbb{HP}^2$ with these two principal E_8-bundles generates the final free summand.

- The $\mathbb{Z}/2$ summand that we know is present independent of any unresolved differentials is generated by h_1p_7, so as usual lifts to $S^1_{nb}\times\mathbb{RP}^7$.
- If $d_2(c) \neq 0$, there is an additional $\mathbb{Z}/2$ summand represented in the E_∞-page by b. We will discuss this summand, and its generator X_8, in §2.2.2.

(9) $\Omega_9^{\xi^{\mathrm{het}}} \cong (\mathbb{Z}/2)^{\oplus 2} \oplus (\mathbb{Z}/2)^{\oplus 2}$, and if the differentials in (D3) vanish, there is an additional $\mathbb{Z}/2 \oplus \mathbb{Z}/2$ summand.

- Two of the $\mathbb{Z}/2$ summands come from $\Omega_9^{\mathrm{Spin}} \cong (\mathbb{Z}/2)^{\oplus 2}$, where they are represented by the generators $\mathbb{HP}^2 \times S^1_{nb}$ and $B \times S^1_{nb}$, with ξ^{het}-structure induced from the corresponding generators in $\Omega_8^{\xi^{\mathrm{het}}}$.
- The other two $\mathbb{Z}/2$ summands that are present no matter the value of the undetermined differentials are in the image of the homomorphism $\iota\colon \widetilde\Omega_9^{\mathrm{String}}(B\mathbb{Z}/2) \to \Omega_9^{\xi^{\mathrm{het}}}$. The generator of the summand in lower Adams filtration has image in the E_∞-page equal to $h_1^2p_7$, so we obtain $S^1_{nb} \times S^1_{nb} \times \mathbb{RP}^7$.
The summand in higher Adams filtration has nonzero image in the group $\widetilde\Omega_9^{\mathrm{Spin}}(B\mathbb{Z}/2) \cong \mathbb{Z}/2\oplus\mathbb{Z}/2$, by inspection of Figure 4. The two generators of $\widetilde\Omega_9^{\mathrm{Spin}}(B\mathbb{Z}/2)$ can be taken to be $\mathbb{HP}^2\times\mathbb{RP}^1$ and $B\times\mathbb{RP}^1$;

to determine which we get, compose further with the Atiyah-Bott-Shapiro [**ABS64**] map $\widetilde{\Omega}_9^{\mathrm{Spin}}(B\mathbb{Z}/2) \to \widetilde{ko}_9(B\mathbb{Z}/2) \cong \mathbb{Z}/2$, which sends $[\mathbb{HP}^2 \times \mathbb{RP}^1] \mapsto 0$ and $[B \times \mathbb{RP}^1]$ to the generator. The image of the map of Adams spectral sequence in Figure 4 is contained in the summand whose image under the Atiyah-Bott-Shapiro map is nonzero, the image of our generator in $\widetilde{\Omega}_9^{\mathrm{Spin}}(B\mathbb{Z}/2)$ is bordant to $B\times\mathbb{RP}^1$; finally, since B and $\mathbb{RP}^1$ are both string, we can take $B\times\mathbb{RP}^1$ as our last generator in this dimension.

- If $d_2(h_1c) = 0$, there is another $\mathbb{Z}/2$ summand whose image in the E_∞-page is h_1b. Thus as usual it lifts to $S^1_{nb} \times X_8$, where X_8 is the manifold we describe in §2.2.2.
- If $d_2(c) = 0$, there is another $\mathbb{Z}/2$ summand whose image in the E_∞-page is c. We were unable to find a manifold X_9 representing this generator. Because c is in Adams filtration 0, corresponding to the mod 2 cohomology class D_1D_2x, if X_9 exists then one can detect it by showing $\int_{X_9} D_1D_2x = 1$.

(10) $\Omega_{10}^{\xi^{\mathrm{het}}} \cong (\mathbb{Z}/2)^{\oplus 3} \oplus \mathbb{Z}/2$, together with potentially another $\mathbb{Z}/2$ summand if the differentials in (D3) vanish.

- Three of the $\mathbb{Z}/2$ summands in $\Omega_{10}^{\xi^{\mathrm{het}}}$ come from $\Omega_{10}^{\mathrm{Spin}} \cong (\mathbb{Z}/2)^{\oplus 3}$. Their generators are known to be $B \times S^1_{nb} \times S^1_{nb}$, $\mathbb{HP}^2 \times S^1_{nb} \times S^1_{nb}$, and a *Milnor hypersurface* X_{10}, defined to be a smooth degree-$(1,1)$ hypersurface in $\mathbb{CP}^2 \times \mathbb{CP}^4$. Milnor [**Mil65**, §3] showed that X_{10} generates the last $\mathbb{Z}/2$ summand in $\Omega_{10}^{\mathrm{Spin}}$.
- The next $\mathbb{Z}/2$ summand is detected by the maps $\widetilde{\Omega}_{10}^{\mathrm{String}}(B\mathbb{Z}/2) \to \Omega_{10}^{\xi^{\mathrm{het}}}$ and $\Omega_{10}^{\xi^{\mathrm{het}}} \to \Omega_{10}^{\mathrm{Spin}}(B\mathbb{Z}/2)$, and by a similar argument to the one we gave for the higher-filtration orange $\mathbb{Z}/2$ summand in degree 9, we may choose $B \times \mathbb{RP}^1 \times S^1_{nb}$ as the generator.
- If $d_2(h_1c) = 0$, then there is an additional $\mathbb{Z}/2$ summand whose image in the E_∞-page is h_1c. Thus we can take $S^1_{nb} \times X_9$ for a manifold representative, though as discussed above we do not know what X_9 is.

(11) We have not determined generators for $\Omega_{11}^{\xi^{\mathrm{het}}}$, nor even its isomorphism type. This is a question whose answer would be useful for anomaly cancellation for the $\mathrm{E}_8 \times \mathrm{E}_8$ heterotic string; see Question 0.3 and §3.2.1. Nonetheless, the Adams argument we gave above implies $\Omega_{11}^{\xi^{\mathrm{het}}}$ contains a $\mathbb{Z}/8$ subgroup, the image of $\iota\colon \widetilde{\Omega}_{11}^{\mathrm{String}}(B\mathbb{Z}/2) \to \Omega_{11}^{\xi^{\mathrm{het}}}$. By comparing with the map $\widetilde{\Omega}_{11}^{\mathrm{String}}(B\mathbb{Z}/2) \to \widetilde{\Omega}_{11}^{\mathrm{Spin}}(B\mathbb{Z}/2)$ as in Figure 4, one learns that the class of $B \times \mathbb{RP}^3$ generates this $\mathbb{Z}/8$.

2.2.2. *X_8, a potentially nonzero class in $\Omega_8^{\xi^{\mathrm{het}}}$.* Though we were unable to determine if the class $b \in E_2^{2,10}$ survives to the E_∞-page, we are able to write down a manifold representative X_8 of the class it determines in $\Omega_8^{\xi^{\mathrm{het}}}$; if b does survive, X_8 should be added to the list of generators above.

DEFINITION 2.67. Let $\mathbb{Z}/2$ act on $S^3 \times S^3 \times S^2$ by the antipodal map on S^2 and the first copy of S^3, and a reflection through a plane on the second S^3. This is a free action; let X_8 denote the quotient, which is a smooth manifold.

X_8 is a generalized Dold manifold of the sort studied by Nath and Sankaran [**NS19**]. Manifolds similar to X_8 frequently appear as generators of bordism groups: see [**FH21a**, §5.5.1] and [**DDHM23**, §14.3.3] for related examples.

LEMMA 2.68. *X_8 admits a string structure, and one can choose a string structure on X_8 so that the induced string structure on $S^3 \times S^3$ is the one induced by the Lie group framing on $S^3 \times S^3 \cong \mathrm{SU}_2 \times \mathrm{SU}_2$.*

PROOF. Adding the normal bundles for $S^{k-1} \hookrightarrow \mathbb{R}^k$ defines an isomorphism

$$T(S^3 \times S^3 \times S^2) \oplus \underline{\mathbb{R}}^3 \xrightarrow{\cong} \underline{\mathbb{R}}^4 \oplus \underline{\mathbb{R}}^4 \oplus \underline{\mathbb{R}}^3. \tag{2.69}$$

To understand TX_8, we will study (2.69) when we introduce the $\mathbb{Z}/2$-action on $S^3 \times S^3 \times S^2$ whose quotient is X_8. Since the outward unit normal vector field on S^k is O_{k+1}-invariant, $\mathbb{Z}/2$ acts trivially on the $\underline{\mathbb{R}}^3$ on the left side of (2.69), since the outward unit normal vector field provides the trivializations of the normal bundles giving that $\underline{\mathbb{R}}^3$ factor. On the right-hand side, $\mathbb{Z}/2$ by the antipodal map on the first factor of S^3, so acts by -1 on each $\underline{\mathbb{R}}$ summand of the first $\underline{\mathbb{R}}^4$. The reflection on the second S^3 factor means $\mathbb{Z}/2$ acts on the second $\underline{\mathbb{R}}^4$ by -1, 1, 1, and 1 on the four $\underline{\mathbb{R}}$ summands. Finally, the antipodal map on S^2 implies $\mathbb{Z}/2$ acts by -1 on the remaining three $\underline{\mathbb{R}}$ summands.

Passing from equivariant vector bundles on $S^3 \times S^3 \times S^2$ to nonequivariant vector bundles on the quotient, (2.69) induces an isomorphism

$$TX_8 \oplus \underline{\mathbb{R}}^3 \xrightarrow{\cong} \sigma^{\oplus 8} \oplus \underline{\mathbb{R}}^3, \tag{2.70}$$

where $\sigma \to X_9$ is pulled back from the tautological line bundle $\sigma \to \mathbb{RP}^2$. The Whitney sum formula implies $\sigma^{\oplus 8} \to \mathbb{RP}^2$ is spin, and since the string obstruction lives in $H^4(\mathbb{RP}^2; \mathbb{Z}) = 0$, $\sigma^{\oplus 8}$ is string. Thus the pullback to X_8 is also string, so TX_8 is string.

For the Lie group framing string structure, use the fact that the involutions on each S^3 summand can be described in terms of Lie groups: since the quotient of $S^3 \cong \mathrm{SU}_2$ by the antipodal map is $\mathbb{RP}^3 \cong \mathrm{SO}_3$, the Lie group framing on S^3 is equivariant for the antipodal map. Compatibility for the reflection comes from the action of a reflection in $\mathrm{Pin}_3^+ \supset \mathrm{SU}_2$. □

PROPOSITION 2.71. *With the string structure described in Lemma 2.68 and the $\mathbb{Z}/2$-bundle $\sigma \to X_8$, $[X_8]$ is linearly independent from $[S^1_{nb} \times \mathbb{RP}^7]$ in*

$$\widetilde{\Omega}_8^{\mathrm{String}}(B\mathbb{Z}/2) \cong \mathbb{Z}/2 \oplus \mathbb{Z}/2,$$

so the image of $[X_8]$ in the E_∞-page for $\Omega_*^{\mathrm{String}}(B\mathbb{Z}/2)$ is the nonzero class in $E_\infty^{2,10} \cong \mathbb{Z}/2$ (perhaps plus a term in lower filtration).

PROOF. Let $f\colon \widetilde{\Omega}_8^{\mathrm{String}}(\mathbb{RP}^2) \to \widetilde{\Omega}_8^{\mathrm{String}}(B\mathbb{Z}/2)$ be the map induced by $\mathbb{RP}^2 \hookrightarrow \mathbb{RP}^\infty \simeq B\mathbb{Z}/2$. The map this induces on Adams spectral sequences is not hard to analyze: Bruner and Rognes [**BR21**, §4.4, Chapter 6, §12.1] run the whole Adams spectral sequence for $\widetilde{tmf}_*(\mathbb{RP}^2)$, using the identification $\Sigma^\infty \mathbb{RP}^2 \simeq \Sigma\mathbb{S}/2$, and as discussed above *tmf*- and *MTString*-homology agree in degrees 14 and below.[18] Likewise, Davis and Mahowald [**DM78**, Table 3.2] compute the E_2-page of the Adams spectral sequence for $\widetilde{\Omega}_*^{\mathrm{String}}(B\mathbb{Z}/2)$ in the range we need, and with their

[18]See also the closely related work of Beaudry, Bobkova, Pham, and Xu [**BBPX22**], who compute $tmf_*(\mathbb{RP}^2)$ using the elliptic spectral sequence.

calculation, h_i-linearity of differentials, and the $2\eta = 0$ trick from the proof of Lemma 2.59 one sees that $\widetilde{\Omega}_8^{\mathrm{String}}(B\mathbb{Z}/2) \cong \mathbb{Z}/2 \oplus \mathbb{Z}/2$. As discussed in §2.2.1, one of the $\mathbb{Z}/2$ summands is detected by $\mathbb{RP}^7 \times S^1_{nb}$, whose image in the E_∞-page is in filtration 1. Consider the map

$$\Psi\colon \mathrm{Ext}_{\mathcal{A}(2)}(\widetilde{H}^*(\mathbb{RP}^2;\mathbb{Z}/2)) \longrightarrow \mathrm{Ext}_{\mathcal{A}(2)}(\widetilde{H}^*(B\mathbb{Z}/2;\mathbb{Z}/2)) \tag{2.72}$$

induced by $\mathbb{RP}^2 \to \mathbb{RP}^\infty \simeq B\mathbb{Z}/2$; we draw this map in Figure 6. Ψ is also the map between the E_2-pages of these two Adams spectral sequences; looking at Figure 6, Ψ is injective in topological degree 8, with image containing the nonzero element of $E_2^{2,10}$ but not the nonzero class in $E_2^{1,9}$. As both of these elements survive to the E_∞-page, this lifts to imply that $f\colon \widetilde{\Omega}_8^{\mathrm{String}}(\mathbb{RP}^2) \to \widetilde{\Omega}_8^{\mathrm{String}}(B\mathbb{Z}/2)$ is injective and that if one wants to find a class in $\widetilde{\Omega}_8^{\mathrm{String}}(B\mathbb{Z}/2)$ linearly independent from $\mathbb{RP}^7 \times S^1_{nb}$, it suffices to find a nonzero class in $\widetilde{\Omega}_8^{\mathrm{String}}(\mathbb{RP}^2)$.

The map $\sigma\colon X_8 \to B\mathbb{Z}/2$ factors through $\mathbb{RP}^2$ by definition, so we are done if we can show X_8, with its map to $\mathbb{RP}^2$, is nonbounding. To do so, consider the transfer map $\Sigma^\infty\mathbb{RP}^2 \to \Sigma^\infty S^2$ associated to the double cover $S^2 \to \mathbb{RP}^2$; this induces on string bordism a map $\widetilde{\Omega}_*^{\mathrm{String}}(\mathbb{RP}^2) \to \widetilde{\Omega}_*^{\mathrm{String}}(S^2)$ sending $(M, f\colon M \to \mathbb{RP}^2)$ to the double cover $M' \to M$ associated to the line bundle $f^*\sigma$, together to the map $M' \to S(\sigma) = S^2$.

The map $\Omega_k^\xi \to \widetilde{\Omega}_{k+\ell}^\xi(S^\ell)$ sending $M \mapsto (M \times S^\ell, \mathrm{proj}_2\colon M \times S^\ell \to S^\ell)$ (where S^ℓ carries the bounding stable framing, which with the ξ-structure on M induces a ξ-structure on $M \times S^\ell$) is always an isomorphism (e.g., check this with the Atiyah-Hirzebruch spectral sequence), and $\Omega_6^{\mathrm{String}} \cong \mathbb{Z}/2 \times \mathbb{Z}/2$ [**Gia71**, §3, §4], generated by $S^3 \times S^3$ with its Lie group framing, because it is represented by h_2^2 in the Adams spectral sequence. Therefore $\widetilde{\Omega}_8^{\mathrm{String}}(S^2) \cong \mathbb{Z}/2$ is generated by $S^3 \times S^3 \times S^2$, with the map to S^2 given by projection onto the third factor. The image of X_8 under the transfer is its double cover, which is $S^3 \times S^3 \times S^2$, with the correct string structure and map to S^2, so $[X_8] \neq 0$ in $\widetilde{\Omega}_8^{\mathrm{String}}(\mathbb{RP}^2)$, which suffices to prove the theorem. □

Finally, by looking at the map $\widetilde{\Omega}_*^{\mathrm{String}}(B\mathbb{Z}/2) \to \Omega_*^{\xi^{\mathrm{het}}}$, we conclude with Corollary 2.73.

COROLLARY 2.73. *Suppose $d_2(c) = 0$ in the Adams spectral sequence for ξ^{het}. Then $[X_8] \neq 0$ in $\Omega_8^{\xi^{\mathrm{het}}}$, and its image in the E_∞-page is the class $b \in E_\infty^{2,10}$ (perhaps plus some elements in lower filtration).*

2.3. ξ^{het} bordism at odd primes.

THEOREM 2.74. *$\Omega_*^{\xi^{\mathrm{het}}}$ has no odd-primary torsion in degrees 11 and below.*

PROOF. This amounts to a direct computation with the Adams spectral sequence. We will go over the case $p = 3$ in detail; for $p = 5, 7$ the story is similar but easier, and for $p \geq 11$ it is trivial because the degrees of the Steenrod powers are too high for the Adams spectral sequence to produce torsion.

First we compute $H^*(B\mathbb{G}^{\mathrm{het}};\mathbb{Z}/3)$ as a module over the Steenrod algebra $\mathcal{A}$ in low degrees in Proposition 2.77, then we do the same for $H^*(MT\xi^{\mathrm{het}};\mathbb{Z}/3)$ in Proposition 2.83. Once we have this, we can run the Adams spectral sequence, and do so in Proposition 2.86.

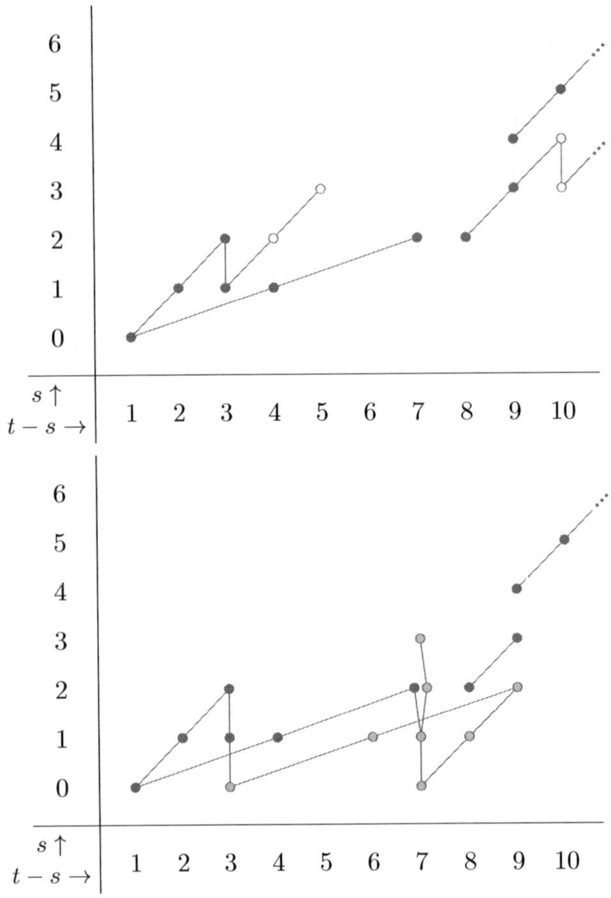

FIGURE 6. The map between the Adams spectral sequences for reduced string bordism induced by the map $\mathbb{RP}^2 \hookrightarrow \mathbb{RP}^\infty \simeq B\mathbb{Z}/2$, which we use in the proof of Proposition 2.71. Top: $\mathrm{Ext}_{\mathcal{A}(2)}(\widetilde{H}^*(\mathbb{RP}^2;\mathbb{Z}/2),\mathbb{Z}/2)$, the E_2-page of the Adams spectral sequence computing $\widetilde{\Omega}_*^{\mathrm{String}}(\mathbb{RP}^2)_2^\wedge$. Filled dots have nonzero image after mapping to $\mathbb{RP}^\infty$; unfilled dots are the kernel. Bruner and Rognes [**BR21**, §4.4, Chapter 6, §12.1] compute these Ext groups and run this Adams spectral sequence; from their work we learn there are no differentials in this range (though there are hidden ν-extensions that do not enter into our argument; see (*ibid.*, Theorem 12.5)). Bottom: $\mathrm{Ext}_{\mathcal{A}(2)}(\widetilde{H}^*(B\mathbb{Z}/2;\mathbb{Z}/2),\mathbb{Z}/2)$, a summand of the E_2-page of the Adams spectral sequence computing $\widetilde{\Omega}_*^{\mathrm{String}}(B\mathbb{Z}/2)_2^\wedge$. Filled dots are in the image of the map from $\mathbb{RP}^2$; gray dots are the cokernel. The E_2-page was computed by Davis and Mahowald [**DM78**, Table 3.2], and during the proof of Proposition 2.71 we argue that there are no differentials or hidden extensions by 2 in the range depicted.

Throughout this subsection, $\mathcal{P}^i$ refers to the ith Steenrod power, a degree-$4i$ operation on mod 3 cohomology, and β is the Bockstein homomorphism for the sequence $0 \to \mathbb{Z}/3 \to \mathbb{Z}/9 \to \mathbb{Z}/3 \to 0$.

LEMMA 2.75. *Let $C \in H^3(K(\mathbb{Z},3);\mathbb{Z}/3)$ denote the mod 3 reduction of the tautological class. Then*

$$H^*(K(\mathbb{Z},3);\mathbb{Z}/3) \cong \mathbb{Z}/3[C, \mathcal{P}^1 C, \beta\mathcal{P}^1 C, \dots]/(C^2, \dots), \tag{2.76}$$

where all missing generators and relations are in degrees 14 and above.

PROOF. This is a standard application of the Serre spectral sequence for the fibration $K(\mathbb{Z},2) \to * \to K(\mathbb{Z},3)$, so we will be succinct. $E_2^{0,*} \cong H^*(K(\mathbb{Z},2);\mathbb{Z}/3) \cong \mathbb{Z}/3[x]$, with $|x| = 2$; by the E_∞-page, all powers of x must be killed by differentials.

The only way to kill x is with a transgressing $d_3\colon E_3^{0,2} \to E_3^{3,0}$. Let $C := d_3(x)$. $C^2 = 0$ follows by graded commutativity. The Leibniz rule for differentials means that when $3 \nmid k$, $d_3(x^k) = \pm x^{k-1}C$, and if $3 \mid k$, $d_3(x^k) = 0$.

So x^3 survives to the E_4-page. The only remaining differential that can kill x^3 is the transgressing $d_7\colon E_7^{0,6} \to E_7^{7,0}$, so $d_7(x^3) \neq 0$; by the Kudo transgression theorem [**Kud56**], because $x^3 = \mathcal{P}^1(x)$, $d_7(x^3) = \mathcal{P}^1 C$. The Leibniz rule then implies $d_7(x^6) = x^3\mathcal{P}^1 C$, so by the E_8-page, everything on the line $p = 0$ in total degree less than 18 has been killed.

Because $d_3(x^3) = 0$, x^2C survives to the E_4-page; the only remaining way for it to support a differential is to have a new class $w \in H^8(K(\mathbb{Z},3);\mathbb{Z}/3)$ such that $d_5(x^2C) = w$. To see that $\beta(\mathcal{P}^1 C) = \pm w$, compare with the analogous spectral sequence for $\mathbb{Z}/9$-valued cohomology to see that $\mathcal{P}^1 C$ is not in the image of the mod 3 reduction map from $\mathbb{Z}/9$ cohomology to $\mathbb{Z}/3$ cohomology.[19] □

PROPOSITION 2.77. *Let $D \in H^4(B\mathrm{E}_8;\mathbb{Z}/3)$ be the mod 3 reduction of the class c from Definition 1.4, and let D_1 and D_2 be the two copies of D in $H^*(B\mathrm{E}_8^2;\mathbb{Z}/3)$ coming from the two factors of $B\mathrm{E}_8$. In degrees 13 and below, the pullback map on $\mathbb{Z}/3$ cohomology induced by $\phi\colon B\mathbb{G}^{\mathrm{het}} \to B\mathrm{Spin} \times B(\mathrm{E}_8^2 \rtimes \mathbb{Z}/2)$ is the quotient ring homomorphism sending $\lambda - D_1 - D_2 \mapsto 0$, $-p_2 - \mathcal{P}^1(D_1 + D_2) \mapsto 0$, and $\beta\mathcal{P}^1(D_1 + D_2) \mapsto 0$.*

Here ϕ is the map we constructed in (1.44) which forgets the B-field.

PROOF. Throw the Serre spectral sequence at the fibration

$$K(\mathbb{Z},3) \longrightarrow B\mathbb{G}^{\mathrm{het}} \longrightarrow B\mathrm{Spin} \times B(\mathrm{E}_8^2 \rtimes \mathbb{Z}/2). \tag{2.78}$$

The base space is not simply connected, so we might have to worry about local coefficients, but this turns out not to be the case, because the $\mathbb{Z}/2$ symmetry swapping the two E_8 factors, which is the origin of the π_1 in the base, acts trivially on the B-field, which gives us the fiber in (2.78).

In order to run the Serre spectral sequence for (2.78), we need to know the cohomology of $B\mathrm{Spin}$ and $B(\mathrm{E}_8^2 \rtimes \mathbb{Z}/2)$. The former is the polynomial ring on the mod 3 reductions of the Pontrjagin classes, which is a theorem of Borel and

[19] Alternatively, one could deduce this Bockstein by setting up the Serre spectral sequence for $K(\mathbb{Z},3) \to * \to K(\mathbb{Z},4)$ and Hill's calculation [**Hil09**, Corollary 2.9, Figure 1(a)] of the low-degree mod 3 cohomology of $K(\mathbb{Z},4)$ as an $\mathcal{A}$-module: C transgresses to a degree-4 class D, and Hill shows $\beta(\mathcal{P}^1 D) \neq 0$, so by the Kudo transgression theorem [**Kud56**], $\beta(\mathcal{P}^1 C) \neq 0$ in $H^*(K(\mathbb{Z},3);\mathbb{Z}/3)$.

Hirzebruch [**BH59**, §30.2]; for the latter, run the Serre spectral sequence for the fibration

$$BE_8^2 \longrightarrow B(E_8^2 \rtimes \mathbb{Z}/2) \longrightarrow B\mathbb{Z}/2. \tag{2.79}$$

Because $H^*(B\mathbb{Z}/2;\mathbb{Z}/3)$ vanishes in positive degrees, this Serre spectral sequence collapses to imply

$$H^*(B(E_8^2 \rtimes \mathbb{Z}/2);\mathbb{Z}/3) \xrightarrow{\cong} H^*(BE_8^2;\mathbb{Z}/3)^{\mathbb{Z}/2}. \tag{2.80}$$

The answer now follows from the Künneth formula, the fact that we can replace BE_8 with $K(\mathbb{Z},4)$ in the range we need by the result of [**BS58**, Theorems IV, V(e)] we mentioned in §2.2, and the mod 3 cohomology of $K(\mathbb{Z},4)$ in low degrees, worked out by Cartan [**Car54**] and Serre [**Ser52**], and stated explicitly by Hill [**Hil09**, Corollary 2.9].

Now back to (2.78) and its Serre spectral sequence. The fibration (2.78) is classified by the degree-4 cohomology class $\lambda - D_1 - D_2$, i.e., it is the pullback of the universal $K(\mathbb{Z},3)$-bundle

$$K(\mathbb{Z},3) \longrightarrow EK(\mathbb{Z},3) \longrightarrow BK(\mathbb{Z},3) \simeq K(\mathbb{Z},4) \tag{2.81}$$

by the map $B\mathrm{Spin} \times B(E_8^2 \rtimes \mathbb{Z}/2) \to K(\mathbb{Z},4)$ classified by $\lambda - D_1 - D_2$.[20] In the Serre spectral sequence for (2.81), the class $C \in E_2^{0,3} = H^3(K(\mathbb{Z},3);\mathbb{Z}/3)$ must transgress to the generator of $E_2^{4,0} = H^4(K(\mathbb{Z},4);\mathbb{Z}/3)$, and this generator pulls back to $\lambda - D_1 - D_2$, enforcing the relation $\lambda - D_1 - D_2 = 0$ in the E_5-page.

The other two pullbacks to zero in the theorem statement then follow from the Kudo transgression theorem [**Kud56**]: $\mathcal{P}^1 C \in E_2^{0,7} = H^7(K(\mathbb{Z},3);\mathbb{Z}/3)$ must transgress to $\mathcal{P}^1(\lambda - D_1 - D_2)$, and analogously for $\beta\mathcal{P}^1 C$. To compute these, we must determine how $\mathcal{P}^1$ acts on the mod 3 reductions of Pontrjagin classes. Shay [**Sha77**] proves a formula for Steenrod powers of Chern classes, which yields the formula for Pontrjagin classes by pullback. Hence, as worked out by Nordström [**Nor**], $\mathcal{P}^1 p_1 = p_2$; then an Adem relation tells us

$$\mathcal{P}^1 p_2 = \mathcal{P}^1\mathcal{P}^1 p_1 = -\mathcal{P}^2 p_1 = p_1^3, \tag{2.82}$$

the last equality because $\mathcal{P}^i$ is the cup product cube on classes of degree $2i$. Thus we see that $\mathcal{P}^1 C$ transgresses to $-p_2 - \mathcal{P}^1(D_1 + D_2)$ and $\beta\mathcal{P}^1 C$ transgresses to $\beta\mathcal{P}^1(D_1 + D_2)$, killing those classes by the E_{10}-page.

Now, the Leibniz rule cleans up the rest of the Serre spectral sequence in total degree at most 13: by the E_{10}-page, everything in this range is concentrated on the line $q = 0$. Therefore on the E_∞-page, the extension question is trivial in this range, and we conclude. □

PROPOSITION 2.83. *Let $\mathcal{M}_3$ denote the quotient of $H^*(MT\xi^{\mathrm{het}};\mathbb{Z}/3)$ by all elements of degree* 14 *or higher, $\mathcal{M}_3^{\mathrm{SO}}$ denote the quotient of $H^*(MTSO;\mathbb{Z}/3)$ by all elements of degree* 14 *or higher, and $C\alpha$ denote the $\mathcal{A}$-module which consists of two $\mathbb{Z}/3$ summands in degrees* 0 *and* 4 *linked by $\mathcal{P}^1$. Then, there is an isomorphism of $\mathcal{A}$-modules*

$$\mathcal{M}_3 \cong \mathcal{M}_3^{\mathrm{SO}} \oplus \Sigma^8 C\alpha \oplus \Sigma^{12}\mathbb{Z}/3. \tag{2.84}$$

[20]This map, and hence also the fibration, is only determined up to homotopy, but any two choices of representative give isomorphic answers.

PROOF. In Proposition 2.77, we discovered that the map $\phi\colon B\mathbb{G}^{\mathrm{het}} \to B\mathrm{Spin} \times B(\mathrm{E}_8^2 \rtimes \mathbb{Z}/2))$ induces a surjection on mod 3 cohomology in degrees 13 and below. As ϕ commutes with the maps down to BO that are part of the definition of these tangential structures, ϕ induces a map on Thom spectra

$$MT\xi^{\mathrm{het}} \to MTSpin \wedge B(\mathrm{E}_8^2 \rtimes \mathbb{Z}/2)_+. \tag{2.85}$$

Both of these tangential structures' maps to BO factor through $B\mathrm{SO}$, so the Thom isomorphism for mod 3 cohomology untwists. The Thom isomorphism is natural for maps of tangential structures, so we conclude that the pullback map on mod 3 cohomology induced by (2.85) is a surjection in degrees 13 and below — and therefore that we can compute Steenrod powers in the cohomology of the latter Thom spectrum. And the map $MTSpin \to MTSO$ is an equivalence away from 2, so we may work with $MTSO$ in place of $MTSpin$. Milnor [**Mil60**, Theorem 4] computed the Steenrod module structure on $H^*(MTSO;\mathbb{Z}/3)$, showing that it is a free $\mathcal{A}/\beta$-module. Using this, we can determine the Steenrod powers of Up_i, where U is the Thom class; and this and the Cartan formula finish the proof. □

PROPOSITION 2.86. *In topological degrees* 12 *and below, the Adams E_2-page computing $(\Omega_*^{\xi^{\mathrm{het}}})_3^\wedge$ consists of h_0-towers concentrated in even topological degrees, and therefore this Adams spectral sequence collapses in degrees* 12 *and below.*

PROOF. The direct-sum decomposition in Proposition 2.83 means that it suffices to prove the statement about h_0-towers for $\mathcal{M}_3^{\mathrm{SO}}$, $\Sigma^8 C\alpha$, and $\Sigma^{12}\mathbb{Z}/3$ separately. As usual, with M an $\mathcal{A}$-module, we write $\mathrm{Ext}(M)$ to denote $\mathrm{Ext}_{\mathcal{A}}^{*,*}(M,\mathbb{Z}/3)$. The first ingredient we need is $\mathrm{Ext}(\mathbb{Z}/3)$ itself; the computation of $\mathrm{Ext}_{\mathcal{A}}(\mathbb{Z}/3)$ in degrees $t-s \le 11$ is due to Gershenson [**Ger63**]; May [**May65, May66**] expanded this computation to $t-s \le 88$. In topological degrees 2 and below, $\mathrm{Ext}(\mathbb{Z}/3)$ consists of a single h_0-tower in topological degree 0, implying the conclusion for $\Sigma^{12}\mathbb{Z}/3$.

Next, we compute $\mathrm{Ext}(C\alpha)$ using the fact that a short exact sequence of $\mathcal{A}$-modules induces a long exact sequence in Ext groups. Specifically, factor $C\alpha$ as an extension of $\mathcal{A}$-modules

$$0 \longrightarrow \Sigma^4\mathbb{Z}/3 \longrightarrow C\alpha \longrightarrow \mathbb{Z}/3 \longrightarrow 0, \tag{2.87}$$

which we draw in Figure 7, left, and compute the corresponding long exact sequence in Ext in Figure 7, right. There is one potentially nonzero boundary map in range: $\partial\colon \mathrm{Ext}_{\mathcal{A}}^{0,4}(\mathbb{Z}/3) \to \mathrm{Ext}_{\mathcal{A}}^{1,4}(\mathbb{Z}/3)$. This map must be nonzero because $\mathrm{Ext}_{\mathcal{A}}^{0,4}(C\alpha) = \mathrm{Hom}_{\mathcal{A}}(C\alpha, \Sigma^4\mathbb{Z}/3) = 0$. We see that in degrees 6 and below, $\mathrm{Ext}(C\alpha)$ consists solely of h_0-towers in even degrees, which implies the part of the corollary statement coming from $\Sigma^8 C\alpha$.

Finally, $\mathcal{M}_3^{\mathrm{SO}}$. Milnor [**Mil60**, Theorem 4] showed that this module coincides with a free $\mathcal{A}/\beta$-module in degrees 13 and below, and proves (*ibid.*, Lemma 5) that the Ext groups of such a module consist solely of h_0-towers in even topological degree. Therefore in topological degrees 12 and below, $\mathrm{Ext}(\mathcal{M}_3^{\mathrm{SO}})$ also consists solely of h_0-towers in even topological degrees. □

This suffices to prove Theorem 2.74 for $p=3$: h_0-towers on the E_∞-page lift to $\mathbb{Z}_3$ (i.e., the 3-adic integers) summands in $(\Omega_*^{\xi^{\mathrm{het}}})_3^\wedge$, so there is no 3-torsion in this range. □

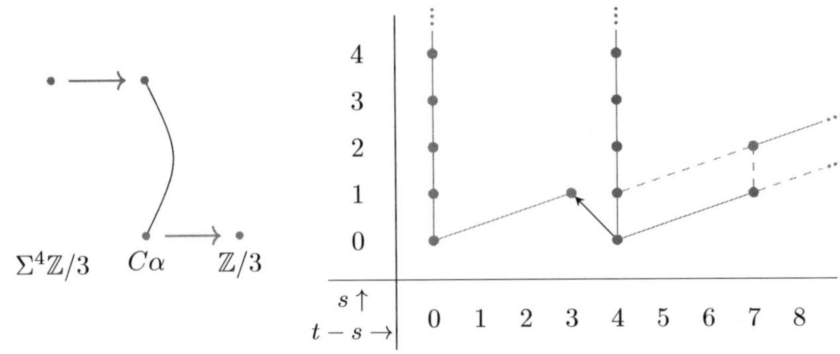

FIGURE 7. Left: the extension (2.87) of $\mathcal{A}$-modules at $p = 3$. Right: the associated long exact sequence in Ext. The dashed gray lines are actions by elements of $\mathrm{Ext}_{\mathcal{A}}(\mathbb{Z}/3)$ that cannot be seen from this long exact sequence and must be deduced another way; we do not need them in this paper, so do not go into the details.

REMARK 2.88. The change-of-rings technique we used at $p = 2$ has an analogue at $p = 3$ for twists of *tmf* (hence also 3-local twisted string bordism in degrees 15 and below, because the Ando–Hopkins–Rezk map [**AHR10**] $MTString_{(3)} \to tmf_{(3)}$ is 15-connected [**HR95, Hil09**]): using Baker and Lazarev's version of the Adams spectral sequence [**BL01**], we can take Ext over the algebra

$$\mathcal{A}^{tmf} := \pi_{-*}\mathrm{Map}_{tmf}(H\mathbb{Z}/3, H\mathbb{Z}/3), \tag{2.89}$$

where $H\mathbb{Z}/3$ is made into a *tmf*-algebra spectrum by the ring spectrum maps $tmf \overset{\tau_{\le 0}}{\to} H\mathbb{Z} \to H\mathbb{Z}/3$, where the first map is the Postnikov 0-connected quotient and the second map is induced from $\mathbb{Z} \twoheadrightarrow \mathbb{Z}/3$. The algebra $\mathcal{A}^{tmf}$ was explicitly calculated by Henriques and Hill, using work of Behrens [**Beh06**] and unpublished work of Hopkins and Mahowald; see Henriques [**DFHH14**, §13.3], Hill [**Hil07**], and Bruner and Rognes [**BR21**, §13] for computations with this Adams spectral sequence.

Just like at $p = 2$, there is a little more work to do apply this spectral sequence to twisted string bordism when the twist does not arise from a vector bundle. We take up this question in joint work with Matthew Yu [**DY23**], where we see how to work over $\mathcal{A}^{tmf}$ for nonvector-bundle twists and that it simplifies the 3-primary computation of $\Omega_*^{\xi^{\mathrm{het}}}$ in degrees relevant to string theory.

2.4. ξ^{CHL} bordism. In this section, we compute the ξ^{CHL} bordism groups. Just like for the ξ^{het} bordism groups, we use the change-of-rings trick from Corollary 2.22 at $p = 2$ and work more directly with the Adams spectral sequence at odd primes. This time, however, we can deduce a lot of information from abstract isomorphisms with the Adams spectral sequences for the string bordism of $B\mathrm{E}_8$, which has been studied by Hill [**Hil09**].

2.4.1. *2-primary computation.*

THEOREM 2.90. *In degrees* 11 *and below, the* 2*-completions of the abelian groups* $\Omega_*^{\xi^{\mathrm{CHL}}}$ *and* $\Omega_*^{\mathrm{String}}(B\mathrm{E}_8)$ *are abstractly isomorphic.*

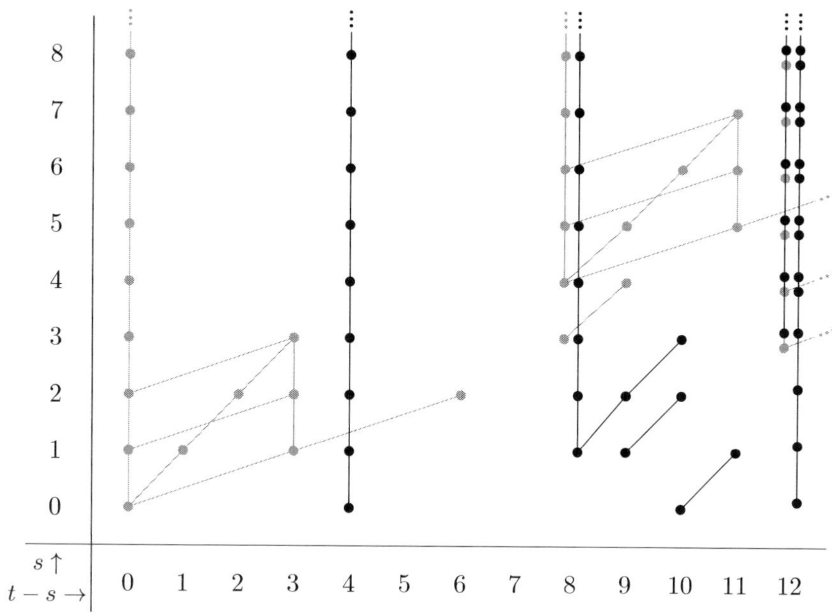

FIGURE 8. The E_2-page for the Adams spectral sequence computing 2-completed ξ^{CHL} bordism. The gray summands correspond to classes with trivial E_8-bundle. See Theorem 2.90 for more information. This figure is adapted from [**Hil09**, Figure 3].

PROOF. By Corollary 2.22, the Adams E_2-page in this range coincides with the Ext of $T(-2c)$ over $\mathcal{A}(2)$. The $\mathcal{A}(2)$-module structure on $T(\mu)$ only depends on the underlying group BG and on μ mod 2, and $2c$ mod $2 = 0$, so as $\mathcal{A}(2)$-modules, $T(-2c) \cong T(0) = H^*(B\mathrm{E}_8;\mathbb{Z}/2)$. So the Adams E_2-page coincides in the range we care about with the E_2-page for $MT\xi^0 = MTString \wedge (B\mathrm{E}_8)_+$. Hill [**Hil09**, Figure 3] computes the E_2-page corresponding for the reduced string bordism of $B\mathrm{E}_8$, which we use to draw the full E_2-page for $\Omega_*^{\xi^{\mathrm{CHL}}}$ in Figure 8.

This is an abstract isomorphism and does not a priori tell us about differentials or extensions. However, quotienting by $\mathbb{T}[1]$ defines a map $\mathbb{G}^{\mathrm{CHL}} \to \mathrm{Spin} \times \mathrm{E}_8$, which induces a map on Adams spectral sequences for Thom spectra of classifying spaces, and this map of Adams spectral sequences is identified with the map induced by $MTString \wedge (B\mathrm{E}_8)_+ \to MTSpin \wedge (B\mathrm{E}_8)_+$, so any differential for the string bordism of $B\mathrm{E}_8$ deduced by pulling back from the Adams spectral sequence for $MTSpin \wedge B\mathrm{E}_8$ remains valid in our Adams spectral sequence for ξ^{CHL} bordism.

Moreover, we can realize the part of $\Omega_*^{\xi^{\mathrm{CHL}}}$ corresponding to the gray summands in Figure 8 by string manifolds with trivial E_8-bundle, so the gray summands split off of the rest of the Adams spectral sequence.

Looking at the black summands in Figure 8, linearity of differentials with respect to the $\mathrm{Ext}_{\mathcal{A}}(\mathbb{Z}/2)$-action on the E_2-page means the only possible nonzero differentials in the range we care about are $d_2\colon E_2^{0,10} \to E_2^{2,11}$ and $d_2\colon E_2^{1,12} \to E_2^{3,13}$. Hill [**Hil09**, §3.3] uses the map to $MTSpin \wedge (B\mathrm{E}_8)_+$ to show that these two differentials are nontrivial, so as we noted above, the same is true for ξ^{CHL} bordism.

As there are no more differentials, and all extensions by 2 in range follow from $\mathrm{Ext}_{\mathcal{A}}(\mathbb{Z}/2)$-action without additional information, we have proven the theorem. $\square$

REMARK 2.91. As described in Remark 1.55, the map $c\colon B\mathrm{E}_8 \to K(\mathbb{Z},4)$ defines a map from ξ^{CHL} structures to $\mathrm{Spin}\langle w_4\rangle$ structures, i.e., the data of a spin structure and a trivialization of w_4. This is the CHL analogue of the passage from ξ^{het} structures to $\xi^{\mathrm{het}'}$ structures from §2.2 — and just as in that case, because c is 15-connected, the induced map $\Omega_k^{\xi^{\mathrm{CHL}}} \to \Omega_k^{\mathrm{Spin}\langle w_4\rangle}$ is an isomorphism for $k \le 14$, so the computations in this section also give $\mathrm{Spin}\langle w_4\rangle$ bordism groups.

An alternate point of view due to [**SSS12**, (2.17)] is that $\mathrm{Spin}\langle w_4\rangle$ structures are twisted string structures in the sense of Corollary 2.12: the trivialization of $w_4(M)$ is equivalent data to a class $\mu \in H^4(M;\mathbb{Z})$ and an identification of 2μ and $\lambda(M)$, so a $\mathrm{Spin}\langle w_4\rangle$-structure is a twisted string structure for the map $-2\colon K(\mathbb{Z},4) \to K(\mathbb{Z},4)$ (corresponding to the classifying space Sati-Schreiber-Stasheff denote $B\mathrm{String}^{2\mathrm{DD}_2}$). See also [**FH21a**, Remark C.18].

The proof of Theorem 2.90 took advantage of an abstract isomorphism, so it tells us nothing about the generators. The elements of $\Omega_*^{\mathrm{String}}(B\mathrm{E}_8)$ coming from $\Omega_*^{\mathrm{String}}(\mathrm{pt})$ are represented by string manifolds with trivial E_8-bundle; these vacuously satisfy the condition $2c = \lambda$, so define classes in $\Omega_*^{\xi^{\mathrm{CHL}}}$ representing the same elements under the abstract isomorphism with $\Omega_*^{\mathrm{String}}(B\mathrm{E}_8)$.

That leaves a few elements left: copies of $\mathbb{Z}$ in degrees 4 and 8, and copies of $\mathbb{Z}/2$ in degrees 9 and 10. We can represent the generator of $\Omega_4^{\xi^{\mathrm{CHL}}} \cong \mathbb{Z}$ by a K3 surface with an E_8-bundle chosen to satisfy the Bianchi identity; it would be interesting to determine generators of $\Omega_k^{\xi^{\mathrm{CHL}}}$ for $k = 8, 9, 10$.

2.4.2. *Odd-primary computation.*

THEOREM 2.92. *For $k \le 12$, $\Omega_k^{\xi^{\mathrm{CHL}}}$ has no odd-primary torsion.*

PROOF. First we show the result for $p = 3$. The mod 3 cohomology, as an $\mathcal{A}$-module, of the string cover $\mathcal{S}(G,\lambda)$ only depends on λ mod 3. Therefore in the CHL case, where $\lambda = 2c$, we might as well work with $\lambda = -c$ — or replacing our $K(\mathbb{Z},4)$ class with its opposite, $\lambda = c$. This string cover corresponds to the universal twist of *MTString* over $K(\mathbb{Z},4)$ from Corollary 2.12, which means that by Theorem 2.11, the Thom spectrum for this twist is *MTSpin* again! That is, the E_2-page of the 3-primary Adams spectral sequence for CHL bordism coincides with the E_2-page for spin bordism — or for oriented bordism, because the forgetful map *MTSpin* $\to$ *MTSO* is a 3-primary equivalence.

Milnor [**Mil60**, Theorem 4] shows that the mod 3 cohomology of *MTSO* is free as an $\mathcal{A}/\beta$-module on even-degree generators, where β is the mod 3 Bockstein; then, he proves [**Mil60**, Theorem 1] that for any spectrum with that property and satisfying a finiteness condition, there is no odd-primary torsion in homotopy. The CHL bordism spectrum satisfies these conditions, so we conclude.

For $p \ge 5$, the argument is essentially the same as in Theorem 2.74. $\square$

3. Consequences in string theory

There are a few different uses of bordism groups in theories of quantum gravity. In this section, we discuss applications and questions raised by the computations in the previous section. Though we stay mostly mathematical, some of what we state in this section is only known at a physical level of rigor.

3.1. The cobordism conjecture. As part of the Swampland program in quantum gravity, McNamara and Vafa [**MV19**] made the following conjecture, a consequence of the generally believed fact that theories of quantum gravity should not have global symmetries:

CONJECTURE 3.1 (McNamara–Vafa cobordism conjecture [**MV19**]). Suppose we have a consistent n-dimensional theory of quantum gravity in which the spacetime backgrounds that are summed over carry a ξ-structure. Then, for $3 \le k \le n-1$, $\Omega_k^\xi = 0$.

The key here is the meaning of "the spacetime backgrounds carry a ξ-structure" — we do not mean just that one could sum over ξ-manifolds, but that ξ is in some to-be-specified sense the maximally general structure for which the theory makes sense. String theorists often work with singular manifolds and even Deligne–Mumford stacks on $\mathcal{M}an$ [**PS05, PS06a, PS06b, DFM11a, DFM11b**], and the notion of ξ-bordism appearing in Conjecture 3.1 is expected to take this into account, as some sort of bordism theory of generalized manifolds.

The tangential structures ξ currently known for various theories of quantum gravity do not satisfy the vanishing criterion in Conjecture 3.1, so there must be additional data or conditions on these theories' backgrounds modifying ξ so as to kill its bordism groups. These modifications often take the form of additional extended objects in the theory.

This leads to a common application of the cobordism conjecture: compute the bordism groups for the tangential structure ξ as we currently understand it, and use any nonvanishing groups as beacons illuminating novel objects in the theory, which one then studies. This idea has been applied in [**MV19, BKRU20, GEMSV20, DH21, MV21, Sch21, ACC22, BC22, BCKM22, DHMT22, MR22, Wit22, DDHM23, MDL23**];[21] in this subsection, we will use our computations from §2 and see what we can learn about the $\mathrm{E}_8 \times \mathrm{E}_8$ heterotic string and the CHL string.

Despite the $k \ge 3$ bound in Conjecture 3.1, modifying ξ to kill classes in Ω_1^ξ and Ω_2^ξ is often physically meaningful, and can predict useful new objects in the theory. This is a common technique in the study of the cobordism conjecture, and we will do this too.

3.1.1. *The* $\mathrm{E}_8 \times \mathrm{E}_8$ *heterotic string.* McNamara and Vafa [**MV19**, §4.5] discussed predictions of their conjecture to the $\mathrm{E}_8 \times \mathrm{E}_8$ heterotic string theory, but after making the simplifying assumption that the gauge $(\mathrm{E}_8 \times \mathrm{E}_8) \rtimes \mathbb{Z}/2$-bundle is trivial; the corresponding tangential structure is then $B\mathrm{String}$. For example, their conjecture must account for $\Omega_3^{\mathrm{String}} \cong \mathbb{Z}/24$, generated by S^3 with its Lie group framing, and they explain how this is trivialized by taking into account the NS5-brane.

With $\Omega_*^{\xi^{\mathrm{het}}}$ in hand, we can predict more objects. Recall the generators we found for ξ^{het}-bordism groups, and our notation for them, from §2.2.1.

[21] Despite all of this work, there are still plenty of already-worked-out computations of bordism groups relevant to various string and supergravity theories whose corresponding defects have not been determined. This includes $\Omega_*^{\mathrm{Spin}}(B\mathrm{E}_8)$ [**Sto86, Edw91**], applicable to the $\mathrm{E}_8 \times \mathrm{E}_8$ heterotic string in the absence of the $\mathbb{Z}/2$ swapping symmetry; Ω_*^{DPin} [**KPMT20**, Appendices E, F], relevant for type I string theory; and $\Omega_*^{\mathrm{m}_c}$ [**FH21a**], useful for the low-energy limit of M-theory.

EXAMPLE 3.2. $\Omega_1^{\xi^{\mathrm{het}}} \cong \mathbb{Z}/2 \oplus \mathbb{Z}/2$, with generators S^1_{nb} and $\mathbb{RP}^1$. McNamara-Vafa already considered S^1_{nb}, but the latter is new. If one allows manifolds with singularities, $\mathbb{RP}^1$ bounds $D^2/(\mathbb{Z}/2)$, i.e., the disc with a principal $\mathbb{Z}/2$-bundle that is singular at the origin, inflated to a singular $\mathbb{G}^{\mathrm{het}}$-bundle via the inclusion $\mathbb{Z}/2 \hookrightarrow \mathbb{G}^{\mathrm{het}}$.

This class corresponds to a 7-brane in the $\mathrm{E}_8 \times \mathrm{E}_8$ heterotic string. The worldvolume of this brane is eight-dimensional, so the link around it in ten-dimensional spacetime is a circle. The monodromy around this circle exchanges the two E_8-bundles. This is exactly the nonsupersymmetric 7-brane recently introduced and discussed by [**KOTY23**].

Related 7-branes in different string theories are studied by Distler, Freed, and Moore [**DFM11a**] and Dierigl, Heckman, Montero, and Torres [**DHMT22**]; the latter study a 7-brane in type IIB string theory, called an R7-brane, which in the cobordism conjecture corresponds to $[\mathbb{RP}^1] \in \Omega_1^{\mathrm{Spin}\text{-}\mathrm{GL}_2^+(\mathbb{Z})}$.

As a way of better understanding Kaidi-Ohmori-Tachikawa-Yonekura's 7-brane, we can try to identify where it is sent under dualities between different string theories. For example, Hořava and Witten [**HW96b, HW96a, Wit96**] identified (a certain limit) of $\mathrm{E}_8 \times \mathrm{E}_8$ heterotic string theory with a theory predicted to be the low-energy limit of a compactification of M-theory on the unit interval. Under this identification, the Kaidi–Ohmori–Tachikawa–Yonekura 7-brane ought to correspond to a defect in M-theory associated to a 2-dimensional bordism class by the cobordism conjecture. Because the passage from M-theory to heterotic string theory requires compactifying on the interval, which is a manifold with boundary, one should use a theory of bordism of compact manifolds which are not necessarily closed.[22] The bordism class should be represented by an interval bundle over $\mathbb{RP}^1$, so we conjecture that the bordism class of the Möbius strip corresponds to the avatar of this brane in M-theory. As a check, M-theory compactified on a Möbius strip is expected to coincide with $\mathrm{E}_8 \times \mathrm{E}_8$ heterotic string theory compactified on $\mathbb{RP}^1$ — they are both predicted to be the CHL string, as we discussed in §1.3, though as usual only a statement about low-energy supergravity limits is known. We will not attempt to fully resolve this question in this paper: among other things, this would require finding "the right" notion of bordism for manifolds with boundary for this application.

Before we leave heterotic/M-duality behind, we point out a notion of bordism of manifolds with boundary, due to Conner and Floyd [**CF66**, §16], for which the Möbius strip is nonbounding; we optimistically conjecture that this is the correct kind of bordism of manifolds with boundary for applications to the cobordism conjecture.

DEFINITION 3.3. Let $\xi_1\colon B_1 \to BO$ and $\xi_2\colon B_2 \to BO$ be tangential structures and $\eta\colon B_1 \to B_2$ be a *map of tangential structures*, i.e., η commutes with the maps ξ_i. A *ξ_2/ξ_1-manifold* is a compact manifold M with ξ_2-structure together with

(1) a ξ_1-structure $\mathfrak{x}$ on ∂M, and
(2) an identification of the ξ_2-structure $\eta(\mathfrak{x})$ on ∂M with the ξ_2-structure induced by taking the boundary on M.

[22] McNamara and Vafa [**MV19**, §5] hint at this generalization, though from the perspective of manifolds with singularities rather than manifolds with boundary.

Conner and Floyd [**CF66**, §16] introduce a notion of bordism for ξ_2/ξ_1-manifolds,[23] which we write $\Omega_*^{\xi_2/\xi_1}$, such that the Thom spectrum corresponding to this notion of bordism is $MT\xi_2/MT\xi_1$, the cofiber of $\eta\colon MT\xi_1 \to MT\xi_2$. This implies the existence of a long exact sequence

$$\cdots \longrightarrow \Omega_k^{\xi_1} \xrightarrow{\eta} \Omega_k^{\xi_2} \xrightarrow{j} \Omega_k^{\xi_1/\xi_2} \xrightarrow{\partial} \Omega_{k-1}^{\xi_1} \longrightarrow \cdots \tag{3.4}$$

where j regards a ξ_2-manifold as a ξ_2/ξ_1-manifold with empty boundary.

LEMMA 3.5. *The class of the Möbius strip M is nonzero in $\Omega_2^{\mathrm{Pin}^+/\mathrm{Spin}}$.*[24]

PROOF. By (3.4), it suffices to prove that $[\partial M] \neq 0$ in Ω_1^{Spin}. The boundary of the Möbius strip is a circle, and for any pin$^+$ structure on M, the boundary circle has the nonbounding spin structure, i.e., is nonzero in Ω_1^{Spin}. This is because if ∂M had the bounding spin structure, one could glue the disc with its standard pin$^+$ structure to M along ∂M and thereby obtain a pin$^+$ structure on $\mathbb{RP}^2$, but $\mathbb{RP}^2$ does not admit a pin$^+$ structure. $\square$

Lemma 3.5 suggests that Conner and Floyd's notion of bordism of manifolds with boundary could be the correct one for our application in heterotic/M-theory duality.

EXAMPLE 3.6. Moving onto higher-codimension objects predicted by higher-dimensional bordism groups, $\Omega_2^{\xi^{\mathrm{het}}}$ is nonzero, but can be generated by products of S^1_{nb} and $\mathbb{RP}^1$. This means that if we trivialize $[\mathbb{RP}^1], [S^1_{nb}] \in \Omega_1^{\xi^{\mathrm{het}}}$ in the sense above, namely by allowing $\mathrm{E}_8 \times \mathrm{E}_8$ heterotic string theory to be defined on singular manifolds whose boundaries are $\mathbb{RP}^1$ and S^1_{nb}, then we can realize our chosen generators of $\Omega_2^{\xi^{\mathrm{het}}}$ as boundaries of singular 3-manifolds: for example, we used $D^2/(\mathbb{Z}/2)$ to realize $\mathbb{RP}^1$ as a boundary, so we can use $S^1_{nb} \times D^2/(\mathbb{Z}/2)$ to realize $S^1_{nb} \times \mathbb{RP}^1$ as a boundary. Thus accounting for $\Omega_2^{\xi^{\mathrm{het}}}$ does not require adding any new kinds of defects or singularities beyond what we used for $\Omega_1^{\xi^{\mathrm{het}}}$.

EXAMPLE 3.7. $\Omega_3^{\xi^{\mathrm{het}}} \cong \mathbb{Z}/8$, generated by $\mathbb{RP}^3$. As in Example 3.2, we can bound $\mathbb{RP}^3$ by $B^4/(\mathbb{Z}/2)$ by allowing a singularity at the origin. This bordism class should correspond to a 5-brane distinct from the NS5-brane.

EXAMPLE 3.8. $\Omega_4^{\xi^{\mathrm{het}}} \cong \mathbb{Z} \oplus \mathbb{Z}/2$. The $\mathbb{Z}/2$ summand is generated by $S^3 \times \mathbb{RP}^1$, where S^3 carries the Lie group framing, so its bordism class can be trivialized using the objects we have already discussed, like in Example 3.6. By Remark 2.64, because $S^3 \times S^1$ is bordant as ξ^{het}-manifolds to S^4 with trivial $\mathbb{Z}/2$-bundle and E_8-bundles with characteristic classes $\pm 1 \in H^4(S^4;\mathbb{Z}) \cong \mathbb{Z}$, this bordism class corresponds to the 4-brane recently found by [**KOTY23**].

The $\mathbb{Z}$ summand in $\Omega_4^{\xi^{\mathrm{het}}}$ is new to us. It is generated by the K3 surface with trivial $\mathbb{Z}/2$-bundle; one E_8-bundle is trivial, and the other cancels $\lambda(\mathrm{K3})$. McNamara and Vafa [**MV19**, §4.2.1] address the K3 surface without data of E_8-bundles or a nontrivial B-field, using it to exhibit a higher-form $\mathbb{T}$-symmetry. Our

[23] Conner and Floyd only consider a few examples of ξ_1 and ξ_2. The works [**Sto68, Ale75, Mit75, RST77, Lau00, Bun15**] consider some more tangential structures.

[24] Strictly speaking, $\mathrm{Pin}^+/\mathrm{Spin}$ is not the correct tangential structure: one should replace Pin^+ with something like $\mathfrak{m}_c$ [**Wit97, Wit16, FH21a**], and should replace Spin with something like ξ^{het}, though $\mathfrak{m}_c$- and pin$^+$ structures on 2-manifolds are equivalent data [**FH21a**, §8.5.1].

K3 surface corresponds to a different bordism class, but McNamara and Vafa's argument still applies: as the K3 surface is believed to be a valid background for $E_8 \times E_8$ heterotic string theory, this higher-form $\mathbb{T}$-symmetry must be broken or gauged in some way. We do not know what this would look like.

$\Omega_5^{\xi^{\mathrm{het}}}$ vanishes and $\Omega_6^{\xi^{\mathrm{het}}} \cong \mathbb{Z}/2$ is generated by $\mathbb{RP}^3 \times S^3$, so as in Example 3.6 we can realize it as a boundary without introducing any new kinds of singularities.

EXAMPLE 3.9. $\Omega_7^{\xi^{\mathrm{het}}} \cong \mathbb{Z}/16$, generated by $\mathbb{RP}^7$. This bordism class is closely analogous to Examples 3.2 and 3.7; this time, we have a 1-brane, i.e., a string.

REMARK 3.10 (Relating bordism classes by compactification[25]). For the cobordism conjecture for type IIB string theory considered on spin-$\mathrm{GL}_2^+(\mathbb{Z})$ manifolds, $[\mathbb{RP}^k] \in \Omega_k^{\mathrm{Spin}\text{-}\mathrm{GL}_2^+(\mathbb{Z})}$ is nonzero for $k = 1$, 3, and 7 [**DDHM23**, §14.3.2], so we would expect these classes to correspond to three different kinds of extended objects, akin to Examples 3.2, 3.7, and 3.9. However, in [**DDHM23**, §7], it is shown that the two higher-codimension objects can be expressed as compactifications of the R7-brane corresponding to $\mathbb{RP}^1$, so there is really only one novel object. We suspect something similar happens here: that in $E_8 \times E_8$ heterotic string theory, the extended objects corresponding to $\mathbb{RP}^3$ and $\mathbb{RP}^7$ can be accounted for using previously known branes and Kaidi, Ohmori, Tachikawa, and Yonekura's 7-brane from Example 3.2 corresponding to $\mathbb{RP}^1$.

From a bordism point of view, we are saying that if we allow singular ξ^{het}-manifolds which locally look like $\mathbb{R}^k \times D^2/(\mathbb{Z}/2)$, it should be possible to not just bound $\mathbb{RP}^1$, but also to bound $\mathbb{RP}^3$ and $\mathbb{RP}^7$. We leave this as a conjecture.

EXAMPLE 3.11. $\Omega_8^{\xi^{\mathrm{het}}}$, which corresponds to codimension-9 objects, is isomorphic to either $\mathbb{Z}^3 \oplus \mathbb{Z}/2$ or $\mathbb{Z}^3 \oplus (\mathbb{Z}/2)^{\oplus 2}$, depending on the fate of the differential (D3). The generators of these four or five summands that we found are:

- $\mathbb{HP}^2$ with two different ξ^{het}-structures, giving two $\mathbb{Z}$ summands;
- the Bott manifold, generating another free summand;
- $\mathbb{RP}^7 \times S^1_{nb}$ generating the $\mathbb{Z}/2$ summand that is present even if (D3) does not vanish; and
- the manifold X_8 that we discussed in §2.2.2, an $S^3 \times S^3$-bundle over $\mathbb{RP}^2$. If the differential (D3) is nonzero, then X_8 bounds as a ξ^{het}-manifold.

$\mathbb{RP}^6 \times S^1_{nb}$ is already accounted for in the sense of Example 3.6, so we focus on the other generators.

Both B and $\mathbb{HP}^2$ are nonbounding in the bordism group $\Omega_8^{\mathrm{Spin}\text{-}\mathrm{Mp}_2(\mathbb{Z})}$, which appears in the study of the cobordism conjecture for type IIB string theory; see [**DDHM23**, §6.9] for a discussion of defects in type IIB corresponding to these bordism classes. Like in Example 3.8, the story in $E_8 \times E_8$ heterotic string theory is presumably not exactly the same, but it may be analogous.

Finally, X_8. Following the arguments in [**MV19**, §4.5] and [**DDHM23**, §7.6, §7.8] the description of X_8 as a fiber bundle over $\mathbb{RP}^2$ with fiber $S^3 \times S^3$ suggests the following string-theoretic construction: use the singular manifold corresponding to the NS5-brane to bound for the first S^3, compactify on the second S^3, and then fiber over $D^3/(\mathbb{Z}/2)$ to make X_8 a boundary of a singular manifold. We do not know whether this is a valid background for the $E_8 \times E_8$ heterotic string; an argument

[25]We thank Markus Dierigl for pointing this out to us.

for or against it could provide an example of a use of Conjecture 3.1 to make a mathematical conjecture for the fate of X_8 based on string-theoretic predictions.

EXAMPLE 3.12. $\Omega_9^{\xi^{\mathrm{het}}}$ corresponds to zero-dimensional objects, i.e., point defects, and is isomorphic to either $(\mathbb{Z}/2)^{\oplus 4}$ or $(\mathbb{Z}/2)^{\oplus 6}$, depending on the fate of (D3). Three of the generators we found in §2.2.1 are of the form S^1_{nb} times a ξ^{het}-manifold, so have already been accounted for in the sense of Example 3.6. The fourth generator is $B \times \mathbb{RP}^1$, so it is also already accounted for.

The remaining two manifolds that might or might not be necessary are $X_8 \times S^1_{nb}$, which as usual is already taken care of, and a manifold X_9 which we did not determine.

3.1.2. *The CHL string.* In Theorems 2.90 and 2.92, we saw that $\Omega_*^{\xi^{\mathrm{CHL}}}$ is abstractly isomorphic to $\Omega_*^{\mathrm{String}}(B\mathrm{E}_8)$. Thus there is a summand corresponding to $\Omega_*^{\mathrm{String}}(\mathrm{pt})$, and as we saw above, these classes can be represented by string manifolds with trivial E_8-bundle. Some of these manifolds were accounted for by [**MV19**, §4.5] in heterotic string theory, e.g., killing S^3 with its nonbounding framing using the fivebrane, and presumably a similar defect is present in the CHL string. McNamara and Vafa leave plenty of string bordism classes' interpretations in terms of defects open to address, and this would be interesting to understand more in the setting of the CHL string.

We also found a few more classes in $\Omega_*^{\xi^{\mathrm{CHL}}}$. For example, $\Omega_4^{\xi^{\mathrm{CHL}}} \cong \mathbb{Z}$, generated by a K3 surface with E_8-bundle chosen to satisfy the Bianchi identity. Like in Example 3.8, this corresponds to some codimension-4 object, though we do not know what it will look like.

3.2. Is the $\mathbb{Z}/2$ symmetry on the $\mathrm{E}_8 \times \mathrm{E}_8$ heterotic string anomalous? Quantum field theories can come with the data of an *anomaly*, a mild inconsistency in which key quantities in the field theory are not defined absolutely without fixing additional data. For example, one wants the partition function of a QFT on a manifold M to be a complex number, but an anomaly signals that the partition function is only an element of a complex line which has not been trivialized. The process of resolving this inconsistency, when necessary, is called *anomaly cancellation.*

Freed and Teleman [**FT14**] describe anomaly cancellation for a broad class of quantum field theories as follows: an n-dimensional quantum field theory Z lives at the boundary of an $(n+1)$-dimensional invertible field theory α, called the *anomaly field theory* of Z. The tangential structures of Z and α match. Anomaly cancellation is the procedure of trivializing α, i.e., establishing an isomorphism from α to the trivial theory.

We think of this from Atiyah and Segal's approach [**Ati88, Seg88**] that field theories are symmetric monoidal functors from (potentially geometric) bordism categories into categories such as $\mathcal{V}ect_{\mathbb{C}}$. The perspective of extended field theory means these are often (∞, n)-categories. If $\mathcal{C}$ and $\mathcal{D}$ are two symmetric monoidal (∞, n)-categories, the (∞, n)-category of symmetric monoidal functors $F\colon \mathcal{C} \to \mathcal{D}$ acquires the symmetric monoidal structure of "pointwise tensor product," specified by the formula

$$(F_1 \otimes F_2)(x) := F_1(x) \otimes_{\mathcal{D}} F_2(x), \tag{3.13}$$

where x is an object, morphism, higher morphism, etc.

Definition 3.14 ([**FM06**, Definition 5.7]). Let $\mathcal{C}$ be a symmetric monoidal (∞, n)-category. An *invertible field theory* is a field theory $\alpha\colon \mathcal{B}ord_n^\xi \to \mathcal{C}$ such that there is another field theory $\alpha^{-1}\colon \mathcal{B}ord_n^\xi \to \mathcal{C}$ such that $\alpha\otimes\alpha^{-1} \simeq \mathbf{1}$, the trivial theory.

The trivial theory $\mathbf{1}\colon \mathcal{B}ord_n^\xi \to \mathcal{C}$ is defined to send all objects to the monoidal unit in $\mathcal{C}$ and all morphisms and higher morphisms to identity morphisms, resp. identity higher morphisms.

Therefore the classification of anomalies follows from the classification of invertible field theories, and anomaly cancellation is an isomorphism from an invertible field theory to $\mathbf{1}$. Freed, Hopkins, and Teleman [**FHT10**] classify invertible *topological* field theories using stable homotopy theory, and Grady and Pavlov [**GP21**, §5] generalize this in the nontopological setting.

In most cases, including the supergravity theories studied in this paper, the QFT under study is unitary, so their anomaly theories have the Wick-rotated analogue of unitarity, *reflection positivity*. Freed and Hopkins [**FH21b**] classify reflection-positive invertible field theories.

Let $I_\mathbb{Z}$ denote the *Anderson dual of the sphere spectrum* [**And69, Yos75**].

Theorem 3.15 ([**FH21b**, Theorem 1.1]). *Let ξ be a tangential structure. There is a natural isomorphism from the group of deformation classes of $(n+1)$-dimensional reflection-positive invertible topological field theories on ξ-manifolds to the torsion subgroup of $[MT\xi, \Sigma^{n+2}I_\mathbb{Z}]$.*

Freed-Hopkins then conjecture (*ibid.*, Conjecture 8.37) that the entire group classifies all reflection-positive invertible field theories, topological or not.

$I_\mathbb{Z}$ satisfies a universal property which leads to the existence of a natural short exact sequence

(3.16)
$$0 \longrightarrow \mathrm{Tors}(\mathrm{Hom}(\Omega_{n+1}^\xi, \mathbb{T})) \longrightarrow [MT\xi, \Sigma^{n+2}I_\mathbb{Z}] \longrightarrow \mathrm{Hom}(\Omega_{n+2}^\xi, \mathbb{Z}) \longrightarrow 0,$$

and this sequence carries physical meaning for the classification of possible anomalies for an n-dimensional QFT Z. For example, $\mathrm{Hom}(\Omega_{n+2}^\xi, \mathbb{Z})$ is a group of $\mathbb{Z}$-valued degree-$(n+2)$ characteristic classes of ξ-manifolds, and the quotient map in (3.16) sends the anomaly field theory of Z to its *anomaly polynomial.* This data can often be computed using perturbative techniques for Z, and is referred to as the *local anomaly.* Consequently, one can use bordism computations to assess what the group of possible anomalies of a QFT is, and whether a specific anomaly field theory is trivializable; see [**FH21a, TY21, DDHM22, LY22, DY22, Tac22, DOS23**] for recent anomaly cancellation theorems in string and supergravity theories using this technique.

3.2.1. *Anomalies for the* $\mathrm{E}_8\times\mathrm{E}_8$ *heterotic string.* For the $\mathrm{E}_8\times\mathrm{E}_8$ heterotic string, the anomaly field theory is an element of the group $[MT\xi^{\mathrm{het}}, \Sigma^{12}I_\mathbb{Z}]$: the free part is noncanonically isomorphic to the free part of $\Omega_{12}^{\xi^{\mathrm{het}}}$, and the torsion part is noncanonically isomorphic to the torsion subgroup of $\Omega_{11}^{\xi^{\mathrm{het}}}$. Though we have not completely determined these groups, $\Omega_{11}^{\xi^{\mathrm{het}}}$ is nonzero, as we showed in Theorem 2.62, so there is the possibility of a nontrivial anomaly to cancel. One generally expects that the anomaly field theory itself is trivial, because physicists have undertaken many consistency checks on $\mathrm{E}_8\times\mathrm{E}_8$ heterotic string theory, but sometimes there is a surprise: in joint work with Dierigl, Heckman, and Montero [**DDHM22**], we found that the anomaly theory for the duality symmetry in type IIB string theory is nonzero, and requires a modification of the theory to be trivialized.

For the $\mathrm{E}_8 \times \mathrm{E}_8$ heterotic string, there has been a fair amount of work already cancelling the anomaly in special cases, but for the full tangential structure ξ^{het}, the question of anomaly cancellation is open. The original work of Green-Schwarz [**GS84**] shows that the anomaly polynomial vanishes, so by (3.16), we only need to look at bordism invariants out of $\Omega_{11}^{\xi^{\mathrm{het}}}$. If one ignores the $\mathbb{Z}/2$ swapping symmetry, the anomaly is known to be trivial: Witten [**Wit86**, §4] showed that the global anomaly is classified by a bordism invariant $\Omega_{11}^{\mathrm{Spin}}(B\mathrm{E}_8) \to \mathbb{C}^\times$, and Stong [**Sto86**] showed that $\Omega_{11}^{\mathrm{Spin}}(B\mathrm{E}_8) = 0$ (see Remark 2.34). Sati [**Sat11a**] studies a closely related question in terms of $\Omega_{11}^{\mathrm{String}}(B\mathrm{E}_8)$.

Recent work of Tachikawa and Yamashita [**TY21**] (see also Tachikawa [**Tac22**] and Yonekura [**Yon22**, §4]) cancels anomalies in a large class of compactifications of heterotic string theory using an ingenious *TMF*-based argument. Their work does not take into account the $\mathbb{Z}/2$ swapping symmetry. It would be interesting to address the full anomaly on ξ^{het}-manifolds, either by directly computing it on generators of $\Omega_{11}^{\xi^{\mathrm{het}}}$ or by adapting Tachikawa and Yonekura's argument. If this symmetry does have a nontrivial anomaly, this would have consequences for the CHL string, either requiring a modification of the theory or showing that it is inconsistent.

3.2.2. *Anomalies for the CHL string.* Anomaly cancellation for the CHL string has been studied less. In Theorems 2.90 and 2.92, we saw that $\Omega_{11}^{\xi^{\mathrm{CHL}}}$ is torsion, so the anomaly polynomial vanishes; and we saw $\Omega_{10}^{\xi^{\mathrm{CHL}}} \cong \mathbb{Z}/2 \oplus \mathbb{Z}/2$, so there is a potential for the anomaly field theory to be nontrivial, which would be interesting to check.

Acknowledgments

I especially want to thank Miguel Montero both for suggesting this project and for many helpful conversations about the material in this paper, and Matthew Yu for many helpful discussions relating to [**DY23**] and other ideas related to this paper. I also want to thank Markus Dierigl and the anonymous referee for helpful comments on a draft. In addition, this paper benefited from conversations with Ivano Basile, Matilda Delgado, Jacques Distler, Dan Freed, Jonathan J. Heckman, Justin Kaidi, Jacob McNamara, Yuji Tachikawa, and Roberto Telléz Domínguez; thank you to all!

Part of this project was completed while AD visited the Perimeter Institute for Theoretical Physics; research at Perimeter is supported by the Government of Canada through Industry Canada and by the Province of Ontario through the Ministry of Research & Innovation.

References

[ABG10] Matthew Ando, Andrew J. Blumberg, and David Gepner, *Twists of K-theory and TMF*, Superstrings, geometry, topology, and C^*-algebras, Proc. Sympos. Pure Math., vol. 81, Amer. Math. Soc., Providence, RI, 2010, pp. 27–63, DOI 10.1090/pspum/081/2681757. MR2681757 ↑249, 250

[ABG⁺14a] Matthew Ando, Andrew J. Blumberg, David Gepner, Michael J. Hopkins, and Charles Rezk, *An ∞-categorical approach to R-line bundles, R-module Thom spectra, and twisted R-homology*, J. Topol. **7** (2014), no. 3, 869–893, DOI 10.1112/jtopol/jtt035. MR3252967 ↑230, 248, 249

[ABG⁺14b] Matthew Ando, Andrew J. Blumberg, David Gepner, Michael J. Hopkins, and Charles Rezk, *Units of ring spectra, orientations and Thom spectra via rigid infinite loop space theory*, J. Topol. **7** (2014), no. 4, 1077–1117, DOI 10.1112/jtopol/jtu009. MR3286898 ↑230, 248, 249, 252

[ABG18] Matthew Ando, Andrew J. Blumberg, and David Gepner, *Parametrized spectra, multiplicative Thom spectra and the twisted Umkehr map*, Geom. Topol. **22** (2018), no. 7, 3761–3825, DOI 10.2140/gt.2018.22.3761. MR3890766 ↑249

[ABP67] D. W. Anderson, E. H. Brown Jr., and F. P. Peterson, *The structure of the Spin cobordism ring*, Ann. of Math. (2) **86** (1967), 271–298, DOI 10.2307/1970690. MR219077 ↑253, 256

[ABP69] D. W. Anderson, E. H. Brown Jr., and F. P. Peterson, *Pin cobordism and related topics*, Comment. Math. Helv. **44** (1969), 462–468, DOI 10.1007/BF02564545. MR261613 ↑245

[ABS64] M. F. Atiyah, R. Bott, and A. Shapiro, *Clifford modules*, Topology **3** (1964), no. suppl, suppl. 1, 3–38, DOI 10.1016/0040-9383(64)90003-5. MR167985 ↑250, 269

[ACC22] David Andriot, Nils Carqueville, and Niccolò Cribiori, *Looking for structure in the cobordism conjecture*, SciPost Phys. **13** (2022), no. 3, Paper No. 071, 38, DOI 10.21468/scipostphys.13.3.071. MR4492333 ↑279

[ADH21] Araminta Amabel, Arun Debray, and Peter J. Haine (editors), *Differential cohomology: Categories, characteristic classes, and connections*, 2021. `arXiv:2109.12250`. 238, 239

[AHR10] Matthew Ando, Michael J. Hopkins, and Charles Rezk, *Multiplicative orientations of KO-theory and of the spectrum of topological modular forms*, 2010. `http://faculty.math.illinois.edu/~mando/papers/koandtmf.pdf`. 250, 276

[Ale75] J. C. Alexander, *On* (U, Sp) *bordism*, Amer. J. Math. **97** (1975), no. 3, 617–625, DOI 10.2307/2373769. MR377944 ↑281

[And69] D. W. Anderson, *Universal coefficient theorems for K-theory*, 1969. `https://faculty.tcu.edu/gfriedman/notes/Anderson-UCT.pdf`. 284

[AS69] M. F. Atiyah and G. B. Segal, *Equivariant K-theory and completion*, J. Differential Geometry **3** (1969), 1–18. MR259946 ↑247

[AS04] Michael Atiyah and Graeme Segal, *Twisted K-theory*, Ukr. Mat. Visn. **1** (2004), no. 3, 287–330; English transl., Ukr. Math. Bull. **1** (2004), no. 3, 291–334. MR2172633 ↑250

[ATD10] Allan Adams, Washington Taylor, and Oliver DeWolfe, *String universality in ten dimensions*, Phys. Rev. Lett. **105** (2010), no. 7, 071601, 4, DOI 10.1103/PhysRevLett.105.071601. MR2684030 ↑233

[Ati61] M. F. Atiyah, *Bordism and cobordism*, Proc. Cambridge Philos. Soc. **57** (1961), 200–208, DOI 10.1017/s0305004100035064. MR126856 ↑251

[Ati88] Michael Atiyah, *Topological quantum field theories*, Inst. Hautes Études Sci. Publ. Math. **68** (1988), 175–186 (1989). MR1001453 ↑283

[Bae02] John C. Baez, *Higher Yang-Mills theory*, 2002. `arXiv:hep-th/0206130`. 228, 236

[Bar06] Tobias Keith Bartels, *Higher gauge theory: 2-bundles*, ProQuest LLC, Ann Arbor, MI, 2006. Thesis (Ph.D.)–University of California, Riverside. MR2709030 ↑237

[BBPX22] Agnès Beaudry, Irina Bobkova, Viet-Cuong Pham, and Zhouli Xu, *The topological modular forms of* $\mathbb{R}P^2$ *and* $\mathbb{R}P^2 \wedge \mathbb{C}P^2$, J. Topol. **15** (2022), no. 4, 1864–1926, DOI 10.1112/topo.12263. MR4584582 ↑270

[BC04] John C. Baez and Alissa S. Crans, *Higher-dimensional algebra. VI. Lie 2-algebras*, Theory Appl. Categ. **12** (2004), 492–538. MR2068522 ↑228

[BC18] Agnès Beaudry and Jonathan A. Campbell, *A guide for computing stable homotopy groups*, Topology and quantum theory in interaction, Contemp. Math., vol. 718, Amer. Math. Soc., [Providence], RI, [2018] ©2018, pp. 89–136, DOI 10.1090/conm/718/14476. MR3869642 ↑230, 245, 257, 258

[BC22] Ralph Blumenhagen and Niccolò Cribiori, *Open-closed correspondence of K-theory and cobordism*, J. High Energy Phys. **8** (2022), Paper No. 37, 25, DOI 10.1007/jhep08(2022)037. MR4462363 ↑279

[BCH19] Francesco Benini, Clay Córdova, and Po-Shen Hsin, *On 2-group global symmetries and their anomalies*, J. High Energy Phys. **3** (2019), 118, 70, DOI 10.1007/jhep03(2019)118. MR3940887 ↑228

[BCKM22] Ralph Blumenhagen, Niccolò Cribiori, Christian Kneißl, and Andriana Makridou, *Dynamical cobordism of a domain wall and its companion defect 7-brane*, J. High Energy Phys. **8** (2022), Paper No. 204, 20, DOI 10.1007/jhep08(2022)204. MR4473075 ↑279

[BdRdWvN82] E. Bergshoeff, M. de Roo, B. de Wit, and P. van Nieuwenhuizen, *Ten-dimensional Maxwell-Einstein supergravity, its currents, and the issue of its auxiliary fields*, *Nucl. Phys. B*, 195:97–136, 1982. 236

[BE23] Daniel Berwick-Evans, *How do field theories detect the torsion in topological modular forms?*, 2023. `arXiv:2303.09138`. 231

[Bea17] Jonathan Beardsley, *Relative Thom spectra via operadic Kan extensions*, Algebr. Geom. Topol. **17** (2017), no. 2, 1151–1162, DOI 10.2140/agt.2017.17.1151. MR3623685 ↑250

[Beh06] Mark Behrens, *A modular description of the $K(2)$-local sphere at the prime 3*, Topology **45** (2006), no. 2, 343–402, DOI 10.1016/j.top.2005.08.005. MR2193339 ↑276

[BF78] A. K. Bousfield and E. M. Friedlander, *Homotopy theory of Γ-spaces, spectra, and bisimplicial sets*, Geometric applications of homotopy theory (Proc. Conf., Evanston, Ill., 1977), Lecture Notes in Math., vol. 658, Springer, Berlin-New York, 1978, pp. 80–130. MR513569 ↑239

[BG23] Lakshya Bhardwaj and Dewi S. W. Gould, *Disconnected 0-form and 2-group symmetries*, J. High Energy Phys. **7** (2023), Paper No. 98, 46, DOI 10.1007/jhep07(2023)098. MR4618042 ↑233

[BH58] A. Borel and F. Hirzebruch, *Characteristic classes and homogeneous spaces. I*, Amer. J. Math. **80** (1958), 458–538, DOI 10.2307/2372795. MR102800 ↑263

[BH59] A. Borel and F. Hirzebruch, *Characteristic classes and homogeneous spaces. II*, Amer. J. Math. **81** (1959), 315–382, DOI 10.2307/2372747. MR110105 ↑274

[BKRU20] Vijay Balasubramanian, Arjun Kar, Simon F. Ross, and Tomonori Ugajin, *Spin structures and baby universes*, J. High Energy Phys. **9** (2020), 192, 30, DOI 10.1007/jhep09(2020)192. MR4190232 ↑279

[BL01] Andrew Baker and Andrej Lazarev, *On the Adams spectral sequence for R-modules*, Algebr. Geom. Topol. **1** (2001), 173–199, DOI 10.2140/agt.2001.1.173. MR1823498 ↑276

[BL04] John C. Baez and Aaron D. Lauda, *Higher-dimensional algebra. V. 2-groups*, Theory Appl. Categ. **12** (2004), 423–491. MR2068521 ↑228, 238

[BN14] Ulrich Bunke and Niko Naumann, *Secondary invariants for string bordism and topological modular forms*, Bull. Sci. Math. **138** (2014), no. 8, 912–970, DOI 10.1016/j.bulsci.2014.05.002. MR3278648 ↑231

[BNV16] Ulrich Bunke, Thomas Nikolaus, and Michael Völkl, *Differential cohomology theories as sheaves of spectra*, J. Homotopy Relat. Struct. **11** (2016), no. 1, 1–66, DOI 10.1007/s40062-014-0092-5. MR3462099 ↑238, 239, 240

[Boa65] J. M. Boardman, *Stable homotopy theory*, 1965. Mimeographed notes. 241, 249

[Bou01] A. K. Bousfield, *On the telescopic homotopy theory of spaces*, Trans. Amer. Math. Soc. **353** (2001), no. 6, 2391–2426, DOI 10.1090/S0002-9947-00-02649-0. MR1814075 ↑239

[BR21] Robert R. Bruner and John Rognes, *The Adams spectral sequence for topological modular forms*, Mathematical Surveys and Monographs, vol. 253, American Mathematical Society, Providence, RI, [2021] ©2021, DOI 10.1090/surv/253. MR4284897 ↑258, 259, 265, 270, 272, 276

[Bro82] Edgar H. Brown Jr., *The cohomology of BSO_n and BO_n with integer coefficients*, Proc. Amer. Math. Soc. **85** (1982), no. 2, 283–288, DOI 10.2307/2044298. MR652459 ↑235

[Bru18] Robert Bruner, *Ext (computer software)*, 2018. `http://www.rrb.wayne.edu/papers/index.html#code`. 258

[Bry93] Jean-Luc Brylinski, *Loop spaces, characteristic classes and geometric quantization*, Progress in Mathematics, vol. 107, Birkhäuser Boston, Inc., Boston, MA, 1993, DOI 10.1007/978-0-8176-4731-5. MR1197353 ↑233, 234

[BS58] Raoul Bott and Hans Samelson, *Applications of the theory of Morse to symmetric spaces*, Amer. J. Math. **80** (1958), 964–1029, DOI 10.2307/2372843. MR105694 ↑242, 254, 274

[BS07] John C. Baez and Urs Schreiber, *Higher gauge theory*, Categories in algebra, geometry and mathematical physics, Contemp. Math., vol. 431, Amer. Math. Soc., Providence, RI, 2007, pp. 7–30, DOI 10.1090/conm/431/08264. MR2342821 ↑228

[BS09] John C. Baez and Danny Stevenson, *The classifying space of a topological 2-group*, Algebraic topology, Abel Symp., vol. 4, Springer, Berlin, 2009, pp. 1–31, DOI 10.1007/978-3-642-01200-6_1. MR2597732 ↑237

[BSCS07] John C. Baez, Danny Stevenson, Alissa S. Crans, and Urs Schreiber, *From loop groups to 2-groups*, Homology Homotopy Appl. **9** (2007), no. 2, 101–135. MR2366945 ↑228, 238

[Bun15] Ulrich Bunke, *The universal η-invariant for manifolds with boundary*, Q. J. Math. **66** (2015), no. 2, 473–506, DOI 10.1093/qmath/hav002. MR3356833 ↑281

[Car50] Henri Cartan, *Notions d'algèbre différentielle; application aux groupes de Lie et aux variétés où opère un groupe de Lie* (French), Colloque de topologie (espaces fibrés), Bruxelles, 1950, Georges Thone, Liège, 1951, pp. 15–27. MR42426 ↑235

[Car54] Henri Cartan, *Sur les groupes d'Eilenberg-Mac Lane. II* (French), Proc. Nat. Acad. Sci. U.S.A. **40** (1954), 704–707, DOI 10.1073/pnas.40.8.704. MR65161 ↑274

[CC21] Hood Chatham and Dexter Chua, *Spectralsequences (computer software)*, 2021. `https://github.com/spectralsequences/sseq/`. 258

[CCG10] Pilar Carrasco, Antonio M. Cegarra, and Antonio R. Garzón, *Nerves and classifying spaces for bicategories*, Algebr. Geom. Topol. **10** (2010), no. 1, 219–274, DOI 10.2140/agt.2010.10.219. MR2602835 ↑237

[CDI19] Clay Córdova, Thomas T. Dumitrescu, and Kenneth Intriligator, *Exploring 2-group global symmetries*, J. High Energy Phys. **2** (2019), 184, front matter + 108, DOI 10.1007/jhep02(2019)184. MR3925280 ↑228

[CDLZ20] Mirjam Cvetič, Markus Dierigl, Ling Lin, and Hao Y. Zhang, *String universality and non-simply-connected gauge groups in 8D*, Phys. Rev. Lett. **125** (2020), no. 21, 211602, 6, DOI 10.1103/physrevlett.125.211602. MR4180436 ↑245

[CDLZ21] Mirjam Cvetič, Markus Dierigl, Ling Lin, and Hao Y. Zhang, *Gauge group topology of 8D Chaudhuri-Hockney-Lykken vacua*, Phys. Rev. D **104** (2021), no. 8, Paper No. 086018, 18, DOI 10.1103/physrevd.104.086018. MR4343418 ↑245

[CDLZ22] Mirjam Cvetič, Markus Dierigl, Ling Lin, and Hao Y. Zhang, *All eight- and nine-dimensional string vacua from junctions*, Phys. Rev. D **106** (2022), no. 2, Paper No. 026007, 44, DOI 10.1103/physrevd.106.026007. MR4471094 ↑245

[CF66] P. E. Conner and E. E. Floyd, *The relation of cobordism to K-theories*, Lecture Notes in Mathematics, No. 28, Springer-Verlag, Berlin-New York, 1966. MR216511 ↑231, 280, 281

[CGH22] Veronica Collazuol, Mariana Graña, and Alvaro Herráez, *E_9 symmetry in the heterotic string on S^1 and the weak gravity conjecture*, J. High Energy Phys. **6** (2022), Paper No. 83, 27, DOI 10.1007/jhep06(2022)083. MR4448344 ↑245

[Che52] Shiing-shen Chern, *Differential geometry of fiber bundles*, Proceedings of the International Congress of Mathematicians, Cambridge, Mass., 1950, vol. 2, Amer. Math. Soc., Providence, RI, 1952, pp. 397–411. MR45432 ↑235

[CHL95] Shyamoli Chaudhuri, George Hockney, and Joseph Lykken, *Maximally supersymmetric string theories in $D < 10$*, Phys. Rev. Lett. **75** (1995), no. 12, 2264–2267, DOI 10.1103/PhysRevLett.75.2264. MR1351447 ↑243

[CHZ11] Qingtao Chen, Fei Han, and Weiping Zhang, *Generalized Witten genus and vanishing theorems*, J. Differential Geom. **88** (2011), no. 1, 1–40. MR2819754 ↑245

[CM83] G. F. Chapline and N. S. Manton, *Unification of Yang-Mills theory and supergravity in ten dimensions*, Phys. Lett. B **120** (1983), no. 1-3, 105–109, DOI 10.1016/0370-2693(83)90633-0. MR689501 ↑236

[CMM22] Peng Cheng, Ilarion V. Melnikov, and Ruben Minasian, *Flat equivariant gerbes: holonomies and dualities*, J. High Energy Phys. **4** (2023), Paper No. 74, 56, DOI 10.1007/jhep04(2023)074. MR4578713 ↑245

[CP95] Shyamoli Chaudhuri and Joseph Polchinski, *Moduli space of Chaudhuri-Hockney-Lykken strings*, Phys. Rev. D (3) **52** (1995), no. 12, 7168–7173, DOI 10.1103/PhysRevD.52.7168. MR1375878 ↑243, 245

[CS85] Jeff Cheeger and James Simons, *Differential characters and geometric invariants*, Geometry and topology (College Park, Md., 1983/84), Lecture Notes in Math., vol. 1167, Springer, Berlin, 1985, pp. 50–80, DOI 10.1007/BFb0075216. MR827262 ↑240

[dBDH$^+$00] Jan de Boer, Robbert Dijkgraaf, Kentaro Hori, Arjan Keurentjes, John Morgan, David R. Morrison, and Savdeep Sethi, *Triples, fluxes, and strings*, Adv. Theor. Math. Phys. **4** (2000), no. 5, 995–1186 (2001), DOI 10.4310/ATMP.2000.v4.n5.a1. MR1868756 ↑233, 243

[DD63] Jacques Dixmier and Adrien Douady, *Champs continus d'espaces hilbertiens et de C^*-algèbres* (French), Bull. Soc. Math. France **91** (1963), 227–284. MR163182 ↑233

[DDHM22] Arun Debray, Markus Dierigl, Jonathan J. Heckman, and Miguel Montero, *The anomaly that was not meant IIB*, Fortschr. Phys. **70** (2022), no. 1, Paper No. 2100168, 31, DOI 10.1002/prop.202100168. MR4375189 ↑284, 285

[DDHM23] Arun Debray, Markus Dierigl, Jonathan J. Heckman, and Miguel Montero, *The Chronicles of IIBordia: Dualities, bordisms, and the Swampland*, 2023. `arXiv:2302.00007`. 246, 270, 279, 282

[DDK$^+$] Arun Debray, Sanath Devalapurkar, Cameron Krulewski, Yu Leon Liu, Natalia Pacheco-Tallaj, and Ryan Thorngren, *A long exact sequence in symmetry breaking: order parameter constraints, defect anomaly matching, and higher Berry phase*. 262

[DFHH14] Christopher L. Douglas, John Francis, André G. Henriques, and Michael A. Hill (eds.), *Topological modular forms*, Mathematical Surveys and Monographs, vol. 201, American Mathematical Society, Providence, RI, 2014, DOI 10.1090/surv/201. MR3223024 ↑276

[DFM11a] Jacques Distler, Daniel S. Freed, and Gregory W. Moore, *Orientifold précis*, Mathematical foundations of quantum field theory and perturbative string theory, Proc. Sympos. Pure Math., vol. 83, Amer. Math. Soc., Providence, RI, 2011, pp. 159–172, DOI 10.1090/pspum/083/2742428. MR2742428 ↑250, 279, 280

[DFM11b] Jacques Distler, Daniel S. Freed, and Gregory W. Moore, *Spin structures and superstrings*, Surveys in differential geometry. Volume XV. Perspectives in mathematics and physics, Surv. Differ. Geom., vol. 15, Int. Press, Somerville, MA, 2011, pp. 99–130, DOI 10.4310/SDG.2010.v15.n1.a4. MR2815726 ↑250, 279

[DH21] Markus Dierigl and Jonathan J. Heckman, *Swampland cobordism conjecture and non-Abelian duality groups*, Phys. Rev. D **103** (2021), no. 6, Paper No. 066006, 12, DOI 10.1103/physrevd.103.066006. MR4248172 ↑279

[DHMT22] Markus Dierigl, Jonathan J. Heckman, Miguel Montero, and Ethan Torres, *IIB string theory explored: reflection 7-branes*, Phys. Rev. D **107** (2023), no. 8, Paper No. 086015, 22, DOI 10.1103/physrevd.107.086015. MR4585955 ↑279, 280

[DK70] P. Donovan and M. Karoubi, *Graded Brauer groups and K-theory with local coefficients*, Inst. Hautes Études Sci. Publ. Math. **38** (1970), 5–25. MR282363 ↑250

[DL23] Joo Davighi, Nakarin Lohitsiri, and Arun Debray, *Toric 2-group anomalies via cobordism*, J. High Energy Phys. **7** (2023), Paper No. 19, 52, DOI 10.1007/jhep07(2023)019. MR4611726 ↑246

[DM78] Donald M. Davis and Mark Mahowald, *The immersion conjecture for RP^{8l+7} is false*, Trans. Amer. Math. Soc. **236** (1978), 361–383, DOI 10.2307/1997791. MR646070 ↑258, 260, 270, 272

[DM82] Donald M. Davis and Mark Mahowald, *Ext over the subalgebra A_2 of the Steenrod algebra for stunted projective spaces*, Current trends in algebraic topology, Part 1 (London, Ont., 1981), CMS Conf. Proc., vol. 2, Amer. Math. Soc., Providence, RI, 1982, pp. 297–342, DOI 10.2307/2374058. MR686123 ↑258

[Don78] Harold Donnelly, *Eta invariants for G-spaces*, Indiana Univ. Math. J. **27** (1978), no. 6, 889–918, DOI 10.1512/iumj.1978.27.27060. MR511246 ↑260

[DOS23] Markus Dierigl, Paul-Konstantin Oehlmann, and Thorsten Schimannek, *The discrete Green-Schwarz mechanism in 6D F-theory and elliptic genera of non-critical strings*, J. High Energy Phys. **3** (2023), Paper No. 90, 50, DOI 10.1007/jhep03(2023)090. MR4564964 ↑284

[Dou06] Christopher L. Douglas, *On the twisted K-homology of simple Lie groups*, Topology **45** (2006), no. 6, 955–988, DOI 10.1016/j.top.2006.06.007. MR2263220 ↑250

[Dug01] Daniel Dugger, *Universal homotopy theories*, Adv. Math. **164** (2001), no. 1, 144–176, DOI 10.1006/aima.2001.2014. MR1870515 ↑238

[DY22] Arun Debray and Matthew Yu, *What bordism-theoretic anomaly cancellation can do for U*, 2022. arXiv:2210.04911. 246, 284

[DY23] Arun Debray and Matthew Yu, *Adams spectral sequences for non-vector-bundle Thom spectra*, 2023. arXiv:2305.01678. 230, 231, 232, 251, 252, 263, 276, 285

[Edw91] Steven R. Edwards, *On the spin bordism of $B(E_8 \times E_8)$*, Illinois J. Math. **35** (1991), no. 4, 683–689. MR1115992 ↑279

[Eve65] Leonard Evens, *On the Chern classes of representations of finite groups*, Trans. Amer. Math. Soc. **115** (1965), 180–193, DOI 10.2307/1994264. MR212099 ↑255

[Fan99] Fuquan Fang, *Topology of Dupin hypersurfaces with six distinct principal curvatures*, Math. Z. **231** (1999), no. 3, 533–555, DOI 10.1007/PL00004742. MR1704992 ↑230

[FFG+20] Anamaría Font, Bernardo Fraiman, Mariana Graña, Carmen A. Núñez, and Héctor Parra De Freitas, *Exploring the landscape of heterotic strings on T^d*, J. High Energy Phys. **10** (2020), 194, 73, DOI 10.1007/jhep10(2020)194. MR4203911 ↑245

[FFG+21] Anamaría Font, Bernardo Fraiman, Mariana Graña, Carmen A. Núñez, and Héctor Parra De Freitas, *Exploring the landscape of CHL strings on T^d*, J. High Energy Phys. **8** (2021), Paper No. 095, 46, DOI 10.1007/jhep08(2021)095. MR4317175 ↑245

[FGN18] Bernardo Fraiman, Mariana Graña, and Carmen A. Nuñez, *A new twist on heterotic string compactifications*, J. High Energy Phys. **9** (2018), 078, front matter+97, DOI 10.1007/jhep09(2018)078. MR3868409 ↑245

[FH13] Daniel S. Freed and Michael J. Hopkins, *Chern-Weil forms and abstract homotopy theory*, Bull. Amer. Math. Soc. (N.S.) **50** (2013), no. 3, 431–468, DOI 10.1090/S0273-0979-2013-01415-0. MR3049871 ↑238, 239

[FH21a] Daniel S. Freed and Michael J. Hopkins, *Consistency of M-theory on non-orientable manifolds*, Q. J. Math. **72** (2021), no. 1-2, 603–671, DOI 10.1093/qmath/haab007. MR4271397 ↑243, 245, 246, 261, 263, 267, 270, 278, 279, 281, 284

[FH21b] Daniel S. Freed and Michael J. Hopkins, *Reflection positivity and invertible topological phases*, Geom. Topol. **25** (2021), no. 3, 1165–1330, DOI 10.2140/gt.2021.25.1165. MR4268163 ↑229, 246, 284

[FHT10] Daniel S. Freed, Michael J. Hopkins, and Constantin Teleman, *Consistent orientation of moduli spaces*, The many facets of geometry, Oxford Univ. Press, Oxford, 2010, pp. 395–419, DOI 10.1093/acprof:oso/9780199534920.003.0019. MR2681705 ↑229, 284

[FK96] Fuquan Fang and Stephan Klaus, *Topological classification of 4-dimensional complete intersections*, Manuscripta Math. **90** (1996), no. 2, 139–147, DOI 10.1007/BF02568299. MR1391206 ↑230

[FM06] Daniel S. Freed and Gregory W. Moore, *Setting the quantum integrand of M-theory*, Comm. Math. Phys. **263** (2006), no. 1, 89–132, DOI 10.1007/s00220-005-1482-7. MR2207325 ↑284

[FPDF21] Bernardo Fraiman and Héctor Parra De Freitas, *Symmetry enhancements in 7d heterotic strings*, J. High Energy Phys. **10** (2021), Paper No. 002, 34, DOI 10.1007/jhep10(2021)002. MR4339924 ↑245

[FPDF22] Bernardo Fraiman and Héctor Parra De Freitas, *Freezing of gauge symmetries in the heterotic string on T^4*, J. High Energy Phys. **4** (2022), Paper No. 007, 21, DOI 10.1007/jhep04(2022)007. MR4428826 ↑245

[Fra11] John Francis, *Integrals on spin manifolds and the K-theory of* $K(\mathbb{Z}, 4)$, 2011. `https://sites.math.northwestern.edu/~jnkf/writ/bspin2011.pdf`. 267

[FSS12] Domenico Fiorenza, Urs Schreiber, and Jim Stasheff, *Čech cocycles for differential characteristic classes: an ∞-Lie theoretic construction*, Adv. Theor. Math. Phys. **16** (2012), no. 1, 149–250. MR3019405 ↑228, 232, 236, 239, 240

[FSS15a] Domenico Fiorenza, Hisham Sati, and Urs Schreiber, *The E_8 moduli 3-stack of the C-field in M-theory*, Comm. Math. Phys. **333** (2015), no. 1, 117–151, DOI 10.1007/s00220-014-2228-1. MR3294945 ↑228, 241, 244

[FSS15b] Domenico Fiorenza, Hisham Sati, and Urs Schreiber, *A higher stacky perspective on Chern-Simons theory*, Mathematical aspects of quantum field theories, Math. Phys. Stud., Springer, Cham, 2015, pp. 153–211. MR3330242 ↑228

[FSS20] Domenico Fiorenza, Hisham Sati, and Urs Schreiber, *Twisted cohomotopy implies M-theory anomaly cancellation on 8-manifolds*, Comm. Math. Phys. **377** (2020), no. 3, 1961–2025, DOI 10.1007/s00220-020-03707-2. MR4121615 ↑243

[FSS21] Domenico Fiorenza, Hisham Sati, and Urs Schreiber, *Twisted cohomotopy implies twisted string structure on M5-branes*, J. Math. Phys. **62** (2021), no. 4, Paper No. 042301, 16, DOI 10.1063/5.0037786. MR4242295 ↑228, 241

[FT14] Daniel S. Freed and Constantin Teleman, *Relative quantum field theory*, Comm. Math. Phys. **326** (2014), no. 2, 459–476, DOI 10.1007/s00220-013-1880-1. MR3165462 ↑283

[FW99] Daniel S. Freed and Edward Witten, *Anomalies in string theory with D-branes*, Asian J. Math. **3** (1999), no. 4, 819–851, DOI 10.4310/AJM.1999.v3.n4.a6. MR1797580 ↑250

[FW10] Fuquan Fang and Jianbo Wang, *Homeomorphism classification of complex projective complete intersections of dimensions 5, 6 and 7*, Math. Z. **266** (2010), no. 3, 719–746, DOI 10.1007/s00209-009-0597-5. MR2719429 ↑230

[GEMSV20] Iñaki García Etxebarria, Miguel Montero, Kepa Sousa, and Irene Valenzuela, *Nothing is certain in string compactifications*, J. High Energy Phys. **12** (2020), Paper No. 032, 105, DOI 10.1007/jhep12(2020)032. MR4239420 ↑279

[Ger63] Hillel H. Gershenson, *Relationships between the Adams spectral sequence and Toda's calculations of the stable homotopy groups of spheres*, Math. Z. **81** (1963), 223–259, DOI 10.1007/BF01111545. MR151976 ↑275

[Gia71] V. Giambalvo, *On ⟨8⟩-cobordism*, Illinois J. Math. **15** (1971), 533–541. MR287553 ↑246, 253, 271

[Gia73a] V. Giambalvo, *Cobordism of line bundles with a relation*, Illinois J. Math. **17** (1973), 442–449. MR321121 ↑245

[Gia73b] V. Giambalvo, *Pin and Pin′ cobordism*, Proc. Amer. Math. Soc. **39** (1973), 395–401, DOI 10.2307/2039653. MR321123 ↑245

[Gia76] V. Giambalvo, *Cobordism of spin manifolds with involution*, Quart. J. Math. Oxford Ser. (2) **27** (1976), no. 106, 241–252, DOI 10.1093/qmath/27.2.241. MR400264 ↑245

[Gir71] Jean Giraud, *Cohomologie non abélienne* (French), Die Grundlehren der mathematischen Wissenschaften, Band 179, Springer-Verlag, Berlin-New York, 1971. MR344253 ↑233

[GMM68] S. Gitler, M. Mahowald, and R. James Milgram, *The nonimmersion problem for RP^n and higher-order cohomology operations*, Proc. Nat. Acad. Sci. U.S.A. **60** (1968), 432–437, DOI 10.1073/pnas.60.2.432. MR227997 ↑260

[GP21] Daniel Grady and Dmitri Pavlov, *The geometric cobordism hypothesis*, 2021. `arXiv:2111.01095`. 284

[Gre06] Robert Greenblatt, *Homology with local coefficients and characteristic classes*, Homology Homotopy Appl. **8** (2006), no. 2, 91–103. MR2246023 ↑244

[GS84] Michael B. Green and John H. Schwarz, *Anomaly cancellations in supersymmetric $D = 10$ gauge theory and superstring theory*, Phys. Lett. B **149** (1984), no. 1-3, 117–122, DOI 10.1016/0370-2693(84)91565-X. MR771086 ↑228, 233, 235, 285

[HHD21] Ruizhi Huang, Fei Han, and Haibao Duan, *String^c structures and modular invariants*, Trans. Amer. Math. Soc. **374** (2021), no. 5, 3491–3533, DOI 10.1090/tran/8311. MR4237954 ↑231

[Hil07] Michael A. Hill, *The 3-local* tmf*-homology of* $B\Sigma_3$, Proc. Amer. Math. Soc. **135** (2007), no. 12, 4075–4086, DOI 10.1090/S0002-9939-07-08937-X. MR2341960 ↑276

[Hil09] Michael A. Hill, *The String bordism of* BE_8 *and* $BE_8 \times BE_8$ *through dimension 14*, Illinois J. Math. **53** (2009), no. 1, 183–196. MR2584941 ↑230, 246, 273, 274, 276, 277

[HJ20] Fabian Hebestreit and Michael Joachim, *Twisted spin cobordism and positive scalar curvature*, J. Topol. **13** (2020), no. 1, 1–58, DOI 10.1112/topo.12122. MR3999671 ↑230, 250

[HLMM22] Brian Henning, Xiaochuan Lu, Tom Melia, and Hitoshi Murayama, *Outer automorphism anomalies*, J. High Energy Phys. **2** (2022), Paper No. 094, 38, DOI 10.1007/jhep02(2022)094. MR4407371 ↑233

[Hoà75] Hoàng Xuân Sính, *Gr-categories*, PhD thesis, Université Paris VII, 1975. 236

[Hov08] Mark Hovey, *The homotopy of* $M\mathrm{String}$ *and* $M\mathrm{U}\langle 6\rangle$ *at large primes*, Algebr. Geom. Topol. **8** (2008), no. 4, 2401–2414, DOI 10.2140/agt.2008.8.2401. MR2465746 ↑246

[HR95] Mark A. Hovey and Douglas C. Ravenel, *The 7-connected cobordism ring at* $p = 3$, Trans. Amer. Math. Soc. **347** (1995), no. 9, 3473–3502, DOI 10.2307/2155020. MR1297530 ↑246, 276

[HS05] M. J. Hopkins and I. M. Singer, *Quadratic functions in geometry, topology, and M-theory*, J. Differential Geom. **70** (2005), no. 3, 329–452. MR2192936 ↑238

[HW96a] Petr Hořava and Edward Witten, *Eleven-dimensional supergravity on a manifold with boundary*, Nuclear Phys. B **475** (1996), no. 1-2, 94–114, DOI 10.1016/0550-3213(96)00308-2. MR1412097 ↑231, 280

[HW96b] Petr Hořava and Edward Witten, *Heterotic and type I string dynamics from eleven dimensions*, Nuclear Phys. B **460** (1996), no. 3, 506–524, DOI 10.1016/0550-3213(95)00621-4. MR1381609 ↑231, 280

[Jen05] Jerome A. Jenquin, *Classical Chern-Simons on manifolds with spin structure*, 2005. `arXiv:math/0504524`. 235

[JFT20] Theo Johnson-Freyd and David Treumann, $\mathrm{H}^4(\mathrm{Co}_0; \mathbf{Z}) = \mathbf{Z}/24$, *Int. Math. Res. Not. IMRN*, (21):7873–7907, 2020. `arXiv:1707.07587`. 235

[Joa04] Michael Joachim, *Higher coherences for equivariant K-theory*, Structured ring spectra, London Math. Soc. Lecture Note Ser., vol. 315, Cambridge Univ. Press, Cambridge, 2004, pp. 87–114, DOI 10.1017/CBO9780511529955.006. MR2122155 ↑250

[Kah84] Bruno Kahn, *Classes de Stiefel-Whitney de formes quadratiques et de représentations galoisiennes réelles* (French), Invent. Math. **78** (1984), no. 2, 223–256, DOI 10.1007/BF01388595. MR767193 ↑255

[KOTY23] Justin Kaidi, Kantaro Ohmori, Yuji Tachikawa, and Kazuya Yonekura, *Nonsupersymmetric heterotic branes*, 2023. `arXiv:2303.17623`. 231, 280, 281

[KPMT20] Justin Kaidi, Julio Parra-Martinez, and Yuji Tachikawa, *Topological superconductors on superstring worldsheets*, SciPost Phys. **9** (2020), no. 1, Paper No. 010, 70, DOI 10.21468/scipostphys. With a mathematical appendix by Arun Debray. MR4196335 ↑279

[KS04] Igor Kriz and Hisham Sati, *M-theory, type IIA superstrings, and elliptic cohomology*, Adv. Theor. Math. Phys. **8** (2004), no. 2, 345–394. MR2086925 ↑245

[KT17] Anton Kapustin and Ryan Thorngren, *Higher symmetry and gapped phases of gauge theories*, Algebra, geometry, and physics in the 21st century, Progr. Math., vol. 324, Birkhäuser/Springer, Cham, 2017, pp. 177–202, DOI 10.1007/978-3-319-59939-7_5. MR3702386 ↑246

[Kud56] Tatsuji Kudo, *A transgression theorem*, Mem. Fac. Sci. Kyūsyū Univ. A **9** (1956), 79–81, DOI 10.2206/kyushumfs.9.79. MR79259 ↑273, 274

[Las63] R. Lashof, *Poincaré duality and cobordism*, Trans. Amer. Math. Soc. **109** (1963), 257–277, DOI 10.2307/1993906. MR156357 ↑241

[Lau00] Gerd Laures, *On cobordism of manifolds with corners*, Trans. Amer. Math. Soc. **352** (2000), no. 12, 5667–5688, DOI 10.1090/S0002-9947-00-02676-3. MR1781277 ↑281

[Lea97] Ian J. Leary, *On the integral cohomology of wreath products*, J. Algebra **198** (1997), no. 1, 184–239, DOI 10.1006/jabr.1997.7151. MR1482981 ↑255

[LMSM86] L. G. Lewis Jr., J. P. May, M. Steinberger, and J. E. McClure, *Equivariant stable homotopy theory*, Lecture Notes in Mathematics, vol. 1213, Springer-Verlag, Berlin, 1986. With contributions by J. E. McClure, DOI 10.1007/BFb0075778. MR866482 ↑248, 249

[LS97] Yves Laszlo and Christoph Sorger, *The line bundles on the moduli of parabolic G-bundles over curves and their sections* (English, with English and French summaries), Ann. Sci. École Norm. Sup. (4) **30** (1997), no. 4, 499–525, DOI 10.1016/S0012-9593(97)89929-6. MR1456243 ↑247

[LSW20] John A. Lind, Hisham Sati, and Craig Westerland, *Twisted iterated algebraic K-theory and topological T-duality for sphere bundles*, Ann. K-Theory **5** (2020), no. 1, 1–42, DOI 10.2140/akt.2020.5.1. MR4078224 ↑251

[LT21] Yasunori Lee and Yuji Tachikawa, *Some comments on 6D global gauge anomalies*, PTEP. Prog. Theor. Exp. Phys. **8** (2021), Paper No. 08B103, 29, DOI 10.1093/ptep/ptab015. MR4303583 ↑246

[Lur09] Jacob Lurie, *Higher topos theory*, Annals of Mathematics Studies, vol. 170, Princeton University Press, Princeton, NJ, 2009, DOI 10.1515/9781400830558. MR2522659 ↑238

[LW23] Matthias Ludewig and Konrad Waldorf, *Lie 2-groups from loop group extensions*, 2023. arXiv:2303.13176. 238

[LY22] Yasunori Lee and Kazuya Yonekura, *Global anomalies in 8d supergravity*, J. High Energy Phys. **7** (2022), Paper No. 125, 31, DOI 10.1007/jhep07(2022)125. MR4458304 ↑267, 284

[Mac71] Saunders MacLane, *Categories for the working mathematician*, Graduate Texts in Mathematics, Vol. 5, Springer-Verlag, New York-Berlin, 1971. MR354798 ↑236

[May65] J. Peter May, *The cohomology of the Steenrod algebra; stable homotopy groups of spheres*, Bull. Amer. Math. Soc. **71** (1965), 377–380, DOI 10.1090/S0002-9904-1965-11302-7. MR185596 ↑275

[May66] J. P. May, *The cohomology of restricted Lie algebras and of Hopf algebras*, J. Algebra **3** (1966), 123–146, DOI 10.1016/0021-8693(66)90009-3. MR193126 ↑258, 275

[McI99] Brett McInnes, *The semispin groups in string theory*, J. Math. Phys. **40** (1999), no. 10, 4699–4712, DOI 10.1063/1.532999. MR1715280 ↑233

[McT14] Carl McTague, *The Cayley plane and string bordism*, Geom. Topol. **18** (2014), no. 4, 2045–2078, DOI 10.2140/gt.2014.18.2045. MR3268773 ↑231

[MDL23] David Martín Velázquez, Davide De Biasio, and Dieter Lüst, *Cobordism, singularities and the Ricci flow conjecture*, J. High Energy Phys. **1** (2023), Paper No. 126, 49, DOI 10.1007/jhep01(2023)126. MR4541045 ↑279

[MG95] Mark Mahowald and Vassily Gorbounov, *Some homotopy of the cobordism spectrum $MO\langle 8\rangle$*, Homotopy theory and its applications (Cocoyoc, 1993), Contemp. Math., vol. 188, Amer. Math. Soc., Providence, RI, 1995, pp. 105–119, DOI 10.1090/conm/188/02237. MR1349133 ↑246

[Mil60] J. Milnor, *On the cobordism ring Ω^* and a complex analogue. I*, Amer. J. Math. **82** (1960), 505–521, DOI 10.2307/2372970. MR119209 ↑275, 278

[Mil65] John W. Milnor, *Remarks concerning spin manifolds*, Differential and Combinatorial Topology (A Symposium in Honor of Marston Morse), Princeton Univ. Press, Princeton, NJ, 1965, pp. 55–62. MR180978 ↑269

[Mit75] G. E. Mitchell, *Bordism of manifolds with oriented boundaries*, Proc. Amer. Math. Soc. **47** (1975), 208–214, DOI 10.2307/2040234. MR350758 ↑281

[MLW50] Saunders MacLane and J. H. C. Whitehead, *On the 3-type of a complex*, Proc. Nat. Acad. Sci. U.S.A. **36** (1950), 41–48, DOI 10.1073/pnas.36.1.41. MR33519 ↑238

[MPDF23] Miguel Montero and Héctor Parra De Freitas, *New supersymmetric string theories from discrete theta angles*, J. High Energy Phys. **1** (2023), Paper No. 91, 47, DOI 10.1007/jhep01(2023)091. MR4536894 ↑245

[MR22] Jacob McNamara and Matthew Reece, *Reflections on parity breaking*, 2022. arXiv:2212.00039. 279

[MS87] Mark Mahowald and Paul Shick, *Periodic phenomena in the classical Adams spectral sequence*, Trans. Amer. Math. Soc. **300** (1987), no. 1, 191–206, DOI 10.2307/2000595. MR871672 ↑258

[Mur96] M. K. Murray, *Bundle gerbes*, J. London Math. Soc. (2) **54** (1996), no. 2, 403–416, DOI 10.1093/acprof:oso/9780199534920.003.0012. MR1405064 ↑228, 234

[Mur10] Michael K. Murray, *An introduction to bundle gerbes*, The many facets of geometry, Oxford Univ. Press, Oxford, 2010, pp. 237–260, DOI 10.1093/acprof:oso/9780199534920.003.0012. MR2681698 ↑234

[MV99] Fabien Morel and Vladimir Voevodsky, $\mathbf{A}^1$*-homotopy theory of schemes*, Inst. Hautes Études Sci. Publ. Math. **90** (1999), 45–143 (2001). MR1813224 ↑238

[MV19] Jacob McNamara and Cumrun Vafa, *Cobordism classes and the Swampland*, 2019. `arXiv:1909.10355`. 229, 232, 279, 280, 281, 282, 283

[MV21] Miguel Montero and Cumrun Vafa, *Cobordism conjecture, anomalies, and the string lamppost principle*, J. High Energy Phys. **1** (2021), Paper No. 063, 46, DOI 10.1007/jhep01(2021)063. MR4258241 ↑245, 279

[Nak61] Minoru Nakaoka, *Homology of the infinite symmetric group*, Ann. of Math. (2) **73** (1961), 229–257, DOI 10.2307/1970333. MR131874 ↑254

[Nor] Johannes Nordström, MathOverflow answer: "Steenrod powers of Pontryagin classes", `https://www.mathoverflow.net/a/272417/`. 274

[NS19] Avijit Nath and Parameswaran Sankaran, *On generalized Dold manifolds*, Osaka J. Math. **56** (2019), no. 1, 75–90. MR3908778 ↑270

[NW13] Thomas Nikolaus and Konrad Waldorf, *Four equivalent versions of nonabelian gerbes*, Pacific J. Math. **264** (2013), no. 2, 355–419, DOI 10.2140/pjm.2013.264.355. MR3089401 ↑237

[PDF23] Héctor Parra De Freitas, *New supersymmetric string moduli spaces from frozen singularities*, J. High Energy Phys. **1** (2023), Paper No. 170, 17, DOI 10.1007/jhep01(2023)170. MR4543647 ↑245

[Pet68] F. P. Peterson, *Lectures on cobordism theory*, Lectures in Mathematics; Department of Mathematics, Kyoto University, vol. 1, Kinokuniya Book Store Co., Ltd., Tokyo, 1968. Notes by M. Mimura. MR234475 ↑246

[PS05] T. Pantev and E. Sharpe, *Notes on gauging noneffective group actions*, 2005. `arXiv:hep-th/0502027`. 279

[PS06a] Tony Pantev and Eric Sharpe, *GLSMs for gerbes (and other toric stacks)*, Adv. Theor. Math. Phys. **10** (2006), no. 1, 77–121. MR2222223 ↑279

[PS06b] Tony Pantev and Eric Sharpe, *String compactifications on Calabi-Yau stacks*, Nuclear Phys. B **733** (2006), no. 3, 233–296, DOI 10.1016/j.nuclphysb.2005.10.035. MR2192505 ↑279

[Qui71] Daniel Quillen, *The Adams conjecture*, Topology **10** (1971), 67–80, DOI 10.1016/0040-9383(71)90018-8. MR279804 ↑256

[Ros89] Jonathan Rosenberg, *Continuous-trace algebras from the bundle theoretic point of view*, J. Austral. Math. Soc. Ser. A **47** (1989), no. 3, 368–381. MR1018964 ↑250

[RST77] Nigel Ray, Robert Switzer, and Larry Taylor, *Normal structures and bordism theory, with applications to* $M\mathrm{Sp}_*$, Mem. Amer. Math. Soc. **12** (1977), no. 193, ix+66, DOI 10.1090/memo/0193. MR461520 ↑281

[RSW22] Dominik Rist, Christian Saemann, and Martin Wolf, *Explicit non-abelian gerbes with connections*, 2022. `arXiv:2203.00092`. 240

[Sat11a] Hisham Sati, *Anomalies of* E_8 *gauge theory on string manifolds*, Internat. J. Modern Phys. A **26** (2011), no. 13, 2177–2197, DOI 10.1142/S0217751X1105333X. MR2803340 ↑251, 285

[Sat11b] Hisham Sati, *Geometric and topological structures related to M-branes II: Twisted string and string*^{c} *structures*, J. Aust. Math. Soc. **90** (2011), no. 1, 93–108, DOI 10.1017/S1446788711001261. MR2810946 ↑228, 241, 245

[Sat11c] Hisham Sati, *Twisted topological structures related to M-branes*, Int. J. Geom. Methods Mod. Phys. **8** (2011), no. 5, 1097–1116, DOI 10.1142/S0219887811005567. MR2838539 ↑251

[Sat12] Hisham Sati, *Twisted topological structures related to M-branes II: twisted Wu and Wu*^{c} *structures*, Int. J. Geom. Methods Mod. Phys. **9** (2012), no. 7, 1250056, 21, DOI 10.1142/S0219887812500569. MR2972513 ↑251

[Sat15] Hisham Sati, *Ninebrane structures*, Int. J. Geom. Methods Mod. Phys. **12** (2015), no. 4, 1550041, 24, DOI 10.1142/S0219887815500413. MR3332238 ↑251

[Sch13] Urs Schreiber, *Differential cohomology in a cohesive infinity-topos*, 2013. arXiv:1310.7930. 238, 239

[Sch21] L. Schlechter, *Computational Methods in String Theory and Applications to the Swampland Conjectures*, PhD thesis, Ludwig–Maximilians–Universität München, 2021. https://edoc.ub.uni-muenchen.de/28084/1/Schlechter_Lorenz.pdf. 279

[Seg70] Graeme Segal, *Cohomology of topological groups*, Symposia Mathematica, Vol. IV (INDAM, Rome, 1968/69), Academic Press, London-New York, 1970, pp. 377–387. MR280572 ↑237

[Seg75] G. B. Sigal, *A classifying space of a topological group in the sense of Gel′fand-Fuks* (Russian), Funkcional. Anal. i Priložen. **9** (1975), no. 2, 48–50. MR375306 ↑237

[Seg88] G. B. Segal, *The definition of conformal field theory*, Differential geometrical methods in theoretical physics (Como, 1987), NATO Adv. Sci. Inst. Ser. C: Math. Phys. Sci., vol. 250, Kluwer Acad. Publ., Dordrecht, 1988, pp. 165–171. MR981378 ↑283

[Ser52] Jean-Pierre Serre, *Sur les groupes d'Eilenberg-MacLane* (French), C. R. Acad. Sci. Paris **234** (1952), 1243–1245. MR46047 ↑274

[Ser53] Jean-Pierre Serre, *Cohomologie modulo 2 des complexes d'Eilenberg-MacLane* (French), Comment. Math. Helv. **27** (1953), 198–232, DOI 10.1007/BF02564562. MR60234 ↑253, 255

[Sha77] P. Brian Shay, mod *p Wu formulas for the Steenrod algebra and the Dyer-Lashof algebra*, Proc. Amer. Math. Soc. **63** (1977), no. 2, 339–347, DOI 10.2307/2041815. MR454974 ↑274

[Sha15] Eric Sharpe, *Notes on generalized global symmetries in QFT*, Fortschr. Phys. **63** (2015), no. 11-12, 659–682, DOI 10.1002/prop.201500048. MR3422349 ↑228

[SI67] Nobuo Shimada and Akira Iwai, *On the cohomology of some Hopf algebras*, Nagoya Math. J. **30** (1967), 103–111. MR215896 ↑257

[SP11] Christopher J. Schommer-Pries, *Central extensions of smooth 2-groups and a finite-dimensional string 2-group*, Geom. Topol. **15** (2011), no. 2, 609–676, DOI 10.2140/gt.2011.15.609. MR2800361 ↑237, 238

[SP18] Christopher J. Schommer-Pries, *Tori detect invertibility of topological field theories*, Geom. Topol. **22** (2018), no. 5, 2713–2756, DOI 10.2140/gt.2018.22.2713. MR3811769 ↑243

[SSS12] Hisham Sati, Urs Schreiber, and Jim Stasheff, *Twisted differential string and fivebrane structures*, Comm. Math. Phys. **315** (2012), no. 1, 169–213, DOI 10.1007/s00220-012-1510-3. MR2966944 ↑228, 232, 236, 238, 239, 241, 244, 278

[Sto68] Robert E. Stong, *Notes on cobordism theory*, Princeton University Press, Princeton, NJ; University of Tokyo Press, Tokyo, 1968. Mathematical notes. MR248858 ↑281

[Sto86] R. E. Stong, *Appendix: calculation of* $\Omega_{11}^{\mathrm{Spin}}(K(Z,4))$, Workshop on unified string theories (Santa Barbara, Calif., 1985), World Sci. Publishing, Singapore, 1986, pp. 430–437. MR849112 ↑254, 261, 267, 279, 285

[Sug99] Shigeki Sugimoto, *Anomaly Cancellations in the Type I D9-D9 System and the USp(32) String Theory*, *Progress of Theoretical Physics*, 102(3):685–699, 09 1999. arXiv:hep-th/9905159. 245

[SW15] Hisham Sati and Craig Westerland, *Twisted Morava K-theory and E-theory*, J. Topol. **8** (2015), no. 4, 887–916, DOI 10.1112/jtopol/jtv020. MR3431663 ↑251

[SY21] Hisham Sati and Aliaksandra Yarosh, *Twisted Morava K-theory and connective covers of Lie groups*, Algebr. Geom. Topol. **21** (2021), no. 5, 2223–2255, DOI 10.2140/agt.2021.21.2223. MR4334511 ↑251

[Tac22] Yuji Tachikawa, *Topological modular forms and the absence of a heterotic global anomaly*, PTEP. Prog. Theor. Exp. Phys. **4** (2022), Paper No. 04A107, 8, DOI 10.1093/ptep/ptab060. MR4404832 ↑284, 285

[Tho62] Emery Thomas, *The torsion Pontryagin classes*, Proc. Amer. Math. Soc. **13** (1962), 485–488, DOI 10.2307/2034967. MR141132 ↑235

[Tho20] Ryan Thorngren, *Topological quantum field theory, symmetry breaking, and finite gauge theory in* $3 + 1D$, *Phys. Rev. B*, 101:245160, Jun 2020. `arXiv:2001.11938`. 246

[TY21] Yuji Tachikawa and Mayuko Yamashita, *Topological modular forms and the absence of all heterotic global anomalies*, Comm. Math. Phys. **402** (2023), no. 2, 1585–1620, DOI 10.1007/s00220-023-04761-2. MR4627327 ↑231, 284, 285

[Wal60] C. T. C. Wall, *Determination of the cobordism ring*, Ann. of Math. (2) **72** (1960), 292–311, DOI 10.2307/1970136. MR120654 ↑251

[Wal13] Konrad Waldorf, *String connections and Chern-Simons theory*, Trans. Amer. Math. Soc. **365** (2013), no. 8, 4393–4432, DOI 10.1090/S0002-9947-2013-05816-3. MR3055700 ↑240

[Wal18] Konrad Waldorf, *A global perspective to connections on principal 2-bundles*, Forum Math. **30** (2018), no. 4, 809–843, DOI 10.1515/forum-2017-0097. MR3894086 ↑240

[Wan08] Bai-Ling Wang, *Geometric cycles, index theory and twisted K-homology*, J. Noncommut. Geom. **2** (2008), no. 4, 497–552, DOI 10.4171/JNCG/27. MR2438341 ↑250

[War20] Garth Warner, *Topics in topology and homotopy theory*, 2020. `arXiv:2007.02467`. 239

[Wei49] André Weil, *Géométrie différentielle des espaces fibres*, In *Œuvres Scientifiques - Collected Papers I*, Springer Collected Works in Mathematics, pages 422–436. Springer Berlin Heidelberg, 1949. 235

[Whi42] George W. Whitehead, *On the homotopy groups of spheres and rotation groups*, Ann. of Math. (2) **43** (1942), 634–640, DOI 10.2307/1968956. MR7107 ↑249

[Wit86] Edward Witten, *Topological tools in ten-dimensional physics*, Workshop on unified string theories (Santa Barbara, Calif., 1985), World Sci. Publishing, Singapore, 1986, pp. 400–429. With an appendix by R. E. Stong. MR849111 ↑231, 242, 254, 285

[Wit96] Edward Witten, *Strong coupling expansion of Calabi-Yau compactification*, Nuclear Phys. B **471** (1996), no. 1-2, 135–158, DOI 10.1016/0550-3213(96)00190-3. MR1401294 ↑231, 280

[Wit97] Edward Witten, *On flux quantization in M-theory and the effective action*, J. Geom. Phys. **22** (1997), no. 1, 1–13, DOI 10.1016/S0393-0440(96)00042-3. MR1441697 ↑243, 245, 281

[Wit16] Edward Witten, *The "parity" anomaly on an unorientable manifold*, *Phys. Rev. B*, 94:195150, Nov 2016. `arXiv:1605.02391`. 243, 281

[Wit22] Edward Witten, *A note on the canonical formalism for gravity*, Adv. Theor. Math. Phys. **27** (2023), no. 1, 311–380. MR4622702 ↑279

[WW19a] Zheyan Wan and Juven Wang, *Adjoint* QCD_4, *deconfined critical phenomena, symmetry-enriched topological quantum field theory, and higher symmetry extension*, Phys. Rev. D **99** (2019), no. 6, 065013, 21, DOI 10.1103/physrevd.99.065013. MR3993294 ↑246

[WW19b] Zheyan Wan and Juven Wang, *Higher anomalies, higher symmetries, and cobordisms I: classification of higher-symmetry-protected topological states and their boundary fermionic/bosonic anomalies via a generalized cobordism theory*, Ann. Math. Sci. Appl. **4** (2019), no. 2, 107–311, DOI 10.4310/AMSA.2019.v4.n2.a2. MR4020366 ↑246

[WWZ19] Zheyan Wan, Juven Wang, and Yunqin Zheng, *Quantum 4d Yang-Mills theory and time-reversal symmetric 5d higher-gauge topoological field theory*, Phys. Rev. D **100** (2019), no. 8, 085012, 61, DOI 10.1103/physrevd.100.085012. MR4032010 ↑246

[Yon54] Nobuo Yoneda, *On the homology theory of modules*, J. Fac. Sci. Univ. Tokyo Sect. I **7** (1954), 193–227. MR68832 ↑257

[Yon22] Kazuya Yonekura, *Heterotic global anomalies and torsion Witten index*, J. High Energy Phys. **10** (2022), Paper No. 114, 38, DOI 10.1007/jhep10(2022)114. MR4498769 ↑285

[Yos75] Zen-ichi Yosimura, *Universal coefficient sequences for cohomology theories of* CW-*spectra*, Osaka Math. J. **12** (1975), no. 2, 305–323. MR388375 ↑284

[Yu21] Matthew Yu, *Symmetries and anomalies of* $(1+1)d$ *theories: 2-groups and symmetry fractionalization*, J. High Energy Phys. **8** (2021), Paper No. 061, 30, DOI 10.1007/jhep08(2021)061. MR4317209 ↑246

Department of Mathematics, Purdue University, 150 N. University Street, West Lafayette, Indiana 47907

Email address: `adebray@purdue.edu`

Contemporary Mathematics
Volume **802**, 2024
https://doi.org/10.1090/conm/802/16000

Generalized Abelian Turaev–Viro and U(1) BF theories

Emil Høssjer, Philippe Mathieu, and Frank Thuillier

This paper is dedicated to our colleague and friend Éric Pilon

ABSTRACT. We explain how it is possible to study U(1) BF theory over a connected closed oriented smooth 3-manifold in the formalism of path integral thanks to Deligne–Beilinson cohomology. We show how we can straightforwardly extend the definition to families of theories in any dimension. We then extend the definition of the Turaev–Viro invariant of a connected closed oriented smooth 3-manifold in an Abelian framework to a family of invariants in any dimension. We show that those invariants can be written as discrete BF theories. We explain how the extensions of the U(1) BF theory we defined can be related to the extensions of the Turaev–Viro invariant we constructed.

1. Introduction

1.1. A bit of history. In 1968, Tullio Regge studied with Giorgio Ponzano the large spins limit of the $6j$-symbols defined by Wigner [**26**], which have the symmetries of a tetrahedron. On Figure 1, the labels of the edges of the tetrahedron are indices of SU(2) representations (the so-called spins). In their work, Ponzano and Regge showed that if we regard those indices as the lengths of the edges of the tetrahedron, then in the large spins limit, the $6j$-symbol associated with the labeling representations is related to the volume V of the tetrahedron according to

$$\begin{Bmatrix} a & b & c \\ d & e & f \end{Bmatrix} \sim \frac{1}{\sqrt{12\pi V}} \cos\left(\sum_{k,l} j_{kl}\theta_{kl} + \frac{\pi}{4}\right), \tag{1.1}$$

θ_{kl} being the dihedral angles of the tetrahedron.

Studying then the case of $3nj$-symbols, Ponzano and Regge considered triangulated polyhedra, more general than tetrahedra, with edges labeled by indices of SU(2) representations. Conducting a formal computation consisting of summing over all the possible labels and taking the high spins limit, they recognized the discretization of the partition function of Euclidean general relativity in dimension 3

2020 *Mathematics Subject Classification.* Primary 57K16; Secondary 81T25, 81T27.

Key words and phrases. Differential cohomology, quantum invariants of manifolds, quantum field theory, gauge theory, U(1) BF theory.

The second author was supported by the NSF grant 1947155 and the JTF grant 6152, then by the NSF Grant 200020-192080 of the Simons Collaboration on Global Categorical Symmetries, the COST Action 21109 - Cartan geometry, Lie, Integrable Systems, quantum group Theories for Applications (CaLISTA), and the NCCR SwissMAP, funded by the Swiss National Science Foundation.

©2024 American Mathematical Society

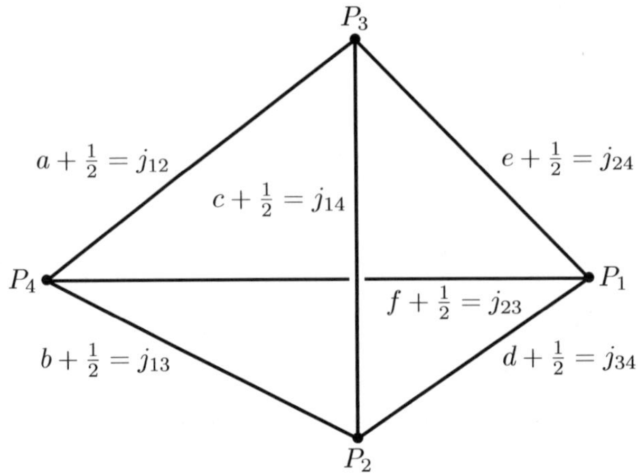

FIGURE 1. Tetrahedron associated with the $6j$-symbol $\begin{Bmatrix} a & b & c \\ d & e & f \end{Bmatrix}$.

for an empty space that Regge had studied in an earlier work in which he expressed General Relativity without coordinates but using a cellular decomposition of the space-time [**27**]. Unfortunately, both quantities are ill-defined. Indeed, the sum over all representations of SU(2) is infinite, and the partition function too, as it consists of a sum over the infinite dimensional space of all the cellular decompositions of space-time.

Those ideas have slept nearly twenty years before emerging back in 1992, when Vladimir Turaev and Oleg Viro defined the so-called state-sum invariant of 3-manifolds, or Turaev–Viro (TV) invariant, using the same idea as Ponzano and Regge of labeled edges of a triangulation of the manifold considered [**32**]. This construction was then rewritten in the formalism of spherical categories by John Barrett and Bruce Westbury [**5**]. In this framework, it is possible to construct the invariant with a finite set of objects, according to the so-called "domination axiom" that the category has to satisfy. The expression of the TV invariant is then formally very similar to the formula of Ponzano and Regge in the case of a realization of the category by representations of $\mathcal{U}_q(\mathfrak{sl}_2(\mathbb{C}))$ (i.e., a quantum deformation of the universal enveloping algebra of $\mathfrak{sl}_2(\mathbb{C})$ at a root of unity q). Contrary to the formula of Ponzano and Regge, the TV invariant is a well-defined quantity, since the domination axiom imposes the sum to be taken over a period of the cyclic representations of $\mathcal{U}_q(\mathfrak{sl}_2(\mathbb{C}))$. It can be therefore considered as a regularization of the formula of Ponzano and Regge.

More recently, in the early years 2000, these works appeared to be particularly useful to regularize some kind of SU(2) BF theories showing up naturally in Loop Quantum Gravity, see, e.g., the works of Laurent Freidel, David Louapre, and Etera Livine [**13–15**].

A first question we may ask is, what would an Abelian version of the TV construction look like, and would it be related in some manner to a U(1) version of the BF theory?[1] These questions were answered between 2015 and 2017 in [**21**–**23**]. This investigation was purely 3-dimensional, but the Abelian TV construction can easily be generalized to any dimension. Likewise, the U(1) BF theory can also be generalized to any dimension. Hence, a natural question is: Is the relation between the Abelian TV invariant and the U(1) BF theory true in any dimension? This question is answered in [**19**, **20**] and the main points of the investigation are presented in this proceeding.

We will first explain what a "higher U(1) BF theory", and second, what a "higher Abelian TV invariant" consist of. Third, we will show how this "higher Abelian TV invariant" can be understood as a discrete BF theory. Finally, we will explain how we can make sense of the claim "BF and TV theories are the same" in this higher dimensional Abelian case.

1.2. A bit of vocabulary. In this proceeding, we will use the following notations and terminology:

- $\mathcal{F}_M$ for the "space of fields" over M, *e.g.* $C^\infty(M)$ (scalar fields), $\Gamma(M,E)$ (sections of a vector bundle E over M), $\mathrm{Conn}_G(M)\,(M)$ (G-connections over M)...
- G for the "structure group" (compact Lie group), *e.g.* most of the time $G = \mathrm{SU}(N)$ or $\mathrm{U}(1)$,
- $\mathcal{G} \subseteq \mathrm{Map}(M,G)$ for the "gauge group", which acts on $\mathcal{F}_M$,
- $S : \mathcal{F}_M \to \mathbb{R}$ for the "action functional" of the theory considered.
- If S is $\mathcal{G}$-invariant, i.e., if $\forall \phi \in \mathcal{F}_M, \forall g \in \mathcal{G}$, $S(g\cdot\phi) = S(\phi)$, then S defines a "gauge theory".
- If $S : \mathcal{F}_M \to \mathbb{R}$ defines a gauge theory, then we may try to consider rather $S : \mathcal{F}_M/\mathcal{G} \to \mathbb{R}$.
- If $e^{2\pi i S}$ is $\mathcal{G}$-invariant, i.e., if $\forall \phi \in \mathcal{F}_M, \forall g \in \mathcal{G}$, $S(g\cdot\phi) = S(\phi) + n$, $n \in \mathbb{Z}$, then S defines a "quantum gauge theory".
- If $S : \mathcal{F}_M \to \mathbb{R}$ defines a quantum gauge theory, then we may try to consider rather $S : \mathcal{F}_M/\mathcal{G} \to \mathbb{R}/\mathbb{Z}$.

2. Generalized U(1) BF theory

2.1. The standard SU(2) case. In this subsection, $M^{(3)}$ is a closed connected oriented smooth 3-manifold. Remark that, since SU(2) is simply connected, then any principal SU(2)-bundle over $M^{(3)}$ is necessarily trivializable.

The first field of the standard SU(2) 3D BF theory is an SU(2)-connection $A^{(1)}$. Since any principal SU(2)-bundle over $M^{(3)}$ is trivializable, $A^{(1)}$ can be regarded as a (global) 1-form with coefficients in $\mathfrak{su}(2)$ (the upper index keeps track of the form degree of the object) such that the group $\mathcal{G}$ acts on $A^{(1)}$ as

$$g\cdot A^{(1)} := g^{-1}A^{(1)}g + g^{-1}\,\mathrm{d}\,g, \tag{2.1}$$

for $g \in \mathcal{G}$ and $A^{(1)} \in \mathrm{Conn}_{\mathrm{SU}(2)}\big(M^{(3)}\big)$.

Recall that the curvature of $A^{(1)}$ is

$$F^{(2)}\left(A^{(1)}\right) = \mathrm{d}\,A^{(1)} + A^{(1)} \wedge A^{(1)} \in \Omega^2\Big(M^{(3)}, \mathfrak{su}(2)\Big) \tag{2.2}$$

[1]Remark that U(1) might be regarded as simpler than SU(2) since U(1) is Abelian, but U(1) is nonsimply connected, contrary to SU(2). Hence, there are nontrivial isomorphism classes of principal U(1)-bundles while principal SU(2)-bundles are all trivializable.

and $F^{(2)}\left(A^{(1)}\right)$ transforms under the action of $\mathcal{G}$ on $A^{(1)}$ as

$$F^{(2)}\left(g\cdot A^{(1)}\right)=g^{-1}F^{(2)}\left(A^{(1)}\right)g, \tag{2.3}$$

for $g\in\mathcal{G}$ and $A^{(1)}\in \mathrm{Conn}_{\mathrm{SU}(2)}\left(M^{(3)}\right)$.

The second field of the standard SU(2) 3D BF theory is a (global) 1-form $B^{(1)}$ with coefficients in $\mathfrak{su}(2)$ such that the group $\mathcal{G}$ acts on $B^{(1)}$ as

$$g\cdot B^{(1)}=g^{-1}B^{(1)}g, \tag{2.4}$$

for $g\in\mathcal{G}$ and $B^{(1)}\in\Omega^1\left(M^{(3)},\mathfrak{su}(2)\right)$.

Then the group $\mathcal{G}$ acts on the pair

$$\left(A^{(1)},B^{(1)}\right)\in \mathrm{Conn}_{\mathrm{SU}(2)}\left(M^{(3)}\right)\times\Omega^1\left(M^{(3)},\mathfrak{su}(2)\right)$$

diagonally, i.e.,

$$g\cdot\left(A^{(1)},B^{(1)}\right)=\left(g\cdot A^{(1)},g\cdot B^{(1)}\right)=\left(g^{-1}A^{(1)}g+g^{-1}\,\mathrm{d}\,g,g^{-1}B^{(1)}g\right), \tag{2.5}$$

for $g\in\mathcal{G}$ and $\left(A^{(1)},B^{(1)}\right)\in \mathrm{Conn}_{\mathrm{SU}(2)}\left(M^{(3)}\right)\times\Omega^1\left(M^{(3)},\mathfrak{su}(2)\right)$.

DEFINITION 2.1. The standard SU(2) 3D BF theory with a coupling constant k is defined by

$$\mathcal{F}_{M^{(3)}}=\mathrm{Conn}_{\mathrm{SU}(2)}\left(M^{(3)}\right)\times\Omega^1\left(M^{(3)},\mathfrak{su}(2)\right) \tag{2.6}$$

as space of fields over which is defined the action functional[2]

$$S_{\mathrm{BF}_k}\left(A^{(1)},B^{(1)}\right)=k\int_{M^{(3)}}\mathrm{Tr}\left(B^{(1)}\wedge F^{(2)}\left(A^{(1)}\right)\right). \tag{2.7}$$

PROPOSITION 2.2. *This action functional defines a gauge theory, i.e.,*

$$S_{\mathrm{BF}_k}\left(g\cdot\left(A^{(1)},B^{(1)}\right)\right)=S_{\mathrm{BF}_k}\left(A^{(1)},B^{(1)}\right), \tag{2.8}$$

for $g\in\mathcal{G}$, $A^{(1)}\in \mathrm{Conn}_{\mathrm{SU}(2)}\left(M^{(3)}\right)$ *and* $B^{(1)}\in\Omega^1\left(M^{(3)},\mathfrak{su}(2)\right)$.

For interesting facts about the SU(2) BF theory, in particular its relation to Alexander–Conway invariant of knots and to the SU(2) Chern–Simons theory, see [**9, 10**].

REMARK 2.3. In this SU(2) case, k is *not* quantized, i.e., k can be any real number.

2.2. The 3D U(1) **case.** In this subsection, $M^{(3)}$ is still a closed connected oriented smooth 3-manifold.

The U(1) case is fundamentally different, as there are in general isomorphism classes of nontrivial bundles.[3] The U(1) case *cannot* be simply regarded as the Abelianization of the SU(2) case.

Moreover, unlike the SU(2) case, we want here our fields $\mathbf{A}^{(1)}$ and $\mathbf{B}^{(1)}$ to be *gauge classes* of U(1)-connections (whence the bold face used for denoting the fields), which means we won't go through the usual gauge fixing procedure [**25**].

Fortunately, it turns out that the space of gauge classes of U(1)-connections is perfectly well-known: This is the first group of Deligne–Beilinson coho-

[2]Whence the name "BF theory" which is *not* an acronyme, but stands for "$B\wedge F$".

[3]Recall that principal U(1)-bundles are classified by the second Čech cohomology $\check{H}^2\left(M^{(3)}\right)$.

mology $H^1_{\mathrm{DB}}\left(M^{(3)}\right)$, also known as second group of differential cohomology $\widehat{H}^2\left(M^{(3)}\right)$.[4] For a description and more references on this topic, see [**4**, **6**]. We recall here a few important facts we will need later on in this paper.

PROPOSITION 2.4. *The group $H^1_{\mathrm{DB}}\left(M^{(3)}\right)$ is described by the following short exact sequence that splits:*

$$0 \to \Omega^1\left(M^{(3)}\right)/\Omega^1_{\mathbb{Z}}\left(M^{(3)}\right) \to H^1_{\mathrm{DB}}\left(M^{(3)}\right) \to \check{H}^2\left(M^{(3)}\right) \to 0 \tag{2.9}$$

where $\Omega^1\left(M^{(3)}\right)$ is the space of 1-forms over $M^{(3)}$, $\Omega^1_{\mathbb{Z}}\left(M^{(3)}\right)$ is the space of closed 1-forms with integral periods over $M^{(3)}$, and $\check{H}^2\left(M^{(3)}\right)$ is the second Čech cohomology.

PROPOSITION 2.5. *Over $H^1_{\mathrm{DB}}\left(M^{(3)}\right)$, there exists*

- *A (symmetric) pairing*

$$\star : \begin{array}{ccc} H^1_{\mathrm{DB}}\left(M^{(3)}\right) \times H^1_{\mathrm{DB}}\left(M^{(3)}\right) & \longrightarrow & \Omega^3\left(M^{(3)}\right)/\Omega^3_{\mathbb{Z}}\left(M^{(3)}\right) \\ \left(\mathbf{A}^{(1)}, \mathbf{B}^{(1)}\right) & \longmapsto & \mathbf{A}^{(1)} \star \mathbf{B}^{(1)} \end{array} \tag{2.10}$$

- *A notion of integral over the elements of $Z_1\left(M^{(3)}\right)$, i.e., the singular 1-cycles of $M^{(3)}$*

$$\int : \begin{array}{ccc} Z_1\left(M^{(3)}\right) \times H^1_{\mathrm{DB}}\left(M^{(3)}\right) & \longrightarrow & \mathbb{R}/\mathbb{Z} \\ \left(z_{(1)}, \mathbf{A}^{(1)}\right) & \longmapsto & \int_{z_{(1)}} \mathbf{A}^{(1)} \end{array} \tag{2.11}$$

PROPOSITION 2.6. *Another notion of integral will be very useful here. This is*

$$\int_{M^{(3)}} : \begin{array}{ccc} \Omega^3\left(M^{(3)}\right)/\Omega^3_{\mathbb{Z}}\left(M^{(3)}\right) & \longrightarrow & \mathbb{R}/\mathbb{Z} \\ \mathbf{L}^{(3)} & \longmapsto & \int_{M^{(3)}} \mathbf{L}^{(3)}. \end{array} \tag{2.12}$$

REMARK 2.7. We want to emphasize that the integrals introduced in the two previous propositions are $\mathbb{R}/\mathbb{Z}$-valued. Said differently, only the complex exponential truly makes sense, and we will see later on that this is anyway what we want to consider in a quantum treatment of the theory for the partition function and the expectation values of observables.

DEFINITION 2.8. The U(1) 3D BF theory with a coupling constant[5] $k \in \mathbb{Z}$ is defined by

$$\mathcal{F}_{M^{(3)}}/\mathcal{G} = H^1_{\mathrm{DB}}\left(M^{(3)}\right) \times H^1_{\mathrm{DB}}\left(M^{(3)}\right) \tag{2.13}$$

as space of fields (modulo gauge transformations) over which is defined the *quantum* action functional

$$S_{\mathrm{BF}_k}\left(\mathbf{A}^{(1)}, \mathbf{B}^{(1)}\right) \underset{\mathbb{Z}}{=} k \int_{M^{(3)}} \mathbf{B}^{(1)} \star \mathbf{A}^{(1)} \tag{2.14}$$

and the *quantum* observables

$$W\left(\mathbf{A}^{(1)}, z^{\mathbf{A}}_{(1)}, \mathbf{B}, z^{\mathbf{B}}_{(1)}\right) = e^{2\pi i \int_{z^{\mathbf{A}}_{(1)}} \mathbf{A}^{(1)}} e^{2\pi i \int_{z^{\mathbf{B}}_{(1)}} \mathbf{B}^{(1)}}, \tag{2.15}$$

---

[4]The convention of degree of $H^1_{\mathrm{DB}}\left(M^{(3)}\right)$ allows to keep track of the form degree of the field.

[5]The coupling constant k has to be an integer since the integral in the action is $\mathbb{R}/\mathbb{Z}$-valued.

where $\mathbf{A}^{(1)}, \mathbf{B}^{(1)} \in H^1_{\mathrm{DB}}\left(M^{(3)}\right)$, $z^{\mathbf{A}}_{(1)}, z^{\mathbf{B}}_{(1)} \in Z_1\left(M^{(3)}\right)$.

REMARK 2.9. If $\mathbf{B}^{(1)} = \mathbf{A}^{(1)}$, which is possible because we chose $\mathbf{A}^{(1)}$ and $\mathbf{B}^{(1)}$ to be both (gauge classes of) U(1)-connections, then we get the quantum action of the U(1) Chern–Simons theory

$$S_{\mathrm{CS}_k}\left(\mathbf{A}^{(1)}\right) = k \int_{M^{(3)}} \mathbf{A}^{(1)} \star \mathbf{A}^{(1)}. \tag{2.16}$$

2.3. The nD U(1) **case.** In this subsection, $M^{(n)}$ is a closed connected oriented smooth n-manifold.

PROPOSITION 2.10. *The short exact sequence* (2.9) *can actually be generalized as*

$$0 \to \Omega^p\left(M^{(n)}\right)/\Omega^p_{\mathbb{Z}}\left(M^{(n)}\right) \to H^p_{\mathrm{DB}}\left(M^{(n)}\right) \to \check{H}^{p+1}\left(M^{(n)}\right) \to 0. \tag{2.17}$$

and it still splits.

REMARK 2.11. The exact sequence (2.17) does make sense for $p = -1$ (provided we set $\Omega^{-1}\left(M^{(n)}\right)/\Omega^{-1}_{\mathbb{Z}}\left(M^{(n)}\right) \cong 0$, which implies then that $H^{-1}_{\mathrm{DB}}\left(M^{(n)}\right) \cong \mathbb{Z}$) and $p = n$. However, in the following, we will *not* consider those two cases for the quantum fields of our BF theory, as they would imply that one of the two fields is actually just a number.

REMARK 2.12. The elements of $H^p_{\mathrm{DB}}\left(M^{(n)}\right)$ may be regarded as "gauge classes of connections over $(p-1)$-gerbes", where "$(p-1)$-gerbes" are structures classified up to isomorphism by $\check{H}^{p+1}\left(M^{(n)}\right)$. Hence, a principal U(1)-bundle is nothing but a 0-gerbe, and what is usually called a "gerbe" is actually a 1-gerbe.

PROPOSITION 2.13. *Over* $H^\bullet_{\mathrm{DB}}\left(M^{(n)}\right)$, *we can define*

- *A pairing*

$$\star : \begin{array}{ccc} H^p_{\mathrm{DB}}\left(M^{(n)}\right) \times H^q_{\mathrm{DB}}\left(M^{(n)}\right) & \longrightarrow & H^{p+q+1}_{\mathrm{DB}}\left(M^{(n)}\right) \\ \left(\mathbf{A}^{(p)}, \mathbf{B}^{(q)}\right) & \longmapsto & \mathbf{A}^{(p)} \star \mathbf{B}^{(q)} \end{array} \tag{2.18}$$

- *A notion of integral over the elements of* $Z_p\left(M^{(n)}\right)$, *i.e., the singular* p*-cycles of* $M^{(n)}$

$$\int : \begin{array}{ccc} Z_p\left(M^{(n)}\right) \times H^p_{\mathrm{DB}}\left(M^{(n)}\right) & \longrightarrow & \mathbb{R}/\mathbb{Z} \\ \left(z_{(p)}, \mathbf{A}^{(p)}\right) & \longmapsto & \int_{z_{(p)}} \mathbf{A}^{(p)} \end{array} \tag{2.19}$$

REMARK 2.14. A particular case of this integral is for $p = n$, in which case $z_{(n)} = M^{(n)}$ (potentially multiplied by an integer) and

$$H^n_{\mathrm{DB}}\left(M^{(n)}\right) \cong \Omega^n\left(M^{(n)}\right)/\Omega^n_{\mathbb{Z}}\left(M^{(n)}\right), \tag{2.20}$$

which is an isomorphism that we implicitly used in (2.12).

REMARK 2.15. The pairing $\star : H^p_{\mathrm{DB}}\left(M^{(n)}\right) \times H^q_{\mathrm{DB}}\left(M^{(n)}\right)$ is almost graded-commutative, i.e., with the convention of degree of $H^\bullet_{\mathrm{DB}}(M)$,

$$\mathbf{A}^{(p)} \star \mathbf{B}^{(q)} = (-1)^{p+q+1} \mathbf{B}^{(q)} \star \mathbf{A}^{(p)} \tag{2.21}$$

for $\left(\mathbf{A}^{(p)}, \mathbf{B}^{(q)}\right) \in H^p_{\mathrm{DB}}\left(M^{(n)}\right) \times H^q_{\mathrm{DB}}\left(M^{(n)}\right)$, while with the convention of degree of $\widehat{H}^\bullet\left(M^{(n)}\right)$, the product is truly graded-commutative.

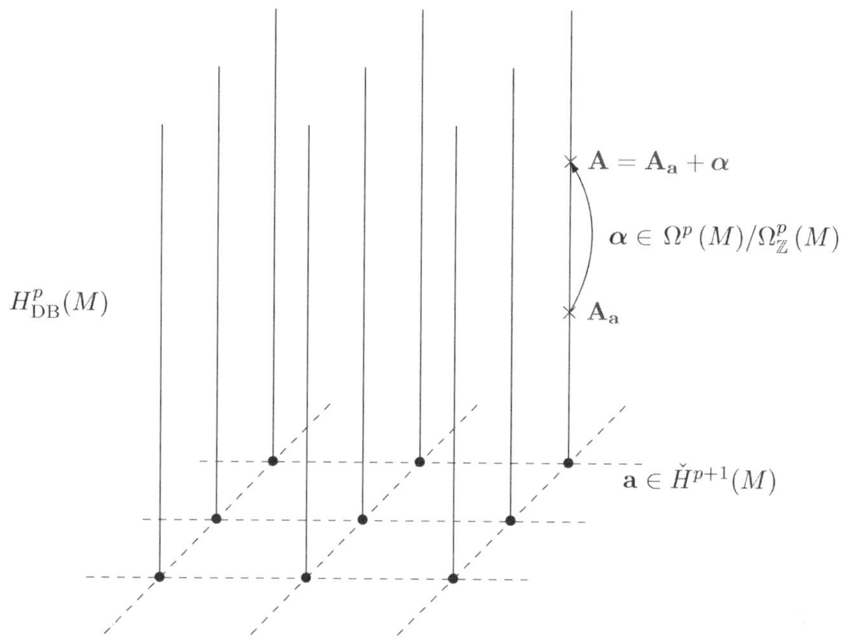

FIGURE 2. Representation of the space of fields $\mathcal{F}_{M^{(n)}}/\mathcal{G} = H^p_{\mathrm{DB}}\left(M^{(n)}\right)$

DEFINITION 2.16. Since we can make sense of Deligne–Beilinson cohomology classes in any degree as well as their product and integral, the U(1) nD BF theory with a coupling constant $k \in \mathbb{Z}$ will be defined by

$$\mathcal{F}_{M^{(n)}}/\mathcal{G} = H^p_{\mathrm{DB}}\left(M^{(n)}\right) \times H^q_{\mathrm{DB}}\left(M^{(n)}\right) \tag{2.22}$$

as space of fields (modulo gauge transformations) over which is defined the *quantum* action functional

$$S_{\mathrm{BF}_k}\left(\mathbf{A}^{(p)}, \mathbf{B}^{(q)}\right) \underset{\mathbb{Z}}{=} k \int_{M^{(n)}} \mathbf{B}^{(q)} \star \mathbf{A}^{(p)} \tag{2.23}$$

and the *quantum* observables

$$W\left(\mathbf{A}^{(p)}, z^{\mathbf{A}}_{(p)}, \mathbf{B}^{(q)}, z^{\mathbf{B}}_{(q)}\right) = e^{2\pi i \int_{z^{\mathbf{A}}_{(p)}} \mathbf{A}^{(p)}} e^{2\pi i \int_{z^{\mathbf{B}}_{(q)}} \mathbf{B}^{(q)}}, \tag{2.24}$$

with $n = p + q + 1$, $\left(\mathbf{A}^{(p)}, \mathbf{B}^{(q)}\right) \in H^p_{\mathrm{DB}}\left(M^{(n)}\right) \times H^q_{\mathrm{DB}}\left(M^{(n)}\right)$, $\left(z^{\mathbf{A}}_{(p)}, z^{\mathbf{B}}_{(q)}\right) \in Z_p\left(M^{(n)}\right) \times Z_q\left(M^{(n)}\right)$ and $k \in \mathbb{Z}$.

Exactly like in the 3D case [**16–18, 21–23**], quantum computations can be performed using the fact that the short exact sequence (2.17) splits, that is, decomposing Deligne–Beilinson cohomology classes as

$$\mathbf{A}^{(p)} = \mathbf{A}^{(p)}_{\mathbf{a}} + \boldsymbol{\alpha}^{(p)} \tag{2.25}$$

where the index $\mathbf{a}$ belongs to $\check{H}^{p+1}\left(M^{(n)}\right)$ and $\boldsymbol{\alpha}^{(p)} \in \Omega^p\left(M^{(n)}\right)/\Omega^p_{\mathbb{Z}}\left(M^{(n)}\right)$, see Figure 2.

We can go even further by writing that, as an Abelian group,

$$\check{H}^{p+1}(M) = \check{F}^{p+1} \oplus \check{T}^{p+1}, \tag{2.26}$$

where $\check{F}^{p+1} \cong \mathbb{Z}^{b^{p+1}}$ is the free part of $\check{H}^{p+1}\left(M^{(n)}\right)$, while

$$\check{T}^{p+1} \cong \mathbb{Z}\big/\zeta_1^{p+1}\mathbb{Z} \oplus \ldots \oplus \mathbb{Z}\big/\zeta_{t^{p+1}}^{p+1}\mathbb{Z}\,, \tag{2.27}$$

with $\zeta_i^{p+1}|\zeta_{i+1}^{p+1}$, is its torsion part, and decomposing $\Omega^p\left(M^{(n)}\right)/\Omega^p_{\mathbb{Z}}\left(M^{(n)}\right)$ as

$$\Omega^p\left(M^{(n)}\right)\Omega^p_{\mathbb{Z}}\left(M^{(n)}\right) = \Omega^p\left(M^{(n)}\right)/\Omega^p_{\text{cl}}\left(M^{(n)}\right) \oplus \Omega^p_{\text{cl}}\left(M^{(n)}\right)/\Omega^p_{\mathbb{Z}}\left(M^{(n)}\right), \tag{2.28}$$

$\Omega^p_{\text{cl}}\left(M^{(n)}\right)$ being the space of closed p-forms, knowing that

$$\Omega^p_{\text{cl}}\left(M^{(n)}\right)\Big/\Omega^p_{\mathbb{Z}}\left(M^{(n)}\right) \cong (\mathbb{R}/\mathbb{Z})^{b^p}\,. \tag{2.29}$$

Hence, we can actually write for any class $\mathbf{A}^{(p)} \in H^p_{\text{DB}}\left(M^{(n)}\right)$,

$$\mathbf{A}^{(p)} = \mathbf{A}^{(p)}_{\mathbf{m}} + \mathbf{A}^{(p)}_{\boldsymbol{\kappa}} + \boldsymbol{\alpha}^{(p)}_0 + \boldsymbol{\alpha}^{(p)}_\perp, \tag{2.30}$$

with $\mathbf{m} \in \check{F}^{p+1}$, $\boldsymbol{\kappa} \in \check{T}^{p+1}$, $\boldsymbol{\alpha}^{(p)}_\perp \in \Omega^p\left(M^{(n)}\right)/\Omega^p_{\text{cl}}\left(M^{(n)}\right)$ and the so-called "zero modes" $\boldsymbol{\alpha}^{(p)}_0 \in \Omega^p_{\text{cl}}\left(M^{(n)}\right)/\Omega^p_{\mathbb{Z}}\left(M^{(n)}\right) \cong (\mathbb{R}/\mathbb{Z})^{b^p}$.

This decomposition is of course not unique, and the computations we want to perform should be independent from the choice of decomposition. The whole point is to find a nice decomposition that makes our computations easy.

Claim 2.17 ([**19**], [**20**], [**16**], [**17**], [**18**]). There exists a decomposition for $\mathbf{A}^{(p)}$ and $\mathbf{B}^{(q)}$ that satisfies the following very convenient properties:

1) $\int_{M^{(n)}} \boldsymbol{\beta}^{(q)}_0 \star \mathbf{A}^{(p)}_{\mathbf{m_A}} \underset{\mathbb{Z}}{=} \mathbf{m_A}\cdot\boldsymbol{\beta}^{(q)}_0$,
2) $\int_{M^{(n)}} \boldsymbol{\beta}^{(q)}_0 \star \mathbf{A}^{(p)}_{\boldsymbol{\kappa}_\mathbf{A}} \underset{\mathbb{Z}}{=} 0$,
3) $\int_{M^{(n)}} \boldsymbol{\beta}^{(q)}_\perp \star \mathbf{A}^{(p)}_{\boldsymbol{\kappa}_\mathbf{A}} \underset{\mathbb{Z}}{=} 0$,
4) $\int_{M^{(n)}} \boldsymbol{\beta}^{(q)}_0 \star \boldsymbol{\alpha}^{(p)}_\perp \underset{\mathbb{Z}}{=} 0$,
5) $\int_{M^{(n)}} \boldsymbol{\beta}^{(q)}_0 \star \boldsymbol{\alpha}^{(p)}_0 \underset{\mathbb{Z}}{=} 0$,
6) $\int_{M^{(n)}} \mathbf{B}^{(q)}_{\mathbf{m_B}} \star \mathbf{A}^{(p)}_{\mathbf{m_A}} \underset{\mathbb{Z}}{=} 0$,
7) $\int_{M^{(n)}} \mathbf{B}^{(q)}_{\mathbf{m_B}} \star \mathbf{A}^{(p)}_{\boldsymbol{\kappa}_\mathbf{A}} \underset{\mathbb{Z}}{=} 0$,
8) $\int_{M^{(n)}} \mathbf{B}^{(q)}_{\boldsymbol{\kappa}_\mathbf{B}} \star \mathbf{A}^{(p)}_{\boldsymbol{\kappa}_\mathbf{A}} \underset{\mathbb{Z}}{=} -Q\left(\boldsymbol{\kappa}_\mathbf{B}, \boldsymbol{\kappa}_\mathbf{A}\right)$,

where $Q : \check{T}^{q+1} \times \check{T}^{p+1} \to \mathbb{Q}/\mathbb{Z}$ is a generalization of the so-called "linking form" [**24**].

Consequence 2.18. The previous properties imply that

$$\int_{M^{(n)}} \mathbf{B}^{(q)} \star \mathbf{A}^{(p)} \underset{\mathbb{Z}}{=} \int_{M^{(n)}} \boldsymbol{\beta}^{(q)}_\perp \star \mathbf{A}^{(p)}_{\mathbf{m_A}} + \int_{M^{(n)}} \mathbf{B}^{(q)}_{\mathbf{m_B}} \star \boldsymbol{\alpha}^{(p)}_\perp + \int_{M^{(n)}} \boldsymbol{\beta}^{(q)}_\perp \star \boldsymbol{\alpha}^{(p)}_\perp$$
$$+ \mathbf{m_B}\cdot\boldsymbol{\alpha}^{(p)}_0 + \mathbf{m_A}\cdot\boldsymbol{\beta}^{(q)}_0 - Q\left(\boldsymbol{\kappa}_\mathbf{B}, \boldsymbol{\kappa}_\mathbf{A}\right). \tag{2.31}$$

Heuristic and motivational calculation. We want to study functional integrals such as

$$\text{“}\, \mathcal{Z}^p_{\text{BF}_k}\left(M^{(n)}\right) = \frac{1}{\mathcal{N}^p_{\text{BF}_k}\left(M^{(n)}\right)} \int_{H^q_{\text{DB}}\left(M^{(n)}\right)\times H^p_{\text{DB}}\left(M^{(n)}\right)} \mathscr{D}\mathbf{B}^{(q)}\,\mathscr{D}\mathbf{A}^{(p)}\, e^{2\pi i S_{\text{BF}_k}\left(\mathbf{A}^{(p)},\mathbf{B}^{(q)}\right)} \,\text{”} \tag{2.32}$$

(the so-called "partition function") and

$$\text{“}\,\left\langle W\left(\mathbf{A}^{(p)}, z^{\mathbf{A}}_{(p)}, \mathbf{B}^{(q)}, z^{\mathbf{B}}_{(q)}\right)\right\rangle \tag{2.33}$$
$$= \frac{1}{\mathcal{N}^p_{\text{BF}_k}\left(M^{(n)}\right)} \int_{H^q_{\text{DB}}\left(M^{(n)}\right)\times H^p_{\text{DB}}\left(M^{(n)}\right)} \mathscr{D}\mathbf{B}^{(q)}\,\mathscr{D}\mathbf{A}^{(p)}$$
$$W\left(\mathbf{A}^{(p)}, z^{\mathbf{A}}_{(p)}, \mathbf{B}^{(q)}, z^{\mathbf{B}}_{(q)}\right) e^{2\pi i S_{\text{BF}_k}\left(\mathbf{A}^{(p)},\mathbf{B}^{(q)}\right)} \,\text{”}$$

(the so-called "expectation value of observables") which are ill-defined integrals.

We will focus here on the partition function (2.32) and try to propose an appropriate definition. If we use in this integral the decomposition and the properties we recalled above, we can write

$$\text{“}\ \mathcal{Z}^p_{\mathrm{BF}_k}\left(M^{(n)}\right) = \frac{1}{\mathcal{N}_{\mathrm{BF}_k}\left(M^{(n)}\right)} \sum_{\boldsymbol{\kappa}_{\mathbf{B}},\boldsymbol{\kappa}_{\mathbf{A}}\in \check{T}^{q+1}\times\check{T}^{p+1}} e^{-2\pi i k Q(\boldsymbol{\kappa}_{\mathbf{B}},\boldsymbol{\kappa}_{\mathbf{A}})} \sum_{\mathbf{m}_{\mathbf{B}},\mathbf{m}_{\mathbf{A}}\in \check{F}^{q+1}\times\check{F}^{p+1}} \int_{(\mathbb{R}/\mathbb{Z})^{b^q}\times(\mathbb{R}/\mathbb{Z})^{b^p}} \mathrm{d}^{b^q}\,\boldsymbol{\beta}_0\,\mathrm{d}^{b^p}\,\boldsymbol{\alpha}_0\, e^{2\pi i k\left(\int_{M^{(n)}}\mathbf{m}_{\mathbf{B}}\cdot\boldsymbol{\alpha}_0^{(p)}+\mathbf{m}_{\mathbf{A}}\cdot\boldsymbol{\beta}_0^{(q)}\right)} \int_{\Omega^q\left(M^{(n)}\right)/\Omega^q_{\mathrm{cl}}\left(M^{(n)}\right)\times\Omega^p\left(M^{(n)}\right)/\Omega^p_{\mathrm{cl}}\left(M^{(n)}\right)} \mathscr{D}\boldsymbol{\beta}^{(q)}_{\perp}\,\mathscr{D}\boldsymbol{\alpha}^{(p)}_{\perp}\, e^{2\pi i k\left(\int_{M^{(n)}}\boldsymbol{\beta}^{(q)}_{\perp}\star\mathbf{A}^{(p)}_{\mathbf{m}_{\mathbf{A}}}+\int_{M^{(n)}}\mathbf{B}^{(q)}_{\mathbf{m}_{\mathbf{B}}}\star\boldsymbol{\alpha}^{(p)}_{\perp}+\int_{M^{(n)}}\boldsymbol{\beta}^{(q)}_{\perp}\star\boldsymbol{\alpha}^{(p)}_{\perp}\right)}\ \text{”} \tag{2.34}$$

and the integral over the topological sector $(\mathbb{R}/\mathbb{Z})^{b^q}\times(\mathbb{R}/\mathbb{Z})^{b^p}$ can be performed and gives

$$\text{“}\ \mathcal{Z}^p_{\mathrm{BF}_k}\left(M^{(n)}\right) = \frac{1}{\mathcal{N}_{\mathrm{BF}_k}\left(M^{(n)}\right)} \sum_{\boldsymbol{\kappa}_{\mathbf{B}},\boldsymbol{\kappa}_{\mathbf{A}}\in \check{T}^{q+1}\times\check{T}^{p+1}} e^{-2\pi i k Q(\boldsymbol{\kappa}_{\mathbf{B}},\boldsymbol{\kappa}_{\mathbf{A}})} \int_{\Omega^q\left(M^{(n)}\right)/\Omega^q_{\mathrm{cl}}\left(M^{(n)}\right)\times\Omega^p\left(M^{(n)}\right)/\Omega^p_{\mathrm{cl}}\left(M^{(n)}\right)} \mathscr{D}\boldsymbol{\beta}^{(q)}_{\perp}\,\mathscr{D}\boldsymbol{\alpha}^{(p)}_{\perp}\, e^{2\pi i k\int_{M^{(n)}}\boldsymbol{\beta}^{(q)}_{\perp}\star\boldsymbol{\alpha}^{(p)}_{\perp}} \sum_{\mathbf{m}_{\mathbf{B}},\mathbf{m}_{\mathbf{A}}\in \check{F}^{q+1}\times\check{F}^{p+1}} \delta_{\mathbf{m}_B}\delta_{\mathbf{m}_A} e^{2\pi i k\left(\int_{M^{(n)}}\boldsymbol{\beta}^{(q)}_{\perp}\star\mathbf{A}^{(p)}_{\mathbf{m}_{\mathbf{A}}}+\int_{M^{(n)}}\mathbf{B}^{(q)}_{\mathbf{m}_{\mathbf{B}}}\star\boldsymbol{\alpha}^{(p)}_{\perp}\right)}\ \text{”} \tag{2.35}$$

$\delta_{\mathbf{m}_B}$ and $\delta_{\mathbf{m}_A}$ being regular Kronecker symbols. Then, we can sum over the topological sector $\check{F}^{q+1}\times\check{F}^{p+1}$, and we obtain

$$\text{“}\ \mathcal{Z}^p_{\mathrm{BF}_k}\left(M^{(n)}\right) = \frac{1}{\mathcal{N}_{\mathrm{BF}_k}\left(M^{(n)}\right)} \sum_{\boldsymbol{\kappa}_{\mathbf{B}},\boldsymbol{\kappa}_{\mathbf{A}}\in \check{T}^{q+1}\times\check{T}^{p+1}} e^{-2\pi i k Q(\boldsymbol{\kappa}_{\mathbf{B}},\boldsymbol{\kappa}_{\mathbf{A}})} \int_{\Omega^q\left(M^{(n)}\right)/\Omega^q_{\mathrm{cl}}\left(M^{(n)}\right)\times\Omega^p\left(M^{(n)}\right)/\Omega^p_{\mathrm{cl}}\left(M^{(n)}\right)} \mathscr{D}\boldsymbol{\beta}^{(q)}_{\perp}\,\mathscr{D}\boldsymbol{\alpha}^{(p)}_{\perp}\, e^{2\pi i k\int_{M^{(n)}}\boldsymbol{\beta}^{(q)}_{\perp}\star\boldsymbol{\alpha}^{(p)}_{\perp}}\ \text{”}. \tag{2.36}$$

We observe here a convenient full decoupling of the two remaining topological sectors, $\Omega^q\left(M^{(n)}\right)/\Omega^q_{\mathrm{cl}}\left(M^{(n)}\right)\times\Omega^p\left(M^{(n)}\right)/\Omega^p_{\mathrm{cl}}\left(M^{(n)}\right)$ and $\check{T}^{q+1}\times\check{T}^{p+1}$. The contribution of the topological sector $\Omega^q\left(M^{(n)}\right)/\Omega^q_{\mathrm{cl}}\left(M^{(n)}\right)\times\Omega^p\left(M^{(n)}\right)/\Omega^p_{\mathrm{cl}}\left(M^{(n)}\right)$ is infinite dimensional and we choose to eliminate it[6] thanks to the normalization,

[6] A lot of authors [**2**, **7**, **28**] extract from this part the Reidemeister torsion of $M^{(n)}$. However, they do not work with the gauge classes of fields. They fix the gauge, and for that they usually introduce a metric, which they want afterwards to get rid of, as the theory is expected to be topological, i.e., partition function and expectation values of observables should not depend on any metric.

i.e., by writing formally

$$\text{“}\, \mathscr{N}^{p}_{\mathrm{BF}_k}\left(M^{(n)}\right) = \int_{\Omega^q(M^{(n)})/\Omega^q_{\mathrm{cl}}(M^{(n)}) \times \Omega^p(M^{(n)})/\Omega^p_{\mathrm{cl}}(M^{(n)})} \mathscr{D}\boldsymbol{\beta}_{\perp}^{(q)}\, \mathscr{D}\boldsymbol{\alpha}_{\perp}^{(p)}\, e^{2\pi i k \int_{M^{(n)}} \boldsymbol{\beta}_{\perp}^{(q)} \star \boldsymbol{\alpha}_{\perp}^{(p)}} . \text{”} \tag{2.37}$$

Therefore, the only factor that remains is the contribution of the sector $\check{T}^{q+1} \times \check{T}^{p+1}$ which we will regard as our *definition* of the partition function $\mathcal{Z}^{p}_{\mathrm{BF}_k}\left(M^{(n)}\right)$.

DEFINITION 2.19. In the following,

$$\mathcal{Z}^{p}_{\mathrm{BF}_k}\left(M^{(n)}\right) := \sum_{\boldsymbol{\kappa}_B, \boldsymbol{\kappa}_A \in \check{T}^{q+1} \times \check{T}^{p+1}} e^{-2\pi i k Q(\boldsymbol{\kappa}_B, \boldsymbol{\kappa}_A)}. \tag{2.38}$$

PROPOSITION 2.20. *For the lens space $L(r,s)$ in dimension 3, we have*

$$\mathcal{Z}^{1}_{\mathrm{BF}_k}\left(L(r,s)\right) = \gcd(r,k) r = |\check{T}^2||\mathrm{Hom}\left(\check{T}^2, \mathbb{Z}/k\mathbb{Z}\right)| \tag{2.39}$$

PROOF. For a lens space $L(r,s)$, we know that $Q = \frac{r}{s}$, so

$$\mathcal{Z}^{1}_{\mathrm{BF}_k}\left(L(r,s)\right) = \sum_{n,m=0}^{r-1} e^{-2\pi i k \frac{s}{r} mn} = \sum_{n,m=0}^{r-1} e^{-2\pi i k' \frac{s}{r'} mn} \tag{2.40}$$

where $k = k' \gcd(r,k)$ and $r = r' \gcd(r,k)$, then

$$\mathcal{Z}^{1}_{\mathrm{BF}_k}\left(L(r,s)\right) = \sum_{n=0}^{r-1} \left(\sum_{m=0}^{r-1} e^{-2\pi i k' \frac{s}{r'} n} \right)^{m} \tag{2.41}$$

and the inner parenthesis is 0 whenever $r' \nmid n$ and r whenever $r' \mid n$, whence

$$\mathcal{Z}^{1}_{\mathrm{BF}_k}\left(M^{(n)}\right) = \gcd(r,k) r = |\check{T}^2||\mathrm{Hom}\left(\check{T}^2, \mathbb{Z}/k\mathbb{Z}\right)|. \tag{2.42}$$

□

PROPOSITION 2.21. *More generally, using the properties of Q, we can show that*

$$\mathcal{Z}^{p}_{\mathrm{BF}_k}\left(M^{(n)}\right) = \prod_{i=1}^{t^{p+1}} \gcd\left(\zeta_i^{p+1}, k\right) \zeta_i^{p+1} = |\check{T}^{p+1}||\mathrm{Hom}\left(\check{T}^{p+1}, \mathbb{Z}/k\mathbb{Z}\right)| \tag{2.43}$$

REMARK 2.22. The Čech homology (respectively cohomology) of $M^{(n)}$ is isomorphic to the simplicial homology (respectively cohomology) of $M^{(n)}$, its singular homology (respectively cohomology) and its cellular homology (respectively cohomology). Hence, from now on, we will denote indifferently those groups

$$H_p\left(M^{(n)}\right) \cong F_p \oplus T_p$$

(respectively $H^p\left(M^{(n)}\right) \cong F^p \oplus T^p$.)

Two other isomorphisms will be very useful in the following:

- $T^{p+1} \cong T_{n-p-1} = T_q$ (Poincaré duality),
- $T^{p+1} \cong T_p$ (universal coefficient theorem).

Later on, we will prefer to write

$$\mathcal{Z}^{p}_{\mathrm{BF}_k}\left(M^{(n)}\right) = |T_p||\mathrm{Hom}\left(T_p, \mathbb{Z}/k\mathbb{Z}\right)|. \tag{2.44}$$

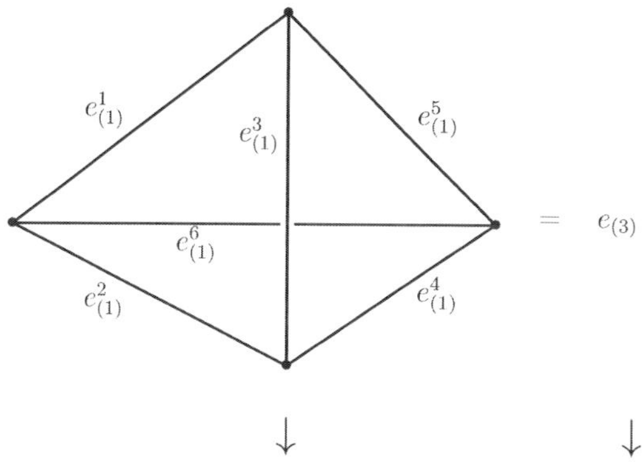

$$\begin{Bmatrix} l^{(1)}\left(e^1_{(1)}\right) & l^{(1)}\left(e^2_{(1)}\right) & l^{(1)}\left(e^3_{(1)}\right) \\ l^{(1)}\left(e^4_{(1)}\right) & l^{(1)}\left(e^5_{(1)}\right) & l^{(1)}\left(e^6_{(1)}\right) \end{Bmatrix}_q = \left|e^{l^{(1)}}_{(3)}\right|$$

FIGURE 3. Quantum $6j$-symbol associated with a tetrahedron in the $\mathcal{U}_q(\mathfrak{sl}_2(\mathbb{C}))$ case.

3. Generalized Abelian TV invariant

3.1. The original $\mathcal{U}_q(\mathfrak{sl}_2(\mathbb{C}))$ 3D case [32]. In this subsection, $M^{(3)}$ is a closed connected oriented smooth 3-manifold.

In this subsection, and in this subsection only, q is a root of unity. Unlike $\mathfrak{sl}_2(\mathbb{C})$, the set of isomorphism classes of representations of $\mathcal{U}_q(\mathfrak{sl}_2(\mathbb{C}))$, quantum deformation of parameter q of the enveloping algebra of $\mathfrak{sl}_2(\mathbb{C})$, is finite. Through misuse of language, we will identify the set of isomorphism classes of representations of $\mathcal{U}_q(\mathfrak{sl}_2(\mathbb{C}))$ with its (finite) indexing set I_q.

Introduce the quantities $w_i = \dim(i)$ and

$$w^2 = \sum_{i \in I} w_i^2$$

that are necessary for the construction of the invariant below. The exact expressions of I_q, w_i and w^2 as a function of q are specified in [**32**] but are irrelevant for what we want to show here.

Consider now a triangulation of $M^{(3)}$, i.e., the collection of a (finite) set of vertices C_0, a (finite) set of edges C_1 connecting those vertices, a (finite) set of triangular faces C_2 bounded by those edges, and a (finite) set of tetrahedra C_3 bounded by those faces.

Define the set of p-labelings to be $C^p_q = \left\{l^{(p)} : C_p \longrightarrow I_q\right\}$. We will be mostly interested here in the case $p = 1$ (labeling of edges). Given a labeling $l^{(1)} \in C^1_q$, each tetrahedron $e_{(3)} \in C_3$ becomes a labeled tetrahedron $e^{l^{(1)}}_{(3)}$, to which we associate a number $|e^{l^{(1)}}_{(3)}| \in \mathbb{C}$, which here is a quantum $6j$-symbol, see Figure 3.

REMARK 3.1. For a closed manifold $M^{(3)}$, any triangular face belongs to two tetrahedra. If, from the viewpoint of one tetrahedron, a given face is labeled by i, j and k, then from the viewpoint of the other tetrahedron to which belongs this face, it is labeled by i^*, j^* and k^*, where i^* means the dual representation of the i-th representation.

Finally, we have Theorem 3.2.

THEOREM 3.2 (V. Turaev, O. Viro [**32**]). *The quantity*

$$\mathcal{Z}_{\mathrm{TV}_q}\left(M^{(3)}\right) = \frac{1}{w^{2|C_0|}} \sum_{l^{(1)} \in C_q^1} \left(\left(\prod_{e_{(1)} \in C_1} w_{l^{(1)}(e_{(1)})} \right) \left(\prod_{e_{(3)} \in C_3} |e_{(3)}^{l^{(1)}}| \right) \right) \in \mathbb{C} \tag{3.1}$$

is independent of the triangulation and defines thus an invariant of $M^{(3)}$.

Barrett and Westbury [**5**] extended the construction to spherical categories, i.e., monoidal categories with a notion of duality. Balsam and Kirillov [**3**] extended in their turn the construction to 3-manifolds provided with a polyhedral decomposition, i.e., a cellular decomposition obtained from a triangulation by removing vertices, edges and faces.

3.2. The 3D $\mathbb{Z}/k\mathbb{Z}$ case. In this subsection, $M^{(3)}$ is still a closed connected oriented smooth 3-manifold provided with a triangulation. For further convenience, let us write this triangulation as a chain complex:

$$C_3 \xrightarrow{\partial_3} C_2 \xrightarrow{\partial_2} C_1 \xrightarrow{\partial_1} C_0 \tag{3.2}$$

where ∂_i is the obvious boundary operator.

The construction in the previous section relies on the fact that I_q is finite, which is made possible thanks to the quantum deformation of the enveloping algebra of $\mathfrak{sl}_2(\mathbb{C})$. But if we want an Abelian version of this construction, we realize immediately that this quantum deformation trick does not work for $\mathfrak{u}(1)$. However, there is at least one other way to extract from U(1) a finite set of representations with a "cutting parameter" k: We may consider $\mathbb{Z}/k\mathbb{Z}$, which is isomorphic to a finite subgroup of U(1).

We will not derive the details from the theory of spherical categories here, but we will refer to [**21, 22**] and recall that

- I_q in the $\mathcal{U}_q(\mathfrak{sl}_2(\mathbb{C}))$ case is replaced by $I_k = \{0, \dots, (k-1)\}$ in the $\mathbb{Z}/k\mathbb{Z}$ case,
- to the chain complex (3.2), we can associate the cochain complex of the p-labelings $C_k^p = \{l^{(p)} : C_p \longrightarrow I_k\}$:

$$C_0 \xrightarrow{\mathrm{d}_0} C_1 \xrightarrow{\mathrm{d}_1} C_2 \xrightarrow{\mathrm{d}_2} C_3 \tag{3.3}$$

  where d is defined by duality as

$$\left(\mathrm{d}^p\, l^{(p)}\right)\left(e_{(p+1)}\right) := l^{(p)}\left(\partial_{p+1}\, e_{(p+1)}\right) \tag{3.4}$$

  for $l^{(p)} \in C_k^p$ and $e_{(p+1)} \in C_{p+1}$. Hence, in particular, $\left(\mathrm{d}^1\, l^{(1)}\right)$ becomes a labeling of the faces,
- the quantum $6j$-symbols associated with a tetrahedron in the $\mathcal{U}_q(\mathfrak{sl}_2(\mathbb{C}))$ case are replaced here by a product of Kronecker symbols modulo k,[7]

[7] What we mean by "Kronecker symbols modulo k" is $\delta_a^{[k]} = \begin{cases} 1 & \text{if } a \equiv 0[k] \\ 0 & \text{otherwise} \end{cases}$.

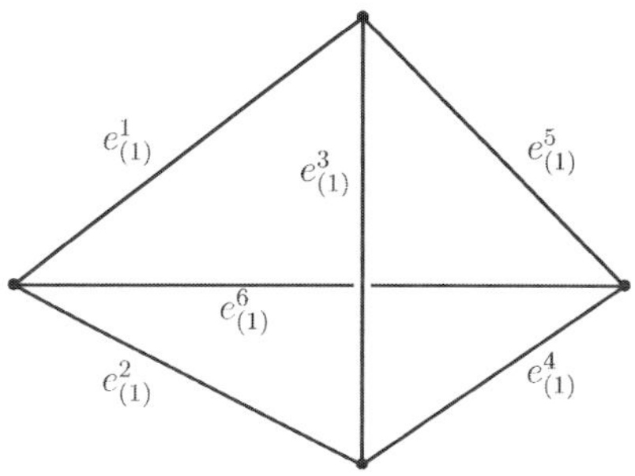

$$= e_{(3)}$$

$$\downarrow \qquad\qquad \downarrow$$

$$\begin{array}{l}\delta^{[k]}_{l^{(1)}\left(e^1_{(1)}\right)+l^{(1)}\left(e^2_{(1)}\right)-l^{(1)}\left(e^3_{(1)}\right)}\ \delta^{[k]}_{l^{(1)}\left(e^3_{(1)}\right)+l^{(1)}\left(e^4_{(1)}\right)-l^{(1)}\left(e^5_{(1)}\right)}\\ \delta^{[k]}_{l^{(1)}\left(e^5_{(1)}\right)-l^{(1)}\left(e^6_{(1)}\right)-l^{(1)}\left(e^1_{(1)}\right)}\ \delta^{[k]}_{l^{(1)}\left(e^6_{(1)}\right)-l^{(1)}\left(e^4_{(1)}\right)-l^{(1)}\left(e^2_{(1)}\right)}\end{array} = \left|e^{l^{(1)}}_{(3)}\right|$$

FIGURE 4. Kronecker symbols modulo k associated with a tetrahedron in the $\mathbb{Z}/k\mathbb{Z}$ case.

each Kronecker symbol representing the colored edges of a face of the tetrahedron considered:

$$\begin{aligned}|e^{l^{(1)}}_{(3)}| = \ &\delta^{[k]}_{l^{(1)}\left(e^1_{(1)}\right)+l^{(1)}\left(e^2_{(1)}\right)-l^{(1)}\left(e^3_{(1)}\right)}\ \delta^{[k]}_{l^{(1)}\left(e^3_{(1)}\right)+l^{(1)}\left(e^4_{(1)}\right)-l^{(1)}\left(e^5_{(1)}\right)}\\ &\delta^{[k]}_{l^{(1)}\left(e^5_{(1)}\right)-l^{(1)}\left(e^6_{(1)}\right)-l^{(1)}\left(e^1_{(1)}\right)}\ \delta^{[k]}_{l^{(1)}\left(e^6_{(1)}\right)-l^{(1)}\left(e^4_{(1)}\right)-l^{(1)}\left(e^2_{(1)}\right)}\end{aligned} \tag{3.5}$$

or, in a more condensed notation:

$$|e^{l^{(1)}}_{(3)}| = \prod_{e_{(2)}\in C_2 \,|\, e_{(2)}\subset \partial_3 e_{(3)}} \delta^{[k]}_{l^{(1)}(\partial_2 e_{(2)})}, \tag{3.6}$$

see Figure 4.

Hence, knowing that for all $i \in I_k$, $w_i = 1$, we can adapt the formula of the TV invariant (3.1) to the Abelian case.[8]

DEFINITION 3.3. We set

$$\begin{aligned}\mathcal{Z}^1_{\mathrm{TV}_k}\left(M^{(3)}\right) &= \frac{1}{k^{|C_0|-1}} \sum_{l^{(1)}\in C^1_k} \prod_{e_{(3)}\in C_3} |e^{l^{(1)}}_{(3)}|\\ &= \frac{1}{k^{|C_0|-1}} \sum_{l^{(1)}\in C^1_k} \prod_{e_{(3)}\in C_3} \prod_{e_{(2)}\in C_2 \,|\, e_{(2)}\subset\partial_3 e_{(3)}} \delta^{[k]}_{l^{(1)}(\partial_2 e_{(2)})} \in \mathbb{C}\end{aligned} \tag{3.7}$$

Note that the gluing condition of two faces along a given edge translates algebraically into the fact that this edge has to carry a representation i from the viewpoint of one face, and i^* from the viewpoint of the other face. Here, the duality is simply $i^* \equiv (k-i)\,[k]$ or $i^* \equiv -i\,[k]$ (strictly speaking, in our settings, $-i \notin I_k$ but $(k-i) \in I_k$.) So in the definition above, each face appears twice, with

[8]The superscript 1 in $\mathcal{Z}^1_{\mathrm{TV}_k}\left(M^{(3)}\right)$ will be explained in the next section.

one orientation and the opposite one. But the Kronecker symbol does not make a difference between the two orientations, so we may simply write:

$$\mathcal{Z}^1_{\mathrm{TV}_k}\left(M^{(3)}\right) = \frac{1}{k^{|C_0|-1}} \sum_{l^{(1)}\in C^1_k} \prod_{e_{(2)}\in C_2} \delta^{[k]}_{l^{(1)}\left(\partial_2 e_{(2)}\right)} \in \mathbb{C}. \tag{3.8}$$

REMARK 3.4. We realize that the assumption that we have *stricto sensu* a triangulation of $M^{(3)}$ is not necessary and that a cellular decomposition is sufficient, although to make sense of formula (3.8), C_2 cannot be empty (but we will come back to this point at the end of this section).

REMARK 3.5. Anticipating one of the next results, let us recall an important fact. The Universal Coefficient Theorem [**8**] claims that, for any Abelian group G,

$$H^p\left(M^{(n)}, G\right) \simeq \mathrm{Hom}\left(H_p\left(M^{(n)}\right), G\right) \oplus \mathrm{Ext}\left(H_{p-1}\left(M^{(n)}\right), G\right), \tag{3.9}$$

i.e., in general, the cohomology with coefficients in G is *not* the group of G-valued homomorphisms over the homology (with coefficients in $\mathbb{Z}$). However, if G is divisible, then $\mathrm{Ext}\left(H_{p-1}\left(M^{(n)}\right), G\right)$ is trivial and as a consequence, *in this case*, the cohomology with coefficients in G *is* the group of G-valued homomorphisms over the homology (with coefficients in $\mathbb{Z}$), i.e.,

$$H^p\left(M^{(n)}, G\right) \cong \mathrm{Hom}\left(H_p\left(M^{(n)}\right), G\right). \tag{3.10}$$

Moreover, the group $\mathbb{Z}/k\mathbb{Z}$ is a divisible group. Hence, the group of cohomology modulo k of degree p is

$$H^p_k := H^p\left(M^{(n)}, \mathbb{Z}/k\mathbb{Z}\right) \simeq \mathrm{Hom}\left(H_p\left(M^{(n)}\right), \mathbb{Z}/k\mathbb{Z}\right). \tag{3.11}$$

CLAIM 3.6. An equivalent definition of $\mathcal{Z}^1_{\mathrm{TV}_k}\left(M^{(3)}\right)$ is

$$\mathcal{Z}^1_{\mathrm{TV}_k}\left(M^{(3)}\right) = \frac{1}{k^{|C_0|-1}} \sum_{l^{(1)}\in C^1_k} \prod_{e_{(2)}\in C_2} \delta^{[k]}_{\left(\mathrm{d}^1 l^{(1)}\right)\left(e_{(2)}\right)} \in \mathbb{C} \tag{3.12}$$

from which we can show that

$$\mathcal{Z}^1_{\mathrm{TV}_k}\left(M^{(3)}\right) = k\frac{|H^1_k|}{|H^0_k|}. \tag{3.13}$$

PROOF. First, denote $Z^p_k = \mathrm{Ker}\,\mathrm{d}^p$ the group of p-cocycles modulo k, $B^p_k = \mathrm{Im}\,\mathrm{d}^{p-1}$ the group of p-coboundaries modulo k. By definition, $H^p_k = Z^p_k/B^p_k$. Then,

$$|H^1_k| = \frac{|Z^1_k|}{|B^1_k|} \tag{3.14}$$

and

$$|B^1_k| = \frac{|C^0_k|}{|Z^0_k|}. \tag{3.15}$$

Now, observe equation (3.6). We realize that, apart from the normalization, we are actually counting the number $|Z^1_k|$ of 1-cocycles modulo k for the cochain complex (3.3). This is of course not an invariant of $M^{(3)}$, but

$$\mathcal{Z}^1_{\mathrm{TV}_k}\left(M^{(3)}\right) = \frac{1}{k^{|C_0|-1}} |Z^1_k| \tag{3.16}$$

$$= \frac{1}{k^{|C_0|-1}} |H^1_k| \cdot |B^1_k| \tag{3.17}$$

$$= \frac{1}{k^{|C_0|-1}} |H^1_k| \cdot \frac{|C^0_k|}{|Z^0_k|} \tag{3.18}$$

$$= \frac{1}{k^{|C_0|-1}} |H^1_k| \cdot \frac{k^{|C_0|}}{|H^0_k|}, \tag{3.19}$$

$$\mathcal{Z}^1_{\mathrm{TV}_k}\left(M^{(3)}\right) = k \frac{|H^1_k|}{|H^0_k|}, \tag{3.20}$$

whence the result. □

CONSEQUENCE 3.7. If $M^{(3)}$ has only one connected component, as we assume here, i.e., if $|H^0_k| = k$, then

$$\mathcal{Z}^1_{\mathrm{TV}_k}\left(M^{(3)}\right) = |H^1_k| \tag{3.21}$$

REMARK 3.8. Taking into account equation (3.21), if $C_2 = \varnothing$, it is reasonable to set

$$\mathcal{Z}^1_{\mathrm{TV}_k}\left(M^{(3)}\right) = 1, \tag{3.22}$$

as the absence of 2-cells means there is no homology in degree 2 and by Poincaré duality, there is no cohomology in degree 1.

REMARK 3.9. We verify *a posteriori* that what we are computing is indeed totally independent from the choice of the cellular decomposition.

EXAMPLE 3.10. Consider the decomposition of S^3 given by Figure 5. Then,

$$\begin{aligned}
\mathcal{Z}^1_{\mathrm{TV}_k}\left(S^3\right) &= \frac{1}{k^{|C_0|-1}} \sum_{l^{(1)} \in C^1_k} \prod_{a=1}^{3} \delta^{[k]}_{(\mathrm{d}^1 l^{(1)})\left(e^a_{(2)}\right)} \\
&= \frac{1}{k^{4-1}} \sum_{l^{(1)} \in C^1_k} \delta^{[k]}_{l^{(1)}\left(\partial_2 e^1_{(2)}\right)} \delta^{[k]}_{l^{(1)}\left(\partial_2 e^2_{(2)}\right)} \delta^{[k]}_{l^{(1)}\left(\partial_2 e^3_{(2)}\right)} \\
&= \frac{1}{k^3} \sum_{l^{(1)} \in C^1_k} \\
&\quad \delta^{[k]}_{l^{(1)}\left(e^1_{(1)}+e^2_{(1)}+e^4_{(1)}+e^5_{(1)}\right)} \delta^{[k]}_{l^{(1)}\left(e^1_{(1)}+e^2_{(1)}+e^3_{(1)}\right)} \delta^{[k]}_{l^{(1)}\left(e^3_{(1)}-e^4_{(1)}-e^5_{(1)}\right)} \\
&= \frac{1}{k^3} \sum_{l^{(1)} \in C^1_k} \delta^{[k]}_{l^{(1)}_k\left(e^1_{(1)}\right)+l^{(1)}\left(e^2_{(1)}\right)+l^{(1)}\left(e^4_{(1)}\right)+l^{(1)}\left(e^5_{(1)}\right)} \\
&\quad \delta^{[k]}_{l^{(1)}\left(e^1_{(1)}\right)+l^{(1)}\left(e^2_{(1)}\right)+l^{(1)}\left(e^3_{(1)}\right)} \delta^{[k]}_{l^{(1)}\left(e^3_{(1)}\right)-l^{(1)}\left(e^4_{(1)}\right)-l^{(1)}\left(e^5_{(1)}\right)} \\
&= \frac{1}{k^3} \sum_{i,j,l,m,n=0}^{k-1} \delta^{[k]}_{i+j+m+n} \delta^{[k]}_{i+j+l} \delta^{[k]}_{l-m-n}
\end{aligned}$$

$$
\begin{aligned}
&= \frac{1}{k^3} \sum_{i,j,m,n=0}^{k-1} \delta^{[k]}_{i+j+m+n} \\
&= \frac{1}{k^3} \sum_{i,j,m=0}^{k-1} 1 \\
&= \frac{1}{k^3} k^3 \\
\mathcal{Z}^1_{\mathrm{TV}_k}\left(S^3\right) &= 1
\end{aligned}
$$

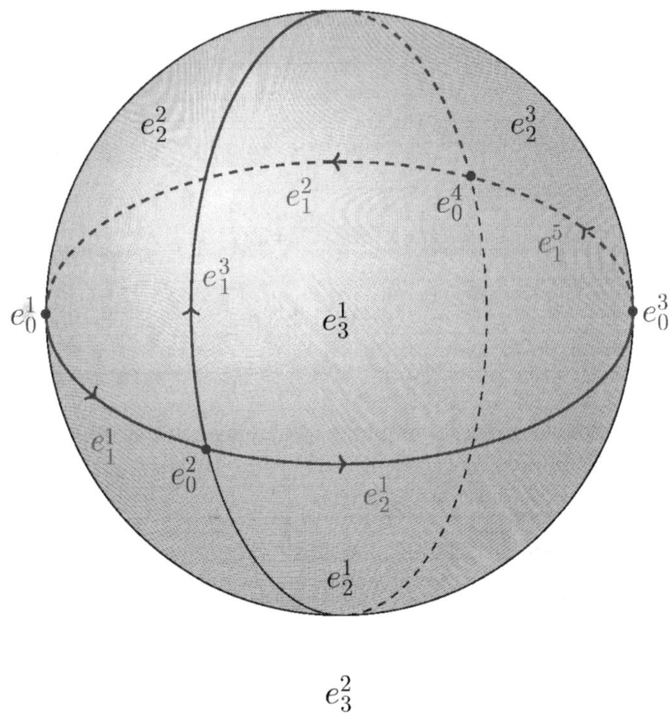

FIGURE 5. A particular cellular decomposition of S^3

Note that this result is trivial from (3.22) if we consider $S^3 = B^3 \underset{\partial B^3}{\cup} \mathrm{pt}$, which is a decomposition that has only one 0-cell and one 3-cell, and no 1-cell nor 2-cell.

REMARK 3.11. If we had followed the standard convention of normalization, i.e., if we had defined $\mathcal{Z}^1_{\mathrm{TV}_k}\left(M^{(3)}\right)$ as

$$
\mathcal{Z}^1_{\mathrm{TV}_k}\left(M^{(3)}\right) = \frac{1}{k^{|C_0|}} \sum_{l^{(1)} \in C^1_k} \left(\prod_{e_{(3)} \in C_3} \delta^{[k]}_{l^{(1)}\left(\partial_2 e_{(2)}\right)} \right) \in \mathbb{C} \tag{3.23}
$$

then we would have had

$$
\mathcal{Z}^1_{\mathrm{TV}_k}\left(S^3\right) = \frac{1}{k} \tag{3.24}
$$

and

$$\mathcal{Z}^1_{\mathrm{TV}_k}\left(S^1 \times S^2\right) = 1. \tag{3.25}$$

This definition is the appropriate one in order for $\mathcal{Z}^1_{\mathrm{TV}_k}\left(M^{(3)}\right)$ to define a TQFT satisfying Atiyah–Segal axioms [**1**, **29**, **30**]. In this framework, the invariant is also defined for compact 3-manifolds with boundary, which is a case we are not considering here. We focus here on *closed* 3-manifolds, and we shifted the number of vertices by 1 in the definition of $\mathcal{Z}^1_{\mathrm{TV}_k}\left(M^{(3)}\right)$ in order to have

$$\mathcal{Z}^1_{\mathrm{TV}_k}\left(S^3\right) = 1 \tag{3.26}$$

for later convenience.

REMARK 3.12. This TV construction (in 3D with a triangulation) may remind the reader of the Dijkgraaf–Witten (DW) construction [**12**]. According to [**31**], for a pair (G, α) where G is a finite group (not necessarily Abelian in general) and α is a 3-cocycle of the classifying space BG of G, then there exists a spherical fusion category $\mathcal{C}$ such that the TV construction for $\mathcal{C}$ matches with the DW construction for the pair (G, α):

$$\mathcal{Z}_{\mathrm{TV}_{\mathcal{C}}}\left(M^{(3)}\right) = \mathcal{Z}_{\mathrm{DW}_{(G,\alpha)}}\left(M^{(3)}\right). \tag{3.27}$$

The spherical fusion category $\mathcal{C}$ is explicitly built from G, α playing the role of associator.

The question here is somewhat the opposite: If we properly follow the categorical construction that leads to (3.8), are we able to find a pair (G, α) such that

$$\mathcal{Z}^1_{\mathrm{TV}_k}\left(M^{(3)}\right) = \mathcal{Z}_{\mathrm{DW}_{(G,\alpha)}}\left(M^{(3)}\right)? \tag{3.28}$$

The answer is positive, with $G = \mathbb{Z}/k\mathbb{Z}$ and α trivial, in the sense that the associativity of the tensor product of the underlying category is not twisted.

Remark that the Kronecker symbols modulo k of formula (3.8) do not appear explicitly in the DW construction. However, this construction consists of summing only over the labels satisfying the condition that, if three labels are associated with the edges of a triangle of the triangulation of M, then their sum must be zero (in additive notations.) If, instead, we want to sum over all the labels, then for $G = \mathbb{Z}/k\mathbb{Z}$, this condition translates into Kronecker symbols modulo k associated with each triangle of the triangulation of M.

3.3. The nD $\mathbb{Z}/k\mathbb{Z}$ case.

In this subsection, $M^{(n)}$ is a closed connected oriented smooth n-manifold.

Clearly, formula (3.13) can be generalized to manifolds of any dimension n by labeling the cells of any dimension p.

DEFINITION 3.13. For all $p \in \{0, \ldots, (n-1)\}$, if $C_{p+1} \neq \varnothing$ then

$$\mathcal{Z}^p_{\mathrm{TV}_k}\left(M^{(n)}\right) := \frac{1}{k^{|C_{p-1}| - |C_{p-2}| + \ldots - \varepsilon_p |C_0| + \varepsilon_p}} \sum_{l_k^{(p)} \in C_k^p} \prod_{a=1}^{|C_{p+1}|} \delta^{[k]}_{\left(\mathrm{d}^p\, l_k^{(p)}\right)\left(e^a_{p+1}\right)}, \tag{3.29}$$

where $\varepsilon_p = (-1)^p$, and

$$\mathcal{Z}^p_{\mathrm{TV}_k}\left(M^{(n)}\right) = 1 \tag{3.30}$$

otherwise.

REMARK 3.14. The normalization in formula (3.29) was chosen in such a way that the quantity $\mathcal{Z}^p_{\mathrm{TV}_k}\left(M^{(n)}\right)$ is truly invariant and satisfies $\mathcal{Z}^p_{\mathrm{TV}_k}(S^n) = 1$ for all $p \in \{0, \dots, (n-1)\}$.

PROPOSITION 3.15. *We can write that*

$$\mathcal{Z}^p_{\mathrm{TV}_k}\left(M^{(n)}\right) =$$

$$\begin{cases} |\mathrm{Hom}(H_p, \mathbb{Z}/k\mathbb{Z})| \dfrac{|\mathrm{Hom}(F_{p-2}, \mathbb{Z}/k\mathbb{Z})| \dots |\mathrm{Hom}(F_0, \mathbb{Z}/k\mathbb{Z})|}{|\mathrm{Hom}(F_{p-1}, \mathbb{Z}/k\mathbb{Z})| \dots |\mathrm{Hom}(F_1, \mathbb{Z}/k\mathbb{Z})|} \dfrac{1}{|\mathbb{Z}/k\mathbb{Z}|} & \text{if } p \text{ is even} \\ |\mathrm{Hom}(H_p, \mathbb{Z}/k\mathbb{Z})| \dfrac{|\mathrm{Hom}(F_{p-2}, \mathbb{Z}/k\mathbb{Z})| \dots |\mathrm{Hom}(F_1, \mathbb{Z}/k\mathbb{Z})|}{|\mathrm{Hom}(F_{p-1}, \mathbb{Z}/k\mathbb{Z})| \dots |\mathrm{Hom}(F_0, \mathbb{Z}/k\mathbb{Z})|} |\mathbb{Z}/k\mathbb{Z}| & \text{if } p \text{ is odd} \end{cases}$$

or

$$\mathcal{Z}^p_{\mathrm{TV}_k}\left(M^{(n)}\right) =$$

$$\begin{cases} |\mathrm{Hom}(T_p, \mathbb{Z}/k\mathbb{Z})| \dfrac{|\mathrm{Hom}(F_p, \mathbb{Z}/k\mathbb{Z})| \dots |\mathrm{Hom}(F_0, \mathbb{Z}/k\mathbb{Z})|}{|\mathrm{Hom}(F_{p-1}, \mathbb{Z}/k\mathbb{Z})| \dots |\mathrm{Hom}(F_1, \mathbb{Z}/k\mathbb{Z})|} \dfrac{1}{|\mathbb{Z}/k\mathbb{Z}|} & \text{if } p \text{ is even} \\ |\mathrm{Hom}(T_p, \mathbb{Z}/k\mathbb{Z})| \dfrac{|\mathrm{Hom}(F_p, \mathbb{Z}/k\mathbb{Z})| \dots |\mathrm{Hom}(F_1, \mathbb{Z}/k\mathbb{Z})|}{|\mathrm{Hom}(F_{p-1}, \mathbb{Z}/k\mathbb{Z})| \dots |\mathrm{Hom}(F_0, \mathbb{Z}/k\mathbb{Z})|} |\mathbb{Z}/k\mathbb{Z}| & \text{if } p \text{ is odd} \end{cases}$$

PROOF. The point of this computation is to iterate the process of the proof of Claim 3.6.

Indeed, from definition (3.29), we understand that we are counting the number $|Z^p_k|$ of p-cocycles modulo k. But

$$|Z^p_k| = |H^p_k||B^p_k| = \frac{|C^{p-1}_k|}{|Z^{p-1}_N|} \tag{3.31}$$

whence

$$|Z^p_k| = \begin{cases} k^{|C_{p-1}|-|C_{p-2}|+\dots-|C_0|} \dfrac{|H^p_k||H^{p-2}_k|\dots|H^0_k|}{|H^{p-1}_k||H^{p-3}_k|\dots|H^1_k|} & \text{if } p \text{ is even,} \\ \\ k^{|C_{p-1}|-|C_{p-2}|+\dots+|C_0|} \dfrac{|H^p_k||H^{p-2}_k|\dots|H^1_k|}{|H^{p-1}_k||H^{p-3}_k|\dots|H^0_k|} & \text{if } p \text{ is odd.} \end{cases} \tag{3.32}$$

But from the universal coefficient theorem [**8**], we can write

$$H^\bullet_k \cong \mathrm{Hom}\,(H_\bullet, \mathbb{Z}/k\mathbb{Z}) \oplus \mathrm{Ext}\,(T_{\bullet-1}, \mathbb{Z}/k\mathbb{Z})\,, \tag{3.33}$$

and moreover

$$\mathrm{Ext}\,(\mathbb{Z}/\zeta\mathbb{Z}\,, \mathbb{Z}/k\mathbb{Z}) \cong \mathbb{Z}/\gcd(\zeta, k)\mathbb{Z} \cong \mathrm{Hom}\,(\mathbb{Z}/\zeta\mathbb{Z}\,, \mathbb{Z}/k\mathbb{Z})\,. \tag{3.34}$$

Hence,

$$H^\bullet_k \cong \mathrm{Hom}\,(F_\bullet, \mathbb{Z}/k\mathbb{Z}) \oplus \mathrm{Hom}\,(T_\bullet, \mathbb{Z}/k\mathbb{Z}) \oplus \mathrm{Hom}\,(T_{\bullet-1}, \mathbb{Z}/k\mathbb{Z})\,, \tag{3.35}$$

from which we obtain that

$$|H^\bullet_k| = |\mathrm{Hom}\,(F_\bullet, \mathbb{Z}/k\mathbb{Z})||\mathrm{Hom}\,(T_\bullet, \mathbb{Z}/k\mathbb{Z})||\mathrm{Hom}\,(T_{\bullet-1}, \mathbb{Z}/k\mathbb{Z})| \tag{3.36}$$

which, injected in equation (3.32), gives by telescoping simplifications the result expected. □

EXAMPLE 3.16. Still with the decomposition on Figure 5 of S^3, we can compute

$$
\begin{aligned}
\mathcal{Z}^0_{\mathrm{TV}_k}\left(S^3\right) &= \frac{1}{k}\sum_{l^0_k\in C^0_k}\prod_{a=1}^{5}\delta^{[k]}_{(\mathrm{d}^0\, l^0_k)(e^a_1)}\\
&= \frac{1}{k}\sum_{l^0_k\in C^0_k}\delta^{[k]}_{l^0_k(\partial_1\, e^1_1)}\,\delta^{[k]}_{l^0_k(\partial_1\, e^2_1)}\\
&\qquad \delta^{[k]}_{l^0_k(\partial_1\, e^3_1)}\,\delta^{[k]}_{l^0_k(\partial_1\, e^4_1)}\,\delta^{[k]}_{l^0_k(\partial_1\, e^5_1)}\\
&= \frac{1}{k}\sum_{l^0_k\in C^0_k}\delta^{[k]}_{l^0_k(e^2_0-e^1_0)}\,\delta^{[k]}_{l^0_k(e^1_0-e^4_0)}\\
&\qquad \delta^{[k]}_{l^0_k(e^4_0-e^2_0)}\,\delta^{[k]}_{l^0_k(e^3_0-e^2_0)}\,\delta^{[k]}_{l^0_k(e^4_0-e^3_0)}\\
&= \frac{1}{k}\sum_{i,j,l,m=0}^{k-1}\delta^{[k]}_{j-i}\,\delta^{[k]}_{i-m}\,\delta^{[k]}_{m-j}\,\delta^{[k]}_{l-j}\,\delta^{[k]}_{m-l},\\
\mathcal{Z}^0_{\mathrm{TV}_k}\left(S^3\right) &= 1,
\end{aligned}
$$

and

$$
\begin{aligned}
\mathcal{Z}^2_{\mathrm{TV}_k}\left(S^3\right) &= \frac{1}{k^{|C_1|-|C_0|+\varepsilon_1}}\sum_{l^2_k\in C^2_k}\prod_{a=1}^{2}\delta^{[k]}_{(\mathrm{d}^2\, l^2_k)(e^a_3)}\\
&= \frac{1}{k^{5-4+1}}\sum_{l^2_k\in C^2_k}\delta^{[k]}_{l^2_k(\partial_2\, e^1_3)}\,\delta^{[k]}_{l^2_k(\partial_2\, e^2_3)}\\
&= \frac{1}{k^2}\sum_{l^2_k\in C^2_k}\delta^{[k]}_{l^2_k(e^1_2+e^2_2+e^3_2)}\,\delta^{[k]}_{l^2_k(e^1_2+e^2_2+e^3_2)}\\
&= \frac{1}{k^2}\sum_{i,j,l=0}^{k-1}\delta^{[k]}_{i+j+l}\\
&= \frac{1}{k^2}\sum_{i,j=0}^{k-1}1\\
&= \frac{1}{k^2}k^2,\\
\mathcal{Z}^2_{\mathrm{TV}_k}\left(S^3\right) &= 1.
\end{aligned}
$$

EXAMPLE 3.17. Consider the following cellular decomposition of S^n: For all $p\in\{0,\dots,n\}$, $e^1_{(p)}, e^2_{(p)} = B^p$, and for all $p\in\{1,\dots,n\}$, $\partial_p\, e^1_{(p)} = -\,\partial_p\, e^2_{(p)} =$

$S^{p-1} = e^2_{(p-1)} + e^1_{(p-1)}$. Then for all $p \in \{0, \dots, (n-1)\}$,

$$\begin{aligned}
\mathcal{Z}^p_{\mathrm{TV}_k}(S^n) &= \frac{1}{k^{2-2+\dots-2\varepsilon_p+\varepsilon_p}} \sum_{l^{(p)} \in C^p_k} \prod_{a=1}^{2} \delta^{[k]}_{(\mathrm{d}^p\, l^{(p)})\left(e^a_{(p+1)}\right)} \\
&= \frac{1}{k} \sum_{l^{(p)} \in C^p_k} \delta^{[k]}_{l^{(p)}\left(\partial_{p+1} e^1_{(p+1)}\right)} \delta^{[k]}_{l^{(p)}\left(\partial_{p+1} e^2_{(p+1)}\right)} \\
&= \frac{1}{k} \sum_{l^{(p)} \in C^p_k} \delta^{[k]}_{l^{(p)}\left(e^1_{(p)} + e^2_{(p)}\right)} \delta^{[k]}_{l^{(p)}\left(-e^1_{(p)} - e^2_{(p)}\right)} \\
&= \frac{1}{k} \sum_{l^{(p)} \in C^p_k} \delta^{[k]}_{l^{(p)}\left(e^1_{(p)}\right) + l^{(p)}\left(e^2_{(p)}\right)} \delta^{[k]}_{-l^{(p)}\left(e^1_{(p)}\right) - l^{(p)}\left(e^2_{(p)}\right)} \\
&= \frac{1}{k} \sum_{i,j=0}^{k-1} \delta^{[k]}_{i+j} \delta^{[k]}_{-i-j} \\
&= \frac{1}{k} \sum_{i,j=0}^{k-1} \delta^{[k]}_{i+j} \\
&= \frac{1}{k} \sum_{i=0}^{k-1} 1, \\
\mathcal{Z}^p_{\mathrm{TV}_k}(S^n) &= 1.
\end{aligned}$$

4. Generalized Turaev–Viro invariant as a discrete BF theory

DEFINITION 4.1. Given a cellular decomposition

$$C_n \xrightarrow{\partial_n} C_{n-1} \xrightarrow{\partial_{n-1}} \dots \xrightarrow{\partial_2} C_1 \xrightarrow{\partial_1} C_0 \tag{4.1}$$

of $M^{(n)}$, we call dual decomposition any cellular decomposition

$$\widetilde{C}_n \xrightarrow{\partial_n} \widetilde{C}_{n-1} \xrightarrow{\partial_{n-1}} \dots \xrightarrow{\partial_2} \widetilde{C}_1 \xrightarrow{\partial_1} \widetilde{C}_0 \tag{4.2}$$

of $M^{(n)}$ such that for any $\widetilde{e}^a_{(p)} \in \widetilde{C}_p$ and $e^b_{(n-p)} \in C_{n-p}$,

$$\widetilde{e}^a_{(p)} \cdot e^b_{(n-p)} = \delta^{ab} \tag{4.3}$$

where $\cdot$ is the number of intersections operator. We define also dual labelings from the dual chain complex, exactly the same way we defined the non-dual labelings, i.e.,

$$\widetilde{C}^p_k = \left\{ \tilde{l}^{(p)} : \widetilde{C}_p \longrightarrow I_k \right\}.$$

CLAIM 4.2. We can write

$$\delta^{[k]}_{\mathrm{d}^p\, l^{(p)}} = \frac{1}{k|\widetilde{C}_q|} \sum_{\widetilde{m}^{(q)} \in \widetilde{C}^q_k} e^{\frac{2\pi i}{k} \widetilde{m}^{(q)} \cdot \mathrm{d}^p\, l^{(p)}}$$

where $n = p+q+1$, and where $l^{(p)}$ is regarded as a vector with $|C_p|$ components, d^p as a matrix of size $|C_p| \times |C_{p+1}|$ and $\widetilde{m}^{(q)}$ as a vector of size $|\widetilde{C}_q| = |C_{n-q}| = |C_{p+1}|$ by construction.

CONSEQUENCE 4.3. We can write

(4.4)
$$\mathcal{Z}^p_{\mathrm{TV}_k}\left(M^{(n)}\right)=\frac{1}{k^{|C_{p+1}|+|C_{p-1}|-|C_{p-2}|+\ldots\pm\varepsilon_p|C_0|\mp\varepsilon_p}}\sum_{l_k^{(p)}\in C_k^p}\sum_{\widetilde{m}_k^{(q)}\in\widetilde{C}_k^q}e^{\frac{2\pi i}{k}\widetilde{m}^{(q)}\cdot \mathrm{d}^p\, l^{(p)}}$$

which can be interpreted as a sort of partition function of a discrete BF theory.

5. Conclusion

PROPOSITION 5.1 (Høssjer, Mathieu, Thuillier).

$$\mathcal{Z}^p_{\mathrm{BF}_k}(M)=\frac{|T_p|}{|\mathrm{Hom}\left(F_p,\,\mathbb{Z}/k\mathbb{Z}\right)|}\frac{|\mathrm{Hom}\left(F_{p-1},\,\mathbb{Z}/k\mathbb{Z}\right)|\ldots}{|\mathrm{Hom}\left(F_{p-2},\,\mathbb{Z}/k\mathbb{Z}\right)|\ldots}\mathcal{Z}^p_{\mathrm{TV}_k}(M). \tag{5.1}$$

REMARK 5.2. We recover the case $n=3$ and $p=1$:

$$\mathcal{Z}^1_{\mathrm{BF}_k}\left(M^{(3)}\right)=\frac{|T_1|}{|\mathrm{Hom}\left(F_1,\,\mathbb{Z}/k\mathbb{Z}\right)|}\mathcal{Z}^1_{\mathrm{TV}_k}\left(M^{(3)}\right). \tag{5.2}$$

REMARK 5.3. The appearance of the $\mathrm{Hom}\left(F_p,\,\mathbb{Z}/k\mathbb{Z}\right)$ may remind us of the ghost, antifield and antighost sectors that appear in Batalin–Vilkovisky formalism [**25**]. An analysis of the normalizing part (2.37) through this formalism may result in some factors that would cancel the $\mathrm{Hom}\left(F_p,\,\mathbb{Z}/k\mathbb{Z}\right)$ factors in (5.2), leading to a better agreement between the Abelian BF and TV theories.

Let us try to give a quick overview of the significance of this result. In the SU(2) 3D case, the TV construction relying on $\mathcal{U}_q(\mathfrak{sl}_2(\mathbb{C}))$ is regarded as a regularization of the SU(2) BF partition function.[9]

In the U(1) 3D case, we are able to propose a non-perturbative definition of the partition function on the one hand, and we are able to build an Abelian realization of a spherical category depending on a "cutting parameter" k, and to use it to build an Abelian TV invariant on the other hand. The two quantities obtained independently turn out to be equal up to a factor that has a cohomological interpretation.

We are able to generalize these Abelian BF and TV constructions to a family of constructions in any dimension. Those constructions are still related, again up to a factor that has a cohomological interpretation.

Remark that formula (5.2) is an equality between sums of exponentials of two different bilinear forms over two different lattices. In [**11**], Florian Deloup and Vladimir Turaev showed a similar equality in a purely algebraic manner, and they also ended up with a factor between the two sums, that can be interpreted as a normalization by the volume of each lattice.

Acknowledgments

The second author thanks Pavel Mnëv, Stephan Stolz, Liviu Nicolaescu, Christopher Schommer-Pries, Justin Beck and Eugene Rabinovich from the Department of Mathematics of the University of Notre Dame, Vivek Shende and Konstantin Wernli from the Center for Quantum Mathematics of Southern Denmark University, Sébastien Leurent from the Institut de Mathématiques de Bourgogne of the Université de Bourgogne, Dominique Manchon and Abel Lacabanne from the

[9] It is often said they are equal, but the point is that the BF partition function is not a well-defined object, so the equality does not really make sense, *stricto sensu*.

Laboratoire Blaise Pascal of the Université Clermont-Auvergne, for their interest and the useful remarks and suggestions they made in order to improve the presentations of this work. Finally, the second author thanks Ralph Kaufmann from the Department of Mathematics of Purdue University for his invitation to the AMS Special Session on Higher Structures in Topology, Geometry and Physics in March 2022.

References

[1] Michael Atiyah, *Topological quantum field theories*, Inst. Hautes Études Sci. Publ. Math. **68** (1988), 175–186 (1989). MR1001453

[2] Scott Axelrod and Isadore Singer, *Chern–Simons perturbation theory*, Proceedings of the XXth International Conference on Differential Geometric Methods in Theoretical Physics, Vol. 1, 2 (New York, 1991), World Sci. Publ., River Edge, NJ, 1992, pp. 3–45. MR1225107

[3] Benjamin Balsam, *Turaev–Viro theory as an extended TQFT*, ProQuest LLC, Ann Arbor, MI, 2012. Thesis (Ph.D.)–State University of New York at Stony Brook. MR3093932

[4] Christian Bär and Christian Becker, *Differential characters*, Lecture Notes in Mathematics, vol. 2112, Springer, Cham, 2014, DOI 10.1007/978-3-319-07034-6. MR3237728

[5] John Barrett and Bruce Westbury, *Invariants of piecewise-linear* 3*-manifolds*, Trans. Amer. Math. Soc. **348** (1996), no. 10, 3997–4022, DOI 10.1090/S0002-9947-96-01660-1. MR1357878

[6] Michel Bauer, Georges Girardi, Raymond Stora, and Frank Thuillier, *A class of topological actions*, J. High Energy Phys. **8** (2005), 027, 35, DOI 10.1088/1126-6708/2005/08/027. MR2165819

[7] Danny Birmingham, Matthias Blau, Mark Rakowski, and George Thompson, *Topological field theory*, Phys. Rep. **209** (1991), no. 4-5, 129–340, DOI 10.1016/0370-1573(91)90117-5. MR1137489

[8] Raoul Bott and Loring Tu, *Differential forms in algebraic topology*, Graduate Texts in Mathematics, vol. 82, Springer-Verlag, New York-Berlin, 1982. MR658304

[9] Alberto Cattaneo, Paolo Cotta-Ramusino, and Maurizio Martellini, *Three-dimensional BF theories and the Alexander-Conway invariant of knots*, Nuclear Phys. B **436** (1995), no. 1-2, 355–382, DOI 10.1016/0550-3213(94)00500-E. MR1316372

[10] Alberto Cattaneo, Paolo Cotta-Ramusino, Jürg Fröhlich, and Maurizio Martellini, *Topological BF theories in* 3 *and* 4 *dimensions*, J. Math. Phys. **36** (1995), no. 11, 6137–6160, DOI 10.1063/1.531238. MR1355902

[11] Florian Deloup and Vladimir Turaev, *On reciprocity*, J. Pure Appl. Algebra **208** (2007), no. 1, 153–158, DOI 10.1016/j.jpaa.2005.12.008. MR2269836

[12] Robbert Dijkgraaf and Edward Witten, *Topological gauge theories and group cohomology*, Comm. Math. Phys. **129** (1990), no. 2, 393–429. MR1048699

[13] Laurent Freidel and Etera Livine, *Ponzano–Regge model revisited. III. Feynman diagrams and effective field theory*, Classical Quantum Gravity **23** (2006), no. 6, 2021–2061, DOI 10.1088/0264-9381/23/6/012. MR2209197

[14] Laurent Freidel and David Louapre, *Ponzano-Regge model revisited. I. Gauge fixing, observables and interacting spinning particles*, Classical Quantum Gravity **21** (2004), no. 24, 5685–5726, DOI 10.1088/0264-9381/21/24/002. MR2107334

[15] Laurent Freidel and David Louapre, *Ponzano-Regge model revisited: II. Equivalence with Chern-Simons*, `arXiv:gr-qc/0410141`.

[16] Enore Guadagnini and Frank Thuillier, *Deligne-Beilinson cohomology and Abelian link invariants*, SIGMA Symmetry Integrability Geom. Methods Appl. **4** (2008), Paper 078, 30, DOI 10.3842/SIGMA.2008.078. MR2470518

[17] Enore Guadagnini and Frank Thuillier, *Three-manifold invariant from functional integration*, J. Math. Phys. **54** (2013), no. 8, 082302, 17, DOI 10.1063/1.4818738. MR3135468

[18] Enore Guadagnini and Frank Thuillier, *Path-integral invariants in Abelian Chern-Simons theory*, Nuclear Phys. B **882** (2014), 450–484, DOI 10.1016/j.nuclphysb.2014.03.009. MR3190476

[19] Emil Høssjer, Philippe Mathieu, and Frank Thuillier, *An extension of the* U(1) *BF theory, Turaev–Viro invariant and Drinfeld center construction. Part I: Quantum fields, quantum currents and Pontryagin duality*, to be submitted, `arXiv:2212.12872`.

[20] Emil Høssjer, Philippe Mathieu, and Frank Thuillier, *An extension of the* U(1) *BF theory, Turaev–Viro invariant and Drinfeld center construction. Part II: Manifold invariants*, in preparation.

[21] Philippe Mathieu and Frank Thuillier, *A reciprocity formula from Abelian BF and Turaev–Viro theories*, Nuclear Phys. B **912** (2016), 327–353, DOI 10.1016/j.nuclphysb.2016.05.007. MR3567588

[22] Philippe Mathieu and Frank Thuillier, *Abelian BF theory and Turaev–Viro invariant*, J. Math. Phys. **57** (2016), no. 2, 022306, 31, DOI 10.1063/1.4942046. MR3462973

[23] Philippe Mathieu and Frank Thuillier, *Abelian Turaev–Virelizier theorem and $U(1)$ BF surgery formulas*, J. Math. Phys. **58** (2017), no. 10, 102301, 8, DOI 10.1063/1.4986850. MR3707599

[24] Françoise Michel and Claude Weber, *Higher-dimensional knots according to Michel Kervaire*, EMS Series of Lectures in Mathematics, European Mathematical Society (EMS), Zürich, 2017, DOI 10.4171/180. MR3699860

[25] Pavel Mnëv, *Quantum field theory: Batalin–Vilkovisky formalism and its applications*, University Lecture Series, vol. 72, American Mathematical Society, Providence, RI, 2019, DOI 10.1090/ulect/072. MR3967709

[26] Giorgio Ponzano and Tullio Regge, *Semiclassical limit of Racah coefficients*, In: F. Bloch (ed) *Spectroscopic and group theoretical methods in physics* **431** (1968), 233–271, North-Holland, Amsterdam.

[27] Tullio Regge, *General relativity without coordinates* (English, with Italian summary), Nuovo Cimento (10) **19** (1961), 558–571. MR127372

[28] Albert Schwarz, *The partition function of degenerate quadratic functional and Ray-Singer invariants*, Lett. Math. Phys. **2** (1977/78), no. 3, 247–252, DOI 10.1007/BF00406412. MR676337

[29] Graeme Segal, *The definition of conformal field theory*, Differential geometrical methods in theoretical physics (Como, 1987), NATO Adv. Sci. Inst. Ser. C: Math. Phys. Sci., vol. 250, Kluwer Acad. Publ., Dordrecht, 1988, pp. 165–171. MR981378

[30] Graeme Segal, *Topological structures in string theory*, R. Soc. Lond. Philos. Trans. Ser. A Math. Phys. Eng. Sci. **359** (2001), no. 1784, 1389–1398, DOI 10.1098/rsta.2001.0841. Topological methods in the physical sciences (London, 2000). MR1853627

[31] Vladimir Turaev and Alexis Virelizier, *Monoidal categories and topological field theory*, Progress in Mathematics, vol. 322, Birkhäuser/Springer, Cham, 2017, DOI 10.1007/978-3-319-49834-8. MR3674995

[32] Vladimir Turaev and Oleg Viro, *State sum invariants of 3-manifolds and quantum 6j-symbols*, Topology **31** (1992), no. 4, 865–902, DOI 10.1016/0040-9383(92)90015-A. MR1191386

Faculty of Science, Department of Mathematics and Computer Science, Centre for Quantum Mathematics, Campusvej 55, 5230 Odense, Denmark

Email address: `hossjer@imada.sdu.dk`

Institut für Mathematik, Universität Zürich, Winterthurerstrasse 190, CH-8057 Zürich

Email address: `philippe.mathieu@math.uzh.ch`

Laboratoire d'Annecy de Physique Théorique (LAPTh), 9 Chemin de Bellevue, 74940 Annecy, France

Email address: `frank.thuillier@lapth.cnrs.fr`

SELECTED PUBLISHED TITLES IN THIS SERIES

802 **Ralph M. Kaufmann, Martin Markl, and Alexander A. Voronov, Editors,** Higher Structures in Topology, Geometry, and Physics, 2024

801 **K. A. Brown, T. J. Hodges, M. Vancliff, and J. J. Zhang, Editors,** Recent Advances in Noncommutative Algebra and Geometry, 2024

800 **Avinoam Mann, Louis H. Rowen, David J. Saltman, Aner Shalev, Lance W. Small, and Uzi Vishne, Editors,** Amitsur Centennial Symposium, 2024

799 **Alberto A. Condori, Elodie Pozzi, William T. Ross, and Alan A. Sola, Editors,** Recent Progress in Function Theory and Operator Theory, 2024

798 **Marat V. Markin, Igor V. Nikolaev, and Carsten Trunk, Editors,** Advances in Functional Analysis and Operator Theory, 2024

797 **Sangita Jha, Mrinal Kanti Roychowdhury, and Saurabh Verma, Editors,** Recent Developments in Fractal Geometry and Dynamical Systems, 2024

796 **John Cremona, John Jones, Jennifer Paulhus, Andrew V. Sutherland, and John Voight, Editors,** LuCaNT: LMFDB, Computation, and Number Theory, 2024

795 **Michael A. Jones, David McCune, and Jennifer M. Wilson, Editors,** Mathematical Analyses of Decisions, Voting and Games, 2024

794 **Jean-Pierre Magnot, Editor,** Recent Advances in Diffeologies and Their Applications, 2024

793 **Abba Gumel, Editor,** Mathematical and Computational Modeling of Phenomena Arising in Population Biology and Nonlinear Oscillations, 2024

792 **Shaoming Guo, Zane Kun Li, and Brian Street, Editors,** Recent Developments in Harmonic Analysis and its Applications, 2024

791 **Mee Seong Im, Bach Nguyen, and Arik Wilbert, Editors,** Algebraic and Topological Aspects of Representation Theory, 2024

790 **Peter Crooks and Alexandru I. Suciu, Editors,** Compactifications, Configurations, and Cohomology, 2023

789 **I. S. Krasil′shchik, A. B. Sossinsky, and A. M. Verbovetsky, Editors,** The Diverse World of PDEs, 2023

788 **I. S. Krasil′shchik, A. B. Sossinsky, and A. M. Verbovetsky, Editors,** The Diverse World of PDEs, 2023

787 **Hemen Dutta, Editor,** Mathematical Modelling, 2023

786 **Hemen Dutta, Editor,** Mathematical Modelling, 2023

785 **A. Leroy and S. K. Jain, Editors,** Algebra and Coding Theory, 2023

784 **Dinh-Liem Nguyen, Loc Hoang Nguyen, and Thi-Phong Nguyen, Editors,** Advances in Inverse Problems for Partial Differential Equations, 2023

783 **Alla Detinko, Michael Kapovich, Alex Kontorovich, Peter Sarnak, and Richard Schwartz, Editors,** Computational Aspects of Discrete Subgroups of Lie Groups, 2023

782 **Galina Filipuk, Alberto Lastra, and Sławomir Michalik, Editors,** Recent Trends in Formal and Analytic Solutions of Diff. Equations, 2023

781 **Emanuel Indrei, Diego Marcon, and Levon Nurbekyan, Editors,** Geometric and Functional Inequalities and Recent Topics in Nonlinear PDEs, 2023

780 **Erik Koelink, Stefan Kolb, Nicolai Reshetikhin, and Bart Vlaar, Editors,** Hypergeometry, Integrability and Lie Theory, 2022

779 **Samuele Anni, Valentijn Karemaker, and Elisa Lorenzo García, Editors,** Arithmetic, Geometry, Cryptography, and Coding Theory 2021, 2022

778 **Carlos Galindo, Alejandro Melle Hernández, Julio José Moyano-Fernández, and Wilson A. Zúñiga-Galindo, Editors,** p-Adic Analysis, Arithmetic and Singularities, 2022

For a complete list of titles in this series, visit the
AMS Bookstore at **www.ams.org/bookstore/conmseries/**.